Applied Mechanical Design

Applied Mechanical Design

Ammar Grous

iSTE

WILEY

First published 2018 in Great Britain and the United States by ISTE Ltd and John Wiley & Sons, Inc.

ISTE Ltd
27-37 St George's Road
London SW19 4EU
UK

www.iste.co.uk

John Wiley & Sons, Inc.
111 River Street
Hoboken, NJ 07030
USA

www.wiley.com

Library of Congress Control Number: 2017962687

British Library Cataloguing-in-Publication Data
A CIP record for this book is available from the British Library
ISBN 978-1-84821-822-2

Contents

Preface

This book is designed for students in specialized schools and in science and technology universities, and also for professionals in the industrial sector. Its content is based on case studies with given solutions, which are targeted and discussed. This content is drawn from the author's own classes in applied mechanical design. It gives an overview of all the necessary elements of knowledge, and the methods for analyzing and selecting materials. The book is written as a didactic tool to guide readers in their approach to design, making heavy reference to industry.

The construction of the book is founded primarily on intuition where Arts and Industry are concerned. University courses in Technology offer solutions to the problems arising in design, but do not always present an exhaustive list of the numerous steps that need to be followed at the implementation stage. Sometimes, solutions put forward in mechanical design refer to diagrams without saying anything of the "how", the "why" or the "where"… The thought process surrounding design is far more subtle, much deeper and, in many respects, considerably more complex than it might, at first glance, appear to be. The approach to design must begin with concise methods which enable us to:

– clearly set out the problem (or problems) at hand;

– compare succinct analytical solutions;

– make informed, and duly documented, choices and selections.

In this book, we examine a protocol which qualifies and quantifies the requirements, beginning with a clearly written set of Functional Specifications (FS). The methods and analyses herein aim to satisfy the expressed need.

In particular, it is absolutely crucial to have, at our disposal, mathematical and physical tools to calculate the mechanical resistance of materials. Knowledge of how materials behave is fundamentally important to enlightened design; in other words, the results of calculations, even fairly accurate, constitute simple pedagogical mathematical exercises which have no bearing on the final decision taken on a project. The case studies presented here are the result of the author's research and targeted work in mechanical design. They are taken directly from the author's teaching materials which have, in turn, given rise to manuals and a wide range of publications, including manufacturer catalogs on an international scale.

At the end of the book, readers will find case studies which have previously been discussed in the author's classes, with a view to finding solutions. They are invited to make use of these materials for an appropriate illuminating study, in keeping with the workshop tools and lab tools available to them. In applied mechanical design, there is no single solution, but instead a range of possible solutions in view of the means and methods being used.

Ammar GROUS
December 2017

Introduction

In design, modeling is an indispensable step. In the preliminary stages of a design project, it is essential to make sketches (outlines), and combine these with a rough graphic-analytical model. When the preliminary sketches appear viable, we move on to other factors of progressive design. The principles of the calculations include combinations of properties to find the best possible performances that the project can deliver. The mathematical model is founded on exact values of the stresses at work, and of the strains undergone by the components, bounded by the operational limitations. In this book, we present concrete examples with different geometrical properties, under stress from a variety of types of loading. Beams, bars, discs and cylinders (to mention only a few classic forms) already have working hypotheses concerning their use, set out in specialized publications. It is wise to draw upon this pre-existing body of work in putting together real-world projects. The important thing is to be aware that the information is out there, and that judicious use can be made of it. The formulae presented herein are taken from the specialist literature, cited in the bibliographies at the end of each chapter. The tables, standards, formulae and other are presented here as an illustrative guide. Under no circumstances should this book be thought of as exhaustive; readers must refer, for more detailed information, to the aforementioned specialized publications.

Historically, the design of products such as mechanisms and machines has been at the heart of engineering sciences and techniques. The development of computer tools and computer-aided design (CAD) has greatly contributed, with the help of the standards brought into force, to better presentation of the graphical results of sketches of definitions and products. Optimization methods, new workflow schemes and mathematical tools employed in mechanical construction can easily be used by project designers. On the basis of a clear set of technical specifications, the designer can achieve their objectives in the shortest possible time and with the lowest possible material cost. A clearly presented preliminary project is a good guarantee for the production of a safe, well-documented design.

There is no reproducible, prefabricated "recipe" for a good design of mechanical systems. Instead, there are step-by-step, incremental methods which are optimized to deliver the desired results. The current book is founded on experience both in industry and in teaching. It discusses indispensable tools which orientate and guide designers in their search for solutions. It uses a skill-based pedagogical approach, and thus presents the target vectors, including:

1) *Methodology*: Readers of this book will follow the steps, from needs analysis to the hunt for mutually acceptable solutions (agreed between the client and the designer), right through to the presentation of the documented preliminaries for the project. Technical specifications, creative design methods (including FAST: Function Analysis System Technique), Gantt charts and FMECA (Failure Modes, Effects and Criticality Analysis) are used here.

2) *Principles behind the calculations*: The elements of analysis and calculations traditionally employed in construction are the preserve of the mathematical and applied physics tools. The novel aspect of this book is that it brings in a clearly documented pedagogical approach. The analytical approach follows the principles of the calculations. The materials and processes, which are subjects close to mechanical systems, are clearly documented.

3) *Graphical-analytical tools*: The graphical and/or analytical methods discussed herein are used in such a way that engineers, technicians and students can, themselves, use them or draw inspiration from them for their own projects. Logically, the fundamentals of computer-assisted design are examined using the software tools recommended in the industry.

4) *Pedagogical and industrial case studies*: The solved examples given here are taken from the author's own experience in industry and in a university setting. These examples cover a wide range of fields in manufacturing industry (recreational equipment, lifting devices and forms of transport, pedagogical demonstrative mechanisms, etc.). The studies also look at how to make the right choice of materials and structures for the projects at hand.

This book, which is devoted to applied mechanical design, draws on case studies taken from the author's own experience in the professional and university spheres. It is based on a methodology and pedagogical approach which are deliberately painstaking, because they are being used instead of pure mathematics and applied physics to present the subject. It is an arduous task to present a mechanical design book, owing to the multidisciplinarity of the field and the computer tools which are crucially important to run the design calculations quickly. Design is not a singular subject: a complete design project will inevitably require the coming together of a multitude of technical, scientific and technological disciplines, in addition to clarity in the drafting of the dossiers making up the studies.

The first chapter discusses the organization of workflow for projects. Its purpose is to guide the projector–designer through the entire organizational process, from the first general overview sketch to the launch of the idea based on the technical specifications. This part tackles the search for practical solutions which can orientate the projects. This, in a manner of speaking, is the art of conducting a design project.

Chapter 2 sets out the main design tasks, with emphasis being placed on the judicious use of materials and geometric form. Throughout the book's eleven chapters, we present case studies with worked solutions as a didactic approach.

Next, the third chapter presents the principles underpinning the calculations for the elements of machines, materials and structures. It is here that calculation tools prove indispensable for the design analysis. This chapter discusses the need for process analysis and materials analysis. It is also at this stage that the issue of sizing comes into play, and that geometry (GPS: Geometrical Product Specifications) takes on its demonstrative part. We also look at the theory of Hertz contact stress.

Chapter 4 is given over to the particular geometric forms used in applied mechanical design: profiled and incurvate parts (we look at the NURBS method: Non-Uniform Rational Basis Spline).

We move on, in our fifth chapter, to the appropriate use of the principles of calculations for the elements of machines employed in mechanical construction. We touch upon cases of material resistance and give further case studies, with solutions, which can serve as concrete examples to be used in tutorials and other applied mechanical design workshops.

The sixth chapter presents case studies devoted to the noise and vibrations produced by mechanical elements and machine supports. The solution sets demonstrated here could be used to devise other projected cases for tutorials.

Chapter 7 offers further context about a number of cases commonly encountered in welded structures, and presents the calculations used to determine the fatigue of stationary parts and bearings.

The eighth chapter then presents case studies on the brakes and clutch systems used in applied mechanical design.

In the ninth chapter, we discuss bolted mechanical structures, such as spring washers and axles.

Chapter 10 is devoted specifically to the principles underlying the calculations for machine elements made of plastic materials.

The eleventh chapter presents concrete projects that have actually been tackled in the author's classes and tutorials, and the resulting commentated sets of solutions also offered.

The Conclusion and Appendix provide a brief summary, containing a glossary, and tables of reference pertaining to the standards in force in construction and mechanical design. Having come to the end of the above sections of the discussion, this final chapter presents a number of recommendations regarding the drafting and final presentation of projects.

This book is specifically targeted at problem-solving in mechanical design. Its purpose is absolutely not to lay out the fundamental concepts used in mechanics, but it does draw upon those concepts (discussed in depth in other works) to put forward succinct, practical solutions. That is where its strength as a tool for teaching and training lies. Users can refer to an illuminating set of commentated solutions.

1

Case Study-based Design Methodology

1.1. Methodology for designing a project product

"Methodology" should be taken to mean the preparation of an exhaustive file of the steps that must be followed at a research center. That file will contain the data defining the product, right down to the expression defining the need for it.

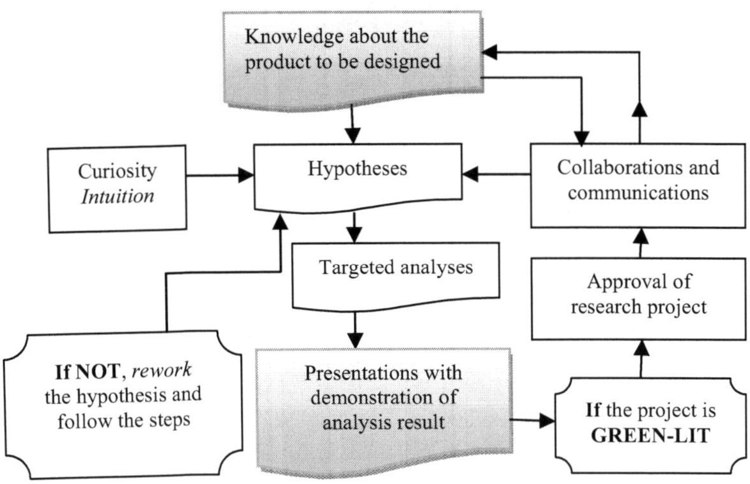

Figure 1.1. *Scientific approach to design*

The scientific approach is named as such because of the dissatisfactions pertaining to fields not explored by the core sciences. If the subject is unique or original, in collaboration between multiple disciplines, new hypotheses may be formulated and patented. Thus, the state of the art will be enriched.

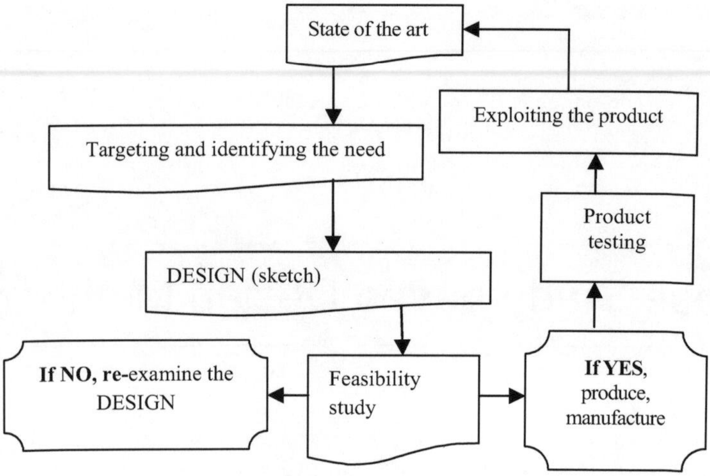

Figure 1.2. *Process of a design project*

Given the multi-disciplinarity (fields of mechanics, pneumatics, hydraulics, electronics, etc.) of subjects which encapsulate design and vast domains of application (the manufacturing industry, agriculture, civil and/or military construction, biomechanics, automotive engineering, aeronautics, etc.) in this book, we present examples found in systems in mechanical and civil engineering. The mathematical foundations are common to the two domains, but the behavior of the materials and structures sometimes differ.

We can already see that design is hugely complex. A designer must have good knowledge of numerous branches of science and technology, in addition to a keen sense of the dual organization of good intuition. The project approach is less stressful because the client will have clearly expressed their requirements at the outset. Solutions will be put forward with the aim of satisfying the requirements, in connection with the technical and economic feasibility. Once the designer has accepted the project to conduct manufacturing studies, tests will be carried out. If the result is a conclusive approval, then the product will be made and potentially released to the public.

1.2. Main players involved in the design process

Design projects are studied in a research hub (RH). Thus, the RH is an entity which is often complex, the size of which is subject to variation from one company to another. The teams employed at RHs are multi-disciplinary, complementary and diverse. The tasks are often trivial and overseen by: the study *engineer*; the study *designer* and the *projector*.

The research engineer is the head of the research project, but delegates some of the work to the projector. In addition to the calculations for which s/he is responsible for (graphic design, core sciences and applied sciences), the research engineer acts as an interface between the administrative and financial management teams.

The study designer, despite what the title suggests, actually deals with more than just designs. His/her role is to finalize the defining designs and drawings in the preliminary stage of the project.

The projector, as the etymology indicates, assists the research engineer in projecting technical tasks, including the defining design drawings. Depending on the structural organization of each company, the projector may sometimes deal with clients, with suppliers and with all the company's separate departments. In addition to the tasks of computer-aided design (CAD), the projector is indispensable in a research center, because the research engineer trusts projector and relies upon him/her for all the work needing to be done.

Specific publications on company structure offer fuller definitions of these hierarchical roles. This book, for its part, is intended to be an instructional design-support tool: sketches, calculations, tests, measurements and checks.

The design workspace evolves around the projector. The below diagram illustrates what we mean by this:

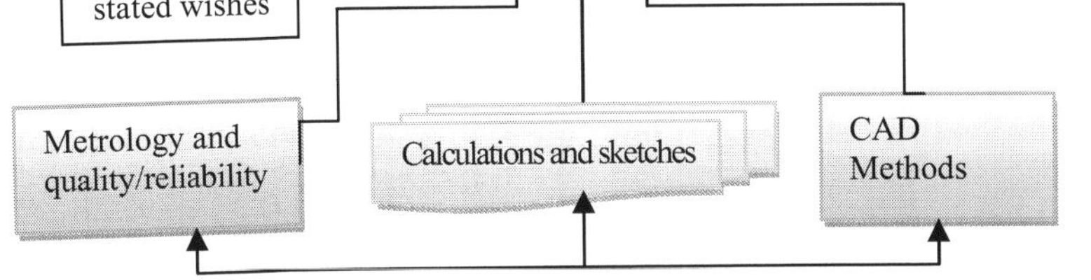

Figure 1.3. *Diagram illustrating the work environment*

The projector is in charge of the needs analysis, the feasibility study for the requirement and the execution of the preliminary ("pre-project") stage. It is worth bearing in mind that there are multiple possible combinations to create a product. The organigrams offered here are simply a few illustrative examples. The staff at the research hub, drawing upon their relevant experiences, would choose possibilities which fit in with their own working methods.

1.3. Conceptualization and creativity

At the initial stage of the design, it is wise to begin with proposals of ideas to help bring the product to life. The team then take an initial overall view of the adopted solutions to lay the foundations for the task of design.

Criteria behind principles of construction		
Human factors	Production	Reliability
Calibration function	Standards	Maintenance
Settings	Materials	Maintainability
Connections	Recycling	Environment
Setups	Components	Handling
Safety/security	...	Applied quality control
Accessibility	etc....	Applied metrology

Table 1.1. *Main construction criteria in design*

1.4. Functional analysis in design: the FAST method

The method known as FAST – Function Analysis System Technique – is a workflow organization method which offers a logical and explicative diagram illustrating the operational and technical functions.

– *Why* does the function need to be performed?

– *How* does the function need to be performed?

– *When* does the function need to be performed?

It is an interrogative method, the answers to which will guide a designer in their choices and methods. Function Analysis System Technique, known by its acronym FAST, is used in functional analysis either to describe (descriptive FAST) or to create (creative FAST). FAST is said to be descriptive if the design solution already exists. This method is used, for example, to carry out a critical study of how a product's technical and/or economic functions are served. The process is as follows:

Prepare for the analysis → Gather and collate the technical functions → Sort the technical functions → Construct a FAST diagram (How? and/or When?) → Write up the technological adopted solution.

Creative FAST is used as follows: Prepare for the analysis → Search for and collect ideas (brainstorming, or "Kaizen") → Order the ideas → Construct a FAST diagram (How? and/or When?) → Write up the technological adopted solution.

Simplified example of a FAST approach in manufacture by machining

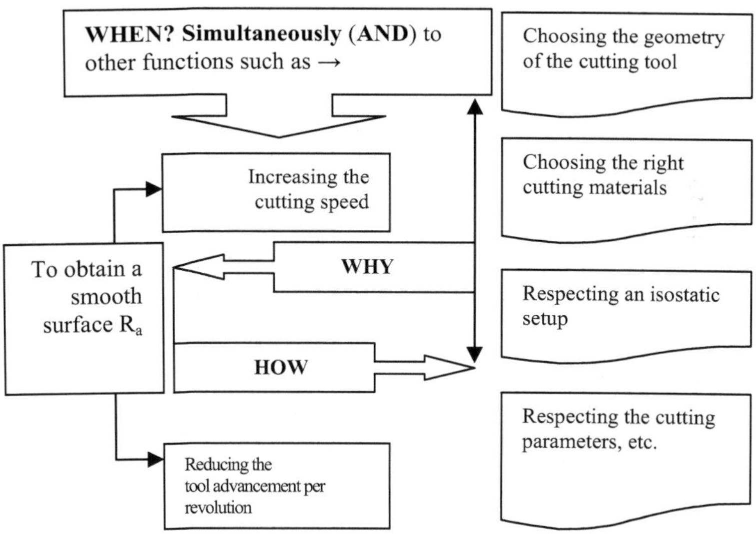

Figure 1.4. *FAST in machining manufacture*

1.4.1. *Decision-support tools in design*

Although decision-support tools can be very helpful if used properly, it is advisable not to blindly accept the accuracy of the proposed solutions. These tools are used at the end of the creative sessions that are an integral part of the preliminary design phase of a project. The purpose of using them is to stabilize the choice of solutions – in other words, to check the validity of the decision and to create a commentated decision dashboard.

Validity of the decision: The questionnaire is designed to extract the true requirements of the product. It then becomes a question of explaining the advantages and disadvantages weighing toward the maintaining of the solution.

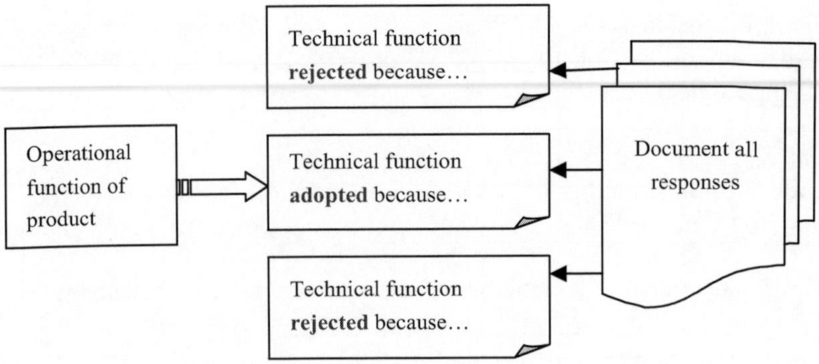

Figure 1.5. *Decision-support system (expert system)*

For example: why use one material instead of another? Why choose that environment? etc. The assessment criteria justify and support the characterization of the product's operational function. Of course, it is crucial to also take account of criteria such as cost, maintenance and even esthetic appearance, etc.

A so-called Bayes table (named after the Scottish statistician Thomas Bayes, 1702-1761), is a multi-criterion table used to formalize a decision. To draw up a Bayes table, we insert the factors contained in relation [1.1], below, in accordance with the following seven steps:

1) Define the objective of the decision to be made

2) Collate the proposed solutions (Sol_1; Sol_2; … Sol_j)

3) Collate the criteria ($Criterion_1$; $Criterion_2$; … $Criterion_j$), assigning each a score between 1 and 3, depending on the case study

4) Weight the criteria (κ_i) with weight scores ranging from 1 to 5

5) Find the value structure for each criterion

6) Carry out an overall evaluation (calculate the total sum)

7) Analyze the results to produce the best solution(s) as a function of the highest weighted score.

Decision dashboard relating to three solutions

$$Score_{Solj} = \sum_{i=1}^{n} N_i \times \kappa_i \quad where \quad \left\{ \begin{array}{l} \kappa_i \text{ is a weighting coefficient } (1 \leq \kappa_i \leq 5) \\ Score_{Solj} \text{ is the score for solution } sol_j \\ N_i \text{ is the criterion score } (1 \leq N_i \leq 3) \\ i \text{ is the criterion number } (1 \leq j \leq n) \end{array} \right\} \quad [1.1]$$

Criteria collected	κ_i coef. Pond.	Solution 1		Solution 2		Solution 3	
		Score	Total	Score	Total	Score	Total
Criterion 1	5	3	5×3=15	1	5×1=5	3	5×3=15
Criterion 2	3	2	3×2=6	2	3×2=6	2	3×2=6
Criterion 3	4	1	4×1=4	2	4×2=8	2	4×2=8
Criterion 4	2	1	2×1=2	2	2×2=4	2	4×2=4
Criterion 5	3	1	3×1=3	1	3×1=3	2	3×2=6
		Weighted total	Σ=30		Σ=26		**Σ=39**
		Minimum score	**1**		**1**		**2**
Conclusion: **Solution 3** is the most advantageous. Hence, it will be given priority.							

Table 1.2. *Example of a decision dashboard*

1.5. Functional specifications (FS)

Before conducting any kind of design study, it is essential to know precisely what the client is looking for. In some cases, the client's expression of their needs may be unclear, lacking detail and sometimes even incomplete. Working closely alongside the client, it is crucial to clearly define the product that is to be designed.

Many methodological tools are put forward in the existing body of literature [FRE 85]; such as the interaction diagram or the APTE method (*Application aux Techniques d'Entreprise* – Application to Business Techniques – taking its name from its parent company, APTE), to cite only those examples. The client may not always be a designer. They will express a requirement, in their own words, which the designer will then have to implement. In order to do so, the designer begins by defining the operational functions of the product being designed.

As the FS cover the entire life cycle of a product, they must satisfy a number of requirements, at the following distinct stages: *design, manufacture, distribution, use, maintenance and end of life – i.e. disposal, recycling or, sometimes, value creation (upcycling)*. Thus, the definition of a product is a true reflection of the set of constraints relating to the environment.

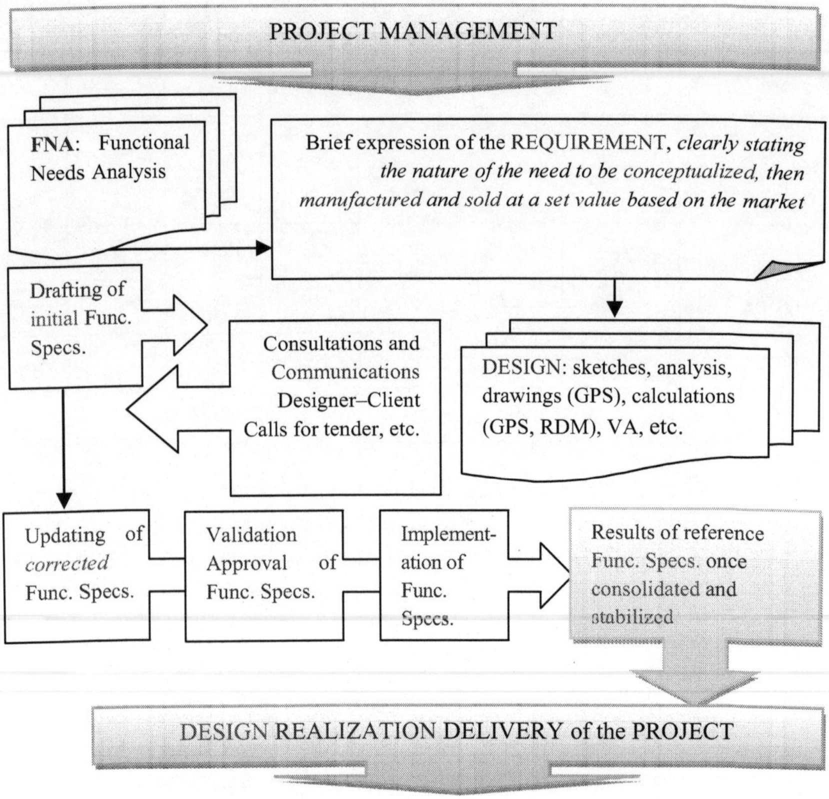

Figure 1.6. *General structure of functional specifications (Func. Specs). GPS: Geometric Product Specification. VA: Value Analysis. For a color version of this figure, see www.iste.co.uk/grous/design.zip*

1.5.1. *Operational functions, using the APTE method or octopus diagram*

The operational functions describe the client's requirement which the designer must serve. To do so, the designer can use a variety of workflow organization methods. For example, s/he might use an octopus diagram in the case, primarily, of designing mechanical systems. An example is given below.

– FP_1: This is the principal function of the product (e.g. a bicycle). It enables the cyclist to move over the ground (link between two media);

– FC_1: Constraint function: withstand disturbances on the cyclable track (road) and be easy to pedal;

– FC_2: Respect the environment;

– FC_3: Look attractive (esthetic value);

– FC_4: Resist external attacking forces (rain, snow, salt, etc.);

– FC_5: Be lightweight, safe and reliable;

– FC_6: Respect the standards and other codes in force;

– FC_7: Be comfortable.

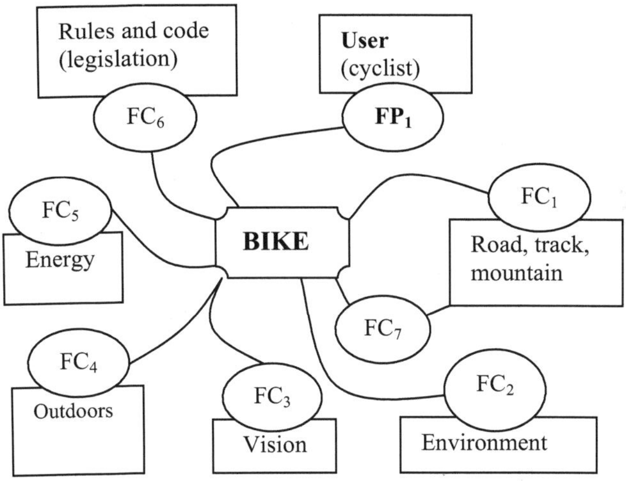

Figure 1.7. *Octopus diagram or APTE principle for a bicycle*

The steps to be taken are generally logical, including:

1) *States*: The states are operational, under construction and undergoing maintenance;

2) *Environment*: Above all, it is important to think about the surroundings of the product we are designing, such as: the physical elements, the user(s), the environment, reliability/safety, energy, etc.

3) *Principal functions*: Hitherto, it has not been customary to insist on the language to be used in technical documents. This, in the author's view, is a mistake, because the language and the expressions must be absolutely free of any ambiguity. A set of functional specifications is a contract between the designer and the client, where sentences need to be structured in such a way as to clearly show the elements of logic: subject, verb and object.

1.5.2. *Linguistic (or syntactical) writing of the functional specifications*

– the label "*subject*" obviously applies to the system to which the design pertains. It needs to be concise, clear and unambiguous;

– the action *verb* (indicative and transitive) must be in the present indicative tense. It is important to avoid verbs which render the statement of the function vague or evasive;

– the direct object *complement* (direct, indirect, circumstantial) applies directly to the user. An indirect object complement is a little different.

With the above completed, the FS can be drafted. The designer adds further information about the characteristics of each function, along with its title. If using an octopus diagram, the number of generated principal functions needs to be taken into account by the designer. The client is free to interject on matters such as F_{NN} (non-negotiable function), F_N (negotiable function) or even F_{FN} (fairly negotiable function).

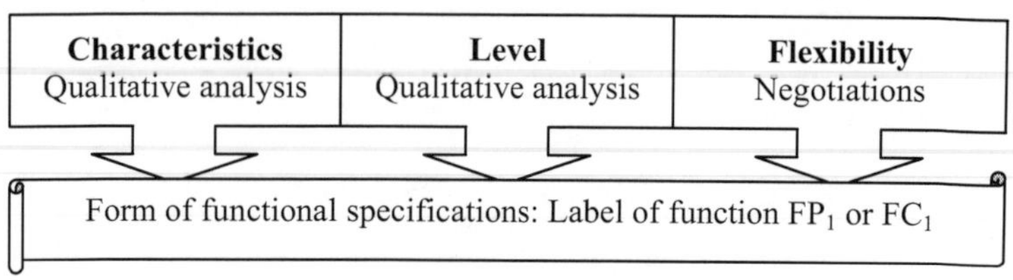

Figure 1.8. *Form of FS*

The designer presents the client with the FS, where all the functions are clearly defined by their characteristics. The levels will be annotated by the client depending on the operational needs for the product. The FS must be signed by both parties in order to formalize the contract. The time allowed for the FS is limited to around half a day's work, and the document is distributed in electronic format. The drawing up of FS constitutes a hugely important step. The design then becomes a matter for specialists. We then seek solutions from a range of specialists with a view of validating the pre-project, using the FMECA design tool.

1.6. Failure Mode Effects and Criticality Analysis

FMECA – Failure Mode Effects and Criticality Analysis – examines all the different types of failure which impact the system's ability to operate. This method, developed by the US Army, has the reference *MIL-P-1629* (1949). It is a reliability method whereby failures are ranked on the basis of their impact on equipment (reliability) and on personnel (safety). For that reason, FMECA is employed widely in aerospace engineering. By the 1970s, the automotive industry (including companies such as Toyota, Nissan, Ford, BMW, Volvo and Peugeot) was using FMECA widely. We can distinguish between different types of FMECA, depending on the domain:

– *product* FMECA can be used from the design of the product to ensure it fulfills the functions for which it was designed;

– *process* FMECA pertains to the realization phase, to verify the impact of the manufacturing processes on the product's conformity;

– *procedural* or *machine* FMECA relates to the equipment used in the manufacture;

– *design* FMECA relates to the design of the product, with the aim being to anticipate potential design failures at the pre-project stage.

We focus here on design FMECA, including the system (the derailleur in a bicycle gear system) and the formation of a research group composed of the maintenance manager, three maintenance agents and two technicians.

i) *failure analysis*: the detailed study of the work orders issued in the wake of the breakdown revealed two modes of failure with their associated effects and causes (see table below);

ii) *criticality calculation*: with the values adopted for F, D and G, it is possible to calculate the criticality attached to each failure of the winder of a hoist;

iii) *actions undertaken*: the highest criticality value, 12, is associated with the wear and tear on the components of the winder. The service decides on preventative action by changing the components of the winder every year.

Ranking of potential failures	
F for FREQUENCY	
1	Max. failure per year (low)
2-3	Max. failure per trimester (average)
4-5	Frequent failures each week
8	Very frequent failures
10	Failure all the time
D for NON-DETECTION	
1	Verification by an operator (naked eye)
2-3	Easy detection by simple examination
4-5	Easy detection by detailed examination
10	Undetectable to the naked eye

G for GRAVITY or UNAVAILABILITY	
1	No stoppage
2-3	No stoppage but sometimes annoyance
4-5	Stoppage > 1 day or nonstandard
8	Dangerous
10	Fatal

Definition of the terms used in Product FMECA (Bicycle)							
System, element or subset	FAILURES			CRITICALITY			
	Modes	**Effects**	**Causes**	**F**	**D**	**G**	$C = F \times D \times G$
Gear change of the hoist	*No gear change*	Winder non compliant	Failure of gear changer	1	2	2	**4**
Idem	Idem	Idem	Defective mechanism	1	2	2	**4**
Idem	Poor function of the winder	Idem	Worn parts of the winder	2	3	2	**12**

ACTIONS UNDERTAKEN	EVOLUTION			
	F	**D**	**G**	$C = F \times D \times G$
Changing of the **winder** every year	1	3	2	**6**
The **Risk Priority Number** is written: $C = N \times R \times P = F \times D \times G$				

Table 1.3. *FMECA for the example of a hoist*

– the criticality represents a mathematical product which evaluates the occurrence and severity, giving us the product: $C = Q \times S$.

The *checks* in design and manufacturing processes constitute mechanisms which prevent the causes at the root of failures. As in design, the clients are separate from the functional specifications. It is manufacturing processes which are affected by possible failures of the product.

– to evaluate the probability found by the check, we use the failure cause tool Statistical Process Control (SPC) [GRO 13];

– when a mechanism, a component or even the whole system ceases to function, *failure* occurs. It is possible that the system may not shut down completely. The performances lack – hence the word "failure".

There are particular ways where failure manifests itself when a system deviates from the specifications. Depending on the kind of failure (abnormal strain, wedging, corrosion, buckling, surpassing of the dimensional tolerance values imposed, etc.), we note a specific *mode* of *failure*.

– the *causes of failure* (upstream) relate to the design and/or manufacture. They can be classified using the 5 M's method [GRO 13] (Machine, Method, Material, Man power/Mind power and Measurement/Medium);

– the *effects of failure* (downstream) are the symptoms of alteration of the components. The so-called risk priority number ($RPN = S \times O \times D$) expresses the severity (gravity) of the occurrence of detection. The RPN is used for high-quality products.

Advantages of FMECA	**Drawbacks to FMECA**
Managing the risks by means of preventative actions	Requires profound knowledge inherent to the question at hand, from design to delivery of the product. This requires a *pluridisciplinary* team. Thus, FMECA is a "cumbersome" method

1.7. PERT method

Program Evaluation and Review Technique (PERT – 1958) was initially introduced in order to study the manufacture programs by ordering numerous tasks, in the form of a grid. In view of their dependencies and chronology, these tasks lead to the obtaining of a finished product. This method is limited, as it contains neither the concepts of duration nor the date. PERT is constructed around the following elements:

a) listing the main operations in the project, placing emphasis on the temporary durations between them;

b) constructing a coherent, succinct graph;

c) quantifying the "at the earliest" and the "at the latest" dates, based on the steps inherent to each operation;

d) targeting the critical operational pathway, in connection with the above.

If necessary, it is possible to construct a Gantt chart. In order to gain a fuller understanding of the logic of PERT, let us look at a simple illustrative example.

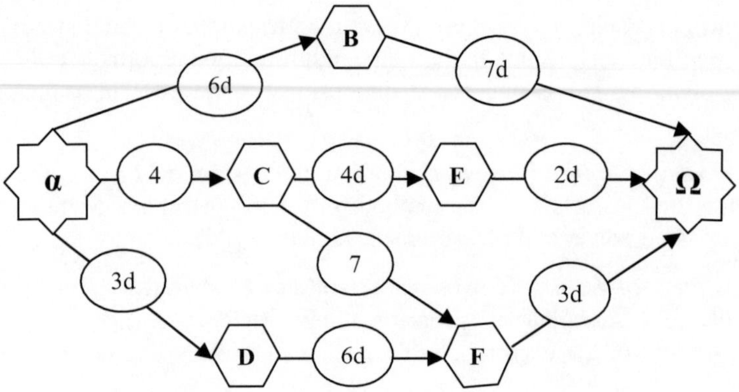

Figure 1.9. *PERT graph for the example (d = days)*

1.7.1. *Logic of construction of the graph per level of operations*

Level I: (α represents the starting point in constructing the network. As we have no previous history to go on, the level I tasks – i.e. C and D – are barred to define level II. Understandably, we first need to finish those steps before we can move onto step F.

Level II: Tasks B, E and F are at level II. Thus, we have to complete them before we are able to begin the final step (**Ω**). **Ω** represents the completion of the construction of the grid, which corresponds to the last day.

i) we must create a fictitious task to model the condition of anteriority C and E, which pre-date task B. The order of the tasks proceeds from left to right;

ii) "level" should be taken to mean the positions of each vertex illustrating the start of the relative tasks – i.e. τ_{prob1}, τ_{prob2}, etc. … τ_{probi}.

– the numbering proceeds from left to right (from → to …) – see Figure 1.9;

– the "*at the earliest*" tasks are performed *from left to right* by adding together the durations of each tasks. The highest value is considered to be at the intersections;

– the "*at the latest*" tasks are performed *from right to left* by successively subtracting the durations of the tasks, beginning with the final task (**Ω**). The lowest value is considered to be at the intersections.

iii) we read the *critical pathway* {**α** → C → F → **Ω**}, going *via* those tasks which correspond to the *earliest* realization date, which corresponds to the *latest* realization date.

– τ_{prob} indicates the probable duration of the task;

– τ_{pt} indicates the *earliest possible* date corresponding to the end of a task;

– τ_{ptd} indicates the *latest possible* date corresponding to the end of a task. Note that if ever that duration were to surpass the value indicated by τ_{td}, the project would suffer a serious delay. In that case, the subsequent steps would be automatically shifted.

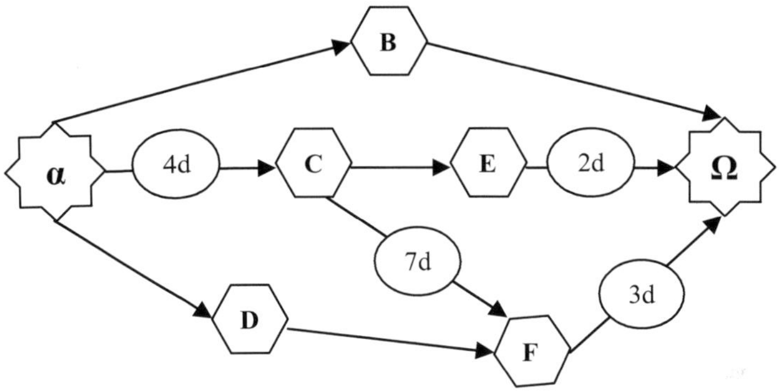

Figure 1.10. *Identification of the critical pathway {α → C → F→ Ω}*

– **Ψ** represents the margin in days thus calculated. $\Psi = \tau_t - \tau_{pt}$ is a tolerance value (in terms of a delay) to represent any possible delay in carrying out a task. For $\Psi = \tau_t - \tau_{pt} = 0$, there is no delay tolerated in the project.

Tasks	Previous tasks	Probable duration τ_{prob} (days)	Earliest possible date τ_{pt} (days)	Latest possible date τ_{td} (days)	Ψ, margin *in days*
Listed at will	Stress history	Durations (days) established	Durations (days) calculated	Durations (days) calculated	Duration $\Psi = \tau_t - \tau_{pt}$
A	-				
B	← A	6 days	6 days	7 days	1 days
C	← A	4 days	4 days	4 days	**0**
D	← A	3 days	3 days	3+2 = 5 days	2 days
E	← C	4 days	4 + 4 = 8 days	7+3+2 = 12 days	4 days
F	← D, C	6 days or 7 days	7 + 4 = 11 days	7+4 = 11 days	**0**
G	← B, E, F	7 days or 2 days or 3 days	7 + 4 + 3 = 14 days	7 + 4 + 3 = 14 days	**0**

Table 1.4. *PERT table explaining the reasoning with an example*

1.7.2. *Statistical approach to the PERT diagram using the Gamma distribution*

There is a statistical approach to representing durations. The chronological order proves to be identical to that which was presented above. To gain an idea of the possible delay tolerance range, we determine an (optimistic) minimum value $\tau_{\text{-}min\text{-}a}$ and a (pessimistic) maximum value $\tau_{\text{-}max\text{-}b}$. Thus, the most probable duration $\tau_{\text{-}prob.}$ indicates the hope of achieving the total duration of the realization. We use $\tau_{\text{-}exp}$ to denote each expected task, with a statistical variance estimator (σ^2) whose statistical relations are as follows. S is a form factor.

$$\begin{cases} \Psi_{\text{margin}} = \tau_{td} - \tau_{pt};\ \tau_{td} = \sum_{\tau=1}^{k} \tau_{prob}\ ;\tau_{\exp} = \left(\tau_{\text{min-}a} - 4\tau_{prob} - \tau_{\text{max-}b}\right)/6 \\ \sigma^2 = \left(\tau_{\text{max-}b} - \tau_{\text{min-}a}\right)^2/6;\ U = \left(P_S - \tau_{\text{max-}b}\right)/\sqrt{\Sigma\sigma^2}\ ;0 < P_s < 1 \end{cases} \quad [1.2]$$

The graphical (statistical) imbrication of PERT is presented as follows:

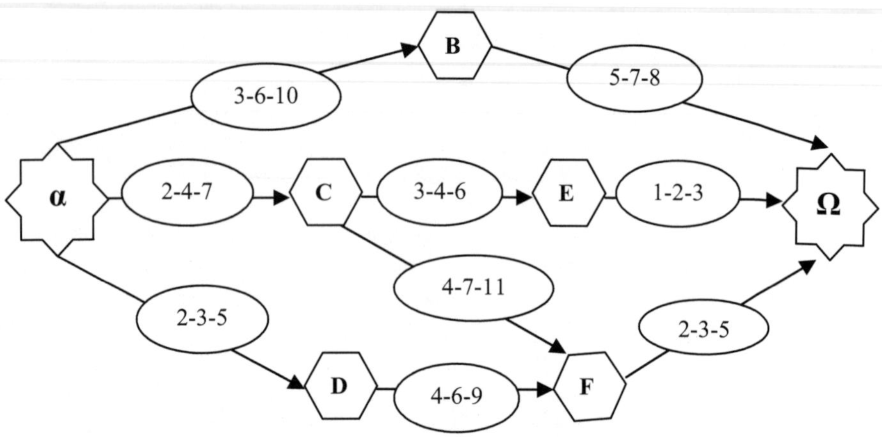

Figure 1.11. *PERT graph, for example, based on statistical durations*

Statistical application to the Gamma distribution for $\lambda = 0.5$ (form factor); $\kappa = 5$ (form factor) and $\tau = 0.5, 0.1, \ldots 20$ gives us:

$$f(\tau) = \left[\frac{\lambda}{(\kappa - 1)!}\right] \cdot \left(e^{-\lambda \cdot \tau}\right) \cdot \left[(\lambda \cdot \tau)^{\kappa - 1}\right]$$

f(τ) =
$63.379 \cdot 10^{-6}$
$789.753 \cdot 10^{-6}$
$3.114 \cdot 10^{-3}$
$7.664 \cdot 10^{-3}$
$14.572 \cdot 10^{-3}$
$23.533 \cdot 10^{-3}$
$33.954 \cdot 10^{-3}$
$45.112 \cdot 10^{-3}$
$56.276 \cdot 10^{-3}$
$66.801 \cdot 10^{-3}$
continuous ...

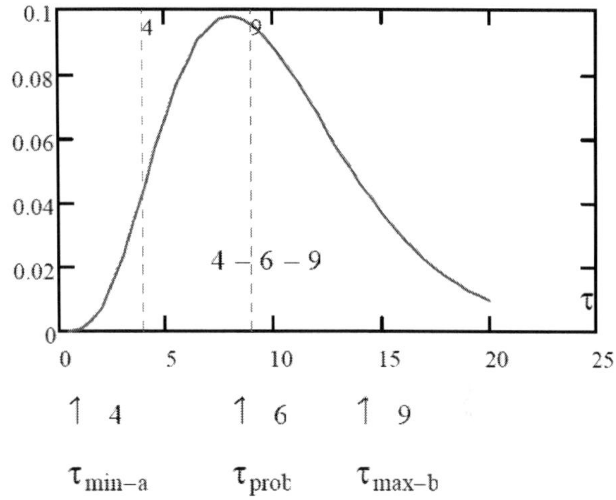

Figure 1.12. *Typical distribution based on the Erlang distribution (Gamma distribution). For a color version of this figure, see www.iste.co.uk/grous/design.zip*

1.8. The Gantt method (Henry Gantt's graph, devised 1910)

This method shows the duration of a task, a manufacturing cycle, an intervention or even expresses the workload of an operator (industrialist). This graph forms the link between the predicted time for a task and the time that is actually spent on it. The purpose is to improve workshop organization. The planning of tasks and activities is followed by deadlines, supported by resources. The Gantt chart helps:

– *adapt* to the set of activities;

– *structure* and *reference* the order in which tasks are to be carried out;

– *facilitate* comprehension – which can sometimes be complex – by simple visual representation;

– *improve* the organization of the workflow by offering an overall view of the structure of the project;

– *plan* the whole duration of the project;

– *dynamize* work by automating the calculations of the dates and durations for the tasks.

A great many project scheduling tools offer models. These software packages propose an approach that is more or less similar to that outlined below:

Step 1: Establish a list of the tasks whose subactivities will be attached, in a clear hierarchical ranking. Any omission or oversight will have repercussions for the state of advancement of the project.

Step 2: Assign the resources and manage the workload. The resources include load-management elements such as production units guided by a simogram.

Step 3: Plan the field of action where the tasks follow the time spread, beginning at the start dates. It is at this stage that the activities can be visually represented on the diagram.

Step 4: At this level, we create the connections between the tasks (arrows between the boxes). It is this connecting function which represents the program's real strong point.

Step 5: We insert the milestones (marking both upstream and downstream in the process). A milestone corresponds to a given step (e.g. the signing of the contract, or the publication of a project handbook). This is a crucially important period in the project. Here, the project gains visibility. Unlike the others, this section has no predetermined duration.

Note that a task cannot begin if a previous one has not yet finished. Similarly, no other task *can commence* if a previous one *has not yet begun*. Also remember that a task may have several previous tasks (see Figure 1.8). In these scenarios, it is imperative that all tasks be completed, so that the subsequent ones can then begin.

It may also be that a task has several subsequent tasks attached to it. In that case, the end of the task conditions the start of the other, subsequent tasks.

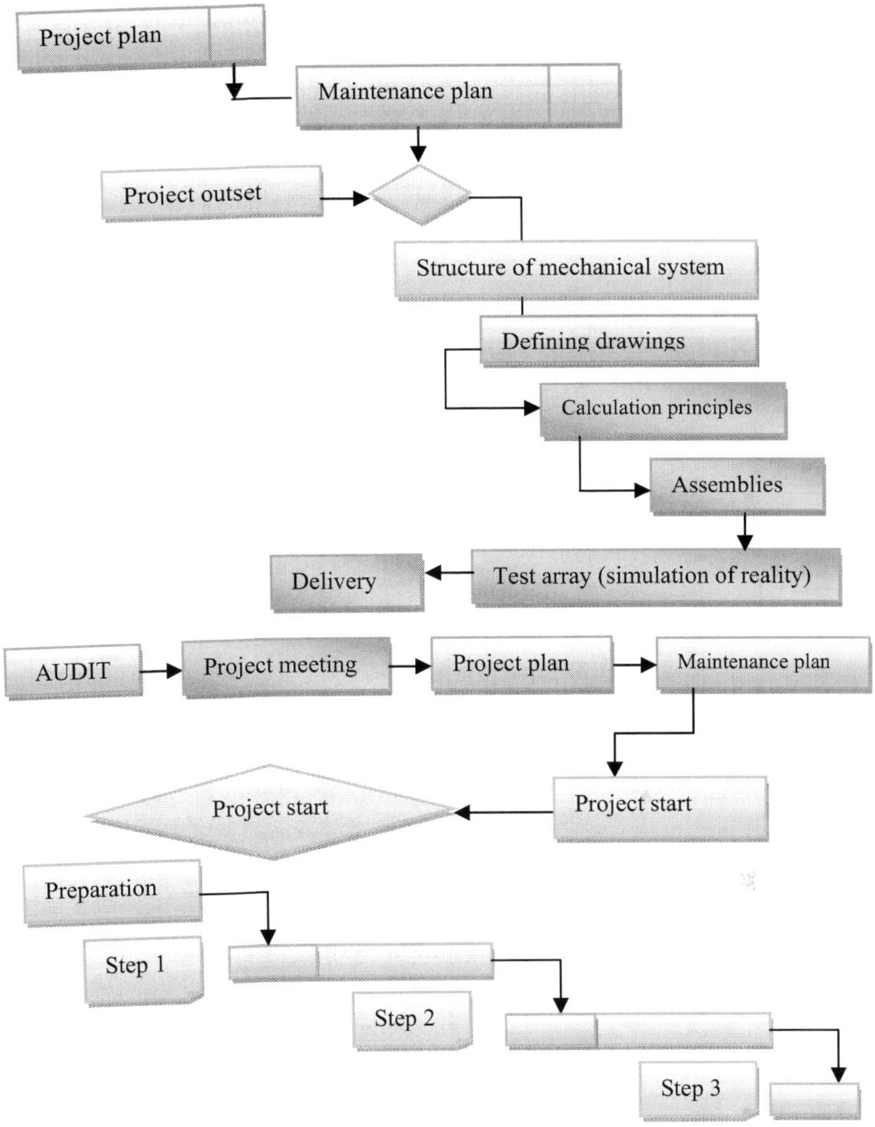

Figure 1.13. *Gantt chart for the example. For a color version of this figure, see www.iste.co.uk/grous/design.zip*

Once the chart has been drawn up, it needs to be followed closely in order to ensure good management of the project, over time. The critical pathway (see Figure 1.10: Identification of the critical pathway $\{\boldsymbol{\alpha} \rightarrow \mathbf{C} \rightarrow \mathbf{F} \rightarrow \boldsymbol{\Omega}\}$) serves to give us a concrete visualization of the succession of tasks, giving an indication of the overall length of the project. Any delay has an impact on the final date of the project.

By clearly visualizing the state of advancement of the tasks, the Gantt chart enables us to assign an appropriate state of advancement to each of the tasks.

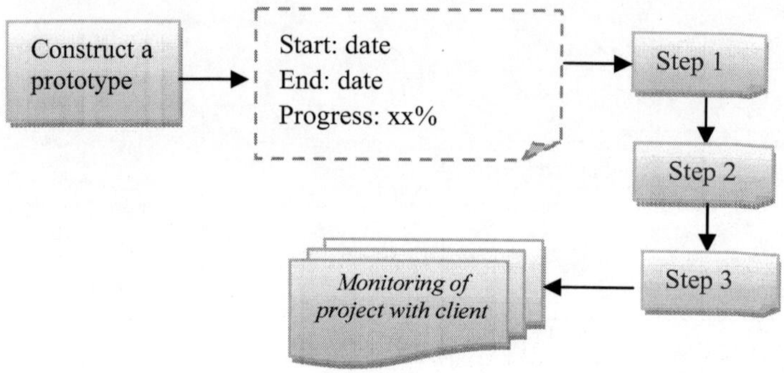

Figure 1.14. *Gantt chart for the example. For a color version of this figure, see www.iste.co.uk/grous/design.zip*

Of course, it is possible to modify the Gantt diagram by planning ahead: it is a dynamic, alterable tool. If, for example, we were to change the duration of a task, the diagram would recalculate the whole duration of the project. That adaptability is useful, and takes account of the interdependence of the created functions. An Excel spreadsheet is perfectly sufficient to create a professional Gantt chart (activity, duration, milestones, resources, etc.).

In the literature specializing in workflow organization, there are numerous interesting and beneficial methods to help save time and better organize the work both at the design stage and at the manufacturing stage. In this book, we shall limit ourselves to the main methods.

1.9. Principal functions of a product

A well-designed product is intended to satisfy the needs expressed by the client. There are numerous methods for analyzing design and manufacture which can be used alongside the FS. These methods are often used at the early stages of a project. The aim is to analyze the optimal path to follow in order to bring a project to fruition, and thus create a competitive finished product that conforms, as closely as possible, to the characteristics included in the FS. In order to do this, it is helpful to clearly define the functions of a product.

A) *Function of a product*: By "function of a product", we mean the action that the product performs in order to achieve its purpose. The verb expressed is in the gerund form. For example: Opening the patient's mouth (separating their jaws) using the mechanism (see case study in Chapter 11) enables the dental hygienist to easily examine their teeth. A product (mechanism) usually serves a range of functions. With this in mind, it is helpful to clearly classify or rank all the functions in the FS and value analysis (VA).

B) *Function of a service*: This function pertains to the use we make of a product. For example, the mechanism in case study 1 serves for (laboratory) demonstrations of the jaws (dental hygiene). It is important not to confuse *service function* and *usage function*. The usage function essentially pertains to the utilitarian aspect of the product. In case study 1, the usage function might not be justified, as the hygienist might be content to use the conventional way of manually separating a patient's jaws. The usage of the designed product is there to make the user's task easy, thus proving the usefulness of the product. There is also what is known as the *esteem function*, which, for its part, is a service function inherent to esthetic beauty and style.

C) *Principal functions (of constraints)*: To begin with, the principal function is the actual reason for creating the product. *Additional functions* could, as and when needed, be added to the principal function – e.g. using a pneumatic actuator rather than a mechanical one. Thus, the idea of a constraint comes into play at the level of mechanical strength, weight, pressure, etc. The *technical functions* are related to the design and manufacture of the product. Here, the client is not directly involved.

The criteria used to assess the aforementioned functions enable us to measure or evaluate the way in which the product's functions are fulfilled. These criteria define the acceptability limits, such as tolerance intervals (see Chapter 3 on GPS), thus proving flexibility, and determine whether a project is approved or rejected on the basis of its metrological quality. For example, choosing H7/h6 quality of assembly instead of H7/p6 could influence a product's flexibility, and thus earn it a less favorable assessment.

1.10. Functional analysis in mechanical design

In mechanical design, the main objective of conducting functional analysis is to: *collect, characterize, rank* and finally *value* all the functions (in the FS) of a product:

– to "collect" means to determine and identify the main functions of the product;

– to "characterize" them is to list the assessment criteria and the flexibility of the functions;

– "ranking" the functions means evaluating the order of importance of the functions considered and adopted;

– To "value" a function is to assign a weighting value to its importance in the project.

i) *External functional needs analysis*: the client (the user) obviously expresses a requirement for a product for which they already have an idea of the function. Thus, the client is best placed to flesh out the details of the product, in the form of service functions and even constraints. The client defines the role of the product. The idea resembles a sort of "black box";

ii) *Internal functional needs analysis*: the client will have already expressed their requirements by an external functional needs analysis. The internal counterpart is a technical analysis which expresses the functional part. It contributes to the formalism of the architecture (a sketch which becomes a drawing defining the product), which expresses the technical functions of the components in the system. It is at this level that the designer contributes his/her opinion of the product.

1.10.1. *Product cost in mechanical design*

Cost is a factor which heavily influences the quality of a product being designed and manufactured. The cost of the product reflects the expenses incurred during the design, manufacture and commercialization routes. In design, that cost is closely related to the amount of time invested in studies (analyses, drawings, calculations, simulations, etc.).

1.10.2. *Creation- and monitoring sheets in mechanical design*

These "sheets" are important, because they contain all the relevant information.

PRE-PROJECT PROPOSAL DOSSIER	
Dossier manager:	Project manager: ....
Date dossier **created**: dd/mm/yy	Dossier last **updated**: dd/mm/yy
Documents by dossier #	References by dossier #
Classification of solutions by dossier	XXX –YY AG 01
Pre-project # 1	XXX –YY AG 02
Pre-project # 2	XXX –YY AG 03
Pre-project # J	XXX –YY AG 0J
Specific notes:	

In that "dossier", the aim is to gather together all the salient ideas from the pre-project stage to constitute an initial classification document.

Solution classification dossier – Calculation principles	
Dossier manager:	Project manager: ....
Reasons: RDM-TMM	Calculations of moments: e.g. pivot shafts
Date dossier **created**: dd/mm/yy	Dossier last **updated**: dd/mm/yy
Dossier Pb # 00 CALCULS 00	Dossier Solution # 00
Dossier Pb # 00 CALCULS 00	Dossier Solution # 00
Dossier Pb # 00 CALCULS 00	Dossier Solution # 00
Solution criteria:	
Metrology/Quality/	
GPS-Tolerance	
Assembly	
Other	
	Criterion used
Explanatory notes:	

The aim of classifying the ideas, by problem, topic or discipline, is to more clearly classify the interests at play in order to assign them to each specialized department.

DOSSIER OF PRE-PROJECT # 1	
Dossier manager	Project manager
Date dossier **created**: dd/mm/yy	Dossier last **updated**: dd/mm/yy
Setting out the problem at hand and the associated calculation principles	
Explanatory drawing with tolerance values, dimensional and geometric constraints	
Drawbacks to the solution	Advantages and relevance of the solution
10-point criteria	Norms and standards
Accuracy-Metrology-Quality control	
Assembly-disassembly	
Materials, characterization	
Points of ease (use, storage, etc.)	
Other	
Specific notes:	

DOSSIER OF PRE-PROJECT # 2	
Dossier manager	Project manager
Date dossier **created**: dd/mm/yy	Dossier last **updated**: dd/mm/yy
Setting out the problem at hand and the associated calculation principles	
Explanatory drawing with tolerance values, dimensional and geometric constraints	
Drawbacks to the solution	Advantages and relevance of the solution
10-point criteria (for example)	Norms and standards
Accuracy-Metrology	
Assembly-disassembly	
Materials	
Points of ease (use, storage, etc.)	
Other	
Specific notes:	

It is wise to ensure that the pre-project is very clearly documented.

DOSSIER OF PRE-PROJECT # 3	
Dossier manager	Project manager
Date dossier **created**: dd/mm/yy	Dossier last **updated**: dd/mm/yy
Setting out the problem at hand and the associated calculation principles	
Explanatory drawing with tolerance values, dimensional and geometric constraints	
Drawbacks to the solution	Advantages and relevance of the solution
10-point criteria (for example)	Norms and standards
Accuracy-Metrology	
Assembly-disassembly	
Materials	
Points of ease (use, storage, etc.)	
Other	
Specific notes:	

This is a logical progression from the above.

Preliminary dossier defining the pre-project	
	| Dossier CODE |
Dossier manager	Project manager
Date dossier **created**: dd/mm/yy	Dossier last **updated**: dd/mm/yy
Exhaustive list of duly indexed documents	
Documents by classification	References
Brief description of the pre-project	Dossier # ....
Drawings (2D and 3D) of the pre-project	Dossier # ....
Sketches of the adopted pre-project	References # ....
Design FMECA	Dossier # ....
Operating instructions (**if** assembly)	Dossier # ....
Safety scores	Dossier # ....
Relevant scores (ranges)	Dossier # ....
Other	
Specific notes:	
	| Dossier CODE |

From the preliminary dossier onwards, we begin to project the sketches and drawings for the adopted pre-project.

FINAL SELECTION OF THE ADOPTED PRE-PROJECT	
	| Dossier CODE |
Dossier manager:	Project manager: ....
Date dossier **created**: dd/mm/yy	Dossier last **updated**: dd/mm/yy
Clearly documented reasons and explanations	
Overall drawing with standard names of the components	
Other	
Specific notes:	

Once the final pre-project has been decided upon, it makes sense to present the defining drawings, the assembly and the manufacturing process. This part is duly documented, based on the norms and other design conventions.

FINAL DESIGN FMECA	
	| Dossier CODE |
Dossier manager:	Project manager: ....
Date dossier **created**: dd/mm/yy	Dossier last **updated**: dd/mm/yy
Function 1: .... Failure mode 1: .... Effect: .... Causes: .... Function 2: .... Failure mode 2: .... Effect: .... Causes: .... Etc. ...	
Note:	

The design FMECA must not contain significant uncertainties. We have restricted ourselves to the example of two functions. The designer may have more functions in mind than the project actually needs.

ASSEMBLY INSTRUCTIONS FOR THE ADOPTED PRE-PROJECT	
	| Dossier CODE |
Dossier manager:	Project manager: ....
Date dossier **created**: dd/mm/yy	Dossier last **updated**: dd/mm/yy
Clearly documented reasons and explanations	
Detailed drawing Assembly drawing	
Other	
Specific notes:	

Justification of the preliminary design	
	| Dossier CODE |
Dossier manager: …………………..	Project manager: …………………..
Date dossier **created**: dd/mm/yy	Dossier last **updated**: dd/mm/yy
Clearly documented reasons and explanations	
Schedule (Gantt, PERT) Sizing (GPS) and precise calculations	
Other	
Specific notes:	

The Gantt schedule clearly sets out the documented reasons for the preliminary project.

Additional dossier(s), if needed	
	| Dossier CODE |
Dossier manager	Project manager
Date dossier **created**: dd/mm/yy	Dossier last **updated**: dd/mm/yy
Clearly documented reasons and explanations	

Table 1.5. *Design dossiers: all the tables are in order*

It can quite easily be observed that the role of design in mechanics is crucially important, on more than one level. Design is based on intuition, calculation and core and applied sciences. It dictates the approach to production in the life cycle of a product or mechanism.

The quick and accurate development of software tools suited for CAD has enabled designers not only to draw better, but also to directly simulate the products of their designs. This considerably reduces the amount of time spent on prototypes. Prototypes make a valuable contribution to actually validating the choices made initially at the design stage, and proving the accuracy of the simulation analyses and calculations. However, for financial reasons, companies with powerful tools at their disposal tend to forego costly prototypes, instead basing their decisions on robust simulations. Powerful simulators yield conclusive results in terms of accidents, failure thresholds, etc.

1.11. Scientific writing on a project

Writing an exhaustive design dossier involves pulling together all the documentation pertaining to the conceptual study. The product must clearly satisfy all the issues raised in the functional specifications. The current chapter is focuses on documentation techniques. It targets the salient points and organization of workflow. As is the case in metrology [GRO 11], *vocabulary* is important in applied mechanical design. Mechanical engineering techniques are not ones in which we would use all our knowledge of mechanics. It is a domain of knowledge of engineering techniques. The universal syllabus for engineer training includes a vast domain of sciences of engineering. Science of Engineering includes solid mechanics, materials, fluid mechanics and thermodynamics, but not necessarily production techniques (Manufacturing), as suggested by the titles of many courses. The author, in his classes in Mechanical Engineering Techniques, has noticed a serious shortfall in Engineering. Sometimes, sadly, people confuse *Machining* and *Mechanical Engineering*. In mechanical design, the projects must be the subject of succinct reports, free of linguistic errors and properly documented. This means presenting and following a product-development process whose objective is stated in the FS from the client. Roughly speaking, the pathway is as follows:

1) Idea for the design project;

2) Study for the design of the project – FS;

3) Preliminary design (systems/subsystems) – pre-project;

4) Detailed design in distinct segments (calculations and drawings);

5) Production of a plan for the manufacture of working prototypes;

6) Validation of the prototype and WRITE-UP (of a report);

7) Confirmed realization of the idea.

1.11.1. *Project process*

The objectives of a development process consist of:

– defining the process: WHO does WHAT, WHEN and HOW to achieve the objective;

– constructing models of one or more systems (sketches and calculations);

– clearly and methodically organizing the progression of the project;

– managing the life cycle of the project from start to finish;

– managing the risks and dangers of the design;

– repeatedly obtaining products of consistent quality.

The activities of development consist of:

– planning and carrying out a feasibility study;

– specifying the theoretical and material needs;

– analyzing (with sketches and drawings based on succinct calculations);

– designing/creating (GPS/ISO technical specification);

– implementing (exact coding if the design is complex);

– testing (components, systems and subsystems);

– integrating and testing (metrology, quality control, esthetics, etc.);

– writing up and delivering;

– maintaining and providing after-sales services.

1.11.2. *Development of the conceptual model*

The development of the conceptual model starts with an idea and looks at how to make it a reality.

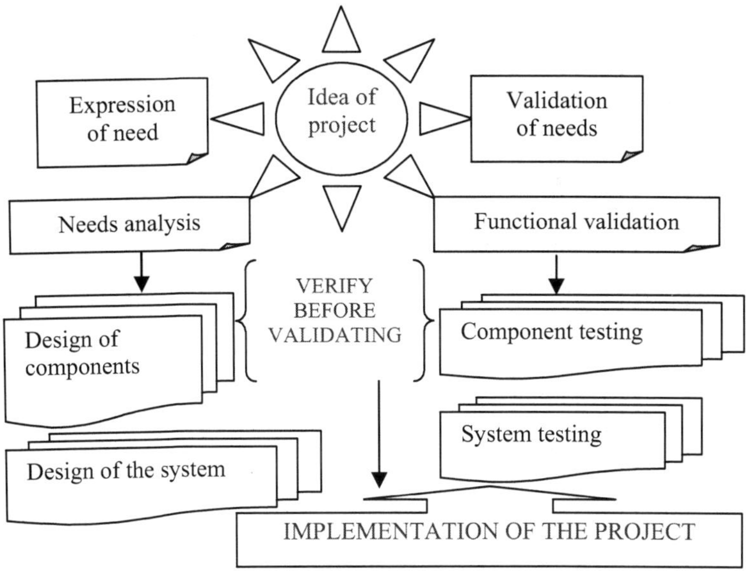

Figure 1.15. *Example of the conceptual model*

1.11.3. *Development (recap) on a spiral model*

There are various ways to present the development of an idea and its path through mechanical design; below is a spiral-type diagram of it.

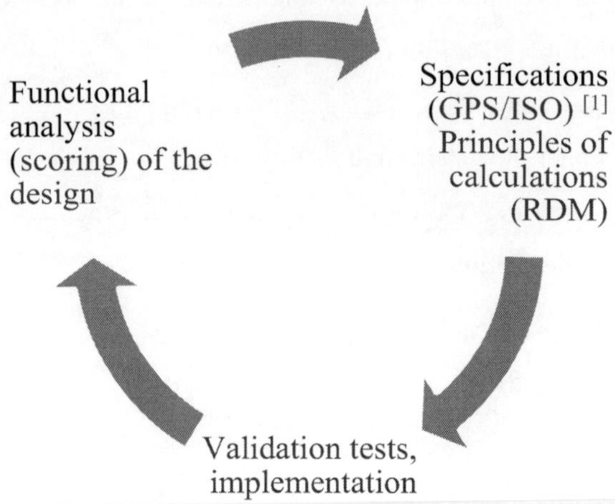

Figure 1.16. *Example of the spiral design model*

1.12. Esthetics of materials in mechanical design

In all future chapters of this book, we shall not touch on the esthetic aspect, focusing instead on the materials. Nevertheless, esthetics plays a not-insignificant role in design products. A well-designed mechanism, machine or even a simple component is functional, because it delivers on its function. If we want to give it a *handsome* and *pleasant finish*, we need to conform to the "*fashion*" in visually pleasing, *attractive* things.

In this book, the materials will be chosen on the basis of rigorous calculations. *Designers* are ordinarily fond of the esthetical dimension. They appeal to fashion sense, where advertising has a definite role to play. It is not true that a mechanical designer does not pay attention to the finish, to the touch and the equilibrium of the products s/he designs. The topics stemming from surface topography (rugosity) and surface treatment hold a twofold interest. Rugosity, for example, is an extremely important requirement in tribology, machining and dimensional metrology [GRO 11]. Surface treatments which are required for protection also produce attractive colors (galvanization, anodizing).

Designers are well-versed in the arts and industry. There is no question, here, of fashioning a meaningless, unsafe or frustrating product. Instead, there is a certain degree of advantage if the product is *pleasing*. An *accomplished* design involves an inventive approach, because it serves a particular need, and the materials used are undeniably a part of that accomplishment.

… MATERIALS AND TEXTURE

In mechanical design (manufacture, assembly and processes), there is an intellectual insouciance in the belief that materials are simply chosen, and that, automatically, their characteristics will conform to the requirements of the design. New materials are *tested*, and often offer more advantageous possibilities (in terms of weight, strength, etc.). Aviation and sports bicycles are enlightening examples of this. From a simple table fork or plastic comb to gargantuan airplanes, it is wise to remain *curious* and *open* to the use of new products.

… AND WHAT OF ENVIRONMENT IN DESIGN?

It is easy to think that the environment and manufacturing techniques can adjust well to time requirements. Many societies are making progress in various technologies and are able to take account of environmental conditions well. There is no incompatibility as long as we remain aware of and receptive to the advantages of methods which take account of the environmental parameter. Certainly, "wild" developments cost us dearly, resulting in natural disasters, health crises and even leading to exodes… All materials contain energy. That energy can be extracted in a variety of ways, and we can even recover emissions and waste to turn them into continuing energy as well. We simply need to remain humble, attentive and imaginative. This is what certain people refer to as eco-energy [GOE 95]. Thus, biodegradability (or its opposite, toxicity) necessitates a certain degree of isolation during the recycling process.

What can a designer do? Alone, not a great deal, other than ponder, research more fully and, above all, listen more closely in order to *attempt to convince others* with powers of persuasion!

1.13. Conclusion

It seems an inalienable necessity that the design process is, first and foremost, iterative: *from concept to concretization, and to the details*. The design is what links parts to the shape, texture, etc. – in a word, the attraction for the *consumer*.

The design steps [FRE 85] take place in accordance with a fundamental principle of operation. The functional decomposition presents the system and subproducts depending on the organization of the workflow [GRO 13], which is the subject of this chapter.

Concretization involves the dimensioning (geometry, stresses and environment) of the components. With the help of optimization techniques, we seek to maximize performances to finally put numbers to the ultimate design. It is at this point that the functional specifications (FS) are said to be concretized.

2

Materials and Geometry in Applied Mechanical Design, Followed by Case Studies

2.1. Introduction to materials in design

Design begins with the idea of a need, and finishes with its succinct realization. Each one of the steps in the design process involves choosing materials and procedures. A designer who is not familiar with the materials is destined for failure. He/she will overlook the greatest evolution of all time: the materials. Before the Stone Age – i.e. around 10,000 BCE – wood, ceramic, glass and natural waste products were used to make weapons (wood and flint). Stone bridges have stood the test of time, and withstood the march of armies. The point is that materials are of prime importance in design: designing and machining the parts without proper appreciation of the materials is tantamount to doing nothing at all.

However material's are not the only influential factor in design; rather, the process of design is guided by the physical form, found by geometrical processes. Nowadays, numerous materials are being "created" all the time. As their characteristics are not well known, designers are wary of using them, or simply refuse to. In this book, we shall use materials which are already well known in the specialized literature [OBE 16]. We do not look at materials science, but rather at the materials themselves. We will be using specialist design software tools, and will, *de facto*, employ the characteristics found in the reference works and specialized tables of the Appendix.

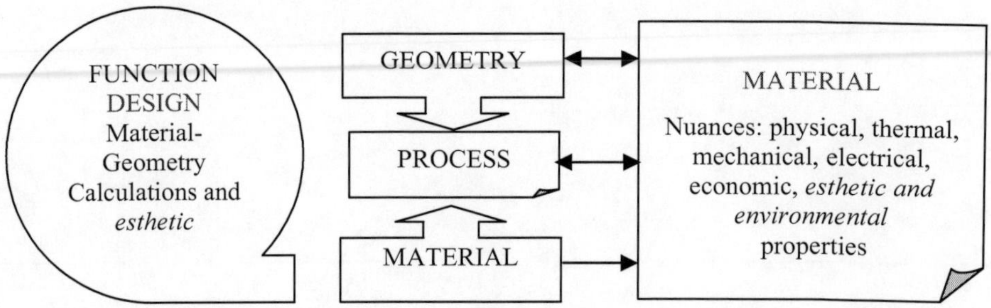

Figure 2.1. *Interaction between material and geometry through the process*

The mechanical components used in our case studies are subject to loads and conduct energy. They are also exposed to wear and tear, fatigue and various types of failure. The components are formed and/or machined to give them a geometry which sometimes has normalized gaps for interchangeability. The question on the mind of any designer catering for a particular set of functional specifications is: what is the right choice out of the vast range of materials available [LEW 90] to serve the expressed requirement? An engineer's rational judgment leads him/her to make multiple choices in connection with the material–geometry couple. The further the engineer advances in his/her design, the more critical the criteria of choice become. At the final stage, the options are fairly restricted, because they are guided by considerations of price and environment. A worthy design requires an innovative use of new *materials* in combination with intelligent exploitation of the technical and esthetic properties.

Whilst design is an iterative process, it is, at the heart of it, intuition-based. We invariably begin with a set of technical specifications – i.e. a definition of the function (hence the requirement for the defining drawings), such as, say, a scooter or even a rocket. Based on knowledge of the materials, physics and experimental data (both successes and failures), we succinctly draw up a pre-project plan. That plan needs to be improved by systematic methods until the project is finally ready to be delivered.

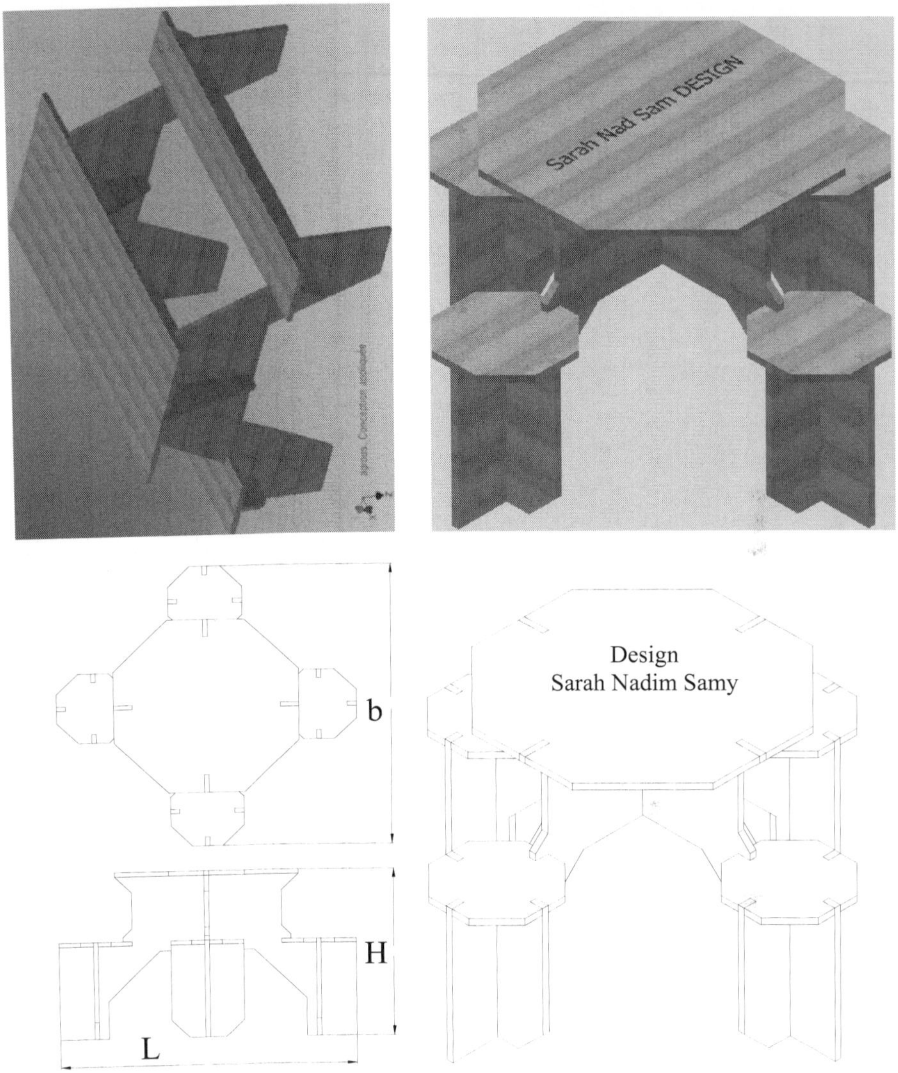

Figure 2.2. *Design of tables based on the material–geometry couple. For a color version of this figure, see www.iste.co.uk/grous/design.zip*

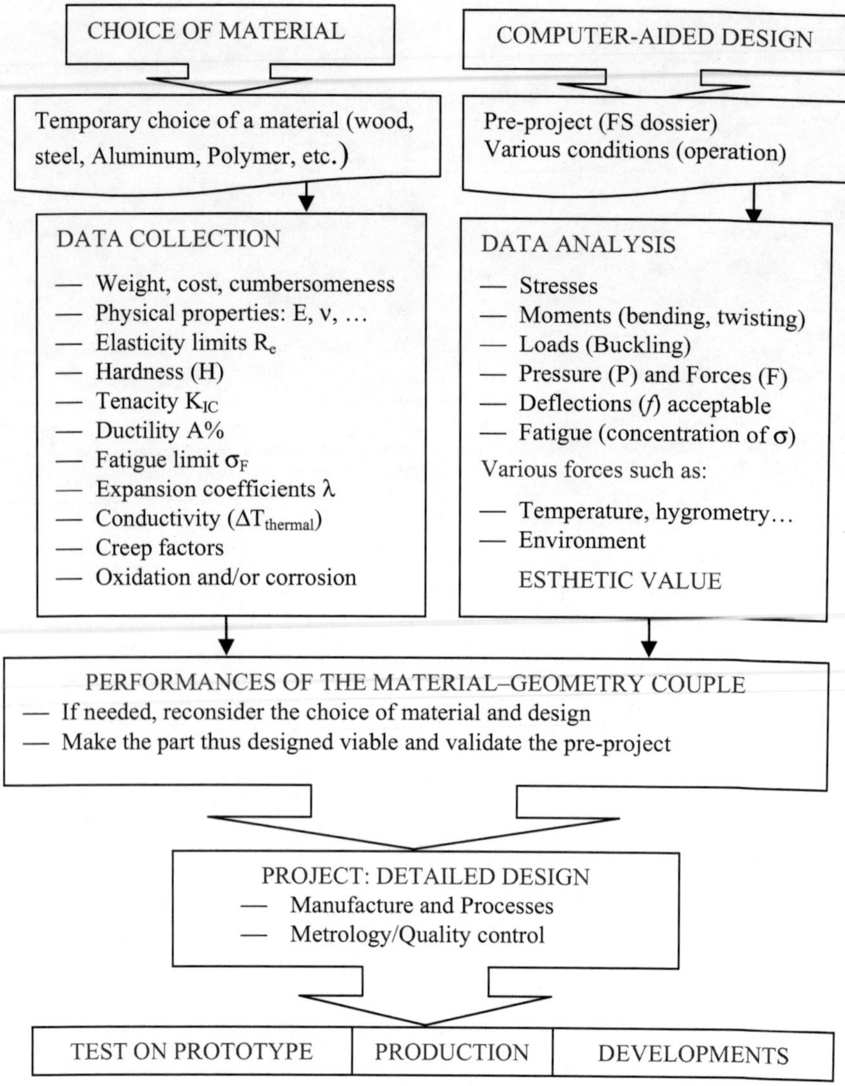

Figure 2.3. *Design approach based on the material–geometry couple*

The choice of materials [FAR 89] is highly important in design. The laws of mechanics are already established. Each material exhibits different strengths and weaknesses, and it is up to the designer to choose wisely in connection with the FS.

For example, in order to sell a racing bike, it needs to look esthetically attractive and lightweight, so requires optimum use of the materials in the design. The performance of the standard section imposes practical quotas for the finesse of the sections. For a given material, those quotas are as follows (Table 2.1).

Wood		Wood	
Name of species	Density ρ expressed in kg/m^3	Name of species	Density ρ expressed in kg/m^3
Acacia	780–820; 0.78–0.82	Maple	560–840; 0.56–0.84
Mahogany	560–850; 0.56–0.85	Ash	840; 0.84
Alder	460–550; 0.46–0.55	Guaiacum	1339; 1.339
Birch	520–730; 0.52–0.73	Beech	800; 0.8
Dry cedar	490	Larch	540–630; 0.54–0.63
Hornbeam	759–900; 0.759–0.9	Elm	540–630; 0.54–063
Chestnut	550–740; 0.55–0.74	Poplar	390; 0.39
White oak	610; 0.61	Red pine	660; 0.66
Holm oak	983; 0.983	Plane tree	650; 0.65
Ebony	1120–1200; 1.12–1.20	Fir tree	450; 0.45

Table 2.1. *Performance of the standard section. Source: [ASH 99]*

Published works dealing with design may seem vague, as they are sometimes confused with publications in mechanical construction. An aptitude for design is almost like Arts and Industry (intuition and technique) in a research hub (RH): choice of material and drawings. We make an initial attempt with one material (chosen on the basis of its strength, cost and sometimes aesthetic value), and we garner the characteristics needed for the calculations in connection with the sketch. We work by means of a functional breakdown, meaning that the product itself constitutes a technical system. This aspect will be discussed in detail over the course of the book, through the lens of our case studies. It must be remembered that the data used here are taken from software packages, standards and handbooks [OBE 16].

2.2. Optimization of mass in mechanical design

In order to make the right choice of lightweight material to ensure the proper stiffness of the structures, let us begin by examining a specific case study:

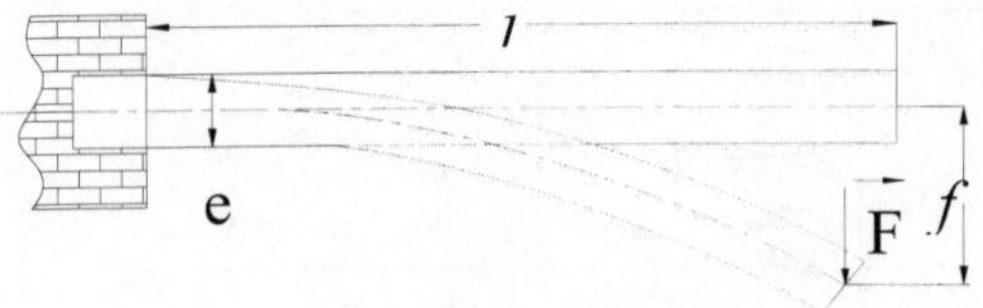

f is the deflection in m or in
E is the Young's modulus in GPa
F is the force in N
e is the thickness in m
l is the length of the beam in m
ρ is the density in Mg/m^3

Figure 2.4. *Anchored beam of thickness (e) and length (l), subjected to a force/load F. For a color version of this figure, see www.iste.co.uk/grous/design.zip*

For: $l = 0.5$ m; F = 500 N; E = 200 GPa; $e_{steel} = 0.1$ m; $\rho = 7.8$ mg/m^3

– *Calculating the deflection (f) of the anchored beam:*

$$f_{steel} = 4l^3 F / Ee^4 = 1.250\times10^{-5}\ m = 4.921\times10^{-4}\ in \qquad [2.1]$$

– *Calculating the mass of the beam while ignoring self-weight:*

$$M_{steel} = l \times e^2 \times \rho = 9.36\times10^{-7}\ Kg = 8.598\times10^{-8}\ lb \qquad [2.2]$$

– *Verification* $e_{steel} = \sqrt{M_{steel}/(l\times\rho)} = 0.1\ m \rightarrow QED:$

Using the expression of [$e_{calculated}$] in that of f_{steel}, we can calculate the deflection in linear elasticity

$$f_{calculated} = \frac{4\,l^5 F}{E} \times \frac{l^2 \rho^2}{M^2} = 1.25\times10^{-5}\ m = 4.921\times10^{-4}\ lb \leftarrow QED \qquad [2.3]$$

– *Verifying the mass of the beam while ignoring self-weight:*

$$M_{calculated} = \sqrt{\frac{4l^5 F}{f_{steel}}}\sqrt{\frac{\rho^2}{E}} = 9.36\times10^{-7}\ kg = 8.598\times10^{-8}\ lb \leftarrow QED \qquad [2.4]$$

CONCLUSION.– To minimize the mass of the beam at a given stiffness F/deflection, we need to choose a material with a minimal $\sqrt{\rho^2/E}$.

Sample Material	$10^3 \sqrt{\rho^2/E}$ **in** $\sqrt{N}/m^3 s^2$	**Density ρ in Kg / m³**	**E in GPa**
Steel	17	7.8	200
Aluminum	10	2.7	69
Wood →	**→ 5.5**	**→ 0.6**	**→ 12**
Concrete	12	2.5	47

Table 2.2. *Main optimization values pertaining to a beam*

DISCUSSION.– In mechanical design, we always expect the structure to withstand a load (a force, F in N) without bending or deforming over a given length (l, in m). In cases where the structure is used for means of transport, for instance a car, an airplane, a boat, a bike, a train, a rocket, or even a simple backpack, it is highly advisable to reduce its *weight* as far as possible. In the example shown in the table above we can see that wood offers a better alternative. Note that wood is used in small constructions (tennis racket handles, vaulting poles, etc.).

2.3. Case study of modeling based on the material–geometry couple

Initial data for a design (modeling) for the legs of a table made of Canadian birch, which has a density of ρ = (0.52-0.73) kg/m³ and a mass of m = (520-730) kg.

– r is the radius of the leg in m;

– l is the length of the table leg in m;

– ρ is the density in Kg/m³;

– J is the moment of inertia of the area of the column (table leg) in m⁴;

– E is the elasticity modulus (Young's modulus) in GPa;

– P is the maximum load that can be applied to the table (four legs) in N;

– M (l) is the mass of the leg in kg or in mg.

Drawing upon the existing body of literature, we can say that this material has the following physical properties:

Nominal density	Density (kg/m³)	Hardness	Bend strength, MPa	Elasticity/rigidity modulus, MPa
0.571	640	4320	94.8	12900

Table 2.3. *Physical properties*

PROJECT QUESTION FOR A TABLE.– For a garden table, we consider: r = 0.01 m, l = 0.35 m 0.40, …, 0.75 m; E = 12.9 GPa; ρ = 0.571 Kg/m^3. We wish to design a table whose legs will not buckle with an optimum choice between the geometry and the material.

ANSWER.– Calculate the minimized mass m (l) applied to the table supported by four legs of length (l) in m.

$$M(l) = \pi \times r^2 / l \times \rho \quad [2.5]$$

M(l) = $\frac{m^4}{kg}$

	0
0	$1.572 \cdot 10^{-3}$
1	$1.375 \cdot 10^{-3}$
2	$1.223 \cdot 10^{-3}$
3	$1.1 \cdot 10^{-3}$
4	$1 \cdot 10^{-3}$
5	$9.17 \cdot 10^{-4}$
6	$8.464 \cdot 10^{-4}$
7	$7.86 \cdot 10^{-4}$
8	$7.336 \cdot 10^{-4}$

QED

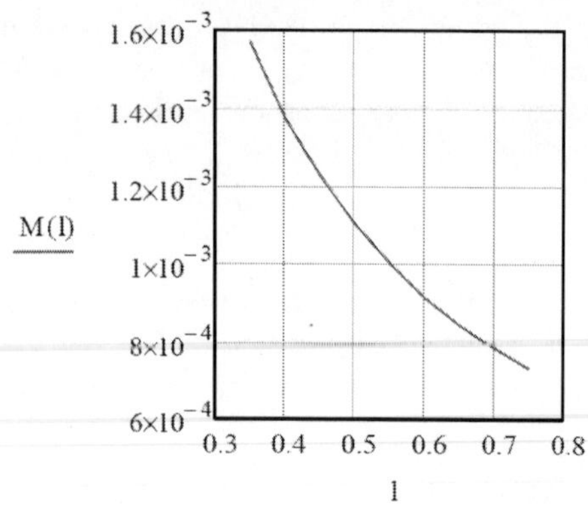

Figure 2.5. *Weight supported by the four table legs. For a color version of this figure, see www.iste.co.uk/grous/design.zip*

The table must be able to hold the critical elastic load P_{crit} without the legs buckling. J is the moment of inertia of the area of the column: $J = \pi r^2/4 = 7.854 \times 10^{-9}, m^4$.

By virtue of Euler's law, we can state the expression of the critical load P_{crit} as follows:

$$P_{critical}^{Euler} = \pi^2 EJ/l^4 = \pi^3 Er^4/4l^4 \ [N] \quad [2.6]$$

$P_{crit}(l) =$ (N)

	0
0	$8.163\cdot10^3$
1	$6.25\cdot10^3$
2	$4.938\cdot10^3$
3	$4\cdot10^3$
4	$3.306\cdot10^3$
5	$2.778\cdot10^3$
6	$2.367\cdot10^3$
7	$2.041\cdot10^3$
8	$1.778\cdot10^3$

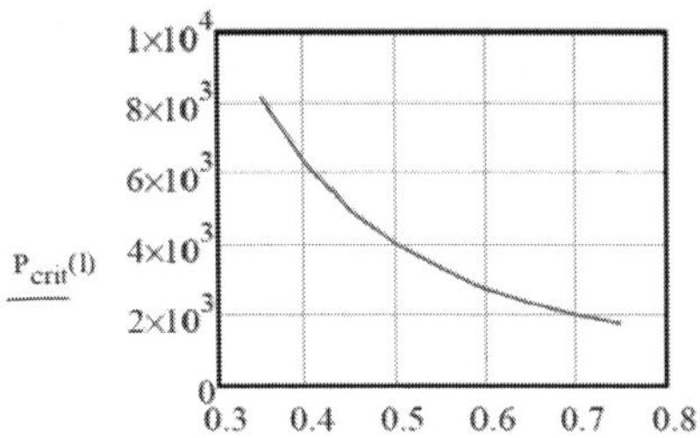

$P_{critical}(l) =$ (·lbf)

	0
0	$1.835\cdot10^3$
1	$1.405\cdot10^3$
2	$1.11\cdot10^3$
3	899.193
4	743.135
5	624.44
6	532.067
7	458.772
8	399.641

Figure 2.6. *Critical buckling load P_{crit}. For a color version of this figure, see www.iste.co.uk/grous/design.zip*

Once we have isolated the variable (r) and substituted it back into the expression M(l), we obtain an expression which verifies the weight initially calculated. Otherwise the model is wrong:

$$M(l) = \pi r^2 / l\rho \text{ or indeed } M(l) \geq \left(4P_{crit}^{Euler}(l)/\pi\right)\left(l^2\right)\rho/\pi\sqrt{E} \quad [2.7]$$

DEMONSTRATION THROUGH CALCULATION.–

The properties of the material are expressed by $\rho/\pi\sqrt{E}$. Thus, we can minimize the mass by choosing materials with the highest possible index: when $\rho = 0.571$ Kg/m³ = 5.71×10^5 Mg/m³, we obtain the index M_1 as follows:

$$M_1 = \frac{\rho}{\sqrt{E}} = 6.29\times10^{-6}\sqrt{GPa}\left(Mg/m^3\right)^{-1} = 6.29 \cong 6\sqrt{GPa}\left(Kg/m^3\right)^{-1} \quad [2.8]$$

$M(l) =$ ($\frac{m^4}{kg}$)

	0
0	$1.572\cdot10^{-3}$
1	$1.375\cdot10^{-3}$
2	$1.223\cdot10^{-3}$
3	$1.1\cdot10^{-3}$
4	$1\cdot10^{-3}$
5	$9.17\cdot10^{-4}$
6	$8.464\cdot10^{-4}$
7	$7.86\cdot10^{-4}$
8	$7.336\cdot10^{-4}$

QED

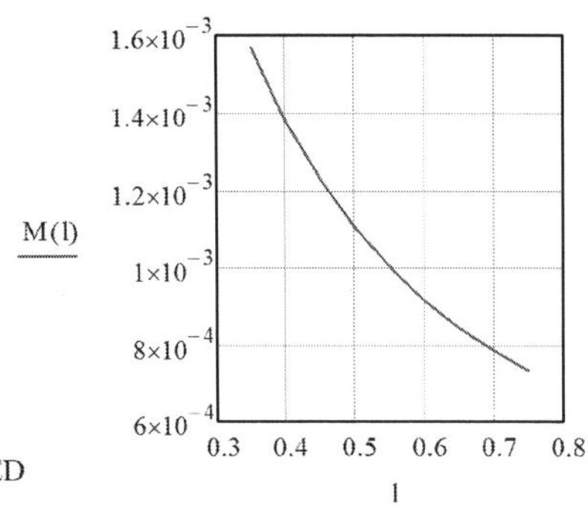

Figure 2.7. *Proof of calculation of the weight supported by four table legs. For a color version of this figure, see www.iste.co.uk/grous/design.zip*

We can calculate the finesse of the table legs by isolating (r) in equation [2.6]. Thus, for an applied P of 56 N (arbitrary choice), we obtain:

$$r(l) = \sqrt[4]{4P/\pi^3} \times \sqrt{l} \times \sqrt[4]{l/E} \ [m] \tag{2.9}$$

r(l) =

	0
0	$2.878 \cdot 10^{-3}$
1	$3.077 \cdot 10^{-3}$
2	$3.263 \cdot 10^{-3}$
3	$3.44 \cdot 10^{-3}$
4	$3.608 \cdot 10^{-3}$
5	$3.768 \cdot 10^{-3}$
6	$3.922 \cdot 10^{-3}$
7	$4.07 \cdot 10^{-3}$
8	$4.213 \cdot 10^{-3}$

m

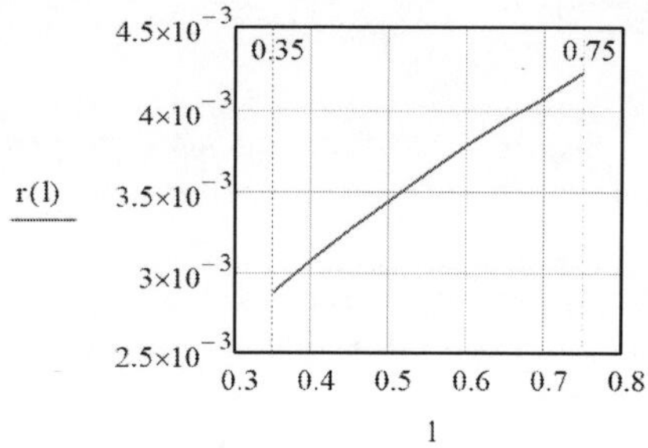

Figure 2.8. *Radii of the four table legs as a function of the lengths chosen. For a color version of this figure, see www.iste.co.uk/grous/design.zip*

The fineness of the legs is maximized by choosing the material with the highest index. Let us set the expression for the fineness index $M_2 = E = 12.9$ GPa $\Rightarrow$ QED

2.4. Geometry by standard sections in strength of materials

In order to achieve an optimized design, it is important to ensure a judicious choice between the cross-sections (areas) and strength is adequately researched, otherwise the design would be subject to caution. Ordinarily, the limits are imposed by the manufacturing constraints and the properties of the material. Those ultimate limits of performance of a given geometry are often twofold, meaning that:

– the limit is empirical (usual cross-section: square, rectangular, circular, etc.) and depends on the geometry of the material;

– the mechanical stability of the geometric sections (see table below):

i) Prismatic forms, as standard sections for profiled parts, are often made by the processes of lamination, stretching, sawing and extrusion. Strength of materials or applied physics manuals will contain the usual relations with the corresponding sections.

ii) In the elastic domain, where we operate, the geometric limits are often the result of the maximum moment of inertia J_{max} in m^4 as a function of the cross-section (or area of the cross-section in m^2). The mathematical expression of the calculations is presented in this simplified form:

$$Log(J) = 1.5 \times Log(A) + Log\left\{\Psi_{geom}\left(1/4\sqrt{\pi}\right)\right\} \quad \left[m^4 10^{-10}\right] \qquad [2.10]$$

– $\Psi_{max}\left(\frac{e}{F}\right)$ is a form factor = 1; A is the area in 10^{-6} m^2

– J_{max} is the maximum moment of inertia of the area in 10^{-10} m^4

To recap with a simple example, here is an indicative table:

Cross-section	Forms	Geometric classes	Nuance and/or processes of the materials	Geometric and/or resistance characteristics
Prismatic cross-sections	Full	Circular Rectangular Complex[1]	Hot laminated Cold laminated, Alu 6061, sawn wood, composites, etc. …	Weight (Kg), Load (N) length, width, height, thickness (m). Properties of cross-section, (S, m^2) moments of inertia (*J*, m^4); modulus of resistance of cross-section (W, m^3) for bending or twisting. Properties of structures (bending rigidity couple, E.I, Nm^2); Bend strength N.m and twist strength. Buckling (Euler P, N)
	Hollow enclosed	Circular Rectangular Complex[1]		
	Hollow open	I-shaped cross-section (double T) U cross-section L cross-section Complex cross-sections [1]		

Table 2.4. *Usual prismatic forms.*[1]*The Finite Element Method is used for the calculation*

Nuances of materials	$\Psi_{max}\left(\frac{e}{F}\right)$	$\Psi_{max}\left(\frac{e}{T}\right)$	$\Psi_{max}\left(\frac{f}{F}\right)$	$\Psi_{max}\left(\frac{f}{T}\right)$
Structural steels	65	25	13	7
Aluminum, 6061	44	31	10	8
GFRP and CFRP	39	26	9	7
Nylons, PE (Polymers)	12	8	5	4
Wood	5	1	3	1
Elastomers	< 6	3	–	–

Table 2.5. *Upper bounds of the form factors. Source: [ASH 99]*

Optimization of the principles of calculation in design is based on in-depth knowledge of geometry and of the materials. The material–geometry couple minimizes costs and improves the design, by making it easier. In mechanics, it is typical to make an initial choice of prospective material in connection with the geometry of the cross-section, and then proceed to shape it. The best geometry–material couple is distinguished by the highest index. A number of examples are given below.

MATERIAL AND GEOMETRY (WEIGHT, VOLUME OR COST) [1]		
Minimum calculation by the geometry–material couple in rigidity/strength		
Mathematical links of Load Geometry Stresses	Strength	Rigidity
Bar under traction (T)	$\sigma_{bending}/\rho$	E/ρ
Beam under bending (F)	$\sqrt[3]{\Psi_f^F \times \sigma_f}/\rho$	$\sqrt{\Psi_f^e \times E}/\rho$
Bar and/or tube under torsion (T)	$\sqrt[3]{\Psi_T^f \times \sigma_f}/\rho$	$\sqrt{\Psi_T^e \times E}/\rho$
Column under compression (e)	σ_f/ρ	$\sqrt{\Psi_F^e \times E}/\rho$
Energy calculation for minimum volume (or weight), case of springs		
Load, geometry and stress	Bending	Twisting
Energy stored as a function of the minimum volume, case of springs	$\dfrac{\left(\Psi_F^f \times \sigma_f\right)^2}{\Psi_F^e \times E}$	$\dfrac{\left(\Psi_T^f \times \sigma_f\right)^2}{\Psi_T^e \times E}$
Energy stored as a function of the minimum weight, case of springs	$\dfrac{\left(\Psi_F^f \times \sigma_f\right)^2}{\Psi_F^e \times E \times \rho}$	$\dfrac{\left(\Psi_T^f \times \sigma_f\right)^2}{\Psi_T^e \times E \times \rho}$

Table 2.6. *Indices for optimization of the geometry –material couple. Source [ASH 99].* [1] *We consider the density ρ by the expression of cost in the index relation $C_m(\rho)$*

Material	E, modulus in GPa	ρ, density in mg/m^3	Ψ_F^e, form factor	$\sqrt{E}/\rho$, mod./density index	Ω_{opt}^1, optimization index
Balsa	4.2 → 5.20	0.17→ 0.24	01→ 02	11	11→ 15
Spruce	9.8 → 11.9	0.36→ 0.44	01→ 02	09	09→ 12
Steel	200 → 210	7.80→ 7.84	25→ 30	1.8	09→ 10
Al 7075 T6	71 → 730	2.80→ 2.82	15→25	03	12→15
CFRP [*]	100 → 160	1.50→ 1.60	10→ 15	07	23→ 28

Table 2.7. *Materials used in design. [*] CFRP = Carbon Fiber Reinforced Polymer*

Design is an area in which we must find the best compromise between the geometry of an object and the material from which it is made. The form index intervals are based on the mean values of the properties. There are always *limitations* to the use of the materials, related to the form factors. These limits are related to the stresses caused by manufacturing and buckling. For instance, wood is relatively inflexible in terms of being shaped. The form factors reach no higher than 5. This is a size restriction. For steels, we can achieve form factors of up to around 50. Thus, the modes of damage present a penalizing parameter, which is the general plasticity (crushing due to the elastic limit and the geometry):

$$F_{buckling-local} = \pi^2 EJ/l^2 \qquad [2.11]$$

$J = \pi r^3.e$ (mm^4) is the moment of inertia, E the elasticity modulus (MPa) of the material in the column (slender tube). The ratio of the buckling equation for F (N) to the area A (mm^2) enables us to isolate the ratio (J/A^2) to express the form factor:

$$\Psi_F^e = 4\pi^2 l/A^2 = r/e \qquad [2.12]$$

It is a dimensionless, so-called form factor. In the knowledge that $\sigma = F/A$ (MPa) expresses the axial stress in the wall of the tube, we deduce the buckling stress:

$$\sigma_{1-buckling} = \sqrt{\pi^2 E\Psi F/4\,l^2} = \pi\sqrt{E\Psi F}/2\,l \quad [MPa] \qquad [2.13]$$

In the design of bikes (suspension forks and frames), mechanical strength and stiffness are two highly important factors to deliver. The forks are subjected to bending forces. In addition, if it were a racing bike, it would need to be as light as possible. Thus, the question of the appropriate choice of material [SHA 93] arises (see table above).

2.4.1. *Choice of materials in design (airplanes and bikes)*

Normally, we look for a material which achieves an optimal combination of the material–geometry couple to minimize weight alongside a specified bending stiffness. The index value to be maximized, then, appears thus:

$$M_{\text{maximized}} = \Psi_F^e \frac{\sqrt{E}}{\rho} \left[\frac{N}{mm^2} \frac{10^3\ mm^3}{Mg} \right] or \left[10^3\ \frac{Nmm}{Mg} \right] \quad [2.14]$$

– $\Psi_{\max}\left(\frac{e}{F}\right)$ is a form factor (given);

– E is the elasticity modulus in MPa or N/mm^2;

– ρ is the density (mass per volume of the material in mg/m^3).

For bicycles, scooters and other similar vehicles (see previous figures), we can model the design of the forks, for example, by choosing a material represented by a beam of length (*l*) to withstand a maximum load P (in N) without failure or strain. The tube can be considered a tube of thickness e (e in mm) and radius r (r in mm). The objective is to minimize the weight on the fork. In order to do so, the optimized couple to offer good strength is based on the best fit of the geometry–material couple represented by an index expressed thus:

$$\Omega^1_{opt\,\text{maximized}} = \sqrt[3]{\Psi_F^f \times \sigma_f} \times \left(\rho^{-1}\right) \quad [2.15]$$

One of the most important parameters to consider in bike design is the strength of the frame and the forks, which undergo bending stress. In professional racing bikes, weight is extremely important. Readers need only refer to the published works specializing in materials to find a wide variety thereof.

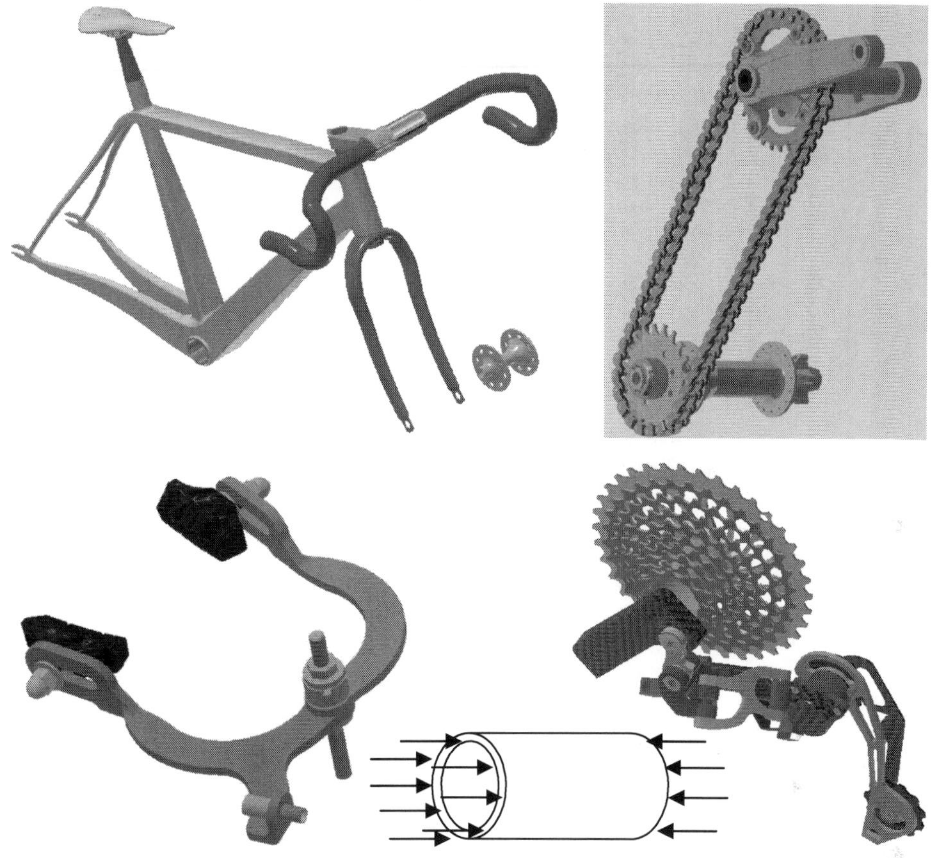

Figure 2.9. *Designed bike parts in 2D and 3D (Inventor Pro). For a color version of this figure, see www.iste.co.uk/grous/design.zip*

The usual forms of the quadratic moments are calculated using the well-known formulae for material strength, some examples of which are given below.

$$E_M = \int_0^{\pi} \frac{M_{bending}^2}{2A\varepsilon E} d\varphi \text{ knowing that } \int_0^{\pi} \frac{P5\sin(\varphi)}{2A\varepsilon E} d\varphi \rightarrow \frac{25\pi P^2}{4A\varepsilon E} \quad [2.16]$$

The following table represents a few examples of cross-sections. Readers may refer to strength of materials publications to find lists of the usual cross-sections.

Usual cross-sections	J_{xx}, in m^4	K, in m^4	W_{axis} m^3 resistance moment	A, area in m^2 V_{axis}, volume in m^3
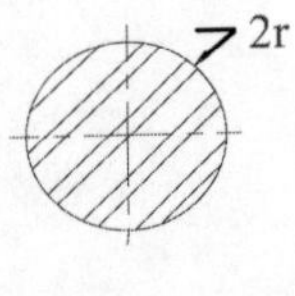	$\frac{\pi}{4}r^4$	$\frac{\pi}{2}r^4$	$\frac{\pi r^3}{4}$	$A=\pi\times r^2$ $V=\frac{\pi\left(r_0^4-r_i^4\right)}{2r_0}\approx 2\pi r^2 e$
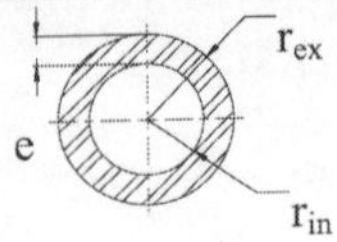	$\frac{\pi\left(r_{ex}^4-r_{in}^4\right)}{4}$ $\approx \pi r^3 e$	$\frac{\pi\left(r_{ex}^4-r_{in}^4\right)}{2}$ $\approx 2\pi r^3 e$	$\frac{\pi\left(r_{ex}^4-r_{in}^4\right)}{4r_{ex}}$ $\approx \pi r^2 e$	$\pi\left(r_{ex}^2-r_{in}^2\right)\approx 2\pi re$ $V=\pi r^3/2$
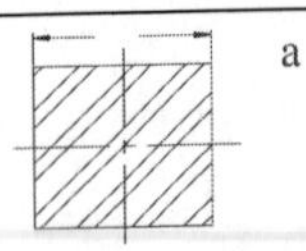	$a^4/12$	$0.14\times a^4$	$a^3/6$	$A=a^2$ $V=0.21\times a^3$
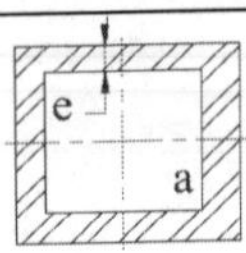	$\frac{2a^3e}{3}$	$a^3 t\left(1-\frac{e}{a}\right)^4$	$\frac{4a^2e}{3}$	$A=4ae$ $V=2a^2e(1-e/a)^2$
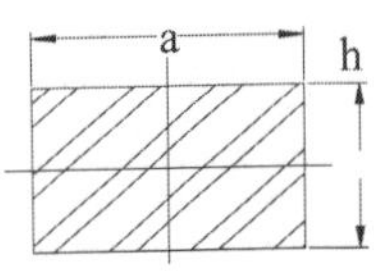	$\frac{bh^3}{12}$	$\frac{hb^3}{3}\left(1-0{,}58\frac{b}{h}\right)$ $(h\succ b)$	$\frac{b\times h^2}{6}$	$A=h.b$ $V=\frac{b^2\times h^2}{3h+1.8b}(h\succ b)$
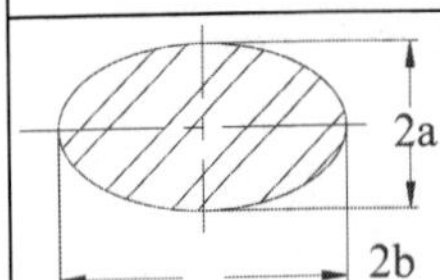	$\frac{\pi a^3 b}{4}$	$\frac{\pi a^3 b}{\left(a^2+b^2\right)}$	$\frac{\pi a^2 b}{4}$	$A=\pi\times(a.b)$ $V=\pi\frac{b\times a^2}{2}(a\prec b)$

Table 2.8. *Moments of the main geometric cross-sections (see Chapter 12)*

By way of example, the following is an illustrative table.

Material	σ_f strength in MPa	ρ, density in mg/m^3	ψ_F^f, form factor	$\frac{\sqrt[3]{\sigma_f}}{\rho}$, index	Ω_{opt}^1, optimization index
Al **6061** T6	$240 \rightarrow 260$	$2.69 \rightarrow 2.71$	$5.5 \rightarrow 6.3$	15	$47 \rightarrow 51$
Spruce	$70 \rightarrow 80$	$0.46 \rightarrow 0.56$	$01 \rightarrow 1.5$	36	$36 \rightarrow 50$
Steel	$770 \rightarrow 990$	$7.82 \rightarrow 7.83$	$7 \rightarrow 8$	12	$44 \rightarrow 48$
CFRP [*]	$300 \rightarrow 450$	$1.50 \rightarrow 1.60$	$4 \rightarrow 4.5$	33	$83 \rightarrow 90$

Table 2.9. *Materials used in cycling (e.g. for the suspension forks). Sources: [WHI 85] and [SHA 93]*

2.4.2. *Form factors Ψ of some usual cross-sections*

Here, we present a few values and expressions of form factors. However, it is preferable to refer to specialized publications to find all the form factors, listed by cross-sectional geometry.

Usual cross-sections	Ψ_F^f	Ψ_T^f	Ψ_F^e	Ψ_T^e
Cross-sectional geometry	Strength		Stiffness	
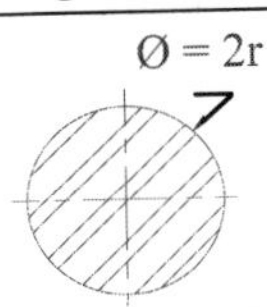	1	1	1	1
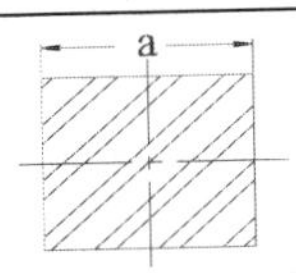	$1.03 = \frac{\pi}{3}$	0.74	$1.18 = \frac{2\sqrt{\pi}}{3}$	0.88
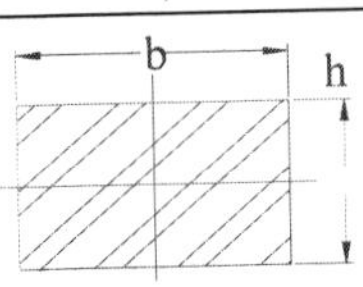	$\frac{2\sqrt{\pi}}{3}\sqrt{\frac{h}{b}}$	$\frac{2\sqrt{\pi}}{3}\frac{\sqrt{b/h}}{1+0.6b/h}$ $h \succ b$	$\frac{\pi h}{3b}$	$\frac{2\pi b}{3h}\left(1-0.58\frac{h}{b}\right)$ $h \succ b$
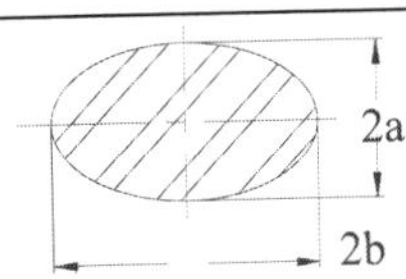	$\sqrt{\frac{a}{b}}$	$\sqrt{\frac{a}{b}}$; $a \prec b$	$\frac{a}{b}$	$\frac{2ab}{\left(a^2+b^2\right)}$

Table 2.10. *Form factors ψ. Source [ASH 99, ASH 00]*

2.4.3. *Form factors in mechanical design [ROA 75]*

The loading of components is axial, torsional (twisting) or flexural (bending). Bars are often subjected to traction, beams to bending, columns to axial compression and shafts to twisting. Here we present a few expressions of the calculations.

$$\text{Elastic bending } \sigma_F = K_1 \times EJ/l^3 \ \ [\text{MPa}] \qquad [2.17]$$

K_1 is a given constant. For example, in the case of simple bending, $K_1 = 3$ with embedding at one end. E is the elasticity modulus in MPa, l is the length in mm and J the moment of inertia. The form factor is dimensionless and is expressed thus:

$$\Psi_F^e = 4\pi \times \frac{J}{A^2} \quad \text{where } J = \int_{\text{section}} y^2 dA \qquad [2.18]$$

$$\text{Elastic torsion } \sigma_T = K \times G/l \ \ [\text{MPa}] \qquad [2.19]$$

The angle of torsion (θ, in degrees) depends on the torsion torque (M) as follows:

$$\theta = M \times J/K \times G \ \ [\text{deg}] \ \textit{and} \ J = \int_{\text{cross-section}} r^2 dA \qquad [2.20]$$

In this case, the form factor Ψ_T^e is dimensionless and is expressed as:

$$\Psi_T^e = 2\pi K / A^2 \qquad [2.21]$$

$$\begin{cases} \text{Bend strength } \sigma_F = M/W = M\gamma_m/J \\ \text{Twist strength } \tau_t = M_\tau/W = M_\tau r_m/J \end{cases} \qquad [2.22]$$

– M_τ is the bending moment in N.m;

– W is the resistance modulus $Q = J/r_m$ (mm^3);

– J is the moment of inertia of the surface in relation to the bending axis;

– A is the area of the cross-section in mm^2;

– G is the shearing moment (rigidity in GPa);

– J is the polar moment of inertia of the circular sections (mm^4 or cm^4).

– K is a constant known as the torsion constant, $K = A^2/2\pi = \pi r^4/2$;

– γ_m = 1/W is a section modulus;

Buckling of columns according to Euler $F_l = \left(\pi^2 EJ_{\min} \times c^2\right)/l^2$ [2.23]

2.5. Case study of design of multi-purpose items

The *time-consuming* aspect of design is pluri-disciplinary work on the specifications of the materials. We make drawings in accordance with the ISO/GPS standards, the calculations, the dynamic system analyses and the behavior of the environment. The esthetic quality is undeniably important to consider. Once we have reached the design stage, we need to interact with the suppliers on the subject of the materials. The choice of manufacturing process will be conditioned by the judicious choice of the materials used.

Metrology and quality control have a crucially important role to play here. We make a prototype product, test it repeatedly and assess its performances relative to the markets. When this stage has been concluded, the product is launched on the market. Thus, the task of the research hub (RH) is not limited to this period alone. The RH will conduct its analyses in connection with retrospection. Obviously, the failure of a moving rod is considerably damaging. With the aim of reducing the forces of inertia and consequently the Hertzian pressures (contact), it is common to use lightweight reds (6061), whose material is also strong, once subjected to its limit stress.

EXAMPLE.– Consider an aluminum 6061 (*Canadian nuance*) rod, which has a density of ρ (Kg/m^3). The area is represented by $A = b\ x\ W$, where b is the thickness of the rod and w is its width. L_{stroke} is the length (mm) from axis to axis. F is the force (N) applied to the rod.

$$m_{rod} = A \times L_{stroke} \times \rho \times [\beta]$$ [2.24]

– A is the rectangular area, given by the expression = $b.W$ (mm^2);

– L_{stroke} is the length of the rod (mm) = Stroke = 2xOA =2R = AB (mm);

– m rod is the mass of the rod in Kg;

– ρ is the density of the material in Kg/m^3;

– β is a factor to take account of the mass of the bore of the platform;

– $\theta(\tau)$ is inward motion $\theta(\tau) = \omega\,\tau$.

In design projects, it is important to solve the two essential problems of *analysis* and *summary*. The two represent a contrary dialectic unit: they are contrary because they are mutual opposites, and united because the two problems need to be addressed using the same mathematical model. Analysis and summary of the mechanisms at play are generally conducted in a number of steps.

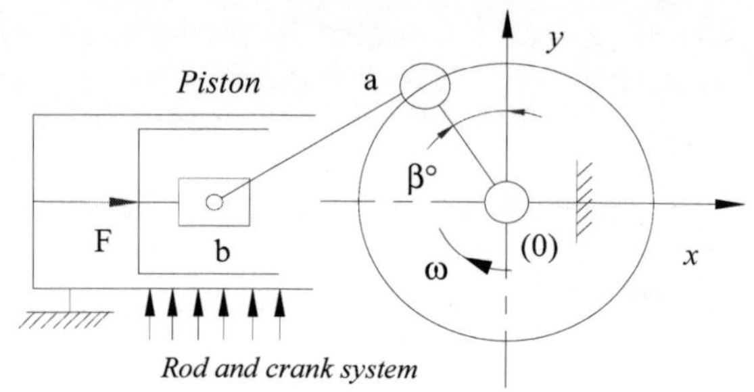

Figure 2.10. *Example of a diagram of the rod and crank system*

The rod is often designed to take account of the fatigue constraint [ROA 75], which is presented as follows: $\sigma_{endurance} \geq F/A$. The safety factor is not taken into account. The mass which satisfies the fatigue constraint is then written:

$$m_{fatigue} = FL_{stroke}\frac{\rho}{\sigma_{end}}[\beta] \;\; where\; M_{performance}1 = \frac{\sigma_{end}}{\rho} \qquad [2.25]$$

The rod is also subjected to buckling stress, which requires a maximum compressive load P where $F \leq P$ in N. By virtue of Euler's buckling relation, we set:

$$\begin{cases} F_{buckling} = \pi^2\, EJ/L_{rod}\;\; N\; or\; lbf; J = b^3W/12,\, mm^4 \rightarrow \\ where\; b = \alpha W \rightarrow m_{Buckling} = \sqrt{12F/\alpha\pi^2}\,L_{rod}^2\,\rho[\beta]/\sqrt{E} \end{cases} \qquad [2.26]$$

– E is the elasticity modulus of the material (in MPa);

– J is the moment of inertia of a rectangular form (in mm^4);

– α is a factor characterizing the transverse cross-section.

We therefore note the expression of the material's index: $M_{performance}2 = \sqrt{E}/\rho$. The table below indicates the parameters used for performance-oriented design [ASH 99].

Nuance of the materials	Elasticity modulus E GPa	ρ, Mg/m^3 density	σ MPa Endurance limit	m_1 mass Kg in σ fatigue	m_2 mass Kg in F, Buckling	m, Kg mass of the rod
Cast iron, FGS	178	7150	250	**0.21**	0.13	0.21
Steel 4041	210	7850	590	0.10	**0.13**	0.13
Al-SiC composite	110	2880	230	**0.09**	0.07	**0.09**
Ti-6-4	115	4400	530	0.06	**0.10**	0.10
USA foundry, Al	70	2700	75	**0.27**	**0.08**	0.27

Table 2.11. *Parameters for performance-oriented design*

Example of calculations of performance for the rods based on the characteristics owing to the peculiarities of the materials: $\mathbf{L_{buckling}}$, the length (150 mm) of the steel rod; **F** the load (50 KN); the coefficients (dimensional aspect factors) $\boldsymbol{\alpha} = 0.5$; $\boldsymbol{\beta} = 1$; E the elasticity modulus of spheroidal graphite cast iron (178 GPa); and $\boldsymbol{\rho}$ the density of the material (7150 Kg/m^3).

SOLUTION.– In fatigue, the performing mass of the rod is calculated using the following formula:

$$m_{fatigue} = FL_{stroke}(\rho\beta)/\sigma_{end} \text{ where } M_{performance}1 = \sigma_{end}/\rho$$

In buckling, the performing mass of the rod is calculated as follows with equation [2.26]:

$$m_{buckling} = \sqrt{12F/\alpha\pi^2} \times L^2 \times \rho\beta/\sqrt{E} \text{ where } M_{performance} = \sqrt{E}/\rho$$

$m(L_{Fatigue}) =$

	0
0	$1.43 \cdot 10^{-7}$
1	$1.502 \cdot 10^{-7}$
2	$1.573 \cdot 10^{-7}$
3	$1.644 \cdot 10^{-7}$
4	$1.716 \cdot 10^{-7}$
5	$1.787 \cdot 10^{-7}$
6	$1.859 \cdot 10^{-7}$
7	$1.93 \cdot 10^{-7}$
8	...

kg

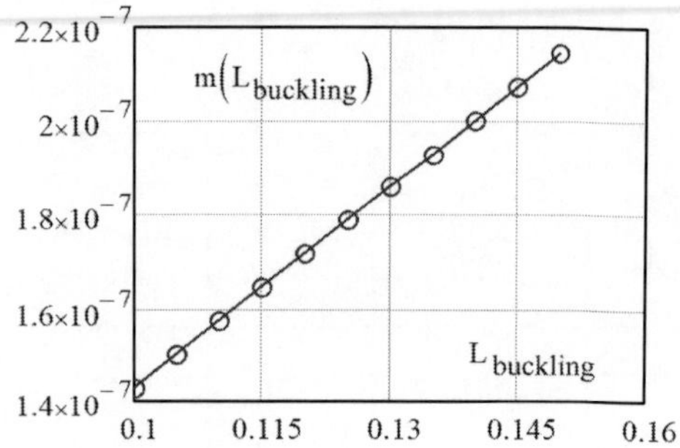

Figure 2.11. *Evolution of mass (with fatigue) as a function of the stroke of the rod. For a color version of this figure, see www.iste.co.uk/grous/design.zip*

$m(L_{buckling}) =$

	0
0	$5.909 \cdot 10^{-8}$
1	$6.515 \cdot 10^{-8}$
2	$7.15 \cdot 10^{-8}$
3	$7.815 \cdot 10^{-8}$
4	$8.509 \cdot 10^{-8}$
5	$9.233 \cdot 10^{-8}$
6	$9.987 \cdot 10^{-8}$
7	...

kg

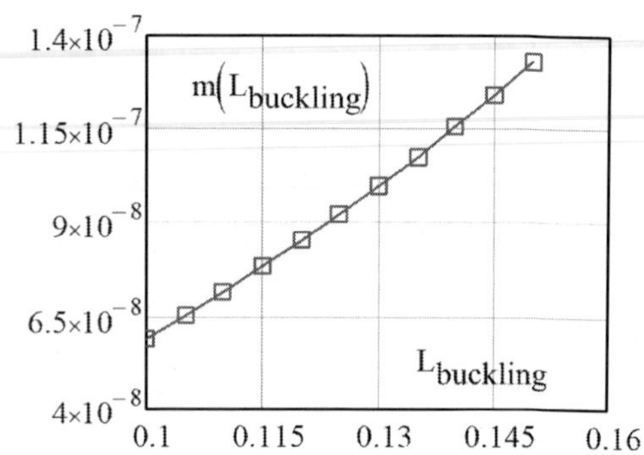

Figure 2.12. *Evolution of mass (with buckling) as a function of the stroke of the rod. For a color version of this figure, see www.iste.co.uk/grous/design.zip*

To avoid fatigue and buckling, we consider the greatest mass between $m_{buckling}$ and $m_{fatigue}$. By equalization two masses, we are able to write the following:

$$\begin{cases} m_{buckling} = \sqrt{\dfrac{12F}{\alpha\pi^2}} L^2 \dfrac{\rho\beta}{\sqrt{E}} \ ; \ M_{perf}2 = \dfrac{\sqrt{E}}{\rho} ; m_{fatigue} = FL_{stroke}\dfrac{\rho\beta}{\sigma_{end}} \\ hence \rightarrow M_{perf}1 = \sigma_{end}/\rho \ and \ M_{perf}2 = \sqrt{12L^2/\left(\alpha F\pi^2\right)} M_{perf}1 \end{cases} \quad [2.27]$$

– the quantity $\sqrt{12L^2/\alpha F\pi^2}$ is known as the coupling constant, and $\left(F/L^2\right)$ represents the structural load on the rod. Thus, the materials which exhibit the best combination in M_{perf2} and $M_{perf\,1}$ will be chosen for the final design project;

– in summary, performance 2 under fatigue is equivalent to performance 1 under buckling, but adjusted by $\sqrt{12L^2/\alpha F\pi^2}$.

NUANCE OF THE MATERIAL	COMMENTS
Magnesium alloys →	Generally, offer good performances
Titanium alloys →	Ti-6-4, which has a high (F/L^2) ratio, is a good choice
Beryllium alloys →	Procedure difficult to implement with (F/L^2)
Aluminum alloys →	Poor performances but are cheaper than Ti and Mg

Many designed rods are made with diverse materials. For example, aluminum and magnesium are often used in recreational car construction. For race cars (with ultra-high-performance engines), titanium and beryllium are the preferred materials. However, rods are difficult to design with *composites*, for behavioral reasons, except if we use continuous fibers.

2.6. Case study of superposed bimetallic materials [ROA 75]

PROBLEM.– Consider a smooth platform made of a bimetallic strip of steel alloyed with titanium measuring ¼ inches in width and 0.030inches in thickness, superposed on another strip of stainless steel 1/4in in width and 0.060 in thickness. Titanium: $E_{titanium} = 17\times10$PSI. ($\lambda$) is the expansion coefficient of titanium = 5.7×10^{-6} in/in °F (Fahrenheit). Stainless steel, $E_{inox} = 28\times10^6$PSI. (λ) is the expansion coefficient of inox = 9.6×10^{-6} in/in °F.

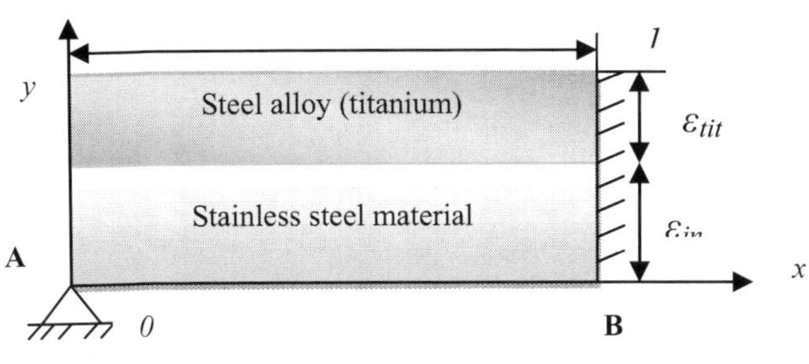

Figure 2.13. *Diagram of the calculation of superposed bi-materials*

– width (l) and thickness (ε) of the respective materials (Titanium, Inox): (l, ε_{tit}, ε_{inox}) = (0.25, 0.03, 0.06) *in* = (6.35, 0.762, 1.524) *mm* =;

– elasticity moduli of the respective materials (Titanium, Inox): (17, 28) $\times 10^6$ *PSI* = (1.172×10^5, 1.931×10^5) MPa;

– expansion coefficient of the materials: (5.7, 9.6).10^{-6}) *in/in* or (5.7, 9.6).10^{-6}) *mm/mm*;

– temperature variation: $\Delta T = 50°F = 10°C$.

QUESTION.– We wish to calculate the maximum bending stress and the required length of bimetal which develops a reaction force of 0.142 (5 oz) of the two in the support (see sketch). The temperature corresponds to 50°F = 10°C.

SOLUTION.– Width and thickness of the materials. Force of reaction on the side of the left terminal (A) $\rightarrow$ $\{l\ \varepsilon_{\text{titanium}}\ \varepsilon_{inox}\} = \{0.32\ 0.042\ 0.072\}\, in =$ $\{8.128\ 1.067\ 1.829\}\, mm$

The elasticity moduli of the materials [ROA 75] are calculated thus:

$$\{E_{\text{titanium}}\ \ E_{inox}\} = \{16.25\ \ 28.35\}\times 10^6\, PSI = \{1.12\ \ 1.955\}\times 10^5\, MPa$$

Expansion coefficient: $\{\lambda_{\text{titanium}}\ \ \lambda_{\text{inox}}\} = \{5.7\ \ 9.6\}\times 10^{-6} = \{5.7\ \ 9.6\}\times 10^{-6}$

The variation in temperature is: $\Delta\tau$ = 54°F = 12.2225°C = 285.372°K. The temperature difference begins at the left end, where $\delta = 0$ in = 0 mm. Let us begin by finding the value of K_1 and then evaluating the equivalent rigidity:

$$K_1 = 4 + 6\frac{\varepsilon_{tit}}{\varepsilon_{inox}} + 46\left(\frac{\varepsilon_{tit}}{\varepsilon_{inox}}\right)^2 + \left(\frac{\varepsilon_{tit}}{\varepsilon_{inox}}\right)^3 \frac{E_{tit}}{E_{inox}} + \frac{\varepsilon_{inox}}{\varepsilon_{tit}}\frac{E_{inox}}{E_{tit}} = 11.37 \qquad [2.28]$$

NOTE.– The following relation is valid only for materials of the same width, as is the case with our design here. The equivalent product of (EJ) would be:

$$EJ = \frac{l\varepsilon_{inox}{}^3\varepsilon_{tit}E_{inox}E_{tit}}{12\varepsilon_{tit}E_{tit}\varepsilon_{inox}E_{inox}}K_1 = 2{,}428\times 10^6\ N.mm^2 = 846.056\ lbf.in^2 \qquad [2.29]$$

Along the entire length of the layers, the temperature of the bimetal, causing incurvation of both layers of the strip thus formed. In order to do so, we first calculate the temperature gradient $[T_2\text{-}T_1]/\tau$ for the two metals together, by what we shall call the equivalent temperature:

$$6\left(\lambda_{inox}-\lambda_{tit}\right)\Delta\tau\left(\varepsilon_{tit}-\varepsilon_{inox}\right)/\varepsilon_{inox}^{2}K_1=2.32227\times10^{-3}\ in^{-1}=9.1428\times10^{-5}\ mm^{-1}$$

$$\begin{cases} R_A=\left[-3\left(L^2-\delta^2\right)/3L^2\right]EJ\times\left[6\left(\lambda_{inox}-\lambda_{tit}\right)\Delta\tau\times\left(\varepsilon_{tit}-\varepsilon_{inox}\right)\right]/\varepsilon_{inox}^{2}K_1\ simplify \\ \rightarrow\left|-5\ lbf/16\right|=0.3329.\text{kg.m}^2/s^2L\rightarrow L=9.431\ in=239{,}545\ mm \end{cases}$$

This is equivalent to the calculation of L carried out manually:

$$L=\frac{-\left(9EJ\Delta\tau\lambda_1\varepsilon_1+9EJ\Delta\tau\lambda_2\varepsilon_2-9EJ\Delta\tau\lambda_1\varepsilon_1-9EJ\Delta\tau\lambda_2\varepsilon_2\right)}{R_A\left(\varepsilon_2^2\times K_1\right)}=9.431\ in=239.545\ mm$$

The *dangerous section* is situated at the point where the bending moment is at its maximum value. Thus, we calculate the maximum $M_{bending}$: $M_{fl}^{max}=R_AL=-2.947\ lbfin=-332.984\ Nmm.$ The bending stress is also at the top of the composite (layer > Titanium):

$$\begin{cases} \sigma_{upper\ layer}^{max}=\dfrac{-6\times M_{fl}^{max}}{l\times\varepsilon_2^2K_1}\left(2+\dfrac{\varepsilon_2}{\varepsilon_1}+\dfrac{E_1\varepsilon_1}{E_2\varepsilon_2}\right)+(-1)\left(\dfrac{\lambda_2-\lambda_1}{K_1}\Delta\tau E_1\right)\times\ldots \\ \ldots\times\left(3\dfrac{\varepsilon_1}{\varepsilon_2}+2\left(\dfrac{\varepsilon_1}{\varepsilon_2}\right)^2-\dfrac{E_2\varepsilon_2}{E_1\varepsilon_1}\right)=3.767\times10^3\ PSI=25{,}972MPa \end{cases} \quad [2.30]$$

The bending stress at the bottom of the composite (lower layer = inox steel)

$$\begin{cases} \sigma_{lower\ layer}^{max}=\dfrac{-6M_{fl}^{max}}{l\varepsilon_2^2K_1}\left(2+\dfrac{\varepsilon_1}{\varepsilon_2}+\dfrac{E_2\varepsilon_2}{E_1\varepsilon_1}\right)+\left(\dfrac{\left(\lambda_2-\lambda_1\right)}{K_1}\Delta\tau E_2\right)\times\ldots \\ \ldots\times\left(3\dfrac{\varepsilon_1}{\varepsilon_2}+2-\dfrac{E_1}{E_2}\left(\dfrac{\varepsilon_1}{\varepsilon_2}\right)^3\right)=-3.151\times10^3\ PSI=-21{,}727\ MPa \end{cases} \quad [2.31]$$

The upper layer of the bimetal is subjected to a positive load. It acts on the lower layer – hence the negative sign of the stress, which means it bends downwards. In both mechanical design and civil engineering construction, such cases are common.

2.7. Curving and incurvate elements by sweeping of sheet metals

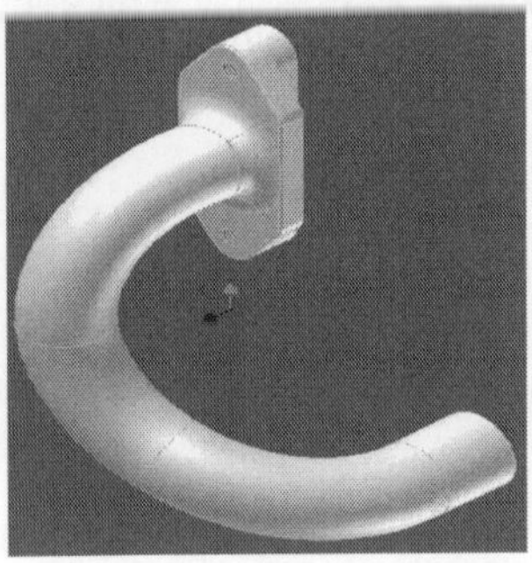

Case studies: hypothesis 1

R = 4.22 in = 107.188 mm
r_{in} = 3.25 in = 82.550 mm
h_{in} = 0.5 in = 12.7 mm
r_{ex} = 4.25 in = 107.950 mm
h_{ex} = 0.62 in = 15.748 mm
h = 1.125 in = 28.575 mm
F = 180 *lbf* = 800.68 N
L = 4.75 in = 120.65 mm

Figure 2.14. *Exhaust elbow designed with sweep function. For a color version of this figure, see www.iste.co.uk/grous/design.zip*

$$\begin{cases} r_n = h/\ln\left(r_{ex}/r_{in}\right) = 4.1936\ in = 106.518\ mm\ and\ \varepsilon = R - r_n \\ \varepsilon = 0.67\ mm\ and\ M_{flexion} = FL = 9.66\times10^4\ Nmm \end{cases}$$

For the section of R = 3.56 in = 90.424 mm, the stress on the internal fiber would be:

$$\begin{cases} Under\ compression \rightarrow \sigma_c = \dfrac{M_{flexion} h_{in}}{A\times\varepsilon\times r_{in}} = 2.696\times10^3\ psi = 18.591\ MPa \\ Under\ traction\ (tension) \rightarrow \sigma_t = \dfrac{M_{flexion} h_{ex}}{A\times\varepsilon\times r_{ex}} = 2557.8 psi = 17.628\ MPa \end{cases} \quad [2.32]$$

HYPOTHESIS 2.– If $r_{ex}1$ = 104.648 mm; $r_{in}1$= 79.248 mm and R_1 = 104.648 mm; L = 120.65 mm

$$\begin{cases} r_{n1} = \dfrac{h}{\ln\left(r_{ex1}/r_{in1}\right)} = 4.0465 in \rightarrow \varepsilon_1 = R_1 - r_{n1} = 0.0735 in \\ = 1.868 mm \rightarrow M_{flexion} = FL = 950\ lbfin = 96600 Nmm \end{cases} \quad [2.33]$$

For the section of R = 88.9 mm, the stress on the internal fiber would be: h_{in1} = 11.582 mm; h_{ex1} = 14.122 mm; A 1.194×10^3 mm^2

NEW DATA.–

$h_0 = 0.95$ in = 24.13 mm; b = 1.75 in = 44.45 mm; c = 0.56 in =14.224 mm

Polar moment of inertia $J = bh_0^3/12 = 0.125\ \left[in^4\right] = 5.204\times10^4\ \left[mm^4\right]$ ↙

Under simultaneous traction and compression and with the above data, we have:

The stress applied $\sigma = M_{flexion}c/J = 3.829\times10^3\,[psi] = 26.403[MPa]$

2.7.1. *Sensible choice of optimizing materials in Palmer micrometers*

Measuring devices [GRO 11] such as micrometers require absolute precision, exactness and fidelity, and undeniable capability. Indubitably, the rigidity and the uncontrollable dimensional variations limit the quality criteria, because of the temperature gradients caused. Designers need to carefully control elastic strain by correcting thermal expansion in normal thermal and environmental conditions ($\Delta\tau \cong 22°C$ and relative humidity RH = 55%).

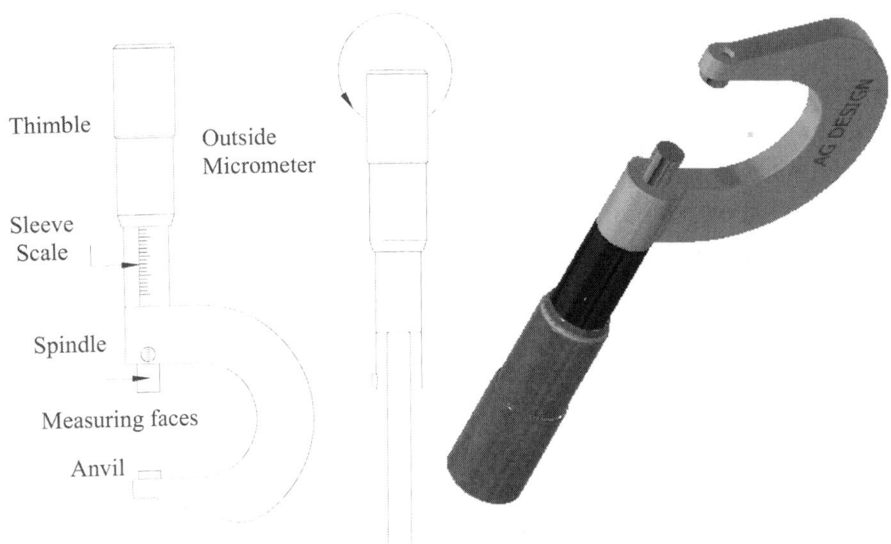

Figure 2.15. *Micrometer with minimized thermal distortion. For a color version of this figure, see www.iste.co.uk/grous/design.zip*

These instruments are often used in machining workshops, where neither the temperature gradients nor the hygrometry are always respected. This causes changes in shape (however minimal), causing distortions to the device [CEB 94]. This distortion cannot be compensated, and results in sensitivity to shock. Excitation engenders noise, which is the source of imprecision and inexactitude. Manufacturers in the field of metrology are aware of the potential for such distortions [CHE 87]. Users are advised of acceptable tolerance intervals and a capability Cp > 1.33. To make the best choice of the appropriate materials for high-precision instruments, it is advisable to refer to the specialized works in the field and to manufacturers' catalogs. In order for the micrometer, in our example, to be able to withstand the flow of heat from the workshops, from the hands of the operator or indeed, sometimes, the heat generated by the instrument itself, it is important to make wise decisions in view of Fourier's now-classic law.

$$Q_{heat} = -\lambda \times \{\partial\tau/\partial x\} \quad [2.34]$$

Q is the incoming heat per unit surface and λ is the heat conductivity. $\partial\tau/\partial x$ is the temperature gradient. We then take note of a certain degree of deformation (acceptable in metrology) assigned to an expansion coefficient (α), written as $\varepsilon = \alpha\Delta\tau = \alpha\left[\tau_0 - \tau\right]$.

Let us set: $$\frac{\partial\varepsilon}{\partial x} = \alpha\frac{\partial\tau}{\partial x} = Q_{heat} \times \left(\frac{\alpha}{\lambda}\right) \quad [2.35]$$

To reduce thermal distortion, the performance index will therefore depend on:

$$i_{performance} \cong \{\alpha/\lambda\} \quad [2.36]$$

Vibration is a real problem which can affect the device's sensitivity. In design, it is suggested to calculate the performance index, as follows:

$$i\left\{\begin{matrix} vibration \\ performance \end{matrix}\right\} \cong \{\sqrt{E}/\rho\} \quad [2.37]$$

A number of specialized publications [ASH 99, LEW 90, FAR 89, ROA 75] put forward formulae for performance indices, thus decreasing thermal distortion.

2.8. Conclusion

In this chapter, our focus has essentially been on the importance of the materials in connection with the geometric forms. This duality is at the very heart of the principles of calculations in design. The drawing stage of the design process poses no problems at all. Choosing the right material for a given application depends on the prices, the environment, safety and lifespan of the product. The complexity of the problem of choosing materials is attributable to the properties of the material, with which it is important to try to find an optimal compromise between: strength, tenacity, weight and geometric form(s). We shall offer further case studies in all the chapters of this book, and specifically in Chapter 11.

The use of crystalline and solid materials (steel, cast iron and other lightweight alloys) is still very widespread in industries such as car building, aviation, heavy industry, etc. Little by little, certain industries, such as aviation and cycling, are beginning to introduce new materials such as fiberglass composites, Kevlar, carbon fibers, etc., and ceramics (benchmark metrology) whose advantageous characteristics are linked to their low density (ρ), their high elasticity modulus (E), their hardness and also their ability to withstand high temperatures, which compensate for their fragility and lack of tenacity.

With such a hugely diverse range of choices of material available to designers, they must be guided by intrinsic criteria such as strength, rigidity, resistance to wear, fatigue, etc., and specific characteristics such as ageing, corrosion, electrical and/or thermal conductivity, etc. Many other criteria, such as the manufacturing processes (machining, press-forming, injection, welding, etc.) also need to be considered.

In particular, we must give consideration to scientific aspects, including the social criteria and the *decision-making structures* inherent to the manufacture locations (countries, tax arrangements, etc.). Those decision-making structures often go hand in hand with an understanding of the technologies and the multiple influences at play, which we shall not discuss here.

3

Geometrical Specification of GPS and ISO Products: Case Studies of Hertzian Contacts

3.1. Introduction

This book is not devoted specifically to functional scoring [GRO 11, GRO 13] (GPS: Geometrical Product Specifications), but adheres to the ISO recommendations on this subject – for example: ISO/TR 14638: 1995; ISO/TS 17450-2: 2005 and ISO 1101: 2012 (Tolerances of form, orientation, location and run-out).

Furthermore, it is important to remember that a mechanism is not the same thing as a machine, as it does not constitute an energy transformer, such as any kind of engine. The best-known mechanism, in mechanism and machine theory (MMT) is the crankshaft (piston) system, whose kinematic chain is characterized by a set of interconnecting solid elements. Figure 3.1 illustrates the closed chain for this system.

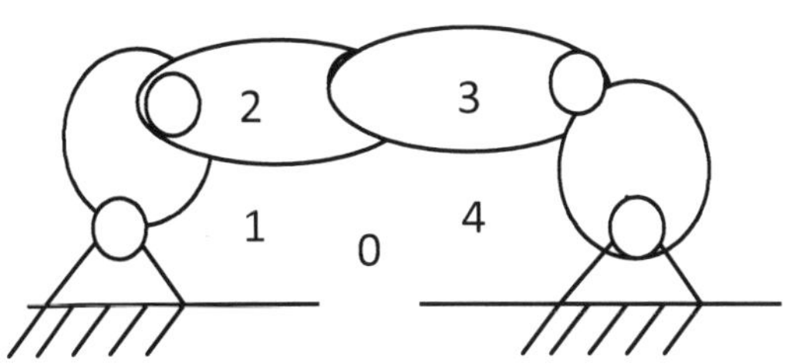

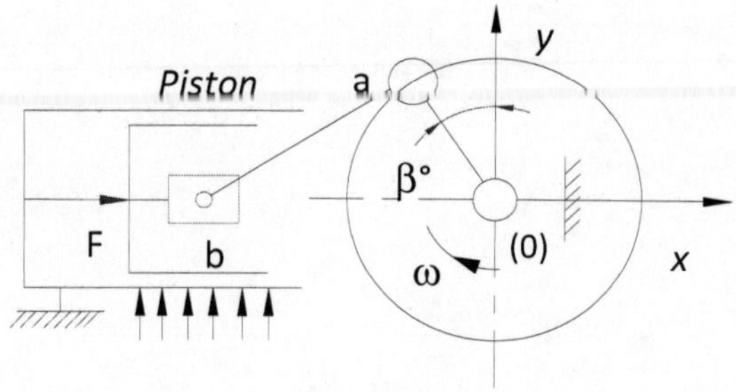

Figure 3.1. *Piston/crankshaft system*

In MMT, we speak of the existence of a continuous open chain (which does not constitute a mechanism as understood in MMT, because it does not carry out transformation of motion) and of complex chains. A complex chain is made up of several contiguous, interlocking closed chains. Ordinarily, we use the symbol (*n*) to denote the number of solid elements in the mechanisms, and (*l*) for the number of connections between those elements. *Nc* is the number of independent closed chains, with $N_c = \{L - n + 1\}$. The degree of hyperstatism of the mechanism is written as follows:

$$h = m_u + m_i + \sum_{i=1}^{n} n_s - 6(n-1) \qquad [3.1]$$

where:

– n is the number of solid elements in the mechanism, including the frame;

– m_u is the mechanism's degree of mobility (useful mobility);

– m_i is the internal mobility of individual elements;

– Σn_s is the number of static unknowns in relation to the connections.

Note that if h = 0, the system is said to be isostatic (the number of unknowns is the same as the number of equations formed). If h ≥ 1, the system is said to be hyperstatic of order h. This brief introduction to MMT is indicative of the importance of studying the logic of the mechanical connections in a system before sketching it out and connecting the solid parts to one another.

3.2. Dimensional and geometrical tolerances in design

Fits for each class	Assembly method	Tolerances	Mean interference	Tolerances for bore	Tolerance for shaft
Clearance fit	Interchangeable	$0.0025\ d^{2/3}$		$0.0025\ d^{1/3}$	$0.0025\ d^{1/3}$
Free fit	Interchangeable	$0.0014\ d^{2/3}$		$0.0013\ d^{1/3}$	$0.0013\ d^{1/3}$
Medium fit	Interchangeable	$0.0009\ d^{2/3}$		$0.0008\ d^{1/3}$	$0.0008\ d^{1/3}$
Normal fit	Selective	0.000	0.000 000	$0.0006\ d^{1/3}$	$0.0004\ d^{1/3}$
Press fit	Selective	0.000	0.00025 d	$0.0006\ d^{1/3}$	$0.0006\ d^{1/3}$
Medium-Forced fit	Selective	0.000	0.00050 d	$0.0006\ d^{1/3}$	$0.0006\ d^{1/3}$
Bracing fit (*)	Selective	0.000	0.00100 d	$0.0006\ d^{1/3}$	$0.0006\ d^{1/3}$

ISO: General GPS table. Main standards for fits of tolerances						
# links	1	2	3	4	5	6
Size ISO →	129 406 286	286 1829	286 8015 14660 1938	1938	13385 10360 463 3670 3611 9493 3599 6906 9121	3650 1938 3670
Distance ISO →	406 129	286			463 3599 6906 7863 13385 10360	

Table 3.1. *Dimensional tolerances in mechanical design*

In design, we use the contact law (Hertz) to determine the pressure on contact between two mechanical assemblies [ROA 75]. The parameters are:

– P_c is the pressure (Hertz pressure) exerted on the contact surface, in MPa or psi;

– T_{interf} is the total interference expressed in inches (in) or (mm);

– d_{in} is the internal diameter, in inches or mm (see Figures 3.2(a) and 3.2(b));

– d_{ex} is the external diameter, expressed in in or in mm (see Figures 3.2(a) and 3.2(b));

– d_c is the contact diameter, expressed in in or in mm (see Figures 3.2(a) and 3.2(b));

– ν_{in} is the Poisson ratio for the internal diameter;

– ν_{ex} is the Poisson ratio for the external diameter;

– E_{in} is the elasticity modulus for the internal diameter;

– E_{ex} is the elasticity modulus for the external diameter.

$$P_{contact} = T_{\text{interference}}^{total} \bigg/ \varnothing_c \left(\frac{d_c^2 + d_{in}^2}{E_{in}\left(d_c^2 - d_{in}^2\right)} + \frac{d_{ex}^2 + d_c^2}{E_0\left(d_{ex}^2 - d_c^2\right)} - \frac{v_{in}}{E_{in}} + \frac{v_{ex}}{E_{ex}} \right) \qquad [3.2]$$

If the diameters are made of the same material, this equation is reduced to:

$$P_{contact} = T_{\text{interference}}^{total} \bigg/ \frac{2 \times d_c^3 \left(d_{ex}^2 - d_c^2\right)}{E\left(d_c^2 - d_{in}^2\right) \times \left(d_{ex}^2 - d_c^2\right)} \qquad [3.3]$$

The tangential stresses on the contact surfaces are determined using Lamé's equations. They tend to be used in situations of fracture under maximum shear:

– σ_{tin} is the tangential stress on the surface at the d_{ex} in MPa or psi;

– σ_{tcex} is the tangential stress on the surface at the d_c for d_{ex};

– σ_{tcin} is the tangential stress on the surface at the d_{in} for d_{in};

– σ_{in} is the tangential stress on the surface at the d_c.

$$\left\{ \begin{aligned} \sigma_{dex} &= P_c \frac{2d_c^2}{d_{ex}^2 - d_c^2} \ ; \ \sigma_{d_c ex} = P_c \frac{d_{ex}^2 + d_c^2}{d_{ex}^2 - d_c^2} \\ \sigma_{din} &= -P_c \frac{2d_c^2}{d_c^2 - d_{in}^2} \ ; \ \sigma_{d_c in} = -P_c \frac{d_c^2 + d_{in}^2}{d_c^2 - d_i^2} \end{aligned} \right\} [MPa] \text{ or } [psi] \qquad [3.4]$$

Thus, the tangential stresses on the contact surfaces are determined using Birnie's equations. These equations are most often used in fracture mechanics:

$$\left\{ \begin{aligned} \sigma_{dex} &= P_c \frac{2d_c^2}{d_{ex}^2 - d_c^2} \ ; \ \sigma_{d_c ex} = P_c \left(\frac{d_{ex}^2 + d_c^2}{d_{ex}^2 - d_c^2} + v_{ex} \right) \\ \sigma_{din} &= -P_c \frac{2d_c^2}{d_c^2 - d_{in}^2} \ ; \ \sigma_{d_c in} = -P_c \left(\frac{d_c^2 + d_{in}^2}{d_c^2 - d_i^2} - v_{in} \right) \end{aligned} \right\} \qquad [3.5]$$

Forces and moments

– μ is the friction coefficient;

– L_{ex} is the contact length of the outer element, expressed in mm or in;

– d_a is the diameter of the shaft in mm or in;

– P_c is the contact pressure (Hertz) in MPa or psi.

$$F_{axial} = \pi\mu d_a L_{ex} P_c \; [N] \text{ or } [lbf], M = \pi\mu d_a^2 L_{ex} P/2 \qquad [3.6]$$

Expansion and/or shrinkage of assemblies

$$\Delta T = I_d / \alpha \times D_{in} \quad °C \text{ or } °F \qquad [3.7]$$

where:

– α, is the thermal expansion in °C or °F;

– I_d is the diametral interference in mm or in;

– D_{in} is the internal diameter of the bore prior to expansion, in mm or in;

– ΔT is the temperature variation in °C or °F.

3.2.1. *Case study of a bicycle wheel hub*

Typically, hubs are designed with almost twice the largest diameter of the bore. Note must also be taken of the shrinkage (or expansion) of the materials while the mechanism is in use. For example, class 8 (ISO) covers this type of system. The issue often encountered here is how to determine the stress (*psi* or *MPa*) and the force (*N*, *lbf*) which allow for this phenomenon of shrinkage of the materials in the system, as it moves:

– diameter of the shaft in contact d_c = 1.25 in = 31.75 mm;
– internal contact diameter d_{in} = 0.00 in = 0.00 mm;
– external diameter of hub, d_{hub}= 2.25 in = 57.15 mm;
– elasticity modulus E = 350×10^5 psi = 2.413×10^5 MPa;
– Poisson ratio of the two materials, ν = 0.3;
– $TO_{Linterference}$ = TO_{interf} = 0.0004 in, 0.0005 in, …, 0.0016 in.

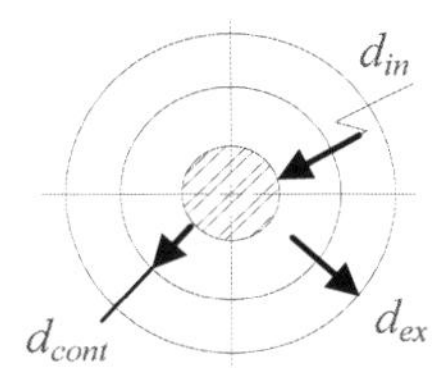

Figure 3.2(a). *Hub*

For the above data, we here propose the best solution. According to ISO, class 8, the bore ranges between 1.000 in and 1.0006 in, and the shaft varies between 1.00010 in and 1.0016 in.

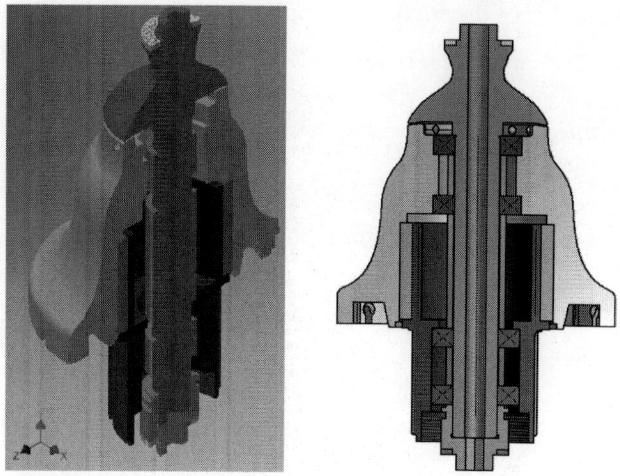

Figure 3.2(b). *Design of a bicycle wheel hub – drawn in Inventor Pro. For a color version of this figure, see www.iste.co.uk/grous/design.zip*

PRE-PROJECT 1.– Before designing the mechanism of the hub, we need to calculate:

1) the contact pressure for a single material (hub, shaft/bore) with the Poisson ratio of $\nu = 0.3$;

2) for class *8* mechanisms under the ISO system, determine the maximum and minimum tangential stress of the system being designed, taking account of the interferences;

3) calculate the axial force F_a and the moment M in N.mm and *lbf.in*;

4) calculate the variation in temperature $\Delta\tau$ of the hub, the shaft and the bore;

5) consider different materials and rerun the calculations for the system – i.e. the maximum and minimum tangential stresses allowable in the final project design.

Sketch and assemble the parts to present the project to the client in graphic and analytical form. Knowing the variation of the contact pressure: Pc = 5×10^3 psi, 1.25×10^4 psi to 1.8×10^4 psi, we calculate the tangential stress using the formula:

T_{interf} = (mm)

	0
0	0.01
1	0.013
2	0.015
3	0.018
4	0.02
5	0.023
6	0.025
7	0.028
8	0.03
9	0.033
10	0.036
11	0.038
12	0.041

P_c = (psi)

	0
0	$5\cdot10^3$
1	$1.25\cdot10^4$

P_c = (MPa)

	0
0	34.474
1	86.184

$P_c(T_{interf})$ = (psi) / $P_c(T_{interf})$ = (MPa)

psi	MPa
$3.8716\cdot10^3$	26.6938
$4.8395\cdot10^3$	33.3672
$5.8074\cdot10^3$	40.0407
$6.7753\cdot10^3$	46.7141
$7.7432\cdot10^3$	53.3876
$8.7111\cdot10^3$	60.0610
$9.6790\cdot10^3$	66.7344
$1.0647\cdot10^4$	73.4079
$1.1615\cdot10^4$	80.0813
$1.2583\cdot10^4$	86.7548
$1.3551\cdot10^4$	93.4282
$1.4519\cdot10^4$	100.1017
$1.5486\cdot10^4$	106.7751

From $P_{contact} = \dfrac{T_{interference}^{total} \times E\left(d_c^2 - d_{in}^2\right)\times\left(d_{ex}^2 - d_c^2\right)}{2\times d_c^3\left(d_{ex}^2 - d_c^2\right)}$, we deduce $\sigma_\tau(P_c) = P_c \dfrac{d_{ex}^2 + d_c^2}{d_{ex}^2 - d_c^2}$ [3.8]

$\sigma_\tau(P_c)$ =

	0
0	$9.4643\cdot10^3$
1	$2.3661\cdot10^4$

$\sigma_{\tau Min} := 9.4643\cdot10^3$ psi

$\sigma_{\tau Max} := 2.3661\cdot10^4$ psi

$\sigma_\tau(P_c)$ =

	0
0	65.254
1	163.135

$\sigma_{\tau Min} = 65.254$ MPa

$\sigma_{\tau Max} = 163.137$ MPa

SOLUTION TO QUESTION 3.–

- d_{ex} = 3 in = 76.2 mm; d_{in} = 0 in = 0 mm; d_{co} = $d_{contact}$ = 1.5 in = 38.1 mm;
- L_{hub} = 4.4 in = 111.76 mm;
- friction coefficient $\mu = 0.12$;
- Poisson ratio ν_{int} (steel) = ν_{ext} (cast iron) = 0.30;
- elasticity modulus, steel, $E_{int} = 25\times10^6$ psi = 1.724×10^5 psi;
- elasticity modulus, cast iron, $E_{int} = 12\times10^6$ psi = 8.618×10^4 psi;
- $\sigma_{\tau contact}$ = 4800 psi = 33.095 MPa (cast iron).

1) Calculation of $\Delta\varepsilon$, the *maximum interference* allowable on the ∅ of contact:

$$\text{From } P_{contact} = \Delta\varepsilon \Big/ d_{co}\left(\frac{d_{co}^2 + d_{int}^2}{E_{int}\left(d_{co}^2 - d_{in}^2\right)} + \frac{d_{ext}^2 + d_{co}^2}{E_0\left(d_{ext}^2 - d_{co}^2\right)} - \frac{\nu_{int}}{E_{int}} + \frac{\nu_{ext}}{E_{ext}}\right)$$

we isolate Δε. The maximum permissible diametral interference is: $\Delta\varepsilon = 8.006\times10^{4}\ in = 0.020335\ mm.$

2) *Axial force*: $F_{ax} = \mu\pi d_{co} L_{hub} P_c = 7.166\times10^3\ lbf = 3.188\times10^4\ N$

Calculation of the *transmitted moment* M, in *lbf.in* and *N.mm*. Put differently, this is the work, in J:

$$M = \mu\pi d_{co}^2 L_{hub} P_c / 2 = 5.374\times10^3\ lbf.in = 6.072\times10^5\ N.mm$$

3) Calculation of the temperature variation due to shrinkage of the materials:

– d_{max} = 2.006 in = 50.952 mm;

– d_{hub} = 3.26 in = 82.804 mm;

– expansion coefficient, α = 0.0000062 (1/Δ°F) = 1.116×10^{-5} (1/Δ°C);

– Poisson ratio, ν_{int} (steel) = ν_{ext} (cast iron) = 0.30;

– expansion (shrinkage) of the material = ε = 0.0078 in = 0.198 mm.

$$\Delta T = \varepsilon / \alpha d_{hub} = 385.9094\ [\Delta°F] = 241{,}394\ [\Delta°C]$$

As $T_{workshop} = 70\ °C = 38.889\ \Delta°C = 455.909\ \Delta°F = 38.889\ °K$

$$\Delta T_{min} = \Delta T + T_{workshop} = 455.909\ \Delta°F = 253.283\ °\Delta C = -19.867\ °C$$

$$N.B. \begin{cases} 20°C = 527.67\Delta°F\ and\ 70°F = 294.261\ \Delta°C \\ 20°C = 293.15\ \Delta°C = 293.15\ \mathrm{K}\ and\ 70°F = 294.261\ K \end{cases}$$

CASE STUDY 2.– On a shaft made of SAE 1018 steel (see Figure 3.2 of the bicycle wheel hub), with diameter d = 3.75 in = 95.25 mm is constructed using press fit, the bore of the layer of thickness ε = 0.54 in = 13.716 mm, with an elasticity modulus E of 2×10^5 MPa = 2.901×10^7 psi. The admissible stress on the steel hub is $[\sigma]_{adm}$ 155 MPa = 22480 psi. The device measures 82.55 mm = 3.25 in in length. The friction between the shaft and the hub is μ = 2/10. This is the peculiarity of the two SAE 1018 materials.

1) What is the value of the interference between the shaft and the tape?

2) Given this press fit, what value of the moment (torque) will be needed?

3) What is the contact pressure of the setup?

4) From the point of view of a designer, discuss the minimum and maximum interferences.

For:

$$\begin{cases} [\sigma]^{hub}_{adm} = 155\ MPa = 2.248\times10^4\ psi; E^{hub}_{shaft} = 2\times10^5\ MPa = 2.901\times10^7\ psi \\ d = 95.25\ mm = 3.75\ in; \varepsilon = 13.716\ mm = 0.54\ in; l = 82.55\ mm = 3.25\ in; \mu = 0.2 \end{cases}$$

SOLUTION WITH DISCUSSION.– The interference resulting from the proposed fit gives rise to the following admissible stress on the tape:

$$I^{fit}_{\max} = [\sigma]_{adm} \times (d/E) = 0.073819\ mm = 0.002906\ in$$

In view of the design of the fit (Figure 3.3), it is possible to calculate the *min max* interference and carry out the calculations for the diameters of the shaft. Hence, we are able to write the minimum interference of the order of $I_{\min}$ = 0.015 mm = 0.0006 in (see the results on the fit sketch produced at the design stage). We conclude that the admissible stress put forward in the FS during the preliminary stage is amply sufficient for the device to last a long time. Let us see what the minimum admissible stress, corresponding to $I_{\min}$, would be.

$$[\sigma]^{\min}_{adm} = I^{ajut}_{\max} \times (d/E) = 32\ MPa = 4.641\times10^3\ psi$$

N.B. This minimum admissible stress is suggested solely for the purposes of conceptual understanding. It need not be calculated, because when the FS is submitted, the materials have already been chosen by the client.

Let us calculate the contact pressure of the press fit with interference (P_{cont}).

Calculation of the moments generated by the press fit

$$\begin{cases} M^{\max}_{I\max} = I^{fit}_{\max} \times \pi \times \varepsilon \times l \times \mu = 1.05\times10^7\ N.mm = 92961\ lbf.in \\ M^{\min}_{I\max} = I^{fit}_{\min} \times \pi \times \varepsilon \times l \times \mu = 2.168\times10^6\ N.mm = 19192\ lbf.in \end{cases}$$

For: $d_{in} = 95.301\ mm = 3.752\ in; d_{ex} = 95.324\ mm = 3.75290\ in;\ d_c = 95.5\ mm = 3.75\ in$

The total interference δ^{press}_{total} is verified by both methods:

$$\delta_{total}^{press} = (-1)\left(I_{\min}^{fit} + I_{\max}^{fit}\right) = -(0.002906 + 0.0006) = -3.506\times10^{-3}\ in = -0.089\ mm$$

$$P_{contact} = \delta \Big/ \frac{2\times d_c^3 \times \left(d_{ex}^2 - d_{in}^2\right)}{E\left(d_c^2 - d_{in}^2\right)\times\left(d_{ex}^2 - d_c^2\right)} = 0.321\ MPa = 46.606\ psi;\ in\ addition \downarrow$$

$$\varphi_{coeff}^{press\ fit} = \frac{I_{\max}^{fit} E\left(d_c^2 - d_{in}^2\right)\left(d_{ex}^2 - d_c^2\right)}{P_{contact} d_c^3\left(d_{ex}^2 - d_{in}^2\right)} simplify \to \text{-}1.6578722190534870907 \lrcorner$$

$$P_{contact} = I_{\max}^{fit} E \frac{\left(d_c^2 - d_{in}^2\right)\left(d_{ex}^2 - d_c^2\right)}{\varphi d_c^3\left(d_{ex}^2 - d_{in}^2\right)} = 0.321\ MPa = 46.606\ psi\ where\ \varphi = -1.658$$

COMMENTS.– According to ANSI B4.1967 (R1974) standard, corresponding to ISO 286, the results obtained from a representation of the system using Inventor Pro (Figure 3.3) enable us to write ∅ 3.75H7/r_6 = ∅ 95.25H7/r_6 or LN1 (H6/n5): recommendation for press fit with interference, middle = -0.00175 in = -0.044 mm (*Press Fit*: fit of the setup with forced tightening (-)). Given this observation, a designer can consider a technique of construction with bracing. Thereby, the parts are heated to begin with, and then allowed to cool in their desired configuration. Finally, the fit will be locked in place (press fit).

3.3. Envelopes and cylinders under pressure (for R/e < 20)

Quite commonly, designers will work on mechanisms with thin or thick envelopes. Unfortunately, confusion can sometimes arise, as the calculations for these mechanisms are run without taking account of the thickness of the wall [ROA 75]. However, it is precisely here that distinctive calculation is demonstrably important. Hereinafter, we shall strictly limit ourselves to employing two cases, which are used in mechanical design. The internal pressure alone, p_{in}:

$$\begin{cases} \sigma_r = \left[\left(p_{in}r_{in}^2 - p_{ex}r_{ex}^2\right) - \left(p_{in} - p_{ex}\right)\left(r_{in}r_{ex}/r\right)^2 \Big/ \left(r_{ex}^2 - r_{in}^2\right)\right] \\ \sigma_{ray} = \left[\left(p_{in}r_{in}^2 - p_{ex}r_{ex}^2\right) + \left(p_{in} - p_{ex}\right)\left(r_{in}r_{ex}/r\right)^2 \Big/ \left(r_{ex}^2 - r_{in}^2\right)\right] \end{cases} \quad [3.9]$$

σ_r and σ_t are the radial and circumferential stresses. The axial stress is:

$$\sigma_{xx} = p_{in} \times r_{in}^2 \Big/ \left(r_{ex}^2 - r_{in}^2\right) \quad [3.10]$$

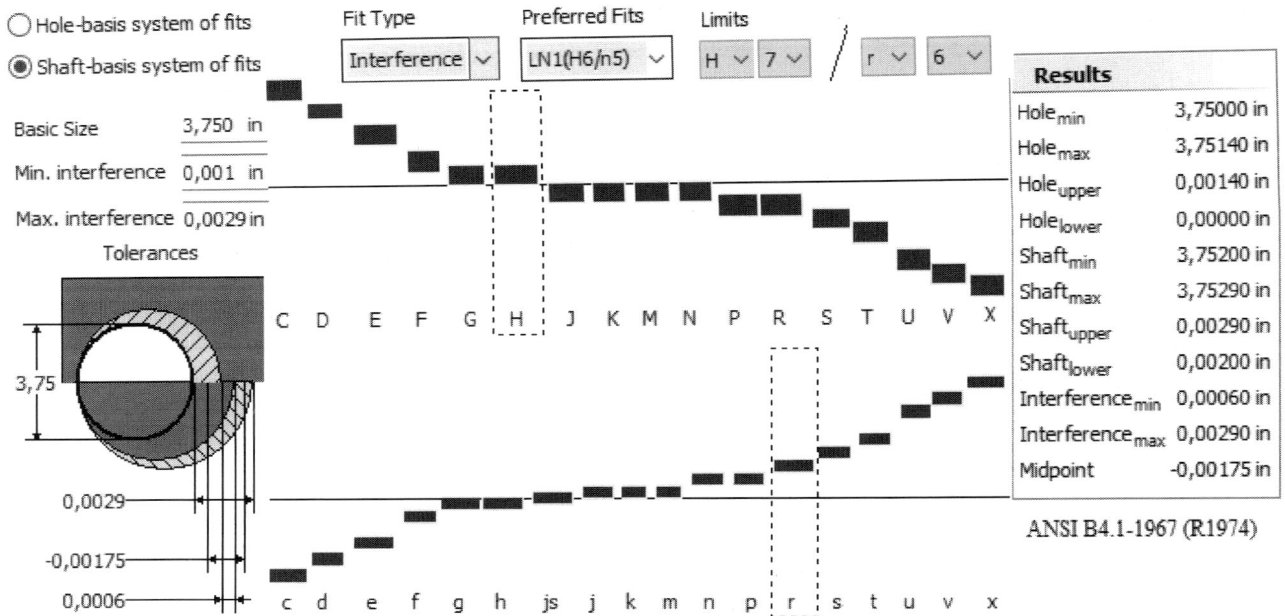

Figure 3.3. *Results of calculation of the press fit with interference for a contact diameter dc = 92.25 mm = 3.75 in. These calculations are performed with the analysis module of the software package Inventor Pro. For a color version of this figure, see www.iste.co.uk/grous/design.zip*

$\sigma > 0$ corresponds to traction and $\sigma < 0$ to compression. The generalized Hooke's law $\varepsilon_t = \delta_r/r = r(\sigma_t - \nu\sigma_r)/E$ can be used to calculate the radial strain.

$$\delta_r = \frac{1}{E\left(r_{ex}^2 - r_{in}^2\right)} r \left\{(1-\nu)\left(p_{in}r_{in}^2 - p_{ex}r_{ex}^2\right) + (1+\nu)\left(p_{in} - p_{ex}\right)\frac{r_{ex}^2 r_{in}^2}{r}\right\} \quad [3.11]$$

If we only look at the internal pressure, the formulae for the stresses at a given point, at a distance r from the center (0), look like this:

$$\sigma_r = \frac{p_{in}r_{in}^2}{r_{ex}^2 - r_{in}^2}\left(1 - \frac{r_{ex}^2}{r}\right) \; ; \; \sigma_t = \frac{p_{in}r_{in}^2}{r_{ex}^2 - r_{in}^2}\left(1 + \frac{r_{ex}^2}{r}\right) \; ; \; \sigma_{xx} = \frac{p_{in}r_{in}^2}{r_{ex}^2 - r_{in}^2} \quad [3.12]$$

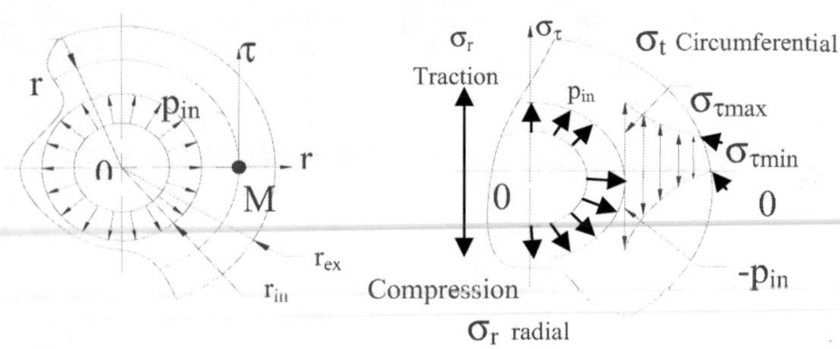

Figure 3.4. *Internal pressure and stresses σ_r and σ_t. For a color version of this figure, see www.iste.co.uk/grous/design.zip*

External pressure alone, p_{ex} [ROA 75]

$$\sigma_r = \frac{p_{ex}r_{ex}^2}{r_{ex}^2 - r_{in}^2}\left(\frac{r_{ex}^2}{r^2} - 1\right) \; and \; \sigma_t = -\frac{p_{ex}r_{ex}^2}{r_{ex}^2 - r_{in}^2}\left(\frac{r_{in}^2}{r} + 1\right) \quad [3.13]$$

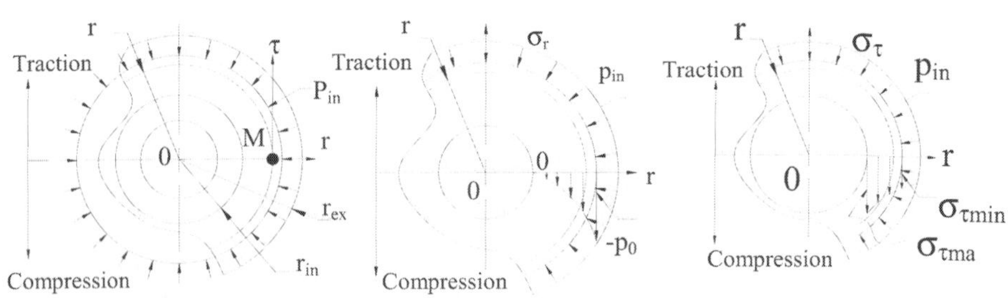

Figure 3.5. *External pressure and stresses σ_r and σ_t. For a color version of this figure, see www.iste.co.uk/grous/design.zip*

NUMERICAL APPLICATION.– Consider a steel cylinder with a Poisson ratio of $\nu = 3/10$; elasticity modulus E = 200 GPa = 29010 psi; internal radius r_{in} = 120 mm = 4.724 in and r_{ex} = 160 mm = 6.299 in. A circumferential stress σ_{limit} is imposed upon it – i.e. an admissible circumferential stress $[\sigma]_{adm}$ = 200 MPa = 2.9×10^7 psi. Our aim is to determine the maximum pressure if only the external pressure is taken into account. The radial displacement then creates an internal pressure, which is evaluated as follows:

SOLUTION.– *internal and external pressures [ROA 75]*

$$\begin{cases} P_{ex} = -[\sigma]_{adm}\left(\dfrac{r_{ex}^2 - r_{in}^2}{2r_{ex}^2}\right) = \text{-}38.505917\ MPa = \text{-}5.585\times10^3\,psi \\ P_{in} = +[\sigma]_{adm}\left(\dfrac{r_{ex}^2 - r_{in}^2}{r_{ex}^2 + r_{in}^2}\right) = 48.984841\ MPa = 7.105\times10^3\,psi \end{cases}$$

If $r \equiv r_{in}$ → maximum radial displacement. Only the internal pressure is considered.

$$\delta_{rad}^{\max} = \varepsilon_t r_{in} = \frac{P_{in} r_{in}}{E}\left(\frac{r_{ex}^2 + r_{in}^2}{r_{ex}^2 - r_{in}^2} + \nu\right) = 0.115\ mm = 4.522\times10^{-3}\ in$$

In industrial constructions, we often find ourselves facing calculations pertaining to rotating cylinders with a *centrifugal effect*: flywheels. In these cases, there are usually no notable pressure values (internal or external). The stresses are circumferential and radial, and are expressed as follows:

$$\sigma_{t1} = \frac{3+\nu}{8}\rho\omega^2\left(r_{ex}^2 + r_{in}^2 + \frac{r_{ex}^2 r_{in}^2}{r^2} - \frac{1+3\nu}{3+\nu}r^2\right) \quad [3.14]$$

$$\sigma_{r1} = \frac{3+\nu}{8}\rho\omega^2\left(r_{ex}^2 + r_{in}^2 - \frac{r_{ex}^2 r_{in}^2}{r^2} - r^2\right) \quad [3.15]$$

The radial strain is expressed thus:

$$\delta_{rad1} = \frac{3+\nu}{8E}\rho\omega^2 r\left(\left(r_{ex}^2 + r_{in}^2\right)(1-\nu) + \frac{r_{ex}^2 r_{in}^2}{r^2}(1-\nu) - \frac{1+\nu^2}{3+\nu}r^2\right) \quad [3.16]$$

3.4. Case study

Two steel rods are fitted against one another, inside a cylinder. The Hertz pressure (contact pressure) is approximately 80 MPa = 1.16×10^4 psi. As the parts are made of carbon steel, the Poisson ratio is ν = 3/10 and the density of the materials ρ = 0.284 lb/in^3 = 7850 kg/m^3.

QUESTION.– Explain the conceptual reason for using hollow cylinders to rotate parts using centrifugal force, such as cement mixers, for example:

– stem 1: R_{ex1} = 48 mm = 2.126 in; R_{in1} = 28 mm = 1.102 in;

– stem 2: $R_{ex}2$ = 28 mm = 2.102 in; R_{in2} = 32 mm = 1.417 in.

We can deduce that r = 0.04 mm, meaning that (0.048 + 0.032)/2 = 0.04 m = 1.575 in, as the two stems can be considered as one and the same part, because of the press fit between them.

SOLUTION.– on the basis of $\sigma_{r1} = \frac{3+\nu}{8}\rho\omega^2\left(R_{ex}^2 + R_{in}^2 - \frac{R_{ex}^2 \times R_{in}^2}{r^2} - r^2\right) = P_{Hertz}$, we calculate:

$$\sigma_{r1} = \frac{3+\nu}{8}\rho\omega^2\left(R_{ex}^2 + R_{in}^2 - \frac{R_{ex}^2 R_{in}^2}{r^2} - r^2\right)\ simplify \rightarrow \frac{0.8206704.Kg.\omega^2}{m} = 8\times10^7\,Pa$$

$$Hence \rightarrow \omega = \sqrt{\frac{m.80000000.\text{Pa}}{0.8206704.Kg}} = 9.873261\times10^3\left[\frac{rad}{s}\right] = 9.428\times10^4\,[RPM]$$

Justification of the use of the hollow cylinder (stem) for this design

When the mechanism achieves this extremely fast rotation, the pressure tends toward 0 and does not impede the motion. In these cases of centrifugal effect, the angular velocity has an almost negligible effect, or one which is limited to speeds of under 10^4 rpm. In the case at hand, ω = 9873.261 rad/s, or 94280 rpm. This is the reason that we choose to use hollow cylinders.

3.5. Rotating cylinders with a full round cross-section: flywheel

DESIGN PROBLEM.– We spin this setup at an unknown angular velocity (ω), but we wish to find out the point beyond which the centrifugal effects disappear without causing notable anomalies in the system. The shaft experiences traction and the wheel compression under the influence of the pressure exerted with 200 MPa. Let us outline the approach taken:

– $\varnothing_{external}$ of the flywheel: d_{fwl} = 24.409 in = 620 mm, and therefore R_{fwl} = 0.31 m;

– $\varnothing_{external}$ of the shaft: d_{shaft} = 4.724 in = 120 mm and therefore R_{shaft} = 0.06 m;

– P_{assy} = Pressure of the assembly of the flywheel on the shaft: P_{assy} = P_{comp} = 200 MPa = 29010 *psi*.

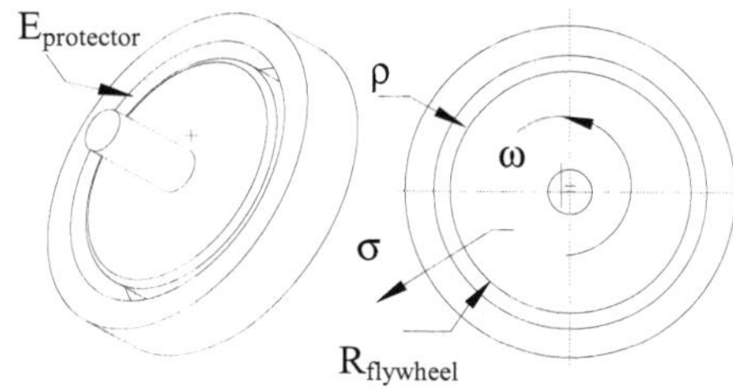

Figure 3.6. *Flywheel with kinetic energy. For a color version of this figure, see www.iste.co.uk/grous/design.zip*

SOLUTION.– By virtue of the mathematical relations governing radial stress, we can posit:

$$\begin{cases} \sigma_{r1} = P_{assy} = \left[(3+\nu)/8\right] \times \rho_{steel}^{steel} \times \omega_{assy}^2 \left(r_{vol}^2 - r_{shaft}^2\right) = 2\times10^8\ Pa\ and \\ \left[(3+\nu)/8\right] \times \rho_{steel}^{steel} \times \omega_{assy}^2 \left(r_{vol}^2 - r_{shaft}^2\right) = 2\times10^8\ Pa \rightarrow \omega_{assembly} \end{cases}$$

$$\begin{cases} 200000000Pa = \left(299.5265625.kg.\omega_{assy}^2\right)/m,\ isolate\ \omega_{assy}^2 = ? \\ \ldots\ \omega_{assy}^2 = \sqrt{\dfrac{m.200000000Pa}{299.5265625.kg}} = 817.141611\left[\dfrac{rad}{s}\right] = 7.803128\times10^3\left[rpm\right] \end{cases}$$

At that frequency = 7803.128 rpm, we can see that there is no noticeable pressure on the shaft/flywheel setup, even at a high rotation frequency (ω).

$$\omega_{\substack{Assembly\\Design}} = \sqrt{\frac{P_{assy}}{\left((3+\nu)/8\right)\rho_{steel}^{steel} \times \left(r_{vol}^2 - r_{shaft}^2\right)}} = 817.141611\frac{1}{s} = 7.803128\times10^3\ rpm$$

To graphically show the limit beyond which centrifugal effects may be experienced, we suggest varying the simulated pressures exerted on the assembly

($P_{assy} = P_{comp} = \sigma_{r1Design}$) between 3.5 MPa and 200 MPa to see the evolution of the rotation frequencies (ω_{Design}) as follows:

$\omega_{Co}(\sigma_{r1A}) =$

	0
0	$1.032 \cdot 10^3$
1	$1.234 \cdot 10^3$
2	$1.407 \cdot 10^3$
3	$1.561 \cdot 10^3$
4	$1.701 \cdot 10^3$
5	$1.83 \cdot 10^3$
6	$1.951 \cdot 10^3$
7	...

·rpm

rpm

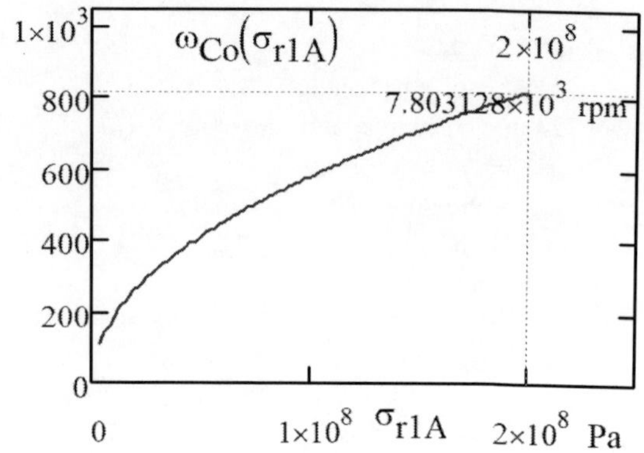

Figure 3.7. *Evolution of the rotation frequency as a function of the exerted pressures. For a color version of this figure, see www.iste.co.uk/grous/design.zip*

3.5.1. *Materials used for flywheels with centrifugal effects*

Flywheels are used to store energy. Lead flywheels are found in children's buggies. In old steam engines, they are made of cast iron. Today, this technique has become popular once more with a goal of storing and recovering energy. The choice of materials is a relevant question.

Designers must consider the possibility of damage being done to the flywheels because of the powerful centrifugal forces to which they are subjected. It is a classic case which arises when the traction stress and/or fatigue exceeds the resistance of the material. The flywheel becomes dislodged, causing additional damage. The strategy is used in techniques of optimization of energy storage systems.

$$E_{kin}^{flywheel} = J\omega^2/2 \text{ where } J = \pi \times \rho \times e \times r^4/2 \qquad [3.17]$$

where:

– J is the moment of inertia of the disk (flywheel) in m^4;
– ρ is the density of the material used for the flywheel in kg/cm^3;
– r is the radius of the flywheel in mm;
– σ_r is the resistance stress in MPa;
– σ_{max} is the maximum stress in MPa;
– e is the thickness of the cross-section of the flywheel in mm;
– m is the mass of the flywheel in kg;
– E_k is the kinetic energy in J;
– ω is the rate of rotation in rad/s;
– ν is the Poisson ratio = 1/3 for solids

Figure 3.8. *Flywheel. For a color version of this figure, see www.iste.co.uk/grous/design.zip*

$$E_{kin}^{flywheel} = \pi\rho r^4 e\omega^2/4 \;;\; m = \pi r^2 e\rho \rightarrow \; E_{kin}^{flywheel}/m = r^2\omega^2/4 \qquad [3.18]$$

When the flywheel is spinning sufficiently quickly, the energy is stored, leading to an increase in stress, resulting from the centrifugal forces. Drawing on the specialized literature, we can therefore posit the following expression for the maximum stress:

$$\sigma_{\max}^{flywheel} = \left(\frac{3+\nu}{8}\right)\rho r^2\omega^2 \;;\; hence \rightarrow \; \frac{E_{cin}^{flywheel}}{m} = \left(\frac{2}{3+\nu}\right)\frac{\sigma_r}{\rho} \qquad [3.19]$$

Among the best candidate materials for use in flywheel design, we choose that which has the best performance index (*kj/kg*) – i.e. that which yields the most energy for a given mass in kg.

$$i_{performance} = \{\sigma_r/\rho\} \; in \; [kj/kg] \qquad [3.20]$$

It is the designer who balances the parameters (r) and (ω). Thus, for a constant speed, at frequency (ω), the energy per volume is written as follows:

$$\left\{E_{kin}^{flywheel}/V\right\} = \rho r^2\omega^2/4 \; hence \rightarrow i\left\{ {Design\max \atop performance} \right\} \cong \{\rho\} \qquad [3.21]$$

The table below presents a few performance indices [LEW 90].

Material	$I_{performance}$ (kj/kg)	Designer's observations
High-resistance (HR) steel	100-200	Inexpensive material
(HR) aluminum alloy	100-200	Inexpensive material
(HR) Mg alloy	100-200	Performance essentially equivalent to that of steels and aluminum alloys
Titanium alloy	100-200	
Cast iron	8-10	(ρ) high density but limited usage speed
Lead alloy	≅ 3	

Table 3.2. *Performance indices as a function of the material*

DESIGNER'S DECISION.– As the energy density in flywheels is considerable, we need to avoid them suddenly becoming loose, which would be catastrophic. Thus, we can surround the disk with a protective screen to contain the explosion and the debris.

3.6. Press fit and thermal effects through bracing

The pressure exerted on a press fit assembly is essentially applied to the shaft/bore systems. Below, we present the formulae and applications relevant to such systems. In children's buggies, it is lead which is used, because it is heavy and soft. Of course, gold, platinum or indeed uranium would work better, but these materials are expensive, and can be quite rare.

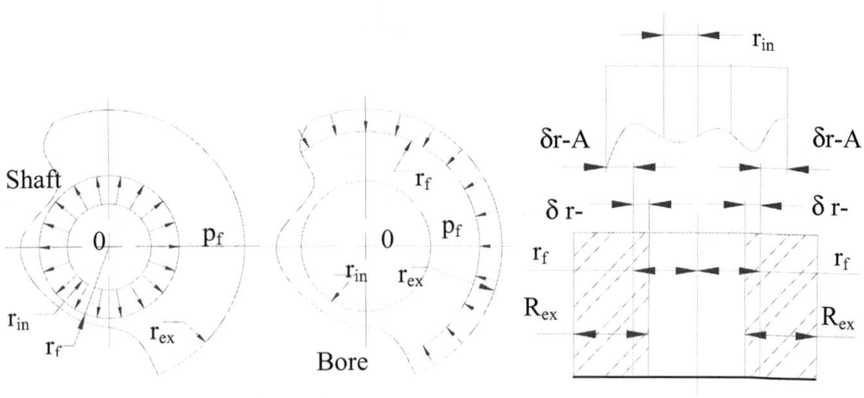

Figure 3.9. *Internal pressure and stresses σ_r and σ_t: press fit. For a color version of this figure, see www.iste.co.uk/grous/design.zip*

The decrease of the radius of the shaft due to the bracing pressure, p_f, is:

$$\delta_{radius}^{shaft} = \varepsilon_{radius}^{shaft} \times r_f = \frac{r_f\left(\sigma_t - \nu_a \sigma_r\right)}{E_a} = -\frac{p_f r_f}{E_a}\left(\frac{r_{ex}^2 + r_f^2}{r_{ex}^2 - r_f^2} - \nu_a\right) \qquad [3.22]$$

The increase in the radius of the container due to the bracing pressure, p_f, is:

$$\delta_{radius}^{bore} = \varepsilon_{radius}^{bore} \times r_f = \frac{r_f\left(\sigma_t - \nu_a \sigma_{bore}\right)}{E_{bore}} = -\frac{p_f r_f}{E_{bore}}\left(\frac{r_{ex}^2 + r_f^2}{r_{ex}^2 - r_f^2} - \nu_{bore}\right) \qquad [3.23]$$

where:

- P_f is the bracing pressure (assembly);
- R_f is the contact radius in assembly;
- ν_L is the of the material from which the container (bore, casing) is made;
- ν_a is the Poisson ratio of the material from which the content (shaft) is made;
- ν_a is the friction coefficient between the two materials (shaft/bore);
- δ_a is the dimensional increase of the radius of the shaft;
- δ_L is the dimensional increase of the radius of the bore;
- l is the length of the assembly;
- S is the safety factor.

We use the term "radial interference" to speak of the influence on the radius, which is presented thus:

$$\delta_{radius}^{rad} = \left\{\delta_{radius}^{bore} - \delta_{radius}^{shaft}\right\} = r\left(\varepsilon_{radius}^{bore} - \varepsilon_{radius}^{shaft}\right) \qquad [3.24]$$

Bracing applies pressure after cooling, and the resulting press fit is expressed as:

$$S_{radius}^{shaft} = 2\delta_{radius} = 2p_f r_f\left(\frac{r_{ex}^2 + r_f^2}{E_{shaft}\left(r_{ex}^2 - r_f^2\right)} + \frac{\nu_{bore}}{E_{bore}} + \frac{r_f^2 + r_{in}^2}{E_{shaft}\left(r_f^2 - r_{in}^2\right)} - \frac{\nu_{shaft}}{E_{shaft}}\right) \qquad [3.25]$$

The duality can be summed up as follows: either we know the value of the press and deduce that of the bracing pressure needed, or the other way around. When the grade of the material is the same for the shaft and the bore, the calculations can be summarized thus:

$$S = 2\delta_{radius} = 4p_f \times r_f^3\left(r_{ex}^2 - r_f^2\right)/E\left(r_{ex}^2 - r_f^2\right)\left(r_f^2 - r_{in}^2\right) \quad [3.26]$$

When the shaft has a solid round cross-section, we set:

$$S = 2\delta_{radius} = 4p_f r_f^3 \times r_{ex}^2/E\left(r_{ex}^2 - r_f^2\right) \quad [3.27]$$

To determine the transmissible forces on the transmission torque:

$$M = Fr_f = 2\pi\mu r_f^2 l p_f\,[N.m]\ because\ F = 2\pi\mu r_f l p_f\,[N] \quad [3.28]$$

These formulae are conceived for evaluating the maximum bracing pressure. When dealing with the preliminary design phase, it is common to begin by finding the minimum bracing pressure in order to make wise and safe choices. This is the reason for the introduction of the safety coefficient at this stage of the design project [ROA 75].

$$p_f^{\min} = \frac{F \times S}{2\pi \times \mu \times r_f \times l} = \frac{M \times S}{2\pi \times \mu \times r_f^2 \times l}\ [MPa\ or\ psi] \quad [3.29]$$

$$\begin{cases} l_{assy} = 125\ mm = 4.921\ in;\ d_{shaft} = 75\ mm = 2.953\ in \rightarrow r_{shaft} = r_{in} = r_f = 0.038\ m;\ \mu = 0.1 \\ d_{shaft} = 120\ mm = 4.724\ in \rightarrow r_{bore} = r_{ex} = 0.06\ mm;\ i = \text{radial interference} = 0.0545\ mm \end{cases}$$

The radial interference is of the order of 5.45/100 mm.

Bracing pressure for a radial interference is 0.0545:

$$\begin{cases} \delta_{radius} = 2p_f r_f\, r_{ex}^2/E\left(r_{ex}^2 - r_f^2\right)\ simplify \rightarrow 0.0000545 \times m = \swarrow \\ \ldots = 9.846\text{e-}13.m.p_f/Pa \rightarrow p_f = 55.351563\ MPa = 8.028 \times 10^3\ psi \end{cases}$$

Axial force necessary for bracing for a radial interference:

$$F = 2\pi \times \mu \times r_f \times l_{assy} \times p_f = 1.63 \times 10^5 [N] = 3.665 \times 10^4 [lbf]$$

We impose a pressure on the assembly $P_{fimposed}$ of the order of 554 MPa:

$$\begin{cases} S = 2\delta_r \text{ and } S = 4p_f\mu \times r_f \times r_{ex}^2 / E\left(r_{ex}^2 - r_f^2\right) = 0.109095\ mm \\ S = 2\delta_r = 1.091 \times 10^{-4} m \rightarrow \delta_r = S/2 = 5.454769 \times 10^{-5}\ m \end{cases}$$

Axial force and moment, necessary for bracing:

$$\begin{cases} F = 2\pi\mu r_f l_{assy} p_{f.imposed} = 1.632 \times 10^6\ N = 3.668 \times 10^5\ lbf \\ M = Fr_f = 2\pi\mu r_f^2 l_{assy} p_{f.imposed} = 6.119 \times 10^4\ N.m = 5.416 \times 10^5\ lbf.in \end{cases}$$

Calculation necessary for bracing for a radial interference:

The thermal effect of bracing on the assembly is evident on a press fit, with interference. When the temperature rises in the room, the strains do likewise. This is characterized by an expansion that is sometimes imperceptible, as it is so small, but nevertheless, it still affects the fit. Each material [ROA 75] has its own expansion value. The strain in directions (x, y, z) is expressed as follows:

$$\varepsilon_{xx} = \varepsilon_{yy} = \varepsilon_{zz} = \alpha\Delta\tau \text{ ; example for } r : \varepsilon_{zz} = \delta_r / r = \alpha_{bore}\Delta\tau \qquad [3.30]$$

The *pressing* or radial strain is written as: $\delta_r = \alpha \times r \times \Delta\tau$

where:

– α_L is the linear coefficient of thermal expansion in 1/°C or 1/°F;

– r is the radius prior to expansion;

– δ_L is the strain sometimes known as radial pressing;

– $\Delta\tau$ is the temperature variation in °C or °F.

3.7. Case study applied to bolted tanks

What follows is a real-world case study in mechanical design. It pertains to the information that can be gleaned from calculations of resistance in connection with the definition drawings and the assembly drawings. Thus, at the end, we need to present a concrete analysis of the stress and resistance criteria in Von Mises' terms.

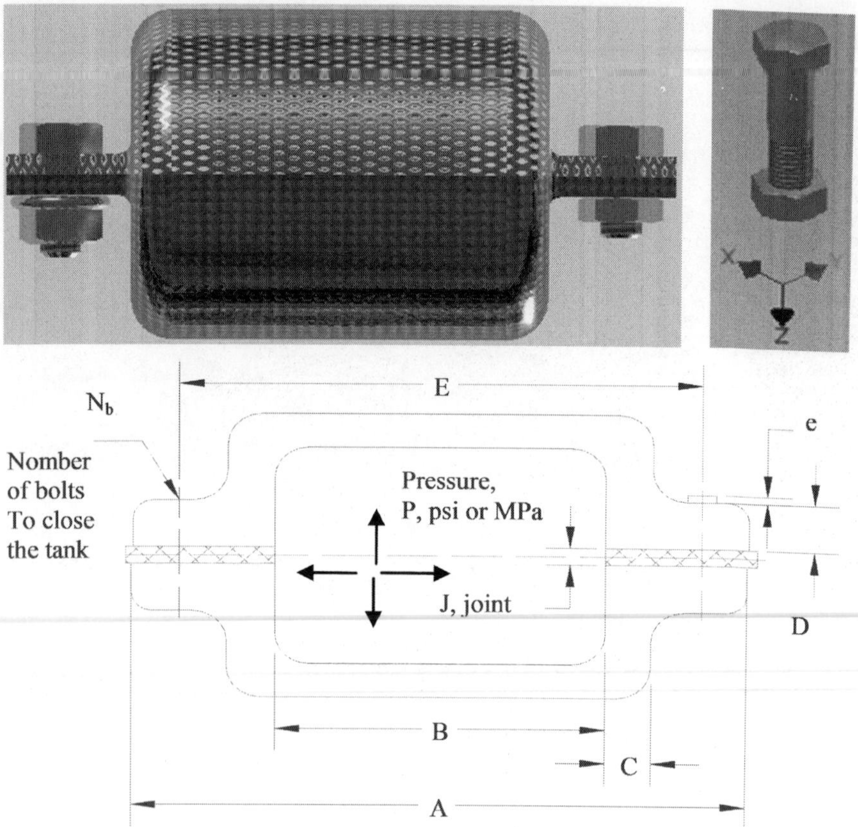

Figure 3.10. *Pressurized tank assembled and locked with bolts. For a color version of this figure, see www.iste.co.uk/grous/design.zip*

OUTLINING THE PROBLEM.– Consider a pressurized cylinder such as a tank (vat) assembled and locked by bolting at the edges of the articulated joints, as illustrated above. It is suggested to use (n) locking bolts. The internal pressure in the tank is exerted by the liquid and vapor contained therein. It ranges from 0 to 25 MPa.

QUESTIONS.–

What is the pressing force needed for each bolt (nut) to ensure complete security – i.e. to ensure there can be no leaks?

What are the dimensions, under the ANSI-USA and ISO systems of norms, for the bolts which can withstand the fatigue due to variable static loading?

Method (Source: Product Engineering, USA)	Coefficients Ktor
Pressing and electronic monitoring thereof	1.0 to 1.5
Direct monitoring with a torque wrench	1.6 to 1.8
Monitoring of power by measuring the elongation of the bolts	1.4 to 1.6
Pneumatic monitoring of power	1.7 to 2.5

Table 3.3. *Factors of pressing moments by different methods*

where:

– internal diameter, d_{in} = 200 mm = 7.874 in;

– maximum power in the vat, P = 25 MPa = 3626 psi;

– elasticity modulus (Young's modulus), E = 2×10^5 MPa = 2.901×10^7 psi.

SOLUTION WITH DISCUSSION.–

Calculating the area A of the cylinder of internal diameter of the cylinder d_{in}:

$$A=\left(\pi d_{in}^2/4\right)=3.142\times10^4\ mm^2=48.695\ in^2$$

Each bolt is subject to the action of the force F_A, expressed thus:

$$F_A=(P\times A/n\times g)=1.001\times10^4\ kg=2.207\times10^4\ lb\rightarrow so\ F_A=10^4\ kg$$

The bolts are considered to have been tightened with a pneumatic wrench – hence the corresponding press factor K_{tor} = 2.5 (from Table 3.3, above).

The load produced by the work on each bolt is:

$$\begin{cases} K_{tor}=2.5\rightarrow P_y=K_{tor}\times F_A=2.502\times10^4\ kg=5.517\times10^4\ lb,\downarrow \\ The\ yield\ \eta=80\%\ is\ then\ P_{uti}=P_y/\eta=3.128\times10^4\ kg=6.896\times10^4\ lb \end{cases}$$

The bolts suggested by the hypothesis adopted at the preliminary phase of the project are *grade 8*. The limiting stress, therefore, is of the order of $[\sigma]_{lim}$ = 1035 MPa = 1.501×105 psi.

The area and nominal diameter of each bolt are:

$$\begin{cases} D_{bolt} = \sqrt{4A_{bolt}/\pi} = 19.434\ mm = 0.7651\ in \rightarrow From\ \sigma_{\lim} = P_{uti}g/A_{bolt} \downarrow \\ calculate\ the\ area\ of\ each\ bolt\ A_{bolt} = P_{uti}g/\sigma_{\lim} = 296.616\ mm^2 = 0.4598\ in^2 \end{cases}$$

In accordance with ANSI/ASME B.1.1, we choose the closest geometric characteristics of bolts. Thus, for a nominal diameter of D = 0.5in, we choose: ½-13-NC. (13 threads per inch). Thick threads are preferred, because of their high resistance. Thus, each bolt is covered with a restrained spring. We calculate the rigidity factor of that spring by this expression: $K = A{\cdot}E/L$

$$\frac{1}{K_B} = \frac{g}{E} \times \left(\frac{0.4D_{nom}}{A_1}\right) + \left(\frac{L_1}{A_1}\right) + \left(\frac{L - L_1}{A_{\min}}\right) + \left(\frac{0.4D_{\min}}{A_{\min}}\right) \qquad [3.31]$$

Where:

– D_{nom} is the nominal diameter of the standardized bolt in (in) or (mm);

– D_{min} is the minimum diameter (in) or (mm);

– A_1 is the area of the non-threaded section of the bolt;

– A_{min} is the area of the threaded section of the bolt at the level of the minor diameter;

– L_1 is the non-threaded length of the bolt; around ¾ (0.75) in;

– L is the implantation length – i.e. the portion of the length which holds the pieces together – at around 2.010 in (ANSI).

When dealing with plastic strain, on the head bolt, material tests (Europe and USA) show that we can write 0.4D and $0.4D_{min}$.

For: L_1 = 0.75 in = 19.05 mm and L = 2.010 in = 51.054 mm

$$For\ D = 0.5\ in = 12.7\ mm \rightarrow A_1 = \left(\pi D^2/4\right) = 126.677\ mm^2 = 0.1963\ in^2$$

The smallest thread diameter of the 0.5-13 UNC thread (large-step series) which is mainly used in nuts and bolts including non-ferrous metals, and particularly to prevent vibrations, is, according to ANSI/ASME B.1.1 or UST (ABC = American-British-Canadian). D_{min} = 0.4056 in, then:

$$D_{\min} = 0.4056\ in = 10.302\ mm \rightarrow A_{\min} = \left(\pi D_{\min}^2/4\right) = 83.359\ mm^2 = 0.1292\ in^2$$

Finally, we obtain the value of K_B as follows:

$$K_B = 1\Big/ \frac{g}{E}\left(\left(\frac{0.4D_{nom}}{A_1}\right)+\left(\frac{L_1}{A_1}\right)+\left(\frac{L-L_1}{A_{\min}}\right)+\left(\frac{0.4D_{\min}}{A_{\min}}\right)\right) = 32690\frac{kg}{mm} = 1.831\times10^6\frac{lb}{in}$$

Calculation of the rigidity of the spring (rate): The calculations giving the rate of rigidity of the spring at the joints are deliberately simplified and reduced at the bolt heads and at the nut. As the bolt and nut are pressed, we see a cylindrical distribution of stress in the vicinity of the head and the pressing nut. In order to carry out calculations in this vicinity, we first need to bind the equivalent area. Then, the design takes place in three distinct steps:

The following three scenarios are possible:

Case 2 (Figure 3.11b) arises when the joint's external diameter is larger than the diameter of the bolt head or the same as the washer, provided that diameter is no greater than $3\times D_a$. When the diameter of the joint D_J is represented by the above expression, (Case 2) or much larger than $3\times D_a$.

$$\left\{\begin{array}{l} CASE\ (a): A_J = \frac{\pi}{4}\times\left(D_{ex}^2 - D_a^2\right) \\ CASE\ (b): A_J = \frac{\pi}{4}\left(D_w^2 - D_n^2\right)+\frac{\pi}{8}\left(\frac{D_J}{D_w}-1\right)\left(\frac{D_wL}{5}-\frac{L^2}{100}\right) \\ CASE\ (c): A_J = \frac{\pi}{4}\times\left(\frac{D_wL}{10}-D_a^2\right) \end{array}\right\} \quad [3.32]$$

Figure 3.11. *Bolts holding the pressurized tank together. For a color version of this figure, see www.iste.co.uk/grous/design.zip*

where:

– D_{ex} is the external diameter in in or mm,

– D_a is the diameter of the bore of the nut (receiving the bolt) in in or mm.

We suppose that D_w is the diameter which supports the bolt to the head of the nut. The design of this bolted assembly enables us to apply *case 2*, meaning that D_w = ¾ in (0.75in) = 19.05mm, which supports the head of the bolt and the nut:

$$CASE\text{ b}: A_J = \frac{\pi}{4}\left(D_w^2 - D_n^2\right) + \frac{\pi}{8}\left(\frac{D_J}{D_w} - 1\right)\left(\frac{D_w L}{5} - \frac{L^2}{100}\right) = 302.716\ mm^2 = 0.4692\ in^2$$

Substituting the value of Aj back into the expression of the degree of rigidity of the spring, we deduce the coefficient for the joint, K_J, as follows:

$$K_J = A_J E / Lg = 1.209 \times 10^5\ kg/mm = 6.772 \times 10^6\ lb/in$$

We now look to determine the part which is not affected by the load on the joint. The load constant K_L, which takes account of the elasticity of the bolt and of the elastic joint, is:

$K_L = K_B / (K_B + K_J) = 0{,}2128$, the portion of work not affected by the load on the joint is presented thus: $F_p = F_p \times K_L = 7880\ kg = 17372\ lb$.

Loss of pressing force due to fixation with bolts:

Experiments have yielded a recommended value of η_1=10%. Thus, the loss of load of the pressing force would be:

$$F_{pressing} = \eta_1 \times F_A = 1001\ kg = 2207\ lb$$

Calculation of the fixation force (bolting) required for the joint:

Based on the fluctuation of the force from 0 *lb* to 8000 *lb*, the minimum fixation force required is 0 lb. Thus, the maximum force would be:

$$F_{ver} = 0\ lb\ and\ F\left\{\begin{matrix}\text{minimum}\\ \text{bolting}\end{matrix}\right\} = K_L\left(F_p + F_K + F_{ver}\right) = 1890\ kg = 4167\ lb$$

The bolt's failure threshold is equivalent, depending on the steel used, to 20,000 lb. It exhibits a static strength which is amply sufficient to hold the joints, with no leaks:

$$R_{fail} = 20{,}000\ lb \succ 4167\ lb \text{ (actually applied)}$$

3.8. Case studies applied to contact stresses (Hertz) in design

Crushing contact between two bodies gives rise to strains under loading. Stresses are experienced in all three spatial directions [σ_x, σ_y, σ_z]. In design (construction, assembly, etc.), these facts are found at play in many elements of machines, such as gear systems, bearings, cams, rails, and so on. In addition, we here present the fundamental issues surrounding Hertzian contact mechanics. In discussing this subject, we touch upon the following:

1) contact area and contact factors (tribology and environment);

2) stresses on contacts and in their vicinity;

3) the resulting strains.

Given Hertz's theory on the subject, we can differentiate the following:

1) so-called point contact, e.g. two spheres in contact and a sphere in contact with a flat surface;

2) lineal contact, e.g. two parallel cylinders and a cylinder in contact with a flat surface, a cogwheel on a gear, a shaft (axle) in a sleeve bearing or roller bearing;

3) so-called surfacic contact, which is two flat surfaces in contact with one another.

Contacts are held to be elastic, in Hertzian theory; the geometry is well known, as a sphere with a given radius R. At the interface, the tribology is found to have no influence, meaning that at the interface, the friction (μ) is null.

Contact takes place with the normal force (F in N or in); the other, external surface forces can be overlooked. In terms of the topography (roughness), the surfaces of the bodies are reasonably smooth, and the materials are isotropic, homogeneous and elastic. In the final analysis, contact is said to be non-conforming when zone a ($a \ll R$) is much smaller than the radius. We can clearly see that the strain is very small to be deemed to have an influence.

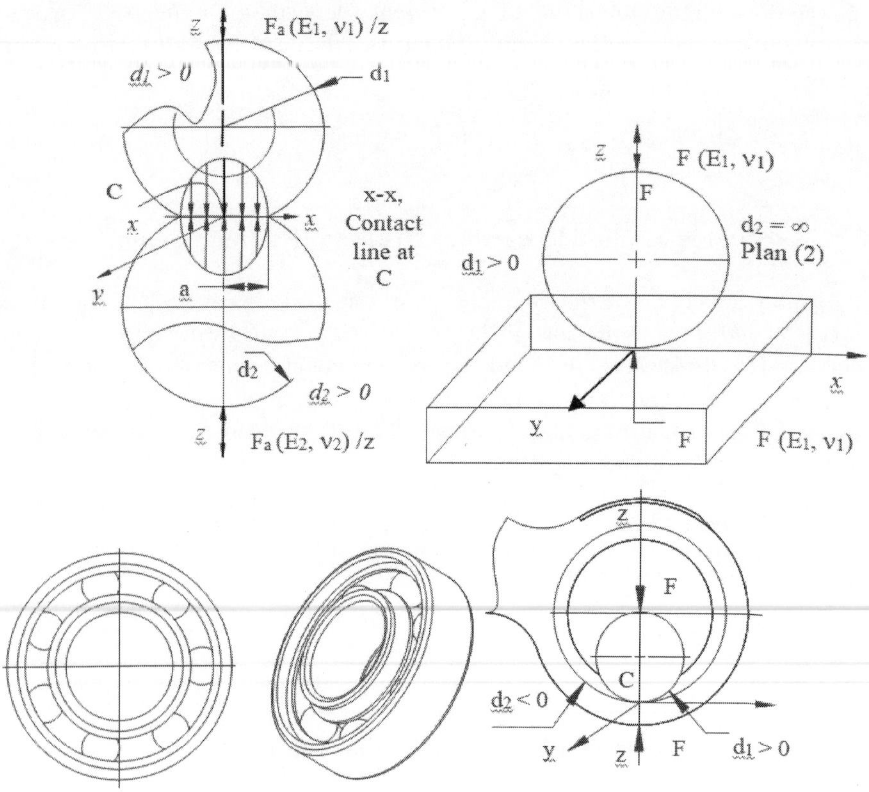

Figure 3.12. *Sphere-to-sphere contact and typical examples of Hertzian contact. For a color version of this figure, see www.iste.co.uk/grous/design.zip*

Below, we present the essentials of contacts between two spheres and two cylinders; these are the cases to which almost all so-called Hertzian contacts relate.

3.8.1. *First case: sphere-to-sphere contact*

Consider two spheres pressed against one another by a force F (in N). Thus, one sphere crushes the other, creating a circular zone, known as C, (see Figure 3.12). The two spheres are deemed to have respective characteristics (1) and (2) [ROA 75]:

– d_1 and d_2 are respectively the diameters of each sphere;

– E_1 and E_2 are respectively the elasticity moduli of the materials from which each sphere is made;

– ν_1 and ν_2 are respectively the Poisson ratios of materials (1) and (2);

– *F* is the load, i.e. the axial force applied in N or *lbf*;

– *C* is the crushing (in mm or in) resulting from the contact under the influence of the axial force F;

– *a* semi-axis (mm or in) of an ellipse. It is the dimension of the crushed zone.

$$C = \sqrt[3]{\frac{3F}{8}\frac{\frac{1-\nu_1^2}{E_1}+\frac{1-\nu_2^2}{E_2}}{\frac{1}{d_1}+\frac{1}{d_2}}} = \left(\frac{3F}{8}\frac{\frac{1-\nu_1^2}{E_1}+\frac{1-\nu_2^2}{E_2}}{\frac{1}{d_1}+\frac{1}{d_2}}\right)^{1/3} \quad [mm\ or\ in] \qquad [3.33]$$

As illustrated above (2D and 3D), contact takes place in the plane (x, y). The force of action passes through the normal (z). The maximum pressure is located in the center. It engenders a semi-elliptical zone (ellipsis of semi-axis a) and is expressed as follows:

$$p_{\max} = 3F/2\pi a^2 \ [Mpa\ or\ psi] \ with\ a\ (mm\ or\ in) \qquad [3.34]$$

NUMERICAL APPLICATION.– For two identical spheres of molybdenum steel (SAE 4040), we have applied a force F = 9.50 N = 2.136 *lbf* on:

– d_1 = 8 mm = 0.315 in and d_2 = 12 mm = 0.472 in;

– $E_1 = E_2 = 2\times10^5$ MPa = 2.901×10^7 psi and $\nu_1 = \nu_2 = 3/10$ (steel on steel).

1) Calculate the value of the depth of the crush zone C in mm and in.

2) The stress σ(a) as a function of the semi-ellipsis (a) formed, in MPa and psi.

3) The maximum pressure at the center of the contact zone (Hertz).

4) Plot the graph of σ(a) as a function of the depth C of crushing.

5) The value of crushing (δ) which characterizes the deflection of the spheres.

SOLUTION.– By virtue of relations [3.33] and [3.34], we can posit the following:

$$\begin{cases} C = \sqrt[3]{(3F/8)\times\left[\left(1-\nu_1^2\right)/E_1+\left(1-\nu_2^2\right)/E_2\right]/(1/d_1+1/d_2)} = 0.054\ mm \\ \text{Max pressure is } P_{\max} = 3F/2pa^2 = 1568\ MPa \end{cases}$$

In view of Mohr's tricircle of the stresses, with respect to the two circles and the coincident point $\sigma_x = \sigma_y$, facilitates $\tau_{xy} = 0$ *and* $\tau_{xz} = \tau_{yz} = (\sigma_x + \sigma_y)/2 = (\sigma_y + \sigma_z)/2$. The shearing stresses become greatest below the surface at values of z ≈ a/2 (see results of simulated calculations below in Figure 3.13):

$$\begin{cases} \sigma_z = \dfrac{-p_{\max}}{1+(z/a)^2},\ z = 0 \rightarrow \sigma_z = -1.568\times10^{-3}\ MPa = -2.568\times10^{3}\ psi \\ \sigma_x(a) = \sigma_y(a) = -45.44\ MPa\ and\ \tau_{\max} = p_{\max}/3 = 522.619\ MPa \end{cases}$$

By varying (a), which is the product of (C), we can calculate the values of the stress σ (a) as follows: z = a = - 0.025a, 0.12a, … 5a. Let us calculate σ(a) and plot its curve:

$$\sigma_x(a) = \sigma_y(a) = -p_{\max}\left(\left(1 - \frac{z}{a}Arctan\left(\frac{1}{z/a}\right)\right)(1+\nu) - \frac{1}{2\left(1+(z/a)^2\right)}\right) \qquad [3.35]$$

The value of the crushing (δ) of the two spheres, according to Hertz's theory, is written as:

$$\delta = 1.040\times\sqrt[3]{F^2\left(\frac{1-\nu_1^2}{E_1}+\frac{1-\nu_2^2}{E_2}\right)^2\Bigg/\left(\frac{1}{d_1}+\frac{1}{d_2}\right)} = \frac{265}{7726013401}m^{5/3} = 3.43\times10^{-8}\times m^{1.667}$$

$\sigma_x(a_v) =$

	0
0	-1.176·10³
1	-910.431
2	-597.213
3	-379.592
4	-235.147
5	-142.307
6	-83.817
7	-47.369
8	...

·MPa

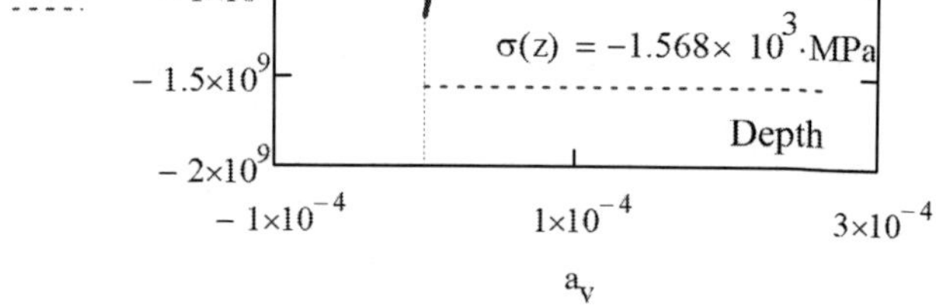

Figure 3.13. *Evolution of stresses as a function of (z) or (a). For a color version of this figure, see www.iste.co.uk/grous/design.zip*

3.8.2. *Second case: contact between two parallel cylinders*

When dealing with two parallel cylinders, in Hertzian contact, we can make the following hypotheses:

– d_1 and d_2 are the respective diameters of the cylinders;

– E_1 and E_2 are respectively the elasticity moduli of the materials of each sphere;

– ν_1 and ν_2 are respectively the Poisson ratios of materials (1) and (2);

– *F* is the load, i.e. the axial force applied in N or *lbf*;

– *C* is the crushing (in mm or in) resulting from the contact under the influence of the axial force F;

– *b* is the semi-axis (mm or in) of an ellipsis. It is the dimension of the crushed zone;

– *Lc* is the *contact length* of the two cylinders.

$$b = \sqrt[2]{\frac{2F}{\pi L}\left(\frac{1-\nu_1^2}{E_1}+\frac{1-\nu_2^2}{E_2}\right)\Big/\left(\frac{1}{d_1}+\frac{1}{d_2}\right)} = [mm\ or\ in] \qquad [3.36]$$

As illustrated below, the contact takes place along L. The force passes along the normal, i.e. along the (z) axis. The maximum pressure is located at the center. It engenders a supposedly rectangular zone of width = 2 *l* and is expressed as follows:

$$p_{\max} = 2F/2\pi(bL)\ [MPa\ or\ psi]\ with\ b\ and\ L\ (mm\ or\ in) \qquad [3.37]$$

The maximum normal stresses are along (z) in the direction of loading (F) and are written as follows:

$$\begin{cases} \sigma_x = -p_{\max}\left(\left(2-\dfrac{1}{\left(1+(z/b)^2\right)}\right)\sqrt{1+(z/b)^2}-\dfrac{2z}{b}\right)\ and\ \sigma_z = \dfrac{p_{\max}}{\sqrt{1+(z/b)^2}} \\ \sigma_y = -2\nu p_{\max}\left(\sqrt{1+(z/b)^2}-z/b\right)\ for\ \dfrac{z}{b}=\dfrac{3}{4} \rightarrow \tau_{\max} = \tau_{zx} = 0.3p_{\max} \end{cases} \qquad [3.38]$$

COMMENT.– Below the surface at the values of z ≈ b/2 (see the results of our simulated calculations below), the shear stresses become greatest. When we are dealing with cases of the same materials, the Poisson ratios balance out, in view of the symmetry of the three main stresses (σ_x, σ_y, σ_z). For example, for (z/*b*) = ¾, we have: $\tau_{\max} = \tau_{zx} = 3/10 \times p_{\max}$ (see Figure b: contact between parallel cylinders).

NUMERICAL APPLICATION.– for two identical molybdenum steel cylinders (SAE 4040), we applied a load of F = 300 N = 67.443 *lbf* to:

– $d_1 = d_2 = 10$ mm = 0.394 in and Lc = 18 mm = 0.709 in, $z = 0.75 \times$ b

– $E_1 = E_2 = 2.07\times10^5$ MPa = 3.002×10^7 psi and $\nu_1 = \nu_2 = 3/10$ (steel on steel)

1) Calculate the value of the crush depth *b* in mm and in in.

2) The maximum pressure at the center of the contact zone (Hertz).

3) The stress σ(a) as a function of the semi-ellipsis (a) formed, in MPa and psi.

4) Plot the graph of σ(a) as a function of the crush depth z.

5) The value of crushing (δ) which characterizes the deflection of the spheres.

SOLUTION WITH DISCUSSION.–

1) Calculation of the value of the crush depth *b* in mm and in:

$$b = \sqrt[2]{\frac{2F}{\pi L_c}\left(\frac{1-\nu_1^2}{E_1}+\frac{1-\nu_2^2}{E_2}\right)\Bigg/\left(\frac{1}{d_1}+\frac{1}{d_2}\right)} = 0.022\ mm = 8.503\times10^{-4}\ in$$

2) Calculation of the maximum pressure at the center of the contact zone (Hertz), z = 0.75×b:

$$p_{\max} = 2F/2\pi b\times L_c = 491.28\ MPa = 71250\ psi$$

3) Calculation of the stresses $\sigma_x(b)$, $\sigma_y(b)$, $\sigma_z(b)$ and τ_{maxi} in MPa and psi:

$$\begin{cases} \sigma_x(b) = -p_{\max}\left[\left(2-1/\left(1+(z/b)^2\right)\right)\sqrt{1+(z/b)^2}-\dfrac{2z}{b}\right] = 147.384\ MPa \\ \sigma_y(b) = -2\nu\, p_{\max}\left(\sqrt{1+(z/b)^2}-z/b\right) = 147.384\ MPa;\ z = 0.75b \\ \sigma_z(b) = -p_{\max}\Big/\sqrt{1+(z/b)^2} = -314.419\ MPa\ where\ z/b = 3/4 \\ \tau_{\max}(b) = \tau_{zx} = 3p_{\max}/10 = 147.384\ MPa,\ where\, z/b = 3/4 = 0.75 \end{cases}$$

By varying (b), which is the previous product, we calculate the values of the stresses $\sigma_x(b)$, $\sigma_y(b)$, $\sigma_z(b)$ and $\tau_{xz}(b) = \tau_{max}(b)$. From Mohr's tri-circle, we set:

$$\begin{cases}\sigma_x\left(b_{sim}\right)=p_{\max}\left(\dfrac{z}{b_{sim}}\right)\left(-\dfrac{2}{10}\right);\ \sigma_y\left(b_{sim}\right)=-p_{\max}\left(\dfrac{z}{b_{sim}}\right)\left(-\dfrac{3}{10}\right)\\ \sigma_z\left(b_{sim}\right)=p_{\max}\left(\dfrac{z}{b_{sim}}\right)\left(-\dfrac{8}{10}\right);\tau_{yz}\left(b_{sim}\right)=\tau_{\max}\left(b_{sim}\right)=-p_{\max}\left(\dfrac{z}{b_{sim}}\right)\left(-\dfrac{3}{10}\right)\end{cases}$$

– b_{sim} is a simulated depth to plot the stress curves as a function of the maximum exerted pressure.

$\sigma_x(b_s) =$	$\sigma_y(b_s) =$	$\sigma_z(b_s) =$	$\tau_{maxi}(b_s) =$
$-7.369\cdot10^3$	$-1.105\cdot10^4$	$-2.948\cdot10^4$	$1.105\cdot10^4$
$-3.685\cdot10^3$	$-5.527\cdot10^3$	$-1.474\cdot10^4$	$5.527\cdot10^3$
$-2.456\cdot10^3$	$-3.685\cdot10^3$	$-9.826\cdot10^3$	$3.685\cdot10^3$
$-1.842\cdot10^3$	$-2.763\cdot10^3$	$-7.369\cdot10^3$	$2.763\cdot10^3$
$-1.474\cdot10^3$	$-2.211\cdot10^3$	$-5.895\cdot10^3$	$2.211\cdot10^3$
...	...	...	...

$\cdot$MPa

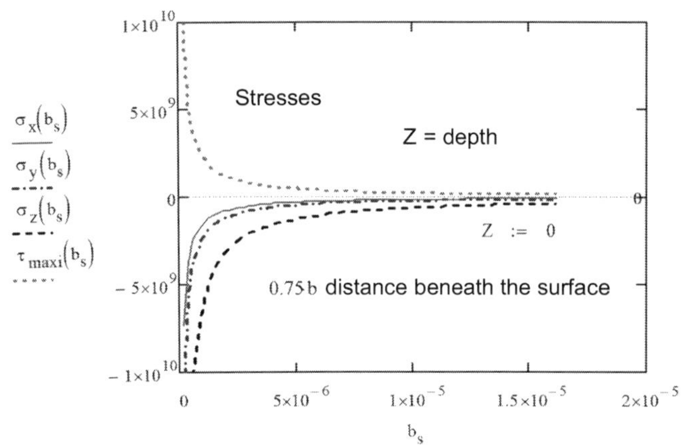

Figure 3.14. *Evolution of stresses as a function of the depth (Z). For a color version of this figure, see www.iste.co.uk/grous/design.zip*

We can see that the highest stress corresponds to $\tau_{xz}(b) = \tau_{max}(b)$. The case of rollers is not greatly different from the former, because they slide, albeit only slightly. We know from mechanics of moving bodies, when two cylinders roll over one another, without sliding. The direction of the shear stresses, under the surface, is reversed. The domain of critical fatigue due to contacts largely deals with this physical phenomenon. In mechanical construction, in power transmission by a gear system, especially, we see sliding which creates further friction, at the root of the superposed tangential and normal stresses. Owing to this phenomenon of forces of friction, the load F is also affected.

3.9. Conclusion

In this chapter, we have presented class simulation case studies, to gain a clearer idea of the issues of choice of materials in design, whilst respecting the GPS/ISO standards. There are many examples which could be drawn from industries, but we have deliberately targeted the most commonplace ones, which require calculations of resistance of materials in connection with judicious choices of the performance criteria.

4

Design of Incurvate Geometries by Sweeping

4.1. Introduction

Parts with an ordinary, skew curve (SC) present cases of unusual calculations in strength of materials [OBE 16, ROA 75]. The 2D or 3D forms created by sweeping (sweep, loft, coil, etc.) are often NURBS (*Non-Uniform Rational B-Splines*), also known as Bézier curves (named for Pierre Bézier). This book is not intended to be a course in complex surfaces. However, such surfaces and other unusual curves do exist in real-world structures, so we cannot ignore the calculations they require.

Usual formulae for calculations on structures:

$$R = r_{in} + h/2;$$

$$\varepsilon = R - \rho_n$$

$$\rho_n = h/\ln\left(r_{ex}/r_{in}\right)$$

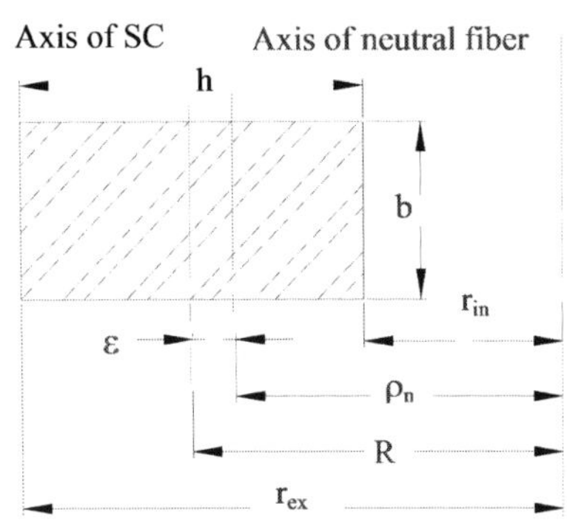

$$R = r_{in} + \frac{d}{2}$$

$$\varepsilon = R - \rho_n$$

$$\rho_n = \frac{\left(\sqrt{r_{in}} + \sqrt{r_{ex}}\right)^2}{4}$$

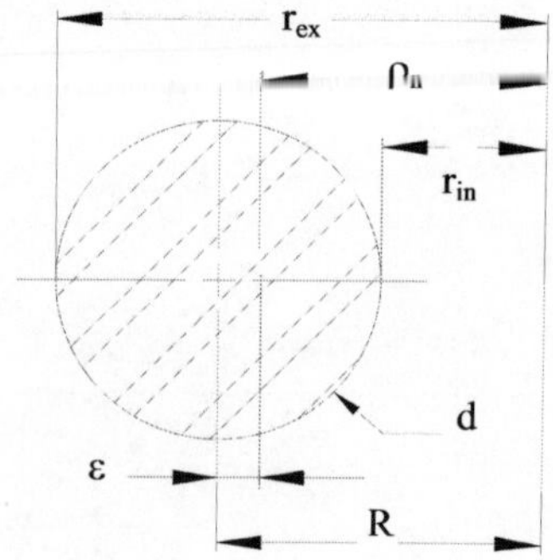

$$R = r_{in} + \frac{\dfrac{h^2 e}{2} + \dfrac{e_{in}^2}{2}\left(b_{in} - e\right)}{he + \left(b_{in} - e\right)e_{in}};$$

$$\varepsilon = R - \rho_n;$$

$$\rho_n = \frac{\left(b_{in} - e\right)e_{in} + h \times e}{\left(b_{in} - e\right)ln\left(\dfrac{r_{in} + e_{in}}{r_{in}}\right) + eln\left(\dfrac{r_{ex}}{r_{in}}\right)}$$

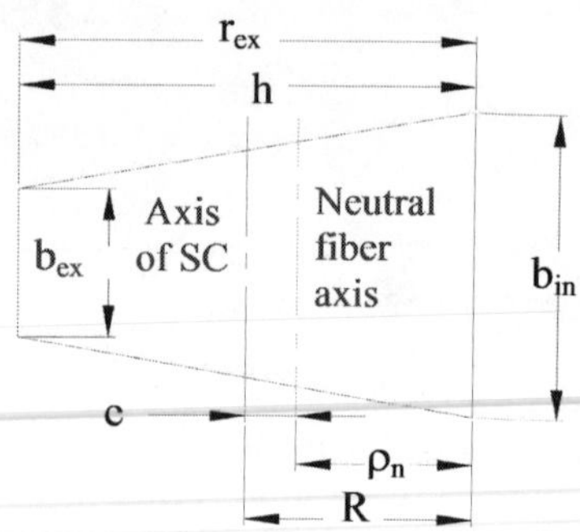

$$R = r_{in} + \frac{\dfrac{h^2 e}{2} + \dfrac{e_{in}^2}{2}\left(b_{in} - e\right) + \left(b_{ex} - e\right)e_{ex}\left(h - \dfrac{1}{2}e_{ex}\right)}{\left(b_{in} - e\right)e_{in} + \left(b_{ex} - e\right)e_{ex} + h \times e}$$

$$\varepsilon = R - \rho_n$$

$$\rho_n = \frac{\left(b_{in} - e\right)e_{in} + \left(b_{ex} - e\right)e_{ex} + h \times e}{b_{in}ln\left(\dfrac{r_{in} + h_{in}}{r_{in}}\right) + ln\left(\dfrac{r_{ex} - e_{ex}}{r_{in} + e_{in}}\right) + b_{ex}ln\left(\dfrac{r_{ex}}{r_{ex} + e_{in}}\right)}$$

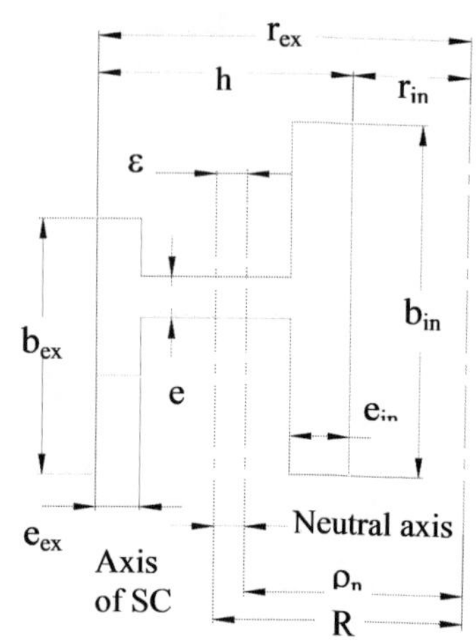

$$R = r_{in} + \frac{h(b_{in} + 2eb_{in})}{3(b_{in} + b_{ex})};$$

$$\varepsilon = R - r_n;$$

$$r_n = \frac{\frac{(b_{in} + b_{ex})}{2} h}{\left(\frac{b_{in} r_{ex} - b_{ex} r_{in}}{h}\right) \times ln\left(\frac{r_{ex}}{r_{in}}\right) - (b_{in} - b_{ex})}$$

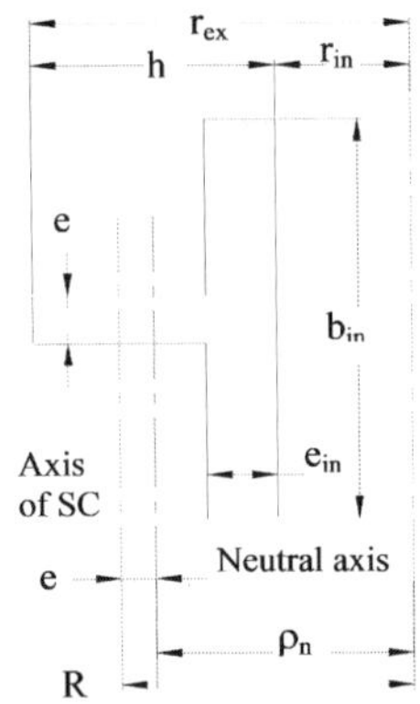

$$R = r_{in} + \frac{\frac{h^2 e}{2} + \frac{e_{in}^2}{2}(b - e) + (b - e)e_{ex}\left(h - \frac{1}{2}e_{ex}\right)}{he + (b - e) + (e_{in} + e_{ex})}$$

$$\rho_n = \frac{(b - e)e(e_{in} + e_{ex})e_{ex} + h \times e}{b\left(\ln\frac{r_{in} + e_{in}}{r_{in}} + \ln\frac{r_{ex}}{r_{ex} - e_{ex}}\right) + e\ln\left(\frac{r_{ex} - e_{ex}}{r_{in} - e_{in}}\right)};$$

$$\varepsilon = R - \rho_n$$

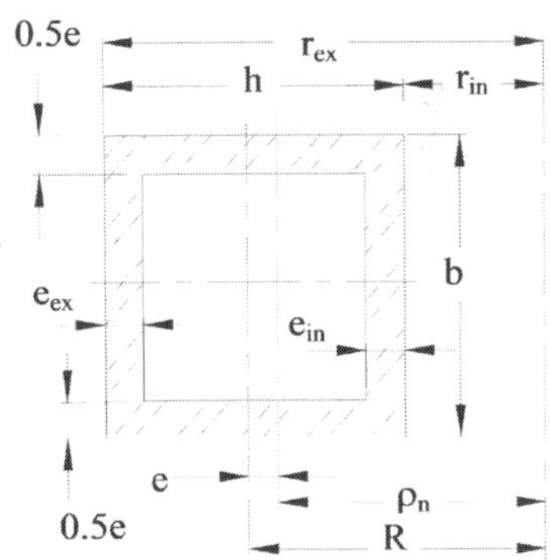

Figure 4.1. *Basic formula for usual geometric models in strength of materials. For a color version of this figure, see www.iste.co.uk/grous/design.zip*

4.2. Case studies

4.2.1. *Case study 1: frame sweeping*

The initial data for a structure (a frame with a round cross-section) swept in accordance with the design sketch are as follows: consider the structure illustrated below (Figure 4.2) (round cross-section frame) with the 2D/3D sweep function on the geometry whose parameters are:

– R is the radius (see figure below) R = 4.75 in = 120.65 mm;

– r_{in} is the internal radius r_{in} = 3.75 in = 95.25 mm;

– r_{ex} is the external radius r_{ex} = 5.75 in = 146.05 mm;

– d is the diameter = 4 in = 101.6 mm;

– A is the area to be checked A = 12.5664 in^2 = 8.107×10^3 mm^2;

– h_{in} is the height, to be calculated (?);

– ε is the excentricity (with respect to the centroid), to be calculated (?);

– L is the length; L = 18 in = 457.2 mm;

– m is the mass = 1856 lb = 841.867 kg;

– F is the force applied (loading); F = 172 *lbf* = 765.094 N.

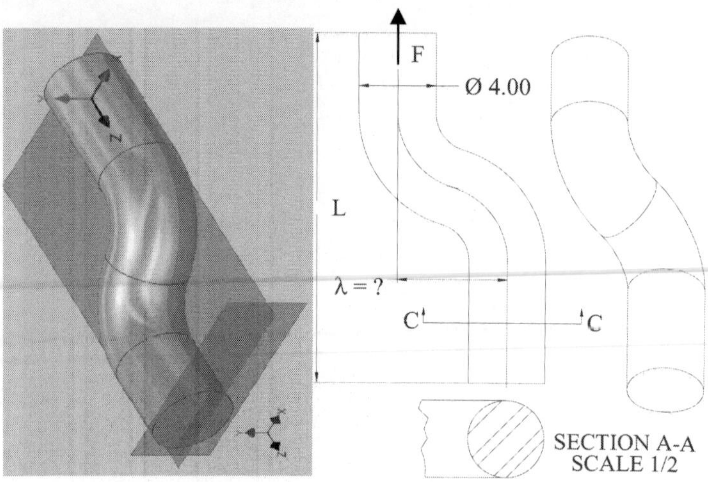

Figure 4.2. *Sweep function giving rise to an arm embedded at one end. For a color version of this figure, see www.iste.co.uk/grous/design.zip*

$$r_{in} = 3.75\ in = 95.25\ mm;\ R = 4.75\ in = 120.65\ mm;\ r_{ex} = 5.75\ in = 146.05\ mm$$

For a *round form*, we set the following equations:

$$\varepsilon = R - r_{in} = 0.0532\ \ in = 1.352\ \ mm\ and\ h_{in} = r_{in} - \varepsilon = 3.6968\ \ in = 93.898\ \ mm$$

For: $d = 4\ in = 101.6\ mm,\ the\ area\ A = \pi\left(d^2/4\right) = 12.564\ in^2 = 8.107\times10^3\ \ mm$

For the *section C-C*: $\begin{cases} M_{bending} = M_f = 1856\left[lbf \times in\right] = 2.097\times10^5\left[N\times mm\right] \\ m = 1856\left[lb\right] = 841.867\left[kg\right] \rightarrow P = m\times g \cong 18560\left[lbf\right] \end{cases}$

In line with the formula for calculating the bending stress on the inner fiber, we set:

$$\begin{cases} \sigma_{bending} = \sigma_f = M_f h_{in} / A\varepsilon r_{in} = 2.735\times10^3 \left[psi\right] = 18.86\left[MPa\right] \\ \sigma_N = P/A = 147.696\, psi = 1.018\, MPa \; ; \left[\sigma\right]_{adm} = 15600\, psi = 107.558\ MPa \end{cases}$$

The bending stress of the tube plus the Pascal stress (P/A) means we can write:

$$\sigma_{total} = \sigma_f + \sigma_N = \left(M_f h_{in} / A\varepsilon r_{in}\right) + P/A = 2883\, psi = 19.878\ MPa$$

Given the simplified expression, we use the software Math CAD Prime to set:

$$\left[\sigma\right]_{adm} = \lambda_{offset}\left(M_f / A\varepsilon r_{in}\right) + P/A \rightarrow simplify\left[10.4987 M_f \lambda_{offset} / mA\varepsilon\right]$$

PROOF.– With the following expression, we can isolate the offset:

$$\begin{cases} \lambda_{offset} = \left\{\left[\sigma\right]_{adm} - \sigma_N\right\} A\varepsilon r_{in} / M_f = 20.8832\ in = 530.433\ mm \rightarrow \text{Proof} \downarrow \\ \lambda_{offset} = \left(1.56\times10^4 - 147.696\right)\dfrac{12.5664\times0.0532\times3.75}{1856} = 20.8832\ in = 530.433\ mm \end{cases}$$

DESIGN DECISION.– On the basis of the above, we would tend to choose $\lambda_{offset}^{calculated} = 21\ in = 533\ mm \rightarrow$. Design, remember, is in no way a matter of isolated drawings, but instead a question of absolute, incontrovertible accuracy, founded on clear calculations. A designer sketches, calculates and then draws, because drawing is the graphic manifestation of what has been calculated and verified; it is the proof and the product of this design decision.

4.2.2. *Case study 2: frame sweeping*

PROBLEM.– We wish to calculate the following functions to enable the frame to safely serve its purpose:

1) The bending moment in all the sections;

2) The compressive stress at the radius r_{in};

3) The traction (tension) stress at the radius r_{ex} (external fiber);

4) Where will the maximum tension be found? Calculate it.

where:

– R is the radius (see Figure 4.3 below); R = 3.55 in = 90.17 mm;

– r_{in} is the internal radius r_{in} = 3.15 in = 80.01 mm;

– r_{ex} is the external radius r_{ex} = 4.15 in = 105.41 mm;

– d is the diameter = 4 in = 101.6 mm;

– A is the area, to be verified A = 2 in^2 = 1.297×10^3 mm^2;

– h is the height; h = 0.85 in = 21.59 mm;

– ε is the excentricity (with respect to the centroid), to be calculated (?);

– L is the length; L = 6.15 in = 156.21 mm;

– F is the force applied (loading); F = 172 *lbf* = 765.094 N.

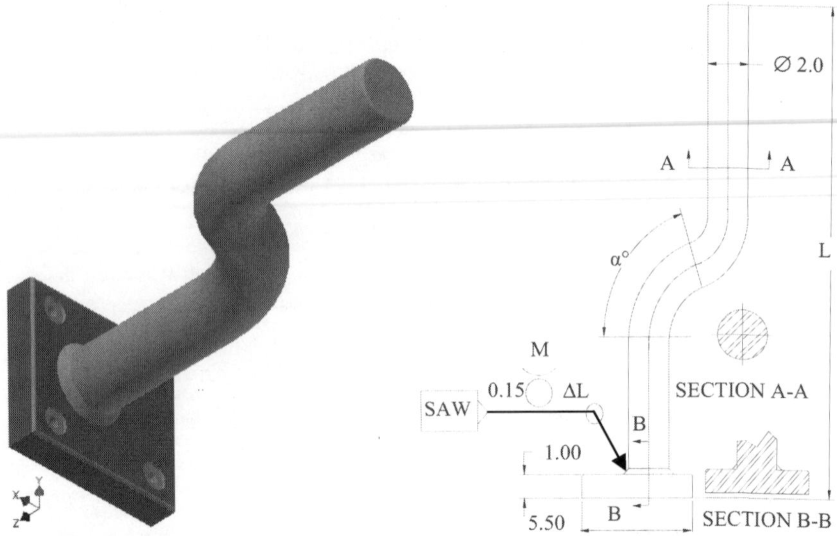

Figure 4.3. *Sweep function giving rise to an arm attached at one end. For a color version of this figure, see www.iste.co.uk/grous/design.zip*

SOLUTION WITH DISCUSSION.–

CASE 1.
$$\begin{cases} From\ r_n = h/\ln(r_{ex}/r_{in});\ \varepsilon = R - r_n;\ and\ R = r_{in} + h/2, \text{ we calculate } \downarrow \\ r_n = 3.0830[in] = 78.308[mm];\ \varepsilon = 0.4670[in] = 11.862[mm] \end{cases}$$

Calculation of the bending moment: $M_{fl} = F \times L = 1058 \times 10^3\ lbf.in = 1.195 \times 10^5\ N.mm.$

For the cross-section of R = 3.55 in, the stress on the internal fiber would be:

$$h_{in} = 0.485\ in = 12.319\ mm; h_{ex} = 0.515\ in = 13.081\ mm; A = 2\ in^2 = 1.29 \times 10^3\ mm^2$$

The internal fiber is subjected to the tension, and the external one to compression:

$$\begin{cases} \text{Internal fiber under tension } \sigma_{inT} = M_{fl} h_{in} / A\varepsilon r_{in} = 174.375\, psi = 1.202\, MPa \\ \text{External fiber under compression } \sigma_{exC} = M_{fl} h_{in} / A\varepsilon r_{in} = 140.544\, psi = 0.969\, MPa \end{cases}$$

CASE 2.– Calculation of the bending moment for F_1 = 190 *lbf* = 845.162 N.mm:

– L is the length; L = 5.56 in = 141.224 mm;

– F is the force applied (loading); F = 190 *lbf* = 845.162 N.

$$\begin{cases} r_{ex} = 4.12\ in = 104.648\ mm;\ r_{in} = 3.12\ in = 79.248\ mm; R = 3.54\ in = 89.916\ mm \\ \rho_n = \dfrac{l}{\ln(r_{ex}/r_{in})} = 3.0573\ in = 78.308\ mm \rightarrow \varepsilon = R - \rho_n = 0.4827\ in = 12.26\ mm \end{cases}$$

Bending moment: $M_{fl}^1 = F_1 \times L = 1056 \times 10^3\ lbf.in = 1.194 \times 10^5\ N.mm$

$$h_{in} = 0.485\ in = 12.319\ mm; h_{ex} = 0.515\ in = 13.081\ mm; A = 2\ in^2 = 1.29 \times 10^3\ mm^2$$

For the cross-section of R = 3.54 in, the internal and external fibers will be subject to traction and compression, respectively:

$$\begin{cases} \text{Internal fiber under traction: } \sigma_{inT} = M_{fl}^1 h_{in} / A\varepsilon r_{in} = 136.792\ psi = 0.943\ MPa \\ \text{External fiber under compression: } \sigma_{exC} = M_{fl}^1 h_{ex} / A\varepsilon r_{ex} = 170.113\ psi = 1.173\ MPa \end{cases}$$

For h_0 = 1 in = 25.4 mm; b = 2 in = 50.8 mm, calculate the polar moment of inertia J:

$$J = bh^3/12 = 0.167\ in^4 = 6.937 \times 10^4\ mm^4,$$

and for c = 0.5 in = 12.7 mm, calculate the normal stress – i.e. the stress of simultaneous compression and traction:

$$\sigma_N = M_{fl}^1 \times c / J = 3.169 \times 10^3 [psi] = 21.851 [MPa]$$

COMMENT.– The above calculation supports the sketch initially projected at the pre-project stage. To prove the sustainability of the design, it is important to respect the geometry in connection with the loads with which we are dealing.

4.2.3. *Case study 3: frame sweeping*

PROBLEM.– Consider a U-shaped flare-fitted die-cut piece (2D-3D), made by sweeping

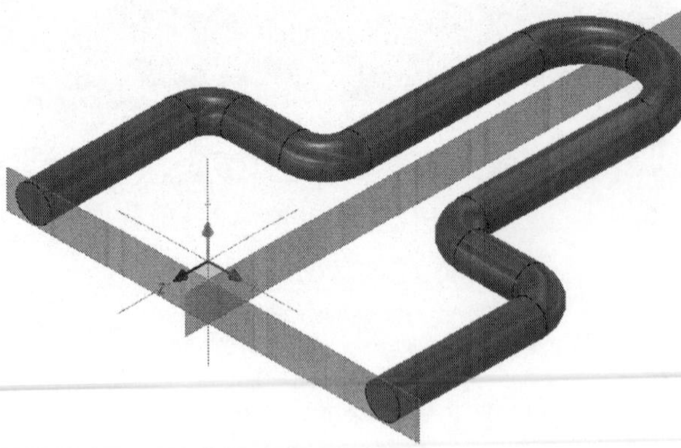

Figure 4.4. *2D and 3D of a swept flared-U piece, with the load concentrated at the center. For a color version of this figure, see www.iste.co.uk/grous/design.zip*

Initial data

– $A = 6.10$ in = 154.94 mm	Middle lever arm
– $c = 0.51$ in = 12.954 mm	½ diameter
– $R = 5.25$ in = 133.35 mm	Radius
– $b = 8.10$ in = 205.74 mm	Length of section of shaped piece
– $h = 1.10$ in = 27.94 mm	Thickness of section of shaped piece
– $r_{in} = 4.10$ in = 104.14 mm	Internal radius of section of piece
– $r_{ex} = 5.10$ in = 129.54 mm	External radius of section of piece
– $L = 22.25$ in = 565.15 mm	Demi-length (lever arm)
– $F_{concentrated} = 450\ lbf = 2.002\times10^3$ N	Force concentrated at the middle of the U
– $F_{app} = 225$ lbf $= 1.001\times10^3$ N	Force applied on each side of middle

SOLUTION WITH DISCUSSION OF THE PROBLEM.–

Polar moment of inertia of the section [α–α] at the level (j-k):

$$J_{xx} = bh^3/12 = 0.898^3\ in^4 = 3.74\times10^5\ mm^4$$

– 1st *section α–α:* calculation of the bending moment and its stress at the level *j-k:*

$$\begin{cases} M_{fl}^{\alpha-\alpha} = F_{app}L = 5.006\times10^3\ lbf.in = 5.656\times10^5\ N.mm \\ \sigma_{fl}^{\alpha-\alpha} = \sigma_{fl}^{j-k} = M_{fl}^{\alpha-\alpha}c/J_{xx} = 2842\ psi = 19.594\ MPa \end{cases}$$

NOTE.– At point *j*, the piece is under tension, and at *k* it is under compression. The shear stress (transverse bending) is zero and points *j and k.*

– 2nd *section β-β:* calculation of the bending moment and its stress at level *m-n.* Calculation of the useful geometric dimensions and the relative bending moment:

$$\begin{cases} l_1 = L - a = 16.15\ in = 410.21\ mm; \rho_n = h/\ln\left(r_{ex}/r_{in}\right) = 5.040\ in = 128.016\ mm \\ \varepsilon = R - \rho_n = 0.210\ in = 5.334\ mm; h_{in} = c - \varepsilon = 0.30\ in = 7.62\ mm; \end{cases}$$

$$A = bh = 5.748\ in^2 = 5748\ mm^2 \rightarrow M_{fl}^{\beta-\beta} = F_{app}l_1 = 3634\ lbf.in = 4.106\times10^5\ N.mm$$

At point (m), we set: $\sigma_{fl}^{m} = M_{fl}^{m}\times h_{in}/A\times\varepsilon\times r_{in} = 142.111[psi] = 0.98[MPa]$

For: $h_{ex} = c + \varepsilon = 0.720\ in = 18.288\ mm \rightarrow$ we calculate the stress at (n).

At point (n), we set: $\sigma_{fl}^{n} = M_{fl}^{n}h_{ex}/A\varepsilon r_{ex} = 274.1789\ psi = 1.890\ MPa \leftarrow tension$

– 3rd *section γ-γ:* at point (ω), note that there is a compressive stress resulting from the bending moment. There is also a direction traction stress. The combination of the two stresses is then calculated by the bending stress, at point (ω), as follows:

$$\begin{cases} l_2 = (L - a - R) = 10.9\ in = 276.86\ mm;\ for\ r_{ex} = 4.02\ in = 102.108\ mm \leftarrow (\gamma-\gamma) \\ r_{in} = 3.02\ in = 76.708\ mm\ \leftarrow (\gamma-\gamma)\ and\ R = 4\ in = 101.60\ mm \rightarrow Calculate: \end{cases}$$

$$\begin{cases} \rho_n = h/\ln(r_{ex}/r_{in}) = 3.8458\ in = 97.684\ mm; \varepsilon = R - \rho_n = 0.1542\ in = 3.916\ mm \\ h_{ex} = c + \varepsilon = 0.664\ in = 16.87\ mm \rightarrow \sigma_{traction} = M_{fl}^{\omega} h_{ex} \big/ A\varepsilon r_{ex} = 185.0495\ psi = 1.276\ MPa \end{cases}$$

The normal stress and the traction stress thus give us the total stress:

$$\begin{cases} \sigma_N = F_{app}/A = 25.2525[psi] = 0.174[MPa] \\ \sigma_C = \sigma_N - \sigma_\tau = 159.797[psi] = 1.102[MPa] \end{cases}$$

At the point (φ), we would have:

$$(\gamma - \gamma) \rightarrow Level\ \varphi \rightarrow h_{in} = (c + \varepsilon) = 0.664\ in = 16.87\ mm$$

The traction stress would be: $\sigma_\tau^\varphi = M_{fl}^\varphi h_{in} \big/ A\varepsilon r_{in} = 392.6223\ psi = 1.276\ \ MPa$

The total traction stress would be: $\sigma_{tot}^U = \sigma_N - \sigma_\tau^U = 417.875[psi] = 1.102[MPa]$

COMMENT AT PRE-PROJECT STAGE.– The maximum stress occurs on the straight portion – that is, the *section α-α* – and its value is 2842 psi = 19.594 MPa. There is traction at point (*k*) and compression at point (j). The maximum value under shearing (*transverse bending*) at points (*j*) and (*k*), which gives a stress of (2842/2) = 1421 psi = 9.797 MPa.

4.2.4. *Case study 4: frame sweeping*

Consider a cold-forged torsion bar of SAE 1018 steel, with a homogeneous circular cross-section (Figure 4.5), subject to loading represented by -1F to 3F, and with a safety factor S = 5/4. The stress intensification factor K_τ = 2, with notch sensitivity. The threads on this shaft are: r = (7/5) in = 1.3607 mm:

– S is the safety factor, S = (5/4) = 1.25;

– d is the diameter of the bar d = (3/4) in = 19.05 mm, so R_b is its radius R_b = (d/2) = 0.375 in = 9.525 mm;

– $r_{threads}$ is the radius of the threads $r_{threads}$ = (1/7) R_b in = 0.0536 in = 1.3607 mm;

– lever arms made to specifications: L = 3.937 in = 100 mm and B = 3.346 in = 85 mm;

– load correction factor other than bending: α = 0.75;

– endurance limit size correction factor: β = 0.85 (d >1/2);

– geometry correction factor, relative to the polished surface state: γ = 0.9, for parts between (1/2 < d < 2) in.

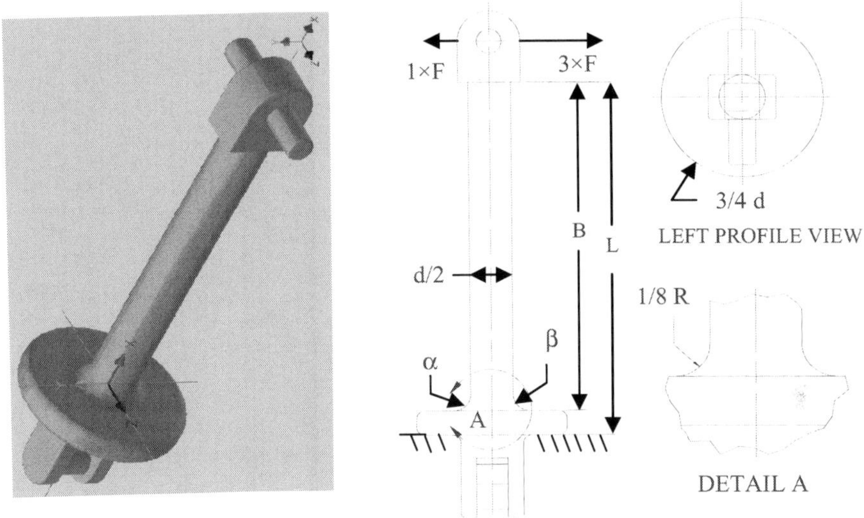

Figure 4.5. *Torsion bar, embedded at one end. For a color version of this figure, see www.iste.co.uk/grous/design.zip*

$$\sigma_{\lim} = 78\times10^3\ psi = 537.791\ MPa\ and\ \sigma_{elas} = 69\times10^3\ psi = 475.738\ MPa$$

The stress intensification factor (SIF), K_f, at the level of the notch sensitivity is $K_f = 1.15$. For a theoretical (*f.i.c*), $K_t = 2$ [ROA 75]:

$$\begin{cases} For\ r_{threads} = (2/5)R_b = (2/5)(d/2) = (2/5)(0.375) = 0.1500\ in = 3.810\ mm \rightarrow \\ K_f = 1 + r_{threads}(K_t - 1) = 1 + 0.15(2-1) = 1.15 \rightarrow\ OK\ because\ 1 \prec K_f \prec 3\ or + \end{cases}$$

Note that for a diameter greater than (½) in = 12.7 mm, we make this assumption:

$$\sigma_{\min}^{\max} \geq 32/\pi d^3 = 32/\pi(1/2)^3 = 81.4873\ psi = 0.5618\ MPa$$

Put differently, in relation to the applied forces, F_α and F_β:

$$\begin{cases} \sigma_{fl}^{\min} \geq 81.48733086305041 1914(M_{fl}) = 81.48733086305041 1914(12F_\alpha)\ psi \\ = 0.56183536876884010(12F_\alpha)\ MPa : Simplify \rightarrow \text{-}325.94932345220164766F_\alpha \end{cases}$$

$$\begin{cases} \sigma_{fl}^{\max} \geq 81.48733086305041 1914\left(M_{fl}\right) = 81.487330863050411914\left(12F_{\alpha}\right)\ psi \\ = 0.5618353368768840 10\left(12F_{\alpha}\right)\ MPa : Simplify \rightarrow \text{-}977.84797035660494297F_{\alpha} \end{cases}$$

$$\sigma_{fl-\alpha}^{Moy} = \frac{1}{2}\begin{bmatrix} (977.84797035660494297)F_{\alpha} + \\ (\text{-}325.94932345220164766F_{\alpha}) \end{bmatrix} simplify \rightarrow 325.94932345220164765F_{\alpha}$$

Remember: $\begin{cases} \sigma_y = \sigma_{elas} = 6.9\times10^4\ psi;\ \sigma_r = \dfrac{\sigma_{\lim}}{2} = 3.9\times10^4\ psi = 268.869\ MPa \\ \sigma_N = \sigma_{fl} = 81.487330863050411914\times M_{fl}\ and\ K_t = 1.15 \end{cases}$

Let us substitute this expression thus: $\left\{\dfrac{1}{S} = \left(\dfrac{\sigma_{ave}}{\sigma_y} + \dfrac{\sigma_N}{\sigma_r\alpha\beta\gamma}\right)\right.$ to isolate F_α:

$$\begin{cases} \dfrac{5}{4} = \left(\dfrac{325.94932345220164765\times F_{\alpha}}{6.9\times10^4} + \dfrac{(1.15)814.87330863050411913\times F_{\alpha}}{3.9\times10^4\times(0.75)\times(0.85)\times(0.9)}\right) \rightarrow \\ \dfrac{466033202508986731 1841\times F_{\alpha}}{10000000000000000000000} = \dfrac{5}{4} \rightarrow F_{\alpha} = 17.166159[lbf] = 76.358880[N] \end{cases}$$

Following the same logic of analysis as for point α, point β has the following effect of F_β:

$$\begin{cases} \sigma_{fl}^{\min} \geq 81.487330863050411914\left(-4F_{\beta}\right) = \text{-}325.94932345220164766\left(F_{\beta}\right)\ psi \\ \sigma_{fl}^{\max} \geq 81.487330863050411914\left(12F_{\beta}\right) = 977.84797035660494297\left(F_{\beta}\right)\ psi \end{cases}$$

$$\sigma_{fl-\beta}^{Ave} = \frac{1}{2}\begin{bmatrix} 977.84797035660494297F_{\beta} + \\ \left(\text{-}325.94932345220164766F_{\beta}\right) \end{bmatrix} simplify \rightarrow 325.94932345220164765F_{\beta} \rightarrow$$

$$\begin{cases} \dfrac{5}{4} = \left(\dfrac{325.94932345220164765\times F_{\alpha}}{6.9\times10^4} + \dfrac{(1.15)814.87330863050411913\times F_{\alpha}}{3.9\times10^4\times(0.75)\times(0.85)\times(0.9)}\right) \rightarrow \\ \dfrac{466033202508986731 1841\times F_{\alpha}}{10000000000000000000000} = \dfrac{5}{4} \rightarrow F_{\alpha} = F_{\beta} = 76.359\ N = 17.166\ lbf \end{cases}$$

4.2.5. *Case study 5: example of a connecting rod of SAE 8650*

A rod of steel alloy 8650, tempered at 1500°F = 815.556°C and cooled to 1010°F = 543.333°C is subjected to the actions of axially reversible loads of F_a = 35,000 *lbf* = 1.557×10^5 N. What is the safety diameter of the rod if the safety coefficient S = 1.75 = 7/4?

where:

– geometry correction factor for F_a, $\alpha = 0.70$ [ROA 75];

– endurance limit size correction factor, $\beta = 0.85$ for d > (3/2) in;

– endurance limit correction factor, $\gamma = 0.8$;

– F_a is the axial force applied to the rod = 35×10^3 lbf = 1.557×10^5 N;

– $T_{cooling}$ = 1010°F = 543.333°C;

–$T_{tempering}$ = 1500°F = 815.556°C.

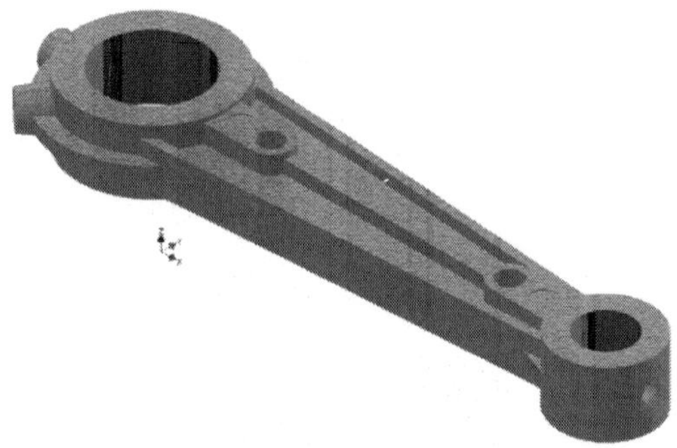

Figure 4.6. *3D drawing of design of SAE 8650 connecting rod. For a color version of this figure, see www.iste.co.uk/grous/design.zip*

Looking at the properties of the material [ROA 75] SAE 8650, we see:

– $\sigma_{lim} = 67\times10^3\ psi = 461.949$ MPa;

– $\sigma_{elastic\ limit} = \sigma_E = 45\times10^3$ psi = 310.264 MPa.

The endurance limit corresponds to half the elastic limit (lim): $\sigma_S = \sigma_u/2 = 3.35\times10^4\ psi = 230.974\ MPa$. The average stress $\sigma_{ave} = 0$. The projected area for: d > (3/2) would be $A_p \succ \pi d^2/4 = 1.767\ in^2 = 38.10\ mm^2$. The normal stress σ_N is then non-zero, and hence: $\sigma_N = F_a/A = 4F_a{}^4/\pi d^2 = 1.981\times10^4\,[psi] = 136.557\,[MPa]$.

The equation is presented in simplified form using Math CAD Prime:

$$\begin{cases} \dfrac{1}{S} = \dfrac{\sigma_{ave}}{\sigma_{\lim}} + \dfrac{\kappa_c\left(4F_a/\pi d^2\right)}{\sigma_s\alpha\beta\gamma} simplify \rightarrow \dfrac{1}{S} = 0 + \dfrac{4F_a}{\sigma_s\alpha\beta\gamma\pi d^2} \rightarrow \\ \dfrac{1}{S} = \left(\dfrac{4\sigma_V}{\sigma_s\alpha\beta\gamma\pi d^2}\right) Solve \rightarrow d = \sqrt{\dfrac{4F_a\kappa_c}{\sigma_s\alpha\beta\gamma\pi}} = 1.678\ in = 42.467\ mm \end{cases}$$

Proof ↓

$$d = \sqrt{4\left(35\times10^3\right)(1)/\pi(0.8)(0.85)(0.7)33500} = 1.678\ in = 42.468\ mm \leftarrow cqfd$$

Thus, we shall choose a diameter of 1.75 in: d_{chosen} = 1.75 in = 44.45 mm, and hence $A = \pi d_{chosen}{}^2/4 = 2.405\ in^2 = 1552\ mm^2$. Therefore, we move on to the design phase with a diameter of (1+3/4) = 1.75 in = 44.45 mm. If the rod rotated at a rate of RPM = 960 rpm, then $\omega = 2\pi$ RPM/60 = 10.528 rad/s, the yield η = 80 %, and by varying the axial force, then $F_{ax} = 15\times10^3\ lbf,\ 16\times10^3\ lbf,\ \ldots,\ 25\times10^3\ lbf$, we obtain the following plots:

$$d\{F_{ax}\} = \sqrt{4F_a\kappa_c/\sigma_s\alpha\beta\gamma\pi}\ ;\ M\{F_{ax}\} = F_{ax}\times d/2; P\{F_{ax}\} = M\times\omega\times\eta \quad [4.1]$$

Presentation and discussion of results:

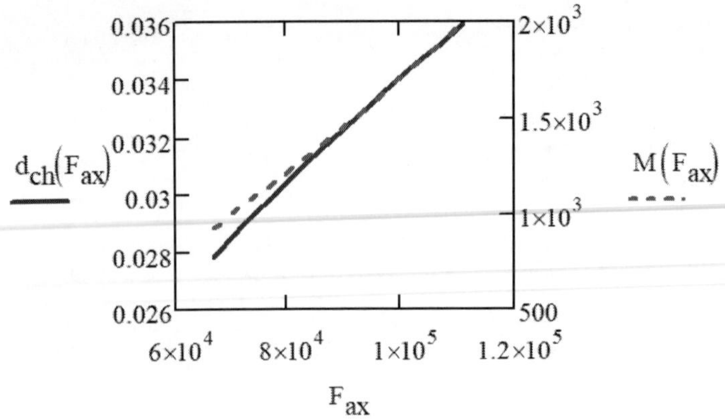

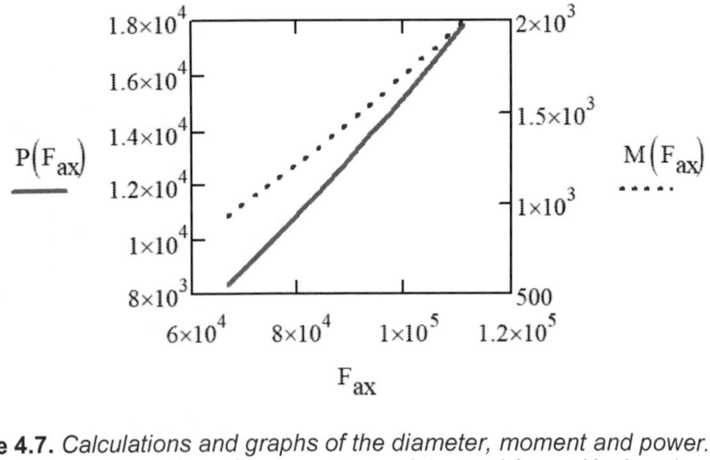

Figure 4.7. *Calculations and graphs of the diameter, moment and power. For a color version of this figure, see www.iste.co.uk/grous/design.zip*

COMMENT.– We can see that the axial loading chosen has a considerable influence on the choice of the diameter (the bore of the rod), the torque (moment) and the power.

$d_{ch}(F_{ax})$	$d_{ch}(F_{ax})$	$M(F_{ax}) =$	$M(F_{ax}) =$	$P(F_{ax}) =$
1.094	27.798	$8.208 \cdot 10^3$	$9.274 \cdot 10^5$	$8.299 \cdot 10^3$ W
1.130	28.709	$9.042 \cdot 10^3$	$1.022 \cdot 10^6$	$9.142 \cdot 10^3$
1.165	29.593	$9.903 \cdot 10^3$	$1.119 \cdot 10^6$	$1.001 \cdot 10^4$
1.199	30.451	$1.079 \cdot 10^4$	$1.219 \cdot 10^6$	$1.091 \cdot 10^4$
1.232	31.285	$1.17 \cdot 10^4$	$1.322 \cdot 10^6$	$1.183 \cdot 10^4$
1.264	32.098	$1.264 \cdot 10^4$	$1.428 \cdot 10^6$	$1.278 \cdot 10^4$
1.295	32.891	$1.36 \cdot 10^4$	$1.536 \cdot 10^6$	$1.375 \cdot 10^4$
1.325	33.665	$1.458 \cdot 10^4$	$1.647 \cdot 10^6$	$1.474 \cdot 10^4$
1.355	34.421	$1.558 \cdot 10^4$	$1.761 \cdot 10^6$	$1.576 \cdot 10^4$
1.384	35.162	$1.661 \cdot 10^4$	$1.877 \cdot 10^6$	$1.68 \cdot 10^4$
1.413	35.887	$1.766 \cdot 10^4$	$1.995 \cdot 10^6$	$1.786 \cdot 10^4$
in	mm	lbf·in	N·mm	W

4.2.6. *Case study 6: swept double elbow*

Initial data for design of spring clip:

– r_{ex} = 4.25 in = 107.95 mm and r_{in} = 3.25 in = 82.55 mm;

– R = 4.75 in = 120.65 mm and d = ¾ in = 0.75 in = 19.05 mm.

1) The maximum normal stress is important to know. The maximum bending stress occurs at the level of the internal fiber of the section with a radius r_{in} = 3.25 in from (α) toward (β) and from (γ) toward (ω). Calculation of the radius of curvature of the internal fiber gives us:

$$\rho_n = \left(\sqrt{r_{ex}} + \sqrt{r_{in}}\right)^2 / 4 = 3.733[in] = 94.825[mm]$$

2) Calculation of the distance from the axis of the neutral fiber to the internal fiber:

$$h_{in} = \rho_n - r_{in} = 0.483\ in = 12.275\ mm$$

3) Calculation of the distance from the axis of the center of gravity to the neutral axis of the neutral fiber:

$$\text{Excentricity} \rightarrow \varepsilon = R - \rho_n = 1.017[in] = 25.825[mm]$$

4) Calculation of the *area (A) of the section*: $A = \pi d^2/4 = 0.442\ in^2 = 285.023\ mm^2$.

5) Calculation of the bending moment and the stress applied to the neutral fiber:

$$M_{fl} = 925\ lbf.in = 1.045\times 10^5\ N.mm \rightarrow \sigma_{fib.\text{int}} = M_{fl} h_{in} / A\varepsilon r_{in} = 306.2069\ psi = 2.111\ MPa$$

6) From (α) to (β): *traction* and from (γ) to (ω): *compression*.

7) The maximum *shear* stress is applied at all points from (α) to (β) and from (γ) to (ω). As the section is symmetrical, the maximum stress at the level of the external fiber also needs to be verified:

$$\tau_{\max} = 0.5\sigma_{fib.\text{int}} = 0.5\left(M_{fl} h_{in} / A\varepsilon r_{in}\right) = 153.1035[psi] = 1.056[MPa]$$

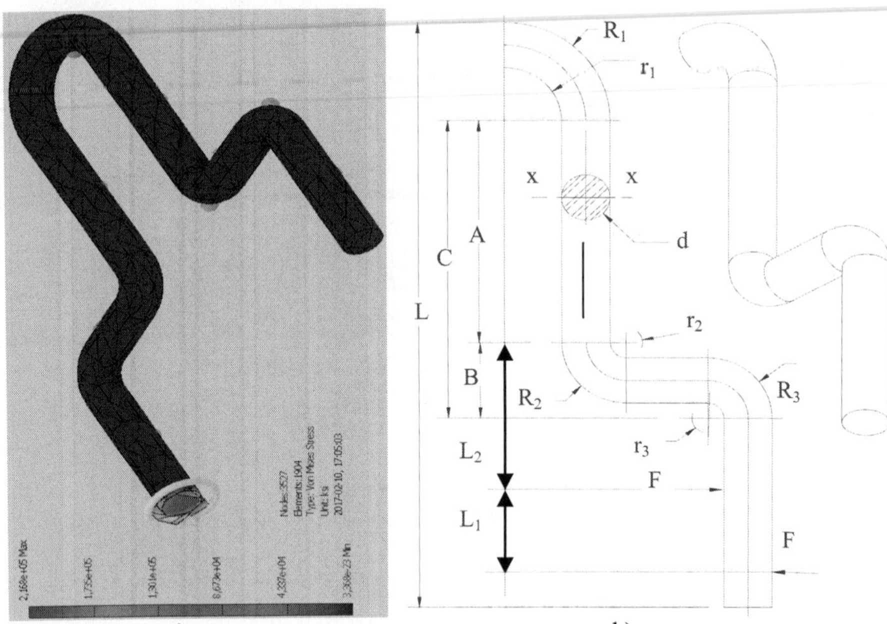

a) b)

Figure 4.8. *Spring clip: incurvate profiled piece by sweeping and inflection. Sketch and 3D design drawing of the spring clip. For a color version of this figure, see www.iste.co.uk/grous/design.zip*

Design drawing of spring clip and discussion of the results.

Looking at the results and the design diagram, we note the following:

– the maximum stress applied to the neutral fiber is fairly low for these data;

– the design of incurvate surfaces requires sustained observation of the curvatures with the aforementioned sweep software functions.

4.2.7. *Case study 7: frame sweeping*

The *C-Clamp* described below must firmly hold a dummy (head and shoulders) to a table. Considering the geometry, calculate the force exerted on the screw if the maximum tension stress in the clamp is limited to 18,000 psi = 124.106 MPa.

Initial data

– r_{in} = 1.05 in = 26.67 mm;

– r_{ex} = 2.05 in = 52.07 mm;

– h = 1.05 in = 26.67 mm;

– e_0 = 0.175 in = 4.445 mm;

– e_{in} = 0.175 in = 4.445 mm;

– b_{in} = 0.8 in = 20.32 mm;

– B = 0.925 in = 23.495 mm;

– L = 2.05 in = 52.07 mm;

– $\sigma_{app} = 18\times10^3$ psi = 124.106 MPa.

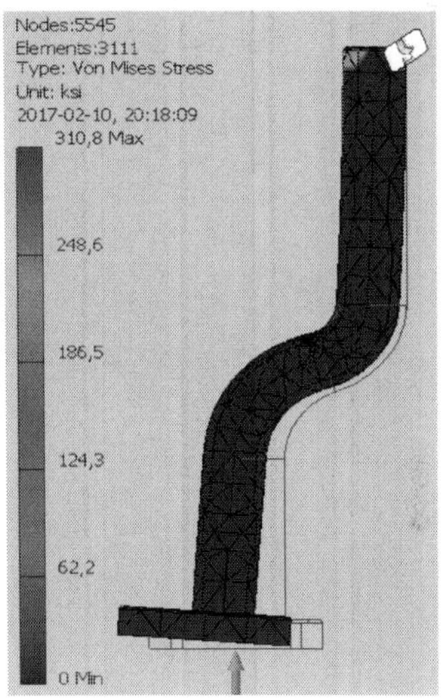

Figure 4.9. *Incurvate profiled piece with sweeping and inflection →. For a color version of this figure, see www.iste.co.uk/grous/design.zip*

SOLUTION WITH DISCUSSION.–

1) The distance R from the center of curvature of the axis of the center of gravity is:

$$R = r_{in} + \left(\frac{h^2 e_0}{2} + \frac{e_{in}^2}{2}\left(b_{in} - e_0\right) \Big/ he_0 + \left(b_{in} - e_0\right)e_{in} \right) = 1.411745\ in = 35.859\ mm$$

2) Using the following formula, we set the expression of the radius of curvature ρ_n:

$$\rho_n = \frac{(b_{in} - e_0)e_{in} + h \times e_0}{(b_{in} - e_0) \times ln((r_{in} + e_{in})/r_{in}) + e_0 \times ln(r_{ex}/r_{in})} = 1.373415\ in = 34.885\ mm$$

3) We can set the expression of the *excentricity* ε, of h_{in} and of the area A of the section:

$$\varepsilon = R - \rho_n = 0.038339\ in = 0.974\ mm\ et\ h_{in} = \rho_n - r_{in} = 0.323415\ in = 8.215\ mm$$

Calculation of the area (A) of the section: $A = (e_0 B) + (e_0 b_{in}) = 0.302\ in^2 = 195\ mm^2$

4) Calculation of the bending moment with respect to the center of gravity (C-G):

$$M_{fl} = F_{max}(L + R) \rightarrow simplify \rightarrow M_{fl} = 0.0879285447761194\ mF_{max}$$

$$As\ (L + R) = 3.461754\ in,\ we\ set \rightarrow M_{fl} = F_{max} \times 3.461754$$

5) Calculation of the *total stress* (bending stress + normal stress) using Math CAD Prime:

$$\begin{cases} \sigma_{total} = \left(\dfrac{M_{fl} \times h_{in}}{A \times \varepsilon \times r_{in}}\right) + \left(\dfrac{F_{max}}{A}\right) factor \rightarrow \\ \rightarrow \dfrac{812032452811014079770000 M_{fl} + 25672929192649275567 7 mF_{max}}{50000000000000000 \times m^3} \end{cases}$$

$$As\ \sigma_{total} = \frac{M_{fl} h_{in}}{A \varepsilon r_{in}} + \frac{F_{max}}{A}\ simplify \rightarrow \left(\frac{M_{fl} h_{in} + F_{max} \varepsilon r_{in}}{A \varepsilon r_{in}}\right)$$

$$\begin{cases} 18 \times 10^3 = \left(\dfrac{F_{max} \times (0.301875)(0.323415)}{(0.301875) \times (0.038339) \times (0.323415)}\right) + \left(\dfrac{F_{max}}{0.301875}\right) factor \rightarrow \\ F_{max} = \dfrac{18000 \times 2000000000000000000}{5879146031720830 68249} = 612.333829\ lbf = 2724\ N \end{cases}$$

6) Calculation of the *bending moment* with respect to the center of gravity:

$$M = F_{max}\left(L+R\right) = 2.12\times10^{3}\left[lbf.in\right] = 2.395\times10^{5}\left[N.mm\right]$$

First verification of the maximum force F_{max}:

$$\frac{F_{max}\times(0.301875)(0.323415)}{(0.301875)\times(0.038339)\times(0.323415)}+\frac{F_{max}}{0.301875}\rightarrow F_{max}=18000\ lbf=80070\ \mathrm{N}$$

Calculation of the *total stress* applied to the clamp:

$$\sigma_{TOT}=\left(M_{fl}h_{in}+F_{max}\varepsilon r_{in}\right)/A\varepsilon r_{in}=5.844\times10^{4}\left[psi\right]=402.948\left[MPa\right]$$

Second verification of the total compressive stress – Software:

$$\sigma_{TOT}=\frac{19448134126740692195 15 J+69927836577249908123\,N.m}{5000000000000\,m^{3}}$$

$$=5.844\times10^{4}\left[psi\right]=402.948\left[MPa\right]\leftarrow QED$$

Third verification of proof on the *maximum force* in connection with the total compressive stress:

$$\sigma_{TOT}=\sigma_{fl}+\sigma_{N}=\frac{M_{fl}h_{in}}{A\times\varepsilon r_{in}}+\frac{F_{max}}{A}=5.844\times10^{4}\left[psi\right]=402.948\left[MPa\right]\leftarrow QED$$

COMMENT.– The stress applied to the external fiber of the section may be greater than that applied to the internal fiber. Hence, we are dealing with a *compressive* stress.

4.3. Conclusion

There are a variety of ways to transform 2D sketches into 3D objects by sweeping (*sweep*, *loft*, *spline*, etc.) to extrude the parts. With regard to the present chapter, we can take away the fact that a graphic design, no matter how elegant, must not be limited to the graphic sciences (2D and 3D). The calculations form the solid backbone of any worthy design in engineering.

5

Principles for Calculations in Mechanical Design: Theory and Problems. Strength of Materials in Constructions

5.1. Essential criteria of constructions in design

The elements of machines, of construction of structures (shafts, connecting rods, etc.) and the support pieces (profiled parts) are not always regular in form. They are not always homogeneous, which poses a significant problem in terms of the reliability of the results of calculations. It is important to fully comprehend the morphologies of the materials which influence the accuracy and faithfulness of the results. Thus, there is an often-found conceptual duality between a *material* and its *geometry*. Ultimately, it is wise to make the choice which is optimal for the purposes of the design project at hand. We may find ourselves dealing with cases where structures are expressed in terms of non-uniform geometric forms (e.g. NURBS). The conventional calculation principles, in such cases, would prove overly simplistic and limiting. For that reason, we turn to the infinitesimal models employed in continuum mechanics. The mathematical tool necessitates minutely detailed representation using the finite element method (FEM), so we use von Mises and/or Tresca yield criteria. Here, we use pre-existing formulae whose homogeneity and consistency have long been established. A *formula* may be homogeneous, but it may nevertheless be *unfit* for use, depending on the case. Hence, a thorough designer must first check that the formulae used are consistent. One of the standard methods for doing so is to examine the units of measurement.

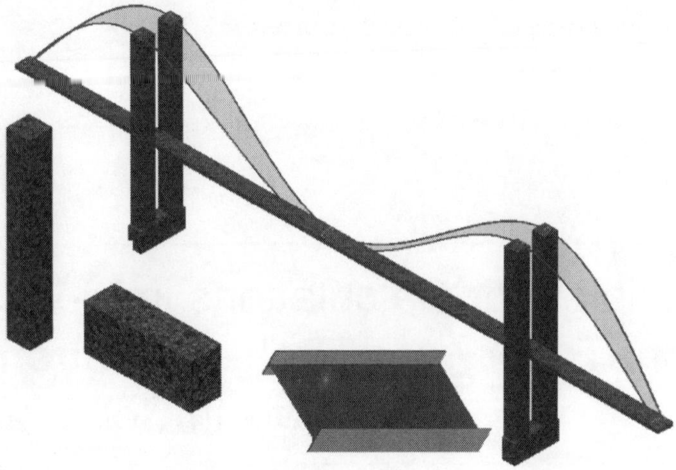

Figure 5.1. *Diagram of a "Cirta" suspension bridge: materials and profiles. For a color version of this figure, see www.iste.co.uk/grous/design.zip*

5.1.1. *Stress intensification in shafts and beams*

When a part (a shaft, a bolt, a beam, etc.) exhibits sudden variations in profile (such as holes, grooves or rounded sections), the relations used in strength of materials (SOM) mean we can write:

$$\sigma_{\max} = K_{\tau} \times \sigma_0;\ with\ \sigma_0 = [F/A = N/A] \leq [\sigma]_{adm} \qquad [5.1]$$

where:

– σ_{max} is the maximum applied stress in psi or MPa;

– A is the area of the section in in^2 or mm^2;

– σ_0 is the applied stress in psi or MPa;

– *F (or N)* is the load applied (force in *N* or *lbf*).

In the vicinity (see Figure 5.2) of the variation of the section (threads, rounded parts, grooves, etc.), the stresses are not evenly distributed. Rather, there is a maximum and a minimum point, written [σ_{min} and σ_{max}]. Only the points nearest to the geometric variation reach the maximum stress, which is known as stress intensification, often represented by the (K_{τ} = *SIF*) stress intensification factor [ROA 75]. It is dependent on the form of the section and the type of change in form. Here is what the above means for each of the standard geometric forms.

5.1.2. *Homogeneous, solid, round sections*

Example: $\begin{cases} r = 10\ mm;\ d = 2r = 20\ mm = 0.787\ in;\ D = 30\ mm; R = 15\ mm = 0.591\ in \\ F_0 = 54 \times 10^3\ N = 1.214 \times 10^4\ lbf;\ A = \pi R^2 = 706.856\ mm^2;\ K_\tau = D/d = 1.5 \end{cases}$

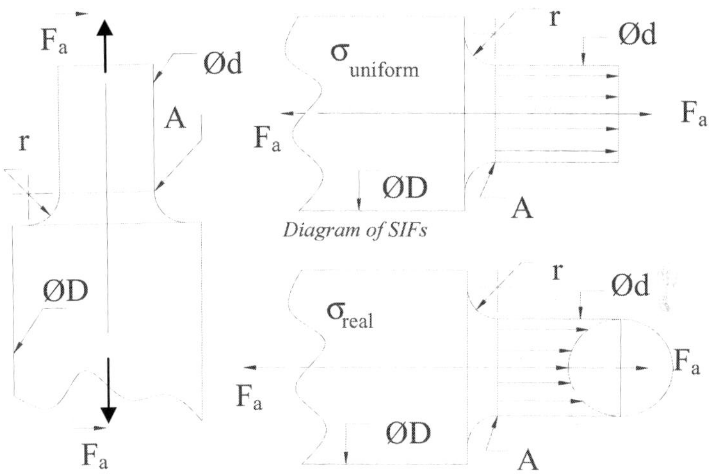

Figure 5.2. *Diagram of stress intensification factors. Round-shouldered shaft. For a color version of this figure, see www.iste.co.uk/grous/design.zip*

– *Without taking account of the SIF*: $\sigma_0 = F_0/A = 76.394\ MPa = 1.108 \times 10^4\ psi$

– *Taking account of the SIF*: $\sigma_{max} = \sigma_0 K_\tau = 114.592\ MPa = 1.662 \times 10^4\ psi$

K_τ is read from the specialized Wöhler curves (S-N). We find $K_\tau = f\,(D/d)$.

5.1.3. *Homogeneous, solid, square sections with recessed section*

Example: $\begin{cases} d_1 = 16\ mm = 0.63\ in; h_1 = 20\ mm; \varepsilon_1 = 10\ mm = 0.394\ in; F_1 = 54 \times 10^3\ N \\ K_{\tau 1} = h_1/d_1 = 1.25;\ \sigma_{M1} = K_{\tau 1} F_1 / d_1 \varepsilon_1 = 421.875\ MPa = 6.119 \times 10^4\ psi \end{cases}$

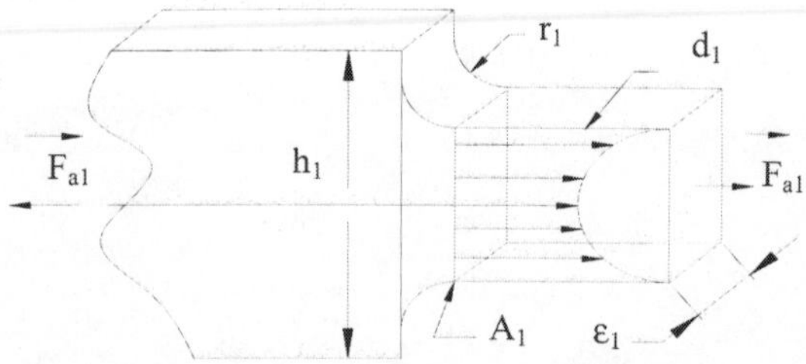

Figure 5.3. *Diagram of stress intensification. Flat piece with recessed section. For a color version of this figure, see www.iste.co.uk/grous/design.zip*

5.1.4. *Homogeneous, hollow, square sections, with no external variation*

$$\text{Example:}\begin{cases} d_3 = 8\ mm = 0.472'';h_3 = 18\ mm = 0.709'';\varepsilon_3 = 10\ mm = 0.394'';F_3 = 54000\ N \\ K_{\tau 3} = \dfrac{h_3}{d_3} = 1.667 \rightarrow \sigma_{M3} = K_{\tau 3}\dfrac{F_3}{(h_3 - d_3)\varepsilon_3} = 1215\ MPa = 1.762\times 10^5\ psi \end{cases}$$

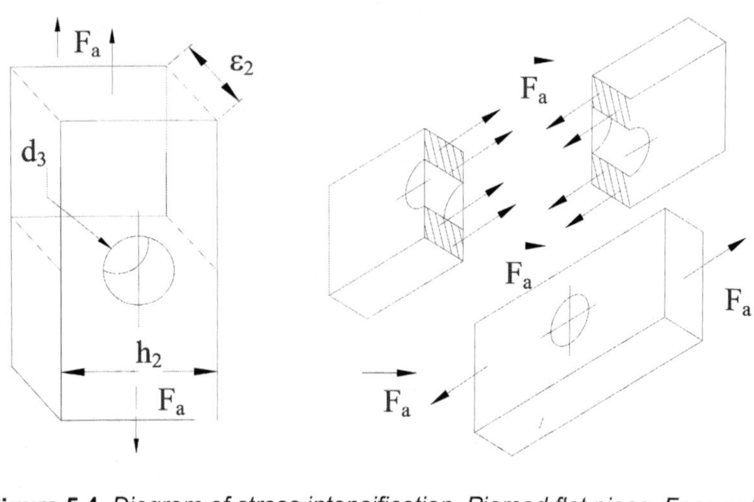

Figure 5.4. *Diagram of stress intensification. Pierced flat piece. For a color version of this figure, see www.iste.co.uk/grous/design.zip*

5.1.5. *Homogeneous, solid, round sections with a shoulder (shouldered shaft)*

$$\text{Example: } \begin{cases} d_4 = 20\ mm = 0.4787'' ; D_4 = 25\ mm = 0.984'' ; F_4 = 54000\ N = 12140\ lbf \\ K_{\tau 4} = \dfrac{D_5}{d_5} = 1.25 \rightarrow \sigma_{M4} = \dfrac{4K_{\tau 4}F_4}{\pi d_4^2} = 241.859\ MPa = 3.116 \times 10^4\ psi \end{cases}$$

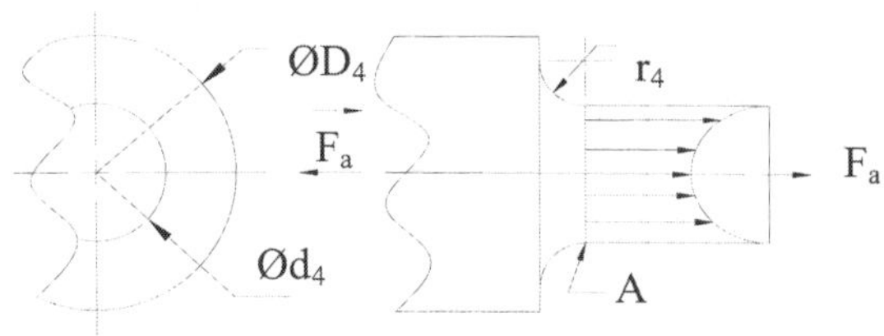

Figure 5.5. *Diagram of stress intensification: flat piece with bottleneck. For a color version of this figure, see www.iste.co.uk/grous/design.zip*

5.1.6. *Homogeneous, solid, rectangular or square sections, with a groove*

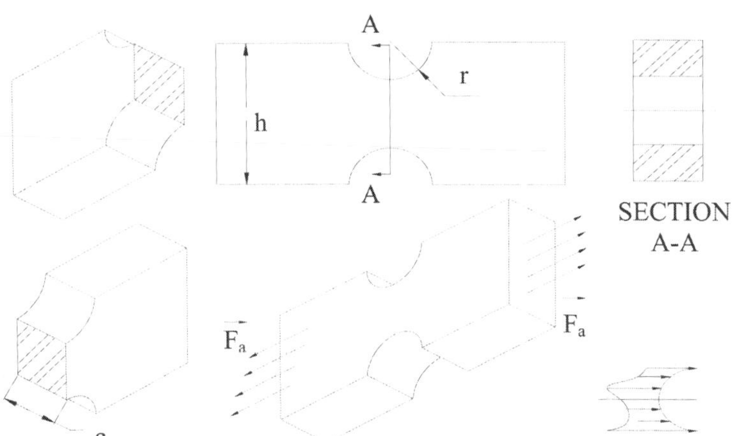

Figure 5.6. *Diagram of stress intensification: flat piece with constriction. For a color version of this figure, see www.iste.co.uk/grous/design.zip*

5.1.7. *Homogeneous, hollow, round and flat sections (pierced flat piece with an axle)*

Example: $$\begin{cases} d_6 = 16\ mm; h_6 = 6\ mm; F_6 = 54000\,N; l_6 = 18\ mm = 0.709"; \varepsilon_6 = 10\ mm \\ K_{\tau 6} = \dfrac{h_6}{l_6} = 0.333 \rightarrow \sigma_{M6} = \dfrac{K_{\tau 6} \times F_6}{(l_6 - d_6)\varepsilon_6} = 241.859\ MPa = 31160\ psi \end{cases}$$

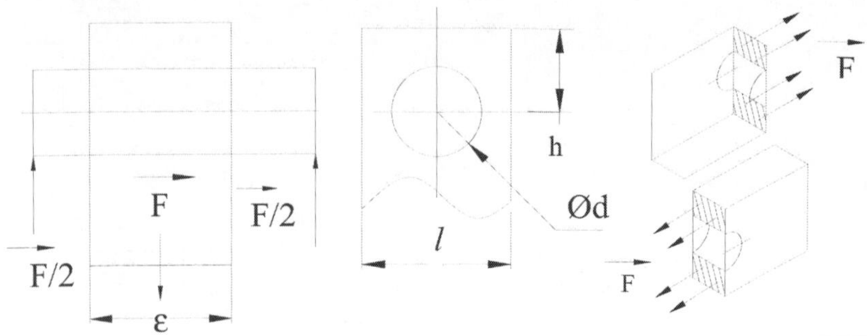

Figure 5.7. *Diagram of stress intensification. Pierced flat piece with an axle. For a color version of this figure, see www.iste.co.uk/grous/design.zip*

5.1.8. *Homogeneous, hollow, round sections (shaft with groove)*

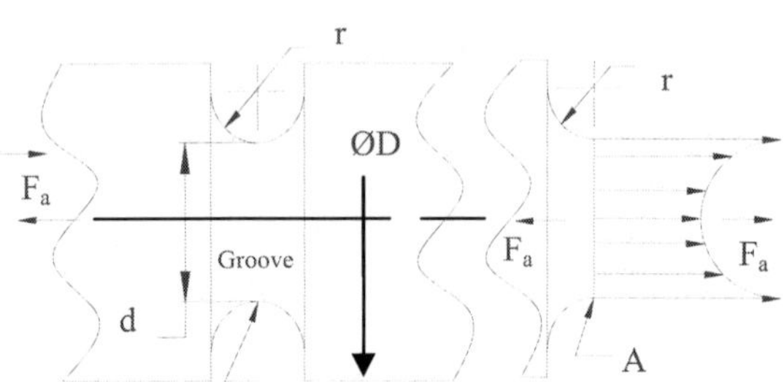

Figure 5.8. *Diagram of stress intensification. Shaft with groove. For a color version of this figure, see www.iste.co.uk/grous/design.zip*

Example: $$\begin{cases} d_7 = 18\ mm = 0.709"; F_7 = 54000\ N; D_7 = 25\ mm = 0.984" \\ K_{\tau 7} = D_7 / d_7 = 1.389 \rightarrow \sigma_{M7} = 4K_{\tau 7}F_7 / \pi d_7^2 = 294.731\ MPa = 42750\ psi \end{cases}$$

Calculations and plotted diagrams: let d = [10 mm, 11 mm, …,18 mm]; and D = 8 mm, F = 54,000, N = 1.214×10^4 *lbf.* Using the formula, let us set: $\sigma_M(d) = K_\tau\left(4F/\pi d^2\right)$.

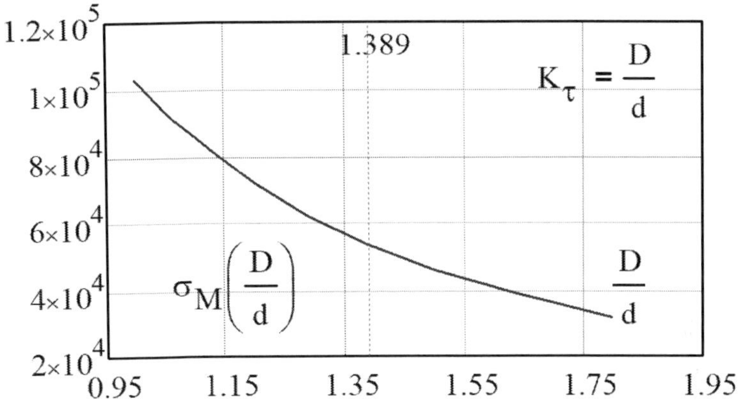

Figure 5.9. *Evolution of stress as a function of the SIF. For a color version of this figure, see www.iste.co.uk/grous/design.zip*

5.2. Principles of calculations for constructions in design

The torsion is sometimes more difficult to calculate than the shear, traction or compression. The calculations bring into play the quadratic moments (*J*, mm^4 or in^4), depending on the geometric form. The most commonly encountered real-world case is that of a screwdriver. The tangential torsion stress (τ, MPa):

– G is the transverse elasticity modulus (also known as the stiffness modulus) in MPa or psi;

– ρ is the radius (distance between the point and the middle line) in mm or in;

– θ is the unit torsion angle (rad/mm or rad/in);

– N is Poisson's ratio;

– E is the elasticity modulus;

– S is the safety coefficient;

– R_{eg} is the elastic shear limit of the material.

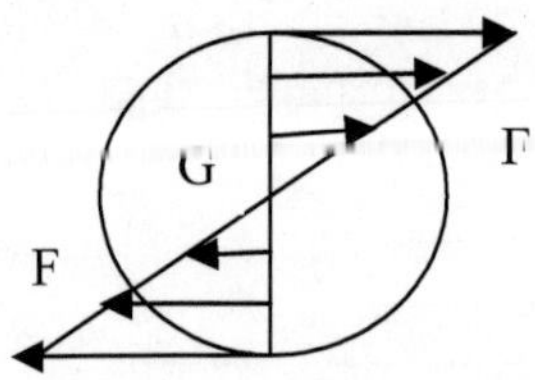

$$\tau = G \times \rho \times (\theta)\ [MPa\ or\ psi]\ where\ G = E/2(1+\nu)$$

$$\tau_{\max} = \rho_{\max}\left(\frac{M_\tau}{J_0}\right) = \rho_{\max}\left(\frac{M_\tau}{J_0/c}\right) \prec [\tau]_{\max}^{admissible}$$

$$[\tau]_{\max} = R_{pg}; \rho_{\max} = c\ ; R_{pg} = R_{eg}/S, R_{eg} \cong R_e/2$$

Figure 5.10. *Stress distribution*

5.2.1. *Example on stress intensifications*

Consider the following: D = 28 mm; d = 20 mm; $K_{\tau s} = K_\tau = (D/d) = 1.4$ and $M_\tau = 25 \times 10^4$ N.mm

$$W_{xx} = J_0/c = \pi d^3/16 = 1.571 \times 10^3\ mm^3 \rightarrow \tau_0 = M_\tau/W_{xx} = 159.155\ MPa = 2.308 \times 10^4\ psi$$

$$Hence\ \tau_{\max} = \tau_0 \times K_{\tau s} = (159.155) \times (1.4) = 222.817\ MPa = 3.232 \times 10^4\ psi$$

In the case of *walls*, the following formulae are used for calculation [ROA 75]:

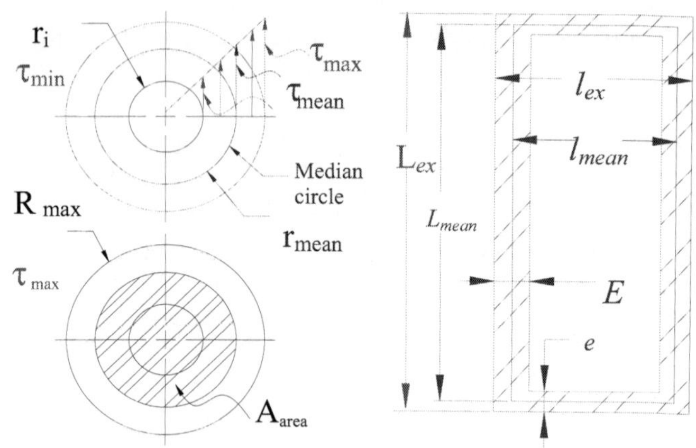

Figure 5.11. *Example: cross-sections of a slender tube and a rectangular beam. For a color version of this figure, see www.iste.co.uk/grous/design.zip*

$$\tau_{mean} = \tau_{med} = M_\tau/(2A\varepsilon)\ [MPa]\ or\ [psi] \qquad [5.2]$$

where:

- A is the internal area to the median line of the section;
- θ is the unit torsion angle;
- α is also a torsion angle (straight section);
- G is the modulus of stiffness;
- L is the length of the beam (bar);
- ε is the thickness of the wall at the point of the mean stress $\tau_{mean} = \tau_{median}$;
- $\oint \frac{ds}{e}$ is the integral on the contour of the median line.

Form	Cross-section	τ_{maxi}	$\theta = \alpha/L$
Square *Side (a)*	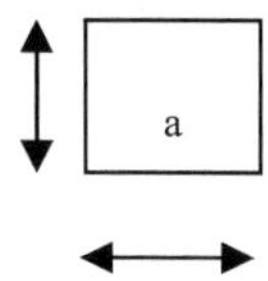	$\frac{4.81 M_\tau}{a^3}$	$\frac{7.10 M_\tau}{Ga^4}$
Rectangle *Width (a)* *Length (b)*	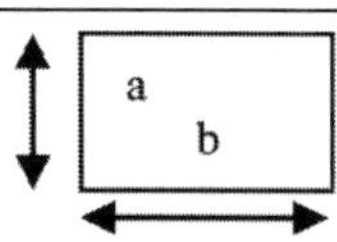	$\frac{M_\tau}{K_1 ab^2}$	$\frac{M_\tau}{K_2 Gab^3}$
Triangle *Side (a)*	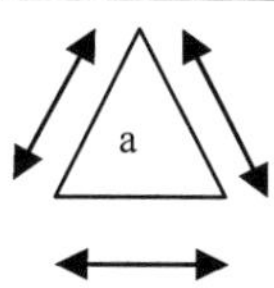	$\frac{20 M_\tau}{a^3}$	$\frac{46 M_\tau}{Ga^4}$
Ellipse *2a short axis* *2b long axis*	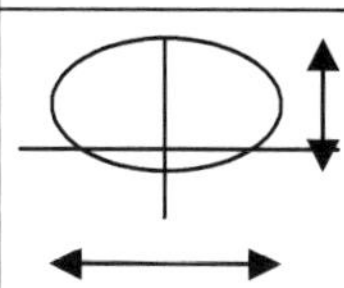 *2b long axis/x* *2a short axis/y*	$\frac{M_\tau}{\pi ab^2}$	$\frac{M_\tau (a^2 + b^2)}{\pi G a^3 b^3}$

Figure 5.12. *Stresses and torsion angles for a number of standard forms*

Values of the coefficients K_1 and K_2 as a function of the geometric ratio (a/b):

a/b	1.0	1.2	1.5	2.0	2.5
K_1	0.208	0.219	0.231	0.246	0.258
K_2	0.1406	0.1661	0.1958	0.229	0.249
a/b	3.0	4.0	5.0	10	∞
K_1	0.267	0.282	0.291	0.312	0.333
K_2	0.263	0.281	0.291	0.312	0.33

Table 5.1. *Coefficients K_1 and K_2. Source: [FAN 14]*

We can see that beyond the ratio (a/b) > 5→$K_1 = K_2$.

By linear interpolation, we are able to write: [a/b ≥ 5 → $K_1 = K_2 = 0.33(1\text{-}063 \times b/a)$].

Certain cross-sections with unusual forms necessitate the use of an appropriate forming process, such as folding or rolling. Ultimately, the various cross-sections can be assimilated to a rectangular bar of width *(a)* and length *(b)*, a simple application of which is shown below:

$$\text{Using } \left\{\tau_{\max} = \frac{M_\tau}{K_1 ab^2} \; and \; \theta = \frac{M_\tau}{K_2 ab^3 G}\right\}, \text{ let us find } \{\tau_{\max} \; and \; \theta\} = ?$$

5.2.2. *Case study on torsion angles*

$a_1 = a_2 = 120$ mm = 4.724 in; $a = a_1 + a_2 = 240$ mm; b = 8 mm; L = 2.56 m = 100.787 in. $[\tau]_{adm} = 45$ MPa = 6527 psi; G = 68 GPa = 9.863×10^6 psi. For (a/b) = 0.033 and $K_1 = K_2$. *Find* the unit torsion angle θ = F (L_{beam}) and the torsion angle of the beam, α = F (L_{beam}). Plot the two graphs separately.

SOLUTION.–

$$K_1 = \frac{a}{b}\left(1 - 0.63\frac{b}{a}\right) = 0.032633 \rightarrow \left\{\tau_{\max} = \frac{M_\tau}{K_1 ab^2} \; and \; \theta = \frac{M_\tau}{K_2 ab^3 G}\right\} = ?$$

From $M_\tau / K_1 ab^2 \prec \tau_{max} \rightarrow M_\tau \leq \tau_{max} \times K_1 ab^2 \rightarrow M_\tau \leq 22.556\ Nm = 199.639\ lbfin$

$$\begin{cases} M_\tau / K_1 ab^2 \prec \tau_{max} simplify \rightarrow \dfrac{1.995e6.M_\tau}{m^3} \leq 45000000.Pa \swarrow \\ M_\tau \leq 22.556\ Nm = 199.639\ lbfin \equiv with\ the\ manual\ calculations \leftarrow QED \end{cases}$$

Calculation of the torsion angles (α) and (θ) as a function of the length of the beam.

By deliberately varying L_{var} between 2 m and 2.5 m with an increment of 1/10 m, let us plot the curves for (α) and (θ):

$$\theta = M_\tau / K_2 ab^3 G = 8.272 \times 10^{-5}\ rad/mm = 2.201 \times 10^{-3}\ rad/in \rightarrow \alpha = ?$$

$$-For\ L = \text{constant} = 2.5\ m \rightarrow \alpha(L) = \theta \times L = 12.133224^\circ = 0.211765\ rad$$

$$-For\ L_{var} = 2\ m, 2.10\ m, ..., 2.50\ m: from\ \alpha(L_{var}) = \theta \times L_{var} \text{ and } \theta(L_{var}) = \alpha / L_{var}$$

We can clearly see from the graph that the unit torsion angle is inversely proportional to the length of the beam. However, the torsion angle (α), for its part, increases in direct proportion to the same variation in length.

RESULTS AND GRAPHS.–

For: $\alpha = 12.277667^\circ \rightarrow \theta(L_{var}) = \alpha / L_{var}$, we have:

$\theta(L_{var}) =$ $\frac{1}{m}$

	0
0	0.107
1	0.102
2	0.097
3	0.093
4	0.089
5	0.086

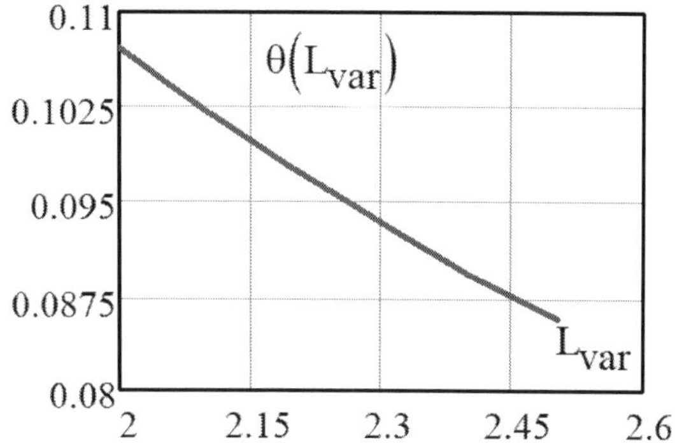

For $\alpha = 12.277667^{\circ} \rightarrow \alpha(L_{\text{var}}) = \theta \times L_{\text{var}}$, *we have*:

α(L) =

	0
0	9.93
1	10.426
2	10.923
3	11.419
4	11.916
5	12.412

·deg

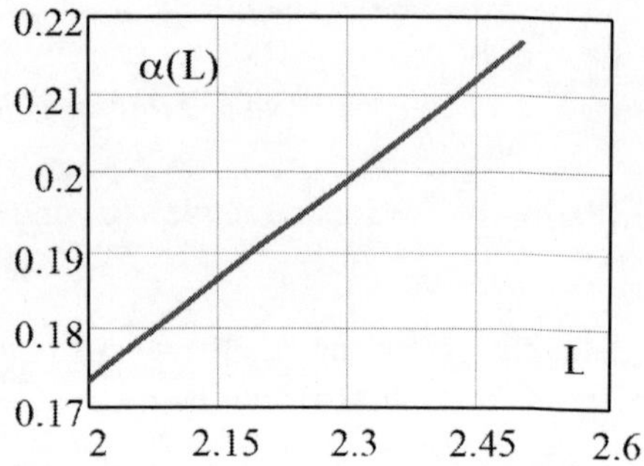

Figure 5.13. *Plots of the torsion angles (α and θ). For a color version of this figure, see www.iste.co.uk/grous/design.zip*

The main formulae used, depending on the form are:

Form	τ_{maxi}	$\theta = \alpha/L$
Flattened circle	$M_\tau / K_3 r^3$	$M_\tau / K_4 G r^4$
Doubly flattened circle	$M_\tau / K_5 r^3$	$M_\tau / K_6 G r^4$
Slotted tube	$\cong \dfrac{3M_\tau}{2\pi r \varepsilon^2}$	$\cong \dfrac{3M_\tau}{2\pi r \varepsilon^3 G}$
Thin-walled rectangle	$\dfrac{M_\tau}{2\varepsilon(a-\varepsilon)(b-\varepsilon)}$	$\dfrac{M_\tau(a+b-2\varepsilon)}{2\varepsilon(a-\varepsilon)^2(b-\varepsilon)^2}$
R = radius; a = width; b = length, ε = thickness, K_1; K_2; K_3; … K_6 = form factors and M = moment.		

Table 5.2. *Stress and torsion angles for various forms*

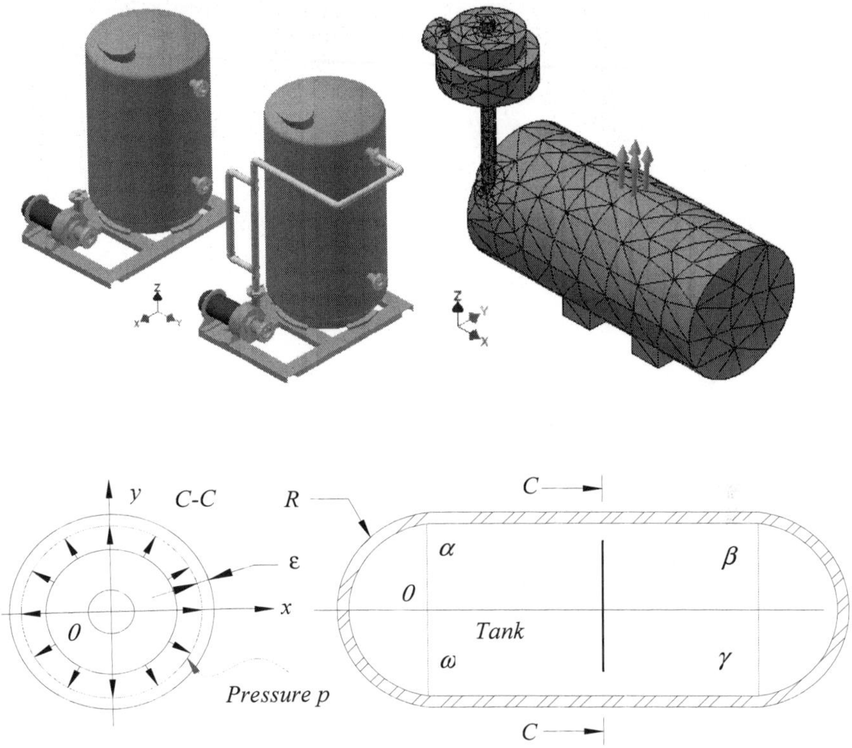

Figure 5.14. *Stress in the envelopes (tanks) under pressure. For a color version of this figure, see www.iste.co.uk/grous/design.zip*

$$\sigma = p_{in}R/2\varepsilon = M_\tau/2A\varepsilon \; [MPa] \; or \; [psi] \qquad [5.3]$$

where:

– A is the area inside the median line of the section in mm^2 or in^2;

– ε is the thickness of the wall in mm or in;

– *R* is the external radius R = D/2, in mm or in;

– *p* is the pressure exerted from the inside (see Figure 5.14) in MPa or psi.

The maximum stress is calculated as: $\tau_{\max} = \dfrac{\sigma}{2} = \dfrac{p_{in}R}{4\varepsilon} = \dfrac{M_\tau}{4A\varepsilon} \; [MPa] \; or \; [psi]$.

The triaxial case where pressure exercises an additional stress would therefore be:

$$\tau_{\max} = (\sigma + p_{in})/2 = p_{in}/2 + p_{in}R/4\varepsilon \left[MPa\right] or \left[psi\right] \qquad [5.4]$$

The stress at a given external point (m) shows that we are dealing with a state of bi-axial stresses where: $\sigma = \sigma_1 = \sigma_2$. Mohr's circle contains it in a single circle, as illustrated by Figure 5.15.

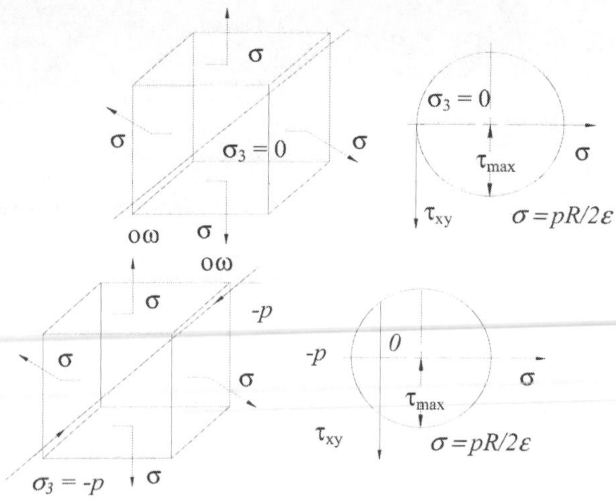

Figure 5.15. *Mohr's circles. For a color version of this figure, see www.iste.co.uk/grous/design.zip*

5.2.3. *Case study: Tresca and von Mises yield criteria*

Calculate the stress exerted by pressure on the wall, p_{in} = 54 bars = 5.4 MPa = 783.204. The radius of the tank R = 1000 mm = 39.37 in over a thickness (ε) = 6 mm = 1.772 in. R_e = 520 MPa = 75420 psi and S = 2. This time, we wish to determine the thickness (ε, mm and in) of the wall required to withstand the exerted pressure. We bring the two design criteria – von Mises and Tresca – into play.

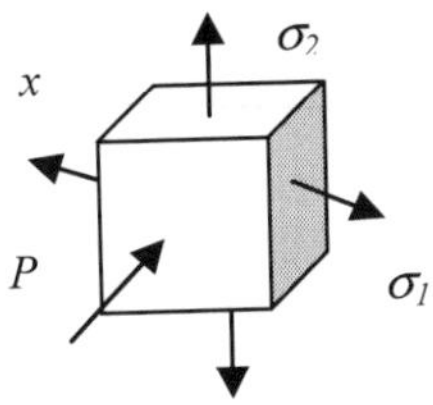

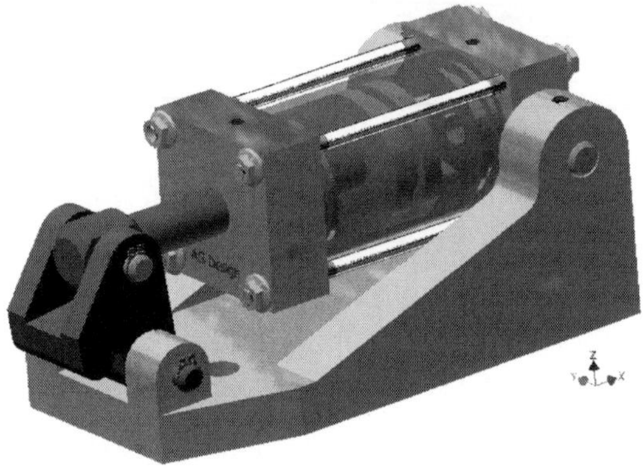

$$\sigma(p_v)=\left(\frac{p_v}{2}\right)+\left(\frac{p_v R}{2\varepsilon}\right)\text{↵}$$

$$\rightarrow simplify \rightarrow\ 83.83p_v$$

Figure 5.16. *Stresses at internal points. Body for internal pressure (jack, vat). For a color version of this figure, see www.iste.co.uk/grous/design.zip*

SOLUTION.–

By calculating the stresses, we are able to write:

$$\sigma_2 = p_{in}R/\varepsilon = 900\ MPa = 130{,}500\ psi\ and\ \sigma_1 = \sigma_2/2 = 450\ MPa = 65270\ psi$$

The maximum pressure appears thus at a point inside the tank (pipe/jack):

$$\tau_{\max} = (p_{in}/2)+(p_{in}R/2\varepsilon) = 452.7\ MPa = 6\ 5658.58\ psi$$

Here is the evolution of the maximum stress at a point inside the tank, as a function of the upward variation in pressure exerted between $P_{variable} = p_v = 2$ MPa, 2.5 MPa, …, 5 MPa. The manual calculation and the calculation done by the software give the same result.

Tresca criterion: according to Tresca's criterion, we see a flow – i.e. plastic deformation – if we have the thickness ε_{Tresca} to design:

$$Calculation\ 1{:}\ \sigma_1 = \sigma_{tot} = p_{in}R/\varepsilon_{Tresca} \geq R_e/S \rightarrow \varepsilon_{Tresca} = 20.7697\ mm = 0.8177\ in$$

$$\begin{cases} \sigma_1 = \sigma_{tot} = \dfrac{p_{in}R}{\varepsilon_{Tresca}} \geq \dfrac{R_e}{S}\ simplify \rightarrow \dfrac{5400000Pa.m}{\varepsilon_{Tresca}} \geq 260000000.\text{P}a \swarrow \\ \rightarrow \varepsilon_{Tresca} = 5400000.Pa.m/260000000.\text{P}a = 20.769231\ mm = 0.817686\ in \end{cases}$$

Deduce the stress from this, using Tresca's criterion as follows:

$$\sigma_1 = \sigma_{tot} = \sigma_{11} = p_{in}R/\varepsilon_{Tresca} = 260\ MPa = 3.771\times10^4\ psi$$

Von Mises criterion: according to the von Mises criterion, a flow (plastic strain) will take place if we have the thickness $\varepsilon_{\text{in-Von-Mises}}$ to design:

$$\begin{cases} \sigma_1 = \sigma_{tot} = \dfrac{p_{in}R\sqrt{3}}{2\varepsilon_{VonMises}} \geq \dfrac{R_e}{S} \, simplify \rightarrow \dfrac{2700000\sqrt{3}.Pa.m}{\varepsilon_{VonMises}} \geq 260000000.Pa \swarrow \\ \rightarrow \varepsilon_{VonMises} = 2700000\sqrt{3}.Pa.m/260000000.Pa = 17.986681\ mm = 0.708137\ in \end{cases}$$

Manual calculation: $\rightarrow \varepsilon_{VonMises} = \left(p_{in}R\sqrt{3}\right)/\left(2\,R_e/S\right) = 17.986681\ mm = 0.708137\ in$

Let us then deduce the stress, using the Von Mises criterion, σ_{flow}, as follows:

$$\sigma_{flow} = \sqrt{\sigma_1^2 + \sigma_2^2 - \sigma_1\sigma_2} = 779.422863\ MPa = 1.13\times10^5\ psi$$

Knowing $R_e/S = 260.000000\ MPa = 3.771\times10^4\ psi$, we can deduce that the resistance condition is broadly met because, according to the von Mises criterion:

$$\sigma_{flow} = \sigma_{tot} = \left(p_{in}R\sqrt{3}/2\varepsilon_{VonMises}\right) \geq R_e/S \rightarrow \text{Proof: } 779.422863\ MPa \succ 260\ MPa$$

$83.83333333333333333\cdot p_v$ = **Calculation run using Math CAD**

	0
0	167.667
1	209.583
2	251.5
3	293.417
4	335.333
5	377.25
6	419.167

MPa

$\sigma(p_v) =$

	0
0	2.432·10⁴
1	3.040·10⁴
2	3.648·10⁴
3	4.256·10⁴
4	4.864·10⁴
5	5.472·10⁴
6	6.079·10⁴

·psi

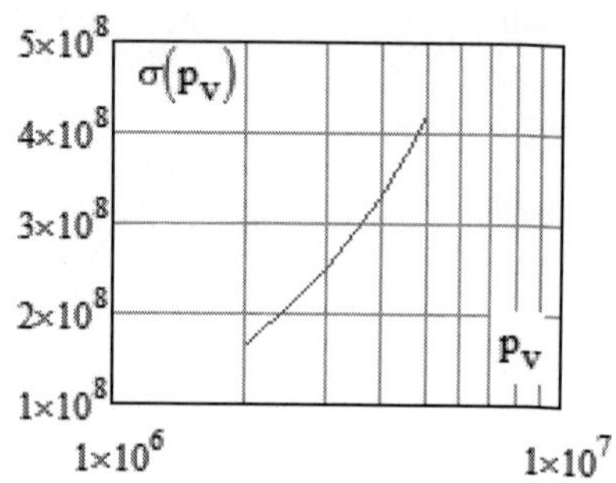

Figure 5.17. *Maximum stress as a function of increasing pressure (semi-log). For a color version of this figure, see www.iste.co.uk/grous/design.zip*

DECISION.– It is precisely at this point that a designer truly comes into their own role, in determining the geometric and loading parameters so that by at least one of the two criteria, the design is completely safe. In terms of *shear*, the *strains* manifest themselves in slippage, written in (γ, rad), of the straight sections of the part.

EXAMPLE.– Consider a parallelepiped with the dimensions (α.β.η.ρ). For the sides $B = \alpha\rho = 44$ mm and $C = \alpha\beta = 84$ mm, $T_\tau = 1200$ N and the stiffness G = 1.2 MPa. Let us calculate (a) in mm and in, and the slip angle (γ), knowing that: $\tan(\gamma) \cong \gamma/a; \tau_{elas} = G\gamma;$ *and* $G = E/2(1+\nu)$.

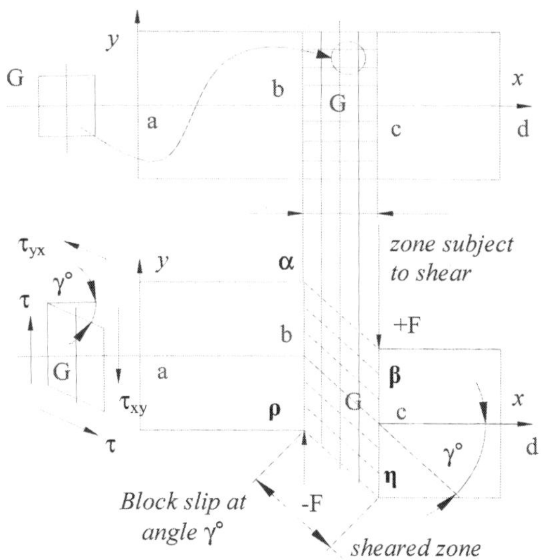

Figure 5.18. *Slip angle (γ in rad or degrees). For a color version of this figure, see www.iste.co.uk/grous/design.zip*

$$\tau = T_\tau / BC = 0.325 \; MPa; \; h\tan(\gamma) \rightarrow a = 13.024 \; mm \rightarrow \gamma = \tau / G = 0.271 \; rad = 15.527°$$

5.3. Pressurized recipients and/or containers

To analyze the characteristics of pressurized recipients, containers and tanks, we can use the shell theorem, in view of their form and the symmetrical application of the loads. We merely need to take the ratio between the thickness of the wall (*e*) and the radius (*r*) to establish the distinction between shells or cylinders with thin or thick walls. If:

– 10e < r, thin wall theory applies;

– 10e > r, thick wall theory applies.

Internal pressure

In the present book, we shall only look at pressurized recipients, whose internal pressures create failures under traction forces. External pressures giving rise to buckling are not studied, in view of the rarity of this type of incident. The equations are as follows.

1) *Thin walls*: circular tensions only.

The maximum stress at the circumference is:

$$\sigma_{\max}^{circumf} = P \times r_{in} / e \qquad [5.5]$$

P represents the internal pressure, r is the internal radius and t the thickness of the wall.

2) *Thick walls*: radial and circular tensions (traction)

The maximum stress at the circumference is:

$$\sigma = P\left(\frac{r_{ex}^2 + r_{in}^2}{r_{ex}^2 - r_{in}^2}\right) \qquad [5.6]$$

r_{ex} represents the external radius, r_{in} the internal radius and P the internal pressure.

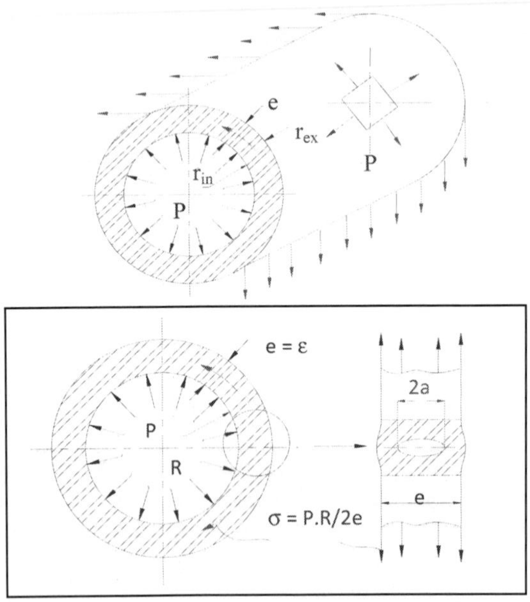

Figure 5.19. *Internal pressure in a cylindrical shell. For a color version of this figure, see www.iste.co.uk/grous/design.zip*

The local stresses on the joint result from intensifications, which may be linked to SIFs. The more sudden the change, the higher the stress will be. For this reason, spherical covers are the best suited.

Example of application to pressurized recipients

Suppose we are designing a recipient whose diameter is *d = 300 mm*, designed to withstand an internal pressure of *P = 3 bars*. The material chosen for it is *polyamide 6*, reinforced with *33 % fiberglass*. For reasons of effectiveness of casing, the walls must have a maximum thickness of *e = 6 mm*. Should the walls be thin or thick?

Initial data: 10×e = 10×6 = 60 mm, and r = 300/2 = 150 mm, e_{real} = 10×e = 10×6 = 60 mm and r = 300/2 = 150 mm → 60 mm < 150 mm. It is therefore possible to envisage thin walls:

$$\sigma_{\max}^{circumf} = \left(P \times r_{in}\right)/e = \left(0{,}3 \times 150\right)/6 = 7.5\ MPa$$

With regard to long-term effects, it is necessary to check that 7.5 MPa are compatible with the material's traction resistance. This pressure of 7.5 MPa is, fortunately, much lower than the material's traction tolerance values (130 MPa at 50 % RH). Hence, this concept can be used for short-term applications. It is helpful to check its suitability to long-term conditions, if necessary. The stress of the spherical form is:

$$\sigma_{\max}^{circumf} = P \times r_{in}/2 \times e = 0.3 \times 150/2 \times 6 = 3.75\ MPa$$

The *spherical form* is the best suited for the extremity (cover). Small pressurized tanks are ordinarily monitored by X-rays, or by an overload test. The aim is to ensure fissuration defaults with diameters equivalent to 2a. The acceptable fissure size is an optimization, choosing the material with the best toughness/stress performance index: $i\left\{{}^{Performance}_{\max}\right\} = K_{Ic}/\sigma_c$.

5.4. Calculation principles and solution method for compound loading

In advanced design, we are often dealing with complex loading – i.e. compound stresses engendered by the hazards of the construction, materials and the environment. To determine the resulting stresses, we use the principle of superposition of forces and work (moments). The stresses resulting from each of the forces are calculated separately, then added together and, finally, verified using specific yield criteria. The aim is to find the stress equivalent to the system being designed. For solid materials, we bring the relative extensions into play: A % and ε_{limit}. Note that, by way of example, for ductile and malleable materials (Alu 6061), we use: A % > 5 and ε_{limit} > 5 %.

For other materials which are even very slightly ductile, or which are less inclined to be malleable, we employ the Tresca and von Mises yield criteria. For fragile materials including A % < 5 and $\varepsilon_{limit} < 5$ %, such as concretes and ceramics, we use the Mohr–Cacquot or Mohr–Coulomb criteria. Specialized publications in SOM offer many more explicit case studies. We shall use pre-existing formulae in SOM without going into the details of their explicative origins. Let us simply note that this book is not, *a priori*, a construction book, although it deliberately does touch on the subject. After all, design is at the heart of all constructions. Design provides the practical tool for the calculation and the organization which goes along with it.

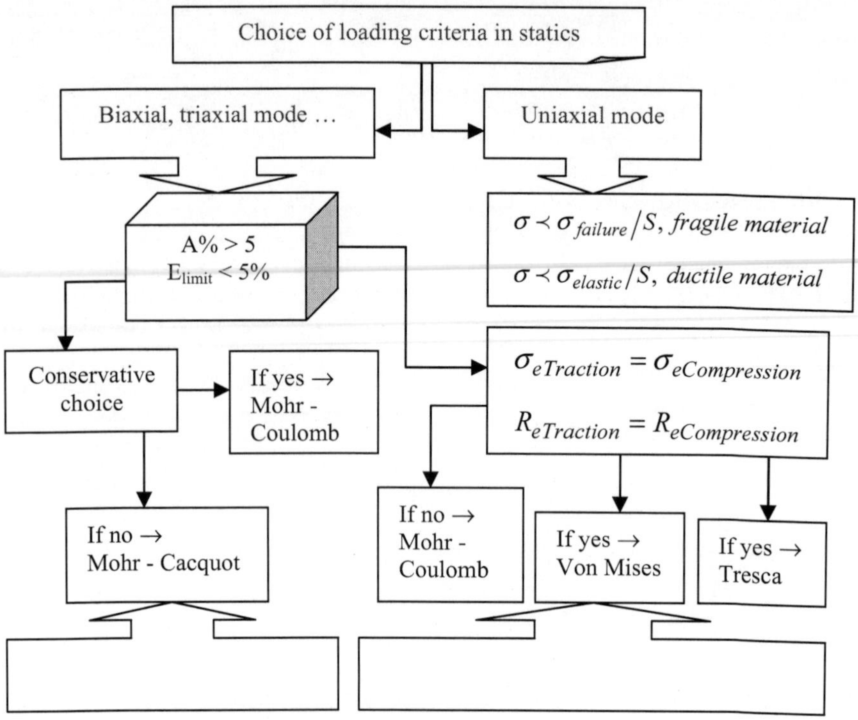

Figure 5.20. *Information on the choice of criteria for the elastic limit*

For ductile materials, we use the Tresca criterion, which is more accurate than the von Mises criterion. In the context with which we are dealing, both criteria yield accurate results. To differentiate between the two, experimental checks must be performed on a case-by-case basis. Certainly, the Tresca criterion is much simpler, which is why it is so frequently used. It allows for a broad safety margin. In the majority of finite-element stress analysis programs, the von Mises criterion is frequently employed, as attested by the software package used in this book (Inventor Pro). As we see it, this tendency to use only the von Mises criterion is rather limiting, because it is imposed/suggested by the program. The von Mises criterion focuses on the flow of ductile materials, although it does not differ greatly from the criterion put forward by Tresca. For instance, to best analyze all the stresses, it would be wise to analyze a regular octahedral element (with eight identical faces). The reason is that the eight faces of an octahedron tend to compress and thereby distort the volume, thus skewing it. In conditions of shear, τ_{oct}, the eight faces tend to better withstand the stress without influencing the volume.

<table>
<tr><td colspan="8">Superpositions of simple forces</td></tr>
<tr><td colspan="3">Calculations of σ</td><td colspan="4">Calculations of τ</td><td>Direction/Analysis</td></tr>
<tr><td>Traction</td><td colspan="2">Bending</td><td>Torsion</td><td colspan="2">Shear</td><td>Bending</td><td>Nature/stress</td></tr>
<tr><td>N</td><td>M_{fly}</td><td>M_{flz}</td><td>M_τ</td><td>Ty</td><td>Tz</td><td>T</td><td>Forces/section</td></tr>
<tr><td>$\sigma_{traction}$</td><td>σ_{fy}</td><td>σ_{fz}</td><td>$\tau_{torsion}$</td><td>τ_y</td><td>τ_z</td><td>τ_T</td><td>Stresses/Point</td></tr>
<tr><td colspan="3">$$\begin{cases} \sigma = \sigma_{trac} + \sigma_{fy} + \sigma_{fz} \\ \sigma = \dfrac{N}{A} + \dfrac{M_{fy}}{J_y} z + \dfrac{M_{fz}}{J_z} z \end{cases}$$</td><td colspan="4">$$\begin{cases} \tau = \tau_{tors} + \tau_y + \tau_z + \tau_T \\ \tau = \dfrac{M_T}{J_0}\rho + \dfrac{T_y}{A} + \dfrac{T_z}{A} + \tau_T \end{cases}$$</td><td>Grouping/addition of stresses [σ, τ] of the same nature</td></tr>
<tr><td colspan="7">Superposition of planar stresses of different natures: →
[σ ≠ 0 τ ≠ 0]
Von Mises: $\sigma_{equiv} = \sqrt{3\tau^2 + \sigma^2} \leq [\sigma]_{adm}$
Tresca: $[\tau]_{adm}^{metals} = \dfrac{[\sigma]_{adm}}{2}$ and $\tau_{eq} = \sqrt{\tau^2 + (\sigma/2)^2} \leq [\tau]_{adm}$</td><td rowspan="2">It is always important to determine the stress equivalent to the stresses (σ, τ). Then, it would be wise to employ the most suitable yield criterion (von Mises or Tresca?) to compare them with the appropriate admissible stress [σ] or [τ]</td></tr>
<tr><td colspan="7">Superposition of planar stresses of the same nature: →
[σ = 0 and τ ≠ 0] or indeed [σ ≠ 0 and τ = 0]. Stresses applied in the same direction are written thus: $\sigma_{eq} = \sigma \leq [\sigma]_{adm}$ or
indeed $\tau_{eq} = \tau \leq [\tau]_{adm}$</td></tr>
</table>

Table 5.3. *SOM problem-solving method*

In mechanical design, this is a major advantage, as all eight faces are identical, leading us to use the von Mises criterion relative to the elasticity limit. Equilibrium in the octahedral element is modeled as follows:

$$\begin{cases} \sigma_{oct} = \left(\sigma_x + \sigma_y + \sigma_z\right)/3 = \left(\sigma_1 + \sigma_2 + \sigma_3\right)/3 \text{ and} \\ \tau_{oct} = \sqrt{\left(\sigma_1 - \sigma_2\right)^2 + \left(\sigma_2 - \sigma_3\right)^2 + \left(\sigma_3 - \sigma_1\right)^2} \end{cases} \quad [5.7]$$

We witness plastic strain – i.e. the flow of the material – when the effective stress (von Mises) is greater than or equal to the elastic limit of the material Re. Given that, in design, it is preferable to stay within the safe zone, we use a value which is within the *safety margin*, and set:

$$\begin{cases} \sigma_{effective} = \sqrt{\left(\sigma_1 - \sigma_2\right)^2 + \left(\sigma_2 - \sigma_3\right)^2 + \left(\sigma_3 - \sigma_1\right)^2} \Big/ \sqrt{2} \prec \left[R_{elas}/S\right] \\ \sigma_{effective} = \left\{\sqrt{\left(\sigma_1 - \sigma_2\right)^2 + \left(\sigma_2 - \sigma_3\right)^2 + \left(\sigma_3 - \sigma_1\right)^2} \Big/ \sqrt{2}\right\} \geq R_{elas} \end{cases} \quad [5.8]$$

5.4.1. *Case study: mechanical fit*

STATEMENT OF THE PROBLEM.–

Consider a swept stem in the form of an (L), embedded at the point (A) on the lever arm (L_{AB}) = 162 mm = 6.378 in, and with the free length L_{BC} = 456 mm = 17.953 in; diameter d = 26 mm = 1.024 in. The force F = 2.8×10^3 N = 629.465 *lbf* is applied at point (C). At point (A), we can envisage using a bearing, represented by the embedment (E). The aim here is to calculate the forces and stresses on the torsion bar and use the Tresca and von Mises criteria, knowing R_e = 850 MPa = 1.233×10^5 *psi*. At the pre-project phase of design, we choose a safety factor S = 1.75 relative to the imposed limit R_e. To do so, let us look at an example, simulated through calculations and modeled with the finite element method.

SOLUTION WITH DISCUSSION.–

1) Calculation of the forces at the anchored point (A). *Torsion torque* M_τ:

$$M_\tau = F_\tau \times L_{BC} = 1.277\times10^6 \ N.mm = 1.13\times10^4 \ lbf.in$$

2) Calculation of the maximum *torsion* stress applied $\tau_{max} = \tau_{xz}$ at point (A):

$$\text{From } W_{xx} = \pi d^3/16 = 3451 \ mm^3 = 0.211 \ in^3 \text{ we calculate } \tau_{xz} = M_\tau / W_{xx} \downarrow$$

This enables us to *calculate the maximum torsion stress*:

$$\tau_{xz} = M_\tau / W_{xx} = M_\tau c / J_0 = 16 M_\tau / \pi d^3 = 369.975478\ MPa = 5.366041 \times 10^4\ psi$$

3) Calculation of the *maximum bending moment* $M_{fl} = M_{flmax}$ *with respect to (A)*:

$$M_{fl} = M_{fl}^{\max} = F_\tau \times L_{BA} = 4.536 \times 10^5\ N.mm = 4.015 \times 10^3\ lbf.in$$

4) Calculation of the *maximum bending stress* $\sigma_{max} = \sigma_{fl} = \sigma_{xx}$ *at point (A)*:

$$W_{zz} = J_z / c = \pi d^3 / 32 = 1.726 \times 10^3\ mm^3 = 0.105\ in^3$$

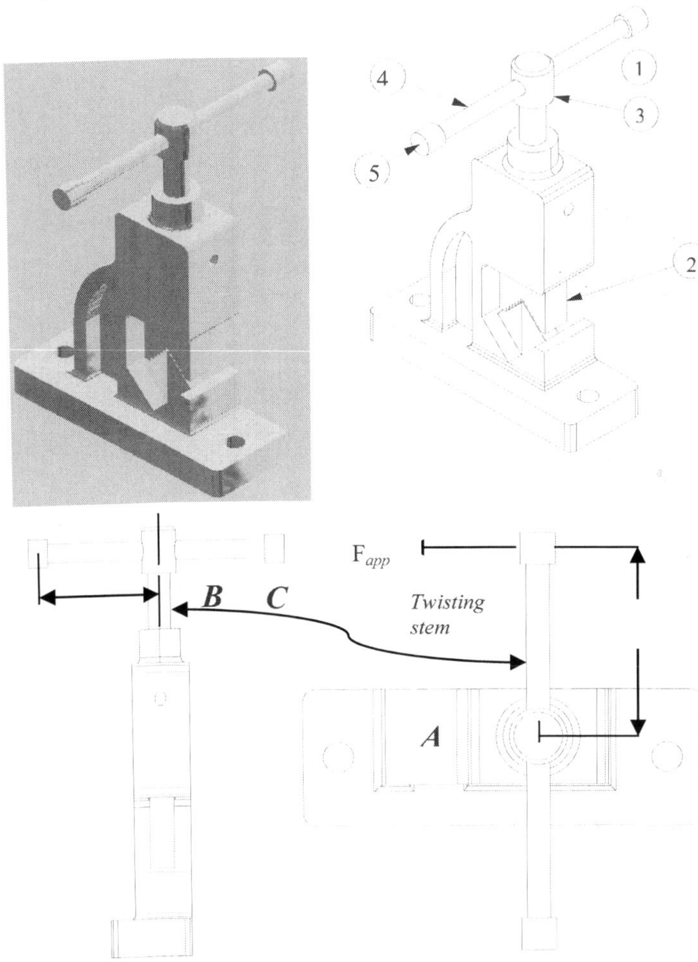

Figure 5.21. *Case study of a stem subject to torsion. For a color version of this figure, see www.iste.co.uk/grous/design.zip*

Maximum bending stress $\sigma_{xx} = \sigma_{fl}^{\max} = M_{fl}/W_{zz} = 262{,}877313\ MPa$

5) Maximum main stresses σ_1 and σ_2 at embedment point (E):

$$\sigma_{(1,2)} = \frac{\sigma_{xx}}{2} \pm \sqrt{\tau_{xz}^2 + \left(\frac{\sigma_{xx}}{2}\right)^2} = \begin{cases} \sigma_{(1)} = 524.068221\ MPa = 7.600967 \times 10^4\ psi \\ \sigma_{(2)} = \text{-}261.190908\ MPa = \text{-}3.788254 \times 10^4\ psi \end{cases}$$

Calculation run with MathCAD Prime:

$$\begin{cases} \sigma_1 = \dfrac{\sigma_{xx}}{2} + \sqrt{\tau_{xz}^2 + \left(\dfrac{\sigma_{xx}}{2}\right)^2} = 524.068221\ MPa = 7.600967 \times 10^4\ psi \checkmark \\ simplify \rightarrow \sigma_1 = 1.3143865668909e8 \times \text{Pa} + 3.9262956442242e8\sqrt{Pa^2} \end{cases}$$

$$\begin{cases} \sigma_2 = \left(\dfrac{\sigma_{xx}}{2} - \sqrt{\tau_{xz}^2 + \left(\dfrac{\sigma_{xx}}{2}\right)^2}\right) = \text{-}261.190908\ MPa = \text{-}3.788254 \times 10^4\ psi \checkmark \\ simplify \rightarrow \sigma_2 = 1.31438656689092e8 \times \text{Pa - } 3.926295644224203e8\sqrt{Pa^2} \end{cases}$$

We see perfect agreement between the two ways of calculating the stresses.

6) *Comparison of the Tresca and von Mises yield criteria:*

$$\begin{cases} \sigma_{flow} = \sqrt{\sigma_1^2 + \sigma_1^2 - \sigma_1 \times \sigma_2}\ simplify \rightarrow 6.92639910088821520 72e8\sqrt{Pa^2} \checkmark \\ \sigma_{flow} = 6.9263991008882152072e8\sqrt{Pa^2} = 692.639910\ MPa = 1.004589 \times 10^5\ psi \end{cases}$$

By manual calculation:

$$\sigma_{flow} = \sqrt{\sigma_1^2 + \sigma_1^2 - \sigma_1 \times \sigma_2} = 692.639910\ MPa = 1.004589 \times 10^5\ psi$$

Application of the *Tresca criterion* in conditions of flow (distortion) of the material. In fact, the stresses are of opposite signs – that is, σ_1 has the opposite sign to σ_2, and:

$\{\sigma_1 - \sigma_2\} = 785.259129\ MPa = 1.138922 \times 10^5\ psi$, or with MathCAD Prime

$$\{\sigma_1 - \sigma_2\}\ Simplify \rightarrow = 7.8525912884484e8Pa = 785.259129\ MPa = 1.138922\times 10^5\ psi$$

Knowing that: R_e = 850 MPa = 1.233×10^5 psi and S = 1.75, in view of the material SAE 1018 steel (0.18 % carbon) of which the part is made, we verify the *von Mises criterion* as follows:

$$\sigma_{flow}^{Von\ Mises} \geq \frac{R_e}{S}\ \text{explicit} \rightarrow\ 692.639910\ MPa \geq 485.714\ MPa,\ because\ \frac{R_e}{S} = \frac{850}{1.75}$$

Confirmation of the von Mises criterion: 692.639910 MPa > 485.714 MPa

Verify the *Tresca criterion:* $\sigma_{flow}^{Tresca} \geq \frac{R_e}{S}$ explicit $\rightarrow$ $785.3\ MPa \geq 485.7\ MPa$

Confirmation of the Tresca criterion: 785.259129 MPa > 485.714 MPa

COMMENTS.– The results, for both the von Mises and Tresca criteria, indicate that there is no distortion of the material with the particular composition (SAE 1018 = 0.18 % carbon) used to manufacture the part. In our case study, we shall simulate the increase in load to over 2.8×10^3 N, to locate the exact point where distortion would occur, and then compare that result to the one found by our FEM model.

7) Calculation of the limit diameters in view of both the von Mises and Tresca yield criteria, to machine the part with the appropriate levels. This is the advantage of assisted mechanical design:

Let us state the problem thus:

$$\sigma_{(1,2)} = \sigma_{xx}/2 \pm \sqrt{\tau_{xz}^2 + (\sigma_{xx}/2)^2} = \{\sigma_{(1)} = ?\ \text{and}\ \sigma_{(2)} = ?\}$$

$$\begin{cases} \sigma_{(1,2)} = \frac{\sigma_{xx}}{2} + \sqrt{\tau_{xz}^2 + \left(\frac{\sigma_{xx}}{2}\right)^2} = \frac{16M_{fl}}{\pi d^3} + \sqrt{\frac{16M_{fl}}{\pi d^3} + \frac{16M_\tau}{\pi d^3}} = \frac{16}{\pi d^3}\left(M_{fl} + \sqrt{M_{fl}^2 + M_\tau^2}\right) \\ \sigma_{(1,2)} = \frac{\sigma_{xx}}{2} - \sqrt{\tau_{xz}^2 + \left(\frac{\sigma_{xx}}{2}\right)^2} = \frac{16M_{fl}}{\pi d^3} - \sqrt{\frac{16M_{fl}}{\pi d^3} + \frac{16M_\tau}{\pi d^3}} = \frac{16}{\pi d^3}\left(M_{fl} - \sqrt{M_{fl}^2 + M_\tau^2}\right) \end{cases}$$

Hence, for the von Mises criterion, there is a risk of failure if:

$$\begin{cases} From\ \sigma_{flow}^{von\ Mises} = \sqrt{\sigma_1^2 + \sigma_2^2 - \sigma_1 \times \sigma_2} = \frac{16}{\pi d_{von\ Mises}^3}\sqrt{4M_{fl}^2 + 3M_\tau^2} \geq \frac{R_e}{S} \swarrow \\ we\ deduct\ d_{von\ Mises} = \sqrt[3]{\frac{16}{\pi\, R_e/S}\sqrt{4M_{fl}^2 + 4M_\tau^2}} \leq 29.265025\ mm = 1.152\ in \end{cases}$$

Confirmation of the limit of the diameter according to the von Mises criterion:

d = 29.265025 mm which is > than the 26 mm prescribed by the hypothesis in the technical specifications (TS). If we were to machine the part to the diameter stated in the TS, we would lose around 3 mm worth of material. The same is true for the Tresca criterion. Thus, for the Tresca criterion, there is a risk of failure if:

$$\begin{cases} From\ \sigma_{flow}^{Tresca} = \left|\sigma_1 - \sigma_2\right| = \dfrac{32}{\pi d_{Tresca}^3}\sqrt{M_{fl}^2 + M_\tau^2} \geq \dfrac{R_e}{S} \rightarrow we\ deduce \\ d_{Tresca} = \sqrt[3]{\left(32/\pi\left(R_e/S\right)\right)\sqrt{M_{fl}^2 + M_\tau^2}} \leq 30.515282\ mm = 1.201\ in \end{cases}$$

Confirmation of the limit of the diameter according to the Tresca criterion:

d = 30.515282 mm = 1.201 in, which is much > than 26 mm, as prescribed by the hypothesis in the TS. Were we to machine the part to 26 mm diameter in accordance with the TS, we would lose around 4.5 mm of material.

8) Value of the ratio between the two criteria:

$$R_{von\ Mises}^{Tresca} = d_{Tresca}/d_{von\ Mises} = 1.042722$$

9) In conditions of *pure bending* ($M_\tau = 0$), we would have the equivalent diameters: $d_{Tresca} = d_{von\ Mises}$

10) In conditions of *pure torsion*, $M_{fl} = 0$, the diameter, according to von Mises, would be:

$$\frac{16 M_\tau \sqrt{3}}{\pi d_{von\ Mises}^3} \leq \frac{R_e}{S} \rightarrow d_{von\ Mises} = \sqrt[3]{\frac{16 M_\tau \sqrt{3}}{\pi R_e/S}} = 28.516151\ mm = 1.226831\ in$$

11) Under *pure torsion*, when $M_{fl} = 0$, the diameter according to Tresca would be:

$$\frac{32 M_\tau}{\pi d_{Tresca}^3} \leq \frac{R_e}{S} \rightarrow d_{Tresca} = \sqrt[3]{\frac{32 M_\tau}{\pi R_e/S}} = 29.916724\ mm = 1.1778241\ in$$

12) Observations: considering this result to be proof of the accuracy of our calculations, we can machine the torsion bar quite safely because, in both cases, the hypothesis stated by the TS is comfortably satisfied.

Value of the ratio between the two criteria: $R^{Tresca}_{von\ Mises} = d_{Tresca}/d_{von\ Mises} = 1.049115$. Note that there is a difference of around 5 % (4.9 %) between the two criteria.

5.4.2. *Case study of a profiled piece stressed under conditions of elasticity*

Calculations of the deflection and the maximum bending moment

Consider the profiled piece with the following design (L, a, ε l) such that:

$-\{L \quad a \quad \varepsilon \quad l\} = \{880 \quad 680 \quad 55.5 \quad 44.5\}\ mm = \{880 \quad 680 \quad 55.5 \quad 44.5\}\ mm;$

– F is the force applied in N, F = 7250 N = 16300 lbf;

– E is the material's elasticity modulus, E = 10^5 MPa = 1.455×10^7 psi;

– moment of inertia of the area of the cross-section, J = 7.2×10^5 mm^4 = 1.73 in^4.

Reactions at the level of the support (A):

$$R_A = F(L-a)/L = 1.648N, M_A = 0 \text{ and } y_A = 0$$

$$\left\{\begin{aligned}\theta_A &= -F(a/6EJL)(2L-a)(L-a) = -2.80114\times10^3\ rad = -\ 0.16° \\ \theta_B &= F(a/6EJL)\times(L^2-a^2) = -4.04609\times10^3\ rad = 0.232°\end{aligned}\right\}$$

The displacements along (OY) would therefore be:

$$\left\{y(x) = \theta_A x + \frac{R_A x^3}{6EJ} - \frac{F}{6EJ}(x-a)^3\ if\,(x \succ a,1,0) = \theta_A + \frac{R_A x^2}{2EJ} - \frac{F}{6EJ}(x-a)^2\ if\,(x \succ a,1,0)\right.$$

The shortening of the neutral surface of the piece is calculated using the following relation. From the initial length, length L1 = 0 mm, let us calculate ΔL as follows:

$$\left\{\begin{aligned}&\Delta L = \frac{1}{2}\int_{L_1}^{L} y(x)dx = \frac{1}{2}\int_{L_1}^{a}\left(\theta_A x + \frac{R_A x^3}{2EJ}\right)dx + \frac{1}{2}\int_{L_1}^{a}\left(\theta_A x + \frac{R_A x^2}{2EJ} - \frac{F}{2EJ}(x-a)^2\right)dx \\ &\to \Delta L \to simplify \to \{0.000002477772.N^2/Pa^2.m_3\} = 2.477772\times10^{-3}\ mm\end{aligned}\right.$$

A is the shortening for positive values of (ΔL), greater than 0. Remember that the distance (c) represents the distance from the neutral fiber, and is therefore written: $c = (\varepsilon/2)$. In addition, the shortening of the neutral surface engenders a displacement which is formulated as follows:

$$\left\{Side\ A \rightarrow |\theta_A| \times c = 0.077732\ mm\ and\ side\ A \rightarrow\ |\theta_B| \times c = 0.112279\ mm\right\}$$

Evaluation of the movement of the lower edges is equivalent to solving the total tension in the bottom fibers of the radius (of the beam) – i.e.:

$$\begin{cases} M_{fl} = F \times L = 6.38 \times 10^6\ N.mm \rightarrow \Delta\varepsilon = \displaystyle\int_{L_0}^{L} \frac{M_{fl} \times c}{EJ} dx = 2.168\ mm\ or\ indeed \swarrow \\ \Delta\varepsilon \rightarrow simplify\left\{3.3916666666666667\text{e-}7.M_{fl} / Pa.N^2\right\} = 2.168\ mm\ (OK) \end{cases}$$

The total relative motion of the lower edges of the relative sections is therefore shifted sideways by the following value: $|\theta_A + \theta_B| \times c - L = 0.187533\ mm =$ 7.383×10^{-3} *in.* This is certainly a small difference, but for a canny designer, it is worth noting it in the technical specifications, and monitoring it closely.

5.5. Buckling of elements of machines, beams, bars, shafts and stems

The specific calculation for power transmission by screwing or bolting obeys the ANSI standards (USA) for English-language standards, and ISO for metric ones. The main parameters which come into play in the design process are presented below.

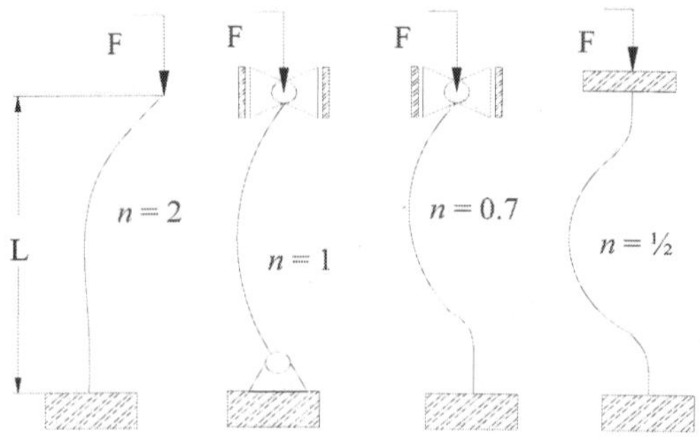

Figure 5.22. *Terminal forms of a threaded stem subject to a force F in (N)*

We have:

– F, the maximum axial force in [N, lb];

– M, the maximum moment in [N.m, lbf.ft];

– S_e, = R_e, the threshold of the elastic limit [MPa, psi];

– E, the elasticity modulus in [MPa, psi];

– $[P]_a$, the admissible pressure in [MPa, psi];

– A_{min}, the minimum area in mm^2 or in^2;

– d, the nominal diameter of the thread in [mm, in];

– d_f, the diameter on the side of the thread in [mm, in];

– d_{min}, the minimum thread diameter in [mm, in];

– H, the height of the screw in [mm, in];

– *p*, the interval of the thread in [mm, in];

– L, the total length subject to the load along the thread in [mm, in];

– K_s, the safety factor [dimensionless];

– μ, the coefficient of friction between the bolt and the screw [dimensionless], depending on the materials in contact;

– *n*, the factor determining the form of the ending of the thread (see Figure 5.22).

Reduced length L_r; minimum cross-section of the area S_{min}, and the *helix angle* α:

$$\left\{L_{reduced} = n \times L;\ \ S_{\min} = \pi d_{\min}^2 / 4;\ \ \alpha = \arctan\left(p/\pi d_s\right)\right\} \quad [5.9]$$

For a quarter turn (fast rotation), the *yield* is expressed as follows:

$$\begin{cases} \eta = \dfrac{\tan(\alpha)}{\tan\left(\alpha + \arctan(\mu_1)\right)} = \dfrac{1 - \mu_1 \times \tan(\alpha)}{1 + \mu_1 \times \text{cotan}(\alpha)},\ \eta \neq 0; if\ \ \alpha \prec \arctan(\mu_1) \\ \rightarrow \eta = 0\ then\ \eta = \dfrac{\tan(\alpha - \arctan(\mu_1))}{\tan(\alpha)} = \dfrac{1 - \mu_1 \text{cotan}(\alpha)}{1 + \mu_1 \tan(\alpha)} \end{cases} \quad [5.10]$$

Calculation of the force and maximum output moment:

$$\left\{ F_{\max}^{axial} = \frac{2M_{\max} 2\pi\eta}{p}; \; M_{\max} = Fp/2\pi\eta, \; Nmm \text{ or } lbfin \leftarrow output \right. \quad [5.11]$$

Slenderness ratio:

$$\lambda_{slenderness} = 4L_{reduced}/d_S \quad [5.12]$$

Calculation of the pressure stress:

$$\sigma_{pressure} = P_t = F/A_{\min} \quad [5.13]$$

Calculation of the torsion stress:

$$\tau_{torsion} = 16M_{torsion}/\left(\pi d_{\min}^3\right) \quad [5.14]$$

Calculation of the *equivalent reduced stress*:

$$\sigma_{reduced} = \sqrt{\sigma_{pressure}^2 + 3\tau_{torsion}^2} \;\; MPa \text{ or } psi \quad [5.15]$$

Calculation of the Rankine critical stress:

$$\sigma_{Rankine} = S_y/\left(1 + \left(S_y \times \lambda^2\right)/\pi^2 E\right) MPa \text{ or } psi \quad [5.16]$$

Euler critical stress:

$$\sigma_{Euler} = \lambda^2 \times E/\lambda^2 \; MPa \text{ or } psi \quad [5.17]$$

Johnson critical stress:

$$\sigma_{Johnson} = S_y\left(1 - \left(\lambda^2 S_y/4\pi^2 E\right)\right) \quad [5.18]$$

Calculation of the pressure applied to the threads:

$$P_{threads} = F/0.75\pi d_s\left(d - d_s\right)H/p \quad [5.19]$$

5.5.1. *Case study: buckling of an I-beam according to AISI specifications*

STATEMENT OF THE PROBLEM.–

The profiled part (I-shaped column) illustrated below is subject to axial compression (x-x and y-y) from the load (P):

– effective length factor, $n = 1$;

– length along (x-x) axis, $L_x = 9$ m;

– length along (y-y) axis, $L_y = 4.5$ m;

– load applied, $P = 7{,}25 \times 10^5$ N;

– stress threshold, $\sigma_y = 250$ MPa;

– elasticity modulus, $E = 2 \times 10^5$ MPa;

– area of the cross-section, $A = 7.5 \times 10^3$ mm²;

– radius of gyration (gyradius) along the y-y axis, $R_y = 50$ mm;

– ratio between the gyradii along the x-x and y-y axes; $r_{xy} = 2$.

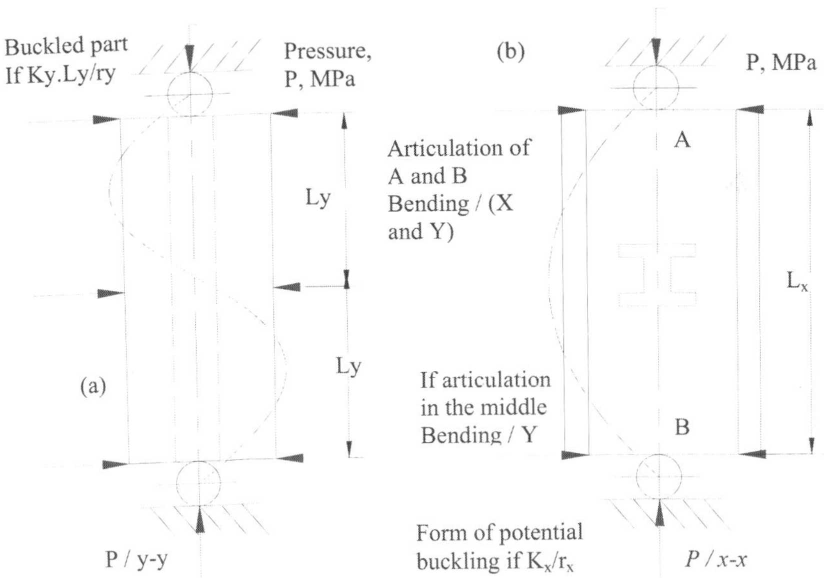

Figure 5.23. *I-beam (or double-T-beam), subject to buckling. For a color version of this figure, see www.iste.co.uk/grous/design.zip*

Step 1. According to the AISC standards, columns under the load P subject to buckling, we need to find the dangerous cross-section (low), as a function of P and the product nL_y: $P = 7.25\times10^5$ N→ $nLy = 4.5$ m and choose a cross section and its properties thus:

Calculation of the gyradii along the X-X axis $r_x = \left[r_{xy} \times r_y\right] = 100\ mm$

Step 2. Calculation of the column's slenderness ratio $C_{slenderness}$:

$$C_{slenderness} = \sqrt{2\pi^2 E/F_y} = 125.664$$

Step 3. Equations for the stress threshold at the point of application of the buckling load (elastic) F_{aelas}:

$$\textit{Elastic buckling}:\ F_{a-elX}\{n,L,r\} = 12\pi^2 E/23(nL/r)^2 = 127.145\ MPa$$

$$\textit{Local buckling}:\ F_{a-locX}\{n,L,r,C_{el}\} = \frac{F_y\left(1-\dfrac{(nL/r)^2}{2C_{el}^2}\right)}{\dfrac{5}{3}+\dfrac{3(nL/r)}{8C_{el}}-\dfrac{(nL/r)^3}{8C_{el}^3}} = 93.386\ MPa$$

If we need to use springs, let us calculate the local or elastic buckling to control the admissible stress F_{aX} with respect to the X-X axis:

$$\left\{\begin{array}{l} \textit{Along } X:\ \textit{If } \{F_{alocX} \prec F_{aelaX}, \text{"local"}, elastic\}, = \textit{along } X = \text{"local"} \\ F_{aX} = \textit{If } \{\textit{along } X = \text{"local"}, F_{alocX}, F_{aelaX}\} = 98.386\ MPa \end{array}\right\}$$

Step 4. Calculation of the local buckling stress F_{alocY} and elastic buckling stress F_{aelaY}, with respect to the Y-Y axis:

$$\left\{\begin{array}{l} \textit{Local buckling}: F_{alocY} = F_{aloc}\left(n, L_y, r_y, C_{el}\right) = 98.386\ MPa \\ \textit{Elastic buckling}: F_{aelaY} = F_{aela}\left(n, L_y, r_y\right) = 127.145\ MPa \end{array}\right\}$$

If we need to use springs, let us calculate the local or elastic buckling to control the admissible stress F_{aX} with respect to the X-X and Y-Y axes:

$$\left\{\begin{array}{l} Along\ Y\text{: } If\ \{F_{alocY} \prec F_Y,"local",elastic\}, = along\ \mathrm{Y} = "local" \\ F_{aY} = If\ \{along\ \mathrm{Y} = "local", F_{alocY}, F_Y\} = 98.386\ MPa \end{array}\right\}$$

Step 5. If we need to use springs, let us calculate the local buckling or the buckling threshold, F_a, to choose the minimum admissible stress F_{aX} with respect to the X-X and Y-Y axes: $F_a = If\ \{F_{aX} \prec F_{aY}, F_{aX}, F_{aY}\} = 98.386\ MPa$.

Step 6. Calculation of the real stress f_a: $F_a = P/A = 96.667\, MPa$. If we use springs to confirm or invalidate the choice of admissible stress, let us set: $Choice\ of\ selection = If\ \{F_a \succ F_{aY},"OK","NG"\}\ choose = "OK"$

Calculation of the percentage over-design (+) or indeed under-design (-). $\% = (F_{aY} - Fa)/F_{aY} = 1.748\% \cong 0.017 \rightarrow$ We choose this profiled piece for the design. If the selection is not acceptable or if the column (profile) is over-designed, we repeat the process from step 1, repeating the calculations, and so on.

5.5.2. *Case study: I-beams and U-beams, homogeneous and isotropic*

Initial data for the I-beam

b = 12 in = 304.8 mm; d = 6 in = 152.4 mm; e = 1 in = 25.4 mm, $A = 2.b.e + e_w.d = 28.5\ in^2 = 18390\ mm^2$.

Distance from the centroid to the endpoints of the beam:

$$\left\{\begin{array}{l} y_1 = (d/2) + e = 4 in = 101.6\ mm \\ y_2 = (d/2) = 6 in = 152.4\ mm \end{array}\right.$$

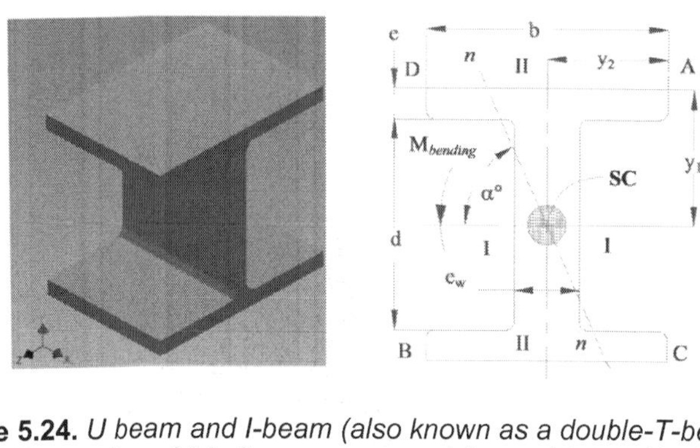

Figure 5.24. *U beam and I-beam (also known as a double-T-beam). For a color version of this figure, see www.iste.co.uk/grous/design.zip*

Moments of inertia:

$$\begin{cases} J_1 = b(d+2e)^3/12 - (d-2e_w)d^3/12 = 309.5\ in^4 = 1.288\times10^8\ mm^4 \\ J_2 = b^3e/6 + e_w^3 d/12 = 288.211.5\ in^4 = 1.200\times10^8\ mm^4 \end{cases}$$

Gyradii with respect to the main central axes:

$$\left\{ r_1 = \sqrt{J_1/A} = 3.295\ in = 80.703\ mm\ ;\ r_2 = \sqrt{J_2/A} = 3.180\ in = 80.773\ mm \right.$$

Yield modulus of the elastic section:

$$\begin{cases} W_1 = \dfrac{e_w d^2}{4} + be(d+e) = 90.75\ in^3 = 1.487\times10^6\ mm^3 \\ W_2 = \dfrac{b^2 e}{2} + \dfrac{e_w^2 d}{4} = 72.884\ in^3 = 1.194\times10^6\ mm^3 \end{cases}$$

Form factors of the elastic section: $\left\{ F_{f1} = \dfrac{W_1 y_1}{J_1} = 1.173\ ;\ F_{f2} = \dfrac{W_2 y_2}{J_2} = 1.516 \right.$

Case study of an I- or double-T-beam (see Figure 5.26)

We have changed the data in this case in order to perform the calculations differently for the purpose of further educational benefit. [$z_a = y_2/2 = 50$ and $y_a = y_1/2$]:

$$\begin{bmatrix} y_a = +120\ mm & y_b = -120\ mm & y_c = -120\ mm & y_d = +120\ mm \\ Z_a = -50\ mm & Z_b = +50\ mm & Z_c = -50\ mm & Z_d = +50\ mm \\ M_{fl} = 5400\ Nm & \theta = 18° & J_y = 121\ cm^4 & J_z = 2225\ cm^4 \end{bmatrix}$$

The respective stresses in (A, B, C and D) are arranged thus:

$$\begin{cases} \sigma_a = (M_{fy}/J_y)Z_a - (M_{fz}/J_z)y_a = -96{,}652\ MPa = -14020\ psi \\ \sigma_b = (M_{fy}/J_y)Z_b - (M_{fz}/J_z)y_b = 96{,}652\ MPa = 14020\ psi \\ \sigma_c = (M_{fy}/J_y)Z_c - (M_{fz}/J_z)y_c = -41{,}256\ MPa = -5984\ psi \\ \sigma_d = (M_{fy}/J_y)Z_d - (M_{fz}/J_z)y_d = 41{,}256\ MPa = 5984\ psi \end{cases}$$

Fiber / plane n-n: $\beta = arc\tan\left((J_z/J_y)\times\tan(\theta)\right) = 1.405\ rad = 84.59° = \begin{pmatrix} 80 \\ 30 \\ 2.053 \end{pmatrix} DMS$

COMMENTS.– The stresses situated to the right of the neutral plane (n-n) are compression stresses, because the results are negative (-96.652 MPa and -41.256 MPa). The stresses situated to the left of the plane n-n are traction stresses (+), (96.652 Mpa and 41.256 MPa). The angle (β) goes from the neutral fiber to the (z) axis.

Case study 2: U beam

The bending moments and the respective stresses in (A and D) are arranged thus:

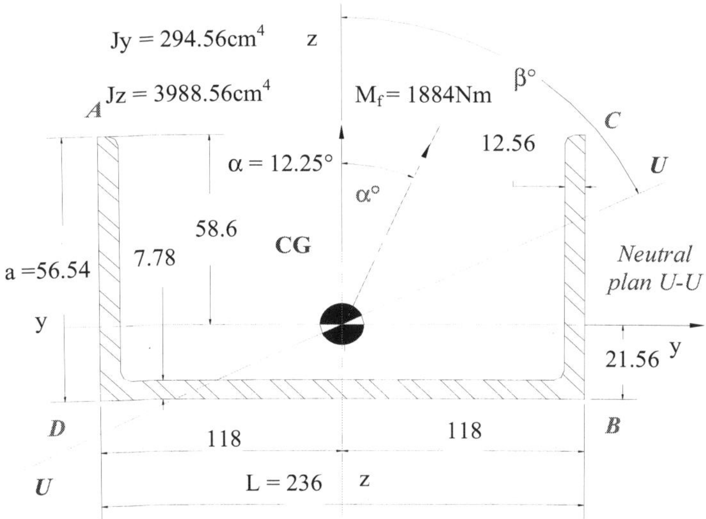

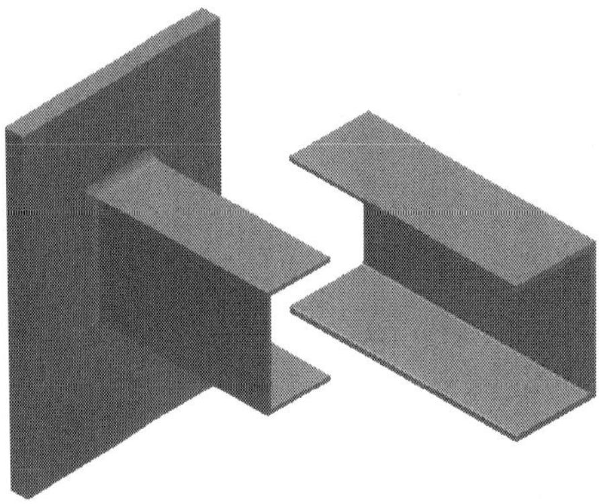

$$\begin{bmatrix} y_a = -118.00\ mm \\ y_b = +118.00\ mm \\ J_y = 294.560\ cm^4 \\ J_z = 3988.56\ cm^4 \\ Z_a = +56.540\ mm \\ Z_b = -21.560\ mm \\ M_{flex} = 1884\ Nm \\ \alpha = 10.027\,\text{deg} \end{bmatrix}$$

Figure 5.25. *U-beams subject to bending. For a color version of this figure, see www.iste.co.uk/grous/design.zip*

$$\begin{cases} M_{fz} = M_{fy} \times \cos(\alpha) = 1.8411 \times 10^3 \; Nm = 1.63 \times 10^4 \; lbfin \\ M_{fy} = M_{fy} \times \sin(\alpha) = 399.7430 \; Nm = 3.538 \times 10^3 \; lbfin \\ \sigma_a = \left(M_{fy}/J_y\right) Z_a - \left(M_{fz}/J_z\right) y_a = +13.119793 \; MPa = 1903 \; psi \\ \sigma_d = \left(M_{fy}/J_y\right) Z_b - \left(M_{fz}/J_z\right) y_b = -8.372709 \; MPa = -1214 \; psi \end{cases}$$

Neutral fiber along the u-u plane:

$$\alpha = arc\tan\left(\tan(\theta)\left(J_z/J_y\right)\right) = 1.243 rad = \begin{pmatrix} 71^\circ \\ 13' \\ 7.154'' \end{pmatrix} DMS$$

COMMENTS.– The stresses to the right of the neutral plane (u-u) correspond to compression stresses and σ_B = (–8.37270!5 MPa) = σ_{max}. The stresses to the left of the neutral plane (u-u) correspond to traction stresses and σ_A = (13.11979: MPa) = σ_{max}.

5.6. Design of stationary and rotating shafts

The design of shafts (profiled pieces) is at the heart of mechanical design. It conforms to very precise calculation requirements. The shaft must satisfy the conditions of resistance and stiffness in the pre-established, clearly defined conditions. As shafts are circular, the usual geometry poses no problem, and it is unnecessary to bring in often-complicated structural calculations, such as the FEM. Ductile materials are governed by the maximum shear stress (stiffness). Here we present the fundamental formulae, knowing that often, the shafts are subject to bending, torsion and sometimes both:

– τ_{xy} is the shear stress (torsion) in psi or MPa (SI);

– σ is the bending stress in psi or MPa (SI);

– M_t is the torsion moment in inch.lb or N.mm (SI);

– M_f is the bending moment in inch.lb or N.mm (SI);

– F_{fa} is the axial load in lb or N (SI);

– d_0 is the external diameter in inch or mm;

– d_i is the internal diameter in inch or mm;

– $K = d_i / d_0$ is the ratio between the diameters;

– K_f is the combined bending load factor (shock and fatigue);

– K_τ is the combined torsion load factor (shock and fatigue);

Loads	K_f	K_τ
Stationary shaft		
Load gradually applied	1.0	1.0
Load suddenly applied	1.5 to 2.0	1.5 to 2.0
Rotating shaft		
Load gradually applied	1.5	1.0
Load suddenly applied (minor shock)	1.5 to 2.0	1.0 to 1.5
Load suddenly applied (major shock)	2.0 to 3.0	1.5 to 3.0

Table 5.4. *Table of load correction factors*

– n = 1: free end; n = 2.25: anchored end; n = 1.6: partially limited end (e.g. bearings);

– k is the radius of gyration, $k = (J/A)^{1/2}$, inch or mm;

– J is the quadratic moment of inertia, inch^4 or mm^4 or cm^4;

– A is the area of the cross-section of the shaft, inch^2 or mm^2;

– σ_s is the compression threshold (stress), *psi* or MPa.

Under the ASME Code, for shafts of construction steel:

– $[\sigma_f]$, admissible stress 8000 psi or MPa for an unkeyed shaft;

– $[\sigma_f]$, admissible stress 6000 psi or MPa for a keyed shaft.

Under bending or compression, the admissible stress is increased by 30 % of the elasticity limit, but never beyond the limit stress for keyed shafts. The value is increased by 25 % for a keyed shaft, in our case. For columns (or profile pieces), the ASME Code describes a factor (α) called the *column-action factor* to express the loading tension:

$$\begin{cases} \alpha = 1/\left(1 - 44\times10^{-4}\,(L/k)\right) \; for \; (L/k) \prec 115 \\ \alpha = \sigma_y\,(L/k)^2 / \pi^2 nE \; for \; (L/k) \succ 115;\; k = \sqrt{J/A} \end{cases} \qquad [5.20]$$

$$\text{Under torsion: } \begin{cases} \text{Solid shaft} \rightarrow \ \tau_{xy} = M_\tau r/J = 16M_\tau/\pi d^3 \\ \text{Hollow shaft} \rightarrow \tau_{xy} = 16M_\tau d_0/\pi\left(d_0^4 - d_i^4\right) \end{cases} \quad [5.21]$$

$$\text{Under bending: } \begin{cases} \text{Solid shaft} \rightarrow \ \sigma_f = M_f r/J = 32M_f/\pi d^3 \\ \text{Hollow shaft} \rightarrow \sigma_f = 32M_f d_0/\pi\left(d_0^4 - d_i^4\right) \end{cases} \quad [5.22]$$

Under axial loads (traction or compression):

$$\begin{aligned} &\text{Solid shaft} \rightarrow \ \sigma_{axial} = 4F_a/\pi d^2 \quad [psi \text{ or } MPa] \\ &\text{Hollow shaft} \rightarrow \sigma_{axial} = 4F_a/\pi\left(d_0^2 - d_i^2\right) \ [psi \text{ or } MPa] \end{aligned} \quad [5.23]$$

The codes set out by ASME (*American Society of Mechanical Engineers*) offer an undeniable advantage when it comes to professional design calculations. These codes introduce shock factors and fatigue factors. These important factors affect the constitution of the classic equations employed in strength of materials. According to the ASME code, the design of a shaft must be:

$$d_0^3 = \frac{16}{\pi.\sigma_s\left(1-K^4\right)}\sqrt{\left(K_f.M_f + \left(\alpha.F_a.d_0 \times \left(1+K^2\right)/8\right)\right)^2 + \left(K_t.M_t\right)^2} \quad [5.24]$$

With no axial load (or with a negligible axial load), the ASME code allows us to write:

$$d^3 = \left(16/\pi.\sigma_s\right)\sqrt{\left(K_f.M_f\right)^2 + \left(K_t.M_t\right)^2} \quad [5.25]$$

5.6.1. *Design (dimensioning) of shafts subjected to rigidity*

These calculations, in general, are run when designing the torsion angle (θ). The calculation of this angle depends particularly on the application at hand. It varies by 0.08 per foot. For tool machines, it varies by 1.0° per foot. It is also calculated using the following expressions:

$$\begin{cases} \text{Solid shaft} \rightarrow \ \theta = 584 \times M_\tau L/Gd^4, \\ \text{Hollow shaft} \rightarrow \theta = 584 \times M_\tau L/G\left(d_0^4 - d_i^4\right) \end{cases} \quad [5.26]$$

where:

– (θ) is the torsion angle, in degrees;

– L is the length of the shaft, in inches or mm;

– M_τ is the torsion moment, in inch.lb or N.mm;

– G is the rigidity modulus (elasticity) in psi or MPa;

– d is the diameter of the shaft in inches or mm.

Standard shaft dimension in accordance with the ASCE (*American Society of Civil Engineers*) code: power transmission shafts in inches:

15/16; 1 3/16; 1 7/16; 1 11/16; 1 15/16; 2 3/6; 2 7/16; 2 15/16; 3 7/16; 3 15/16; 4 7/16; 4 15/16; 5 7/16; 5 15/16

Shafts for machines (tooling machines) in inches:

– from 1/2 inch to 2 1/2 inches in increments of 1/16 inch;

– from 2 5/8 inches to 4 inches in increments of 1/8 inch;

– from 4 1/4 inches to 6 inches in increments of 1/4 inch;

– the standard lengths in case of shock are: 16, 20 and 24 feet (*ft*).

Torsion- and bending moments of transmission shafts

The main parameters which characterize the mechanical design (dimensioning) of the elements of machines are the various moments (torque values). Often, we begin by plotting the graph of the moments (see exercises below) for the shaft under stress. Then, we calculate the dangerous cross-sections where the moment is greatest. For rotating shafts, we determine the rotating moment (torque) thus:

$$M_\tau = h_p 33000 \times 12/2\pi rpm = h_p 63000/rpm \ \left[lbf.in\right] \qquad [5.27]$$

For belt transmissions, the *moment* $M_\tau = \left(T_1 - T_2\right)R$, Nmm:

– T_1 is the tension applied to the belt (side 1);

– T_2 is the tension applied to the belt (side 2);

– R is the radius of the pulley.

For gear transmissions, the moment $M_t = F_\tau \times R$ (Nm). F_τ is the tangential force applied to the primitive radius, $R_p = (m \times Z/2)$ of the gear system in *lb*. Z is the number of teeth and m the modulus of the gear system (normalized). Having presented the fundamental relations of the formulae used in SOM, we shall now present some examples of problems with solutions and discussion, to make the user's task easier.

5.6.2. *Case study 1, solution 1*

Consider a steel shaft with the maximum moment of 6,327,150 N.mm (5600 *lb.in*). The torque of forces applied is T = 271,164 N.mm. (2400 lbf.in). Is the shaft safe if the bending stress applied is of the order of 66 MPa (i.e. 9600 *psi.* = 9600 lbf/in^2)? What is the ideal *torque* to make the shaft safe using this material?

1) Equivalent bending moment $M_e = \left(M + \sqrt{M^2 + T^2} \right) / 2 = 6330054\ Nmm.$

2) Stress applied to the shaft $\sigma_{bending}$ = Mc/J where (c) is the lever arm c = (d/2) mm. J is the polar moment of inertia for a circular form (shaft). J = πd4/ 64 mm^4. For the shaft, the moment is M = M_e:

$$\sigma_{flex} = Mc/J = 53\ MPa = 7672\ psi \to \sigma_{flex} = 53\ MPa \prec \left[\sigma_{flex} \right] = 66\ MPa$$

The bending stress σ_{flex} is lesser than the admissible stress [σ_{flex}]. The shaft is therefore safe for this application. The calculation of the maximum equivalent bending moment gives us:

$$M_{\max}^{equ} = \sigma J/c = 826{,}584\ N.mm \text{ and } M_{\max}^{equ} - M^{equ} = 166{,}039\ N.mm = 1{,}470\ lbf.in$$

This is greater than the current bending moment. Therefore, the shaft is safe. For M = 632,715 N.mm (5600 lbf.in), the ideal torque for the shaft would be:

$$T_{ideal} = M + \sqrt{M^2 + T^2} = 1{,}321{,}089\ N.mm = 11{,}693\ lbf.in$$

This is a case which is applicable to the homogeneous sections of a shaft, for various materials such as steel, aluminum 6061, bronze, brass, etc.

5.6.3. *Case study 2 with solution: shear, moments, slope, elasticity deflection. Applied SOM in mechanics and civil engineering*

Consider:

– the quadratic moment of inertia J = 3.33×10^8 mm^4 (so J = 800 *inch*^{4});

– the length of the piece L = 8.534×10^3 mm (so L = 28 *ft*);

– the elasticity modulus $E = 2.068\times10^5$ MPa (so $E = 30\times10^6$ *lbf/in²*);

– the moment applied $M_0 = 2.305\times10^8$ N.mm (so $M_0 = 17\times10^4$ *lbf.ft*);

– the distance between the load a = 2.438×10^3 mm (a = 8 *ft*) and the left extremity.

SOLUTION.–

Boundary conditions, i.e. reactions (R in N), moments (M in N.mm), slope (θ in degrees) and deflection (y, in mm):

1) Reactions on the left of the cantilever at the free end:

$$\begin{cases} R_A = 0, N = 0\ lbf;\ M_A = 0\ Nmm = 0\ lbfin;\ Slope\ \theta_A = \left[-M_0(L-a)\right]/EJ \\ \ldots\ \theta_A = -1.169^\circ\ and\ y_A = -M_0\left(L^2 - a^2\right)/2EJ = 11.923\ mm = 4.406\ in \end{cases}$$

2) Reactions on the right of the cantilever at the fixed end:

$$R_B = 0\ N\ and\ M_B = M_0 = 2.305\times10^8\ Nmm\ ;\ \theta_B = 0^\circ;\ slope\ \text{and}\ y_B = 0\ mm; deflection$$

3) Transverse shear, bending moment, slope and deflection as a function of (*x*). Graph. (*x*) is the total length of the piece. It varies thus: *x = 0 L; 0.1 L, ... L*. For example, for x = 3.658×10^3 mm, i.e. 12 *ft*.

1) Transverse shear: V(x_1): $V(x) = R_A\ so\ V(x_1) = 0N = 0\ lbf$

2) Bending moment: M(x_1) $M(x) = M_A + R_A x + M_0(x \succ a) = 2.305\times10^8\ Nmm$

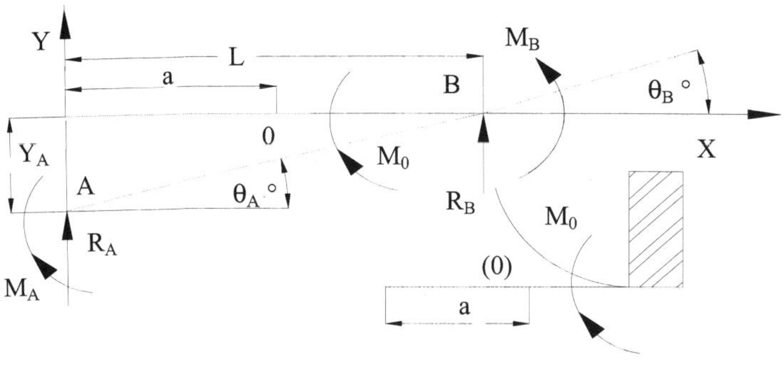

Figure 5.26. *Intensification of moments and cantilever. For a color version of this figure, see www.iste.co.uk/grous/design.zip*

3) Slope, $\theta(x_1)$: $\theta(x)=\theta_A+\frac{M_A x}{EJ}+\frac{R_A x^2}{2EJ}+\frac{M_0}{EJ}(x-a)(x\succ a)\rightarrow\theta(x_1)=-0.935°$

4) Deflection, $y(x_1)$: $\begin{cases} y(x)=y_A+\theta_A x\frac{M_A x^2}{2EJ}+\frac{R_A x^3}{6EJ}+\frac{M_0}{2EJ}(x-a)^2\times(x\succ a) \\ \rightarrow y(x_1)=39.795\ mm=1.567\ inch \end{cases}$

5) Recap of maximum values of the moments and strains: the (-) signs correspond to the direction – that is, the orientation of the vector. The subscript (max) corresponds to the maximum magnitude of the parameters:

$$\begin{cases} M_{\max}=M_0=2.305\times10^8\ N.mm=17\times10^5\ lbf.ft \\ \theta_{\max}=\theta_A=-1.169°\ and\ y_{\max}=y_A=111.923\ mm=4.406\ inch \end{cases}$$

For a = 0, "*val*" corresponds to the maximum achievable values:

$$\theta_{val\max}=\frac{-M_0 L}{E.J}=-1.636°\ y_{val\max}=\frac{M_0 L^2}{2E.J}=121.871\ mm=4.798\ inch$$

6) Plots, V(x); M(x); θ(x) and y(x)

The application to the round, solid sections for these data will give us:

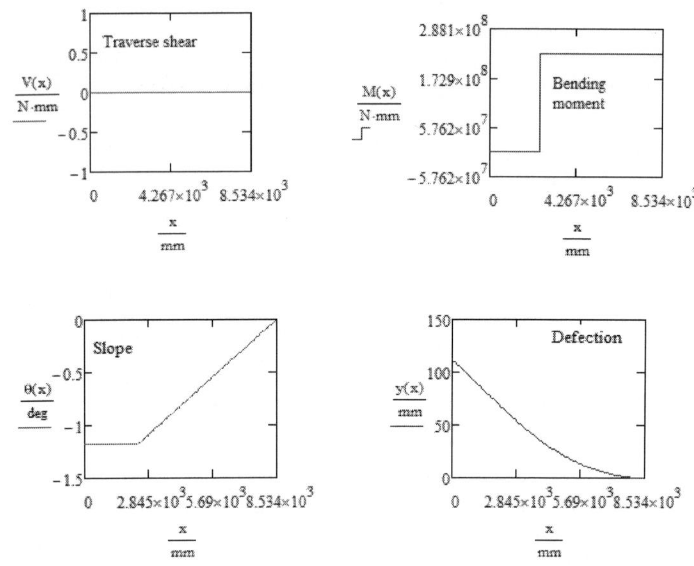

Figure 5.27. *Respective graphs: traverse shear, bending moment, slope and deflection. For a color version of this figure, see www.iste.co.uk/grous/design.zip*

Consider:

– the radius of the bar r = 1.181 in = 29.997 mm;

– the length of the bar L = 5 ft = 1.524 m;

– G is the stiffness modulus $G = 12\times10^6$ psi = 8.274 $\times10^6$ MPa;

– the torsion moment M = 540 lbf.in = 732.142 N.m.

Calculation of the stiffness coefficient K and the maximum torsion stress

$$K = \pi r^4/2 = 3.056\ in^4 = 1.272\times10^6\ mm^4\ \ and\ \ \tau_{max} = 2M/\pi r^3 = 17.267\ MPa = 2504\ psi$$

Calculation of the torsion angle θ

$$\theta = M_\tau L/KG = 0.608° = 0.011 rad$$

Plots of K(r1) and τ_{max}(r) for r1 = (0 in, 1 in, …, 12 in)

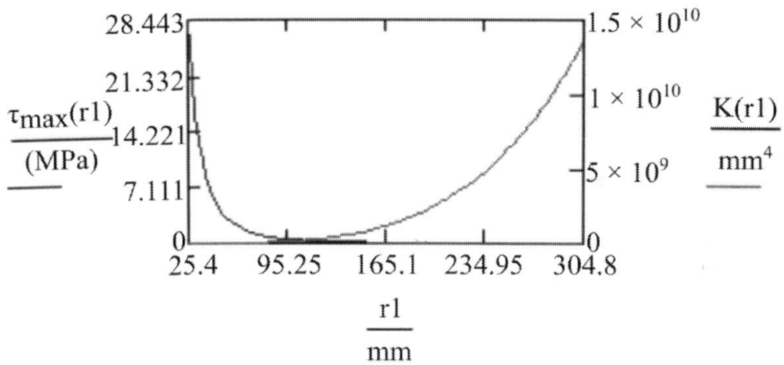

Figure 5.28. *Maximum torsion stress. For a color version of this figure, see www.iste.co.uk/grous/design.zip*

5.7. Power transmission elements: gear systems and pulleys

5.7.1. *Case study*

Mounted on a pulley whose mass is 225 lb, consider a shaft having a diameter of Ø 26 inches, spinning at 450 rpm. The pulley turns at a power of 22 hp. On the shaft is a 22-inch gear system, with a homogeneous mass of 200 lb. The gear system supplies 25 hp of power. What is the load concentrated in the shaft by the pulley and the cogwheel?

Pulley

Frequence of rotation of the shaft	$\omega = 450\ rot/min$;
Radius of the pulley	$r_p = 13$ inch = 330.2 mm;
Power supplied by the pulley	$h_p = 22$ HP = 1.641×10^4W;
Mass of the pulley	$m_p = 220$ lb = 113.398 kg.

Cogwheel

Primitive radius of the wheel	$r_p = 12$ inch = 304.8 mm;
Power supplied by the wheel	$h_p = 22$ Hp = 1.864×10^4 W;
Mass of the pulley	$m_p = 175$ lb = 79.379 kg.

Solution: calculations, method and discussions

1) We begin by calculating the *concentrated load of the pulley*. The concentrated load on the pulley is caused by the pulley itself during the vertical (top to bottom) motions of the belt. We deduce from this that the total concentrated load of the pulley is the sum of the load on the belt and the pulley's own mass. The tension on the pulley on the side where the belt is taut is twice as high as on the loose side of the belt. Thus, the maximum load on the belt is:

$F_p = 3T/r$ *tension of force produced by the load of the belt, so the moment applied to the pulley* $T_p = h_p/2\pi R$, *and hence the force* $F_p = 3T_p/r_p = 711.055\ lbf$

The concentrated load of the pulley is $F_p = m_p g = 961.055\ lbf$.

2) The concentrated load of the cogwheel. The acting force is exercised at the level of the primitive diameter. Thus, the belt drives motion normally. Knowing the moment (torque) and the radius, calculate the force Fg→ $F_g = T/r$.

The moment on the cogwheel: $T_g = h_g/2\pi R$ *Hence the force* $F_g = T_p/r_g =$ 292 lbf.

Thus, the concentrated load on the cogwheel is written: $F_g = m_g \times g = 467\ lbf$.

5.7.2. *Case study: statement of problem 2*

Consider a section of a shaft 5.10 *ft* long, upon which are mounted two bearings supporting a pulley with a load of 220 *lb* in the very center. The 20 HP motion is transmitted from the shaft to the pulley by a key way with an angular speed of 1500 rpm. The belt conveys a force composed of $T = T_1 + T_2 = 1500$ *lb*. $T_1 = 200$ *lb*. We shall agree, in accordance with the ASME code (USA) that the combined shock and fatigue factors – bending and torsion – $K_t = K_b = 1.5$. The stiffness modulus G of the material is 12×10^6 *psi*. What would be the *diameter* $Ø_{shaft}$ *and the torsion angle* θ between the bearings?

We have:

– the rotation speed (RPM) RPM = 150 rpm;

– the power delivered by the pulley h_p 20 HP;

– the shaft length L = 20 inches;

– the torsion modulus in elasticity, psi $G = 12 \times 10^6$ *psi*;

– the bending stress for a keyed shaft $\sigma = 6 \times 10^3$ *psi*;

– the force applied vertically $F_v = 120$ *lb*;

– the force applied horizontally $F_h = 120$ *lb*;

– the correction factor due to shocks and fatigue, combined, under bending, $K_b = 1.5$;

– the correction factor due to shocks and fatigue, combined, under torsion, $K_t = 1.5$.

SOLUTION WITH DISCUSSION.–

Bending moments applied to the shaft (vertically and horizontally), M_v and M_h:

$$\begin{cases} M_v = F_v \times L \;\; \textit{hence the moment } M_v = 3.36 \times 10^3 \; lb.in \\ M_h = F_h \times L \;\; \textit{hence the moment } M_h = 1.624 \times 10^4 \; lb.in \end{cases}$$

Maximum bending and torsion moments applied to the shaft (vertically and horizontally), M_{vmax} and M_{hmax}:

$$\begin{cases} M_{f.Max} = \sqrt{M_v^2 + M_h^2} \;\; \textit{hence the moment } M_{f.Max} = 1.658 \times 10^4 \; lb.in \\ M_{t.Max} = h_p.33000 \times 12 / 2\pi RPM = h_p.63000 / RPM = 8.4 \times 10^3 \; lb.in \end{cases}$$

$$\textit{Diameter of the shaft } d^3 = \frac{16}{\pi 6000}\sqrt{\left(K_f M_{f.Max}\right)^2 + \left(K_t M_{t.Max}\right)^2} \rightarrow d = 2.871 \textit{ in}$$

$$\textit{Torsion angle, } \theta \begin{cases} \theta = 584 \times M_{t.\max} L / GD^4 \textit{ hence the torsion angle } \theta = 0.141° \\ \text{Proof } \textit{of calculations}: \ \theta = 584\left(8400 \times 28 / 12 \times 10^6 \times 3^4\right) = 0.141° \end{cases}$$

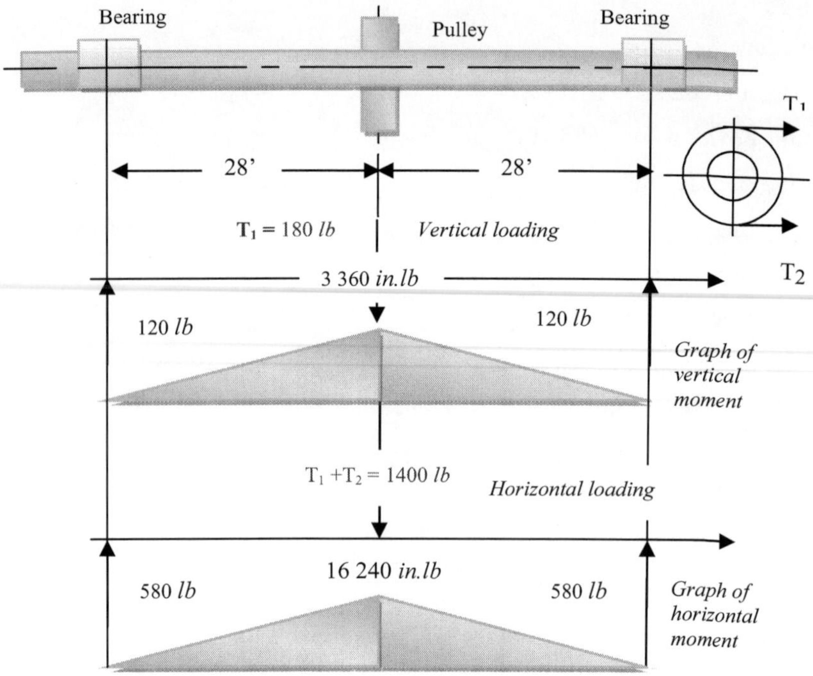

Figure 5.29. *Graph of calculations of moments acting on the shaft. For a color version of this figure, see www.iste.co.uk/grous/design.zip*

DISCUSSION. – The design of mechanical elements is a question which requires great rigorousness. The reason is not always mathematical. It is important not to produce "over-quality" by increasing the value of the diameter.

5.7.3. *Case study: statement of problem 3*

Consider a pulley 24 inches in diameter, mounted on a shaft-pinion 10 inches in diameter. The pulley is driven horizontally by a belt. The pulley's own weight is 300 lb (see Figure 5.30). The tensions on the belt are distributed thus: $T_1 = 1200\ lb$ and $T_2 = 400\ lb$. We accept, in accordance with the ASME code (USA), that the fatigue and combined shock factor under torsion $K_\tau = 1.5$ and bending $K_f = 2.0$. The material's stiffness modulus G is $12\times10^6\ psi$:

– rotation speed (RPM)	RPM = 150 rpm;
– power delivered by the pulley	h_p 22 HP;
– shaft length	L = 40 inch;
– lever arm (L_1), see Figure 5. 30	$L_1 = 9$ inch = 228.6 mm;
– torsion modulus under elasticity, psi	$G = 12\times10^6\ psi$;
– bending stress for a keyed shaft	$\sigma = 6\times10^3\ psi$;
– force applied vertically	$F_1 = F_v = 1324\ lbf = 5.889\times10^3$ N;
– force applied horizontally	$F_2 = F_h = 985\ lbf = 4.381\times10^3$ N;

– correction factor (ASME), shocks and fatigue, combined, under torsion, $K_t = 1.5$;

– correction due to shocks and fatigue, combined, under bending, $K_f = 2$ (ASME Code).

What would $Ø_{shaft}$ and the torsion angle θ be if the length of the diameter is L = 40 inches?

DOCUMENTED SOLUTION.–

1) Determination of the bending moments (M_v and M_h) acting on the shaft:

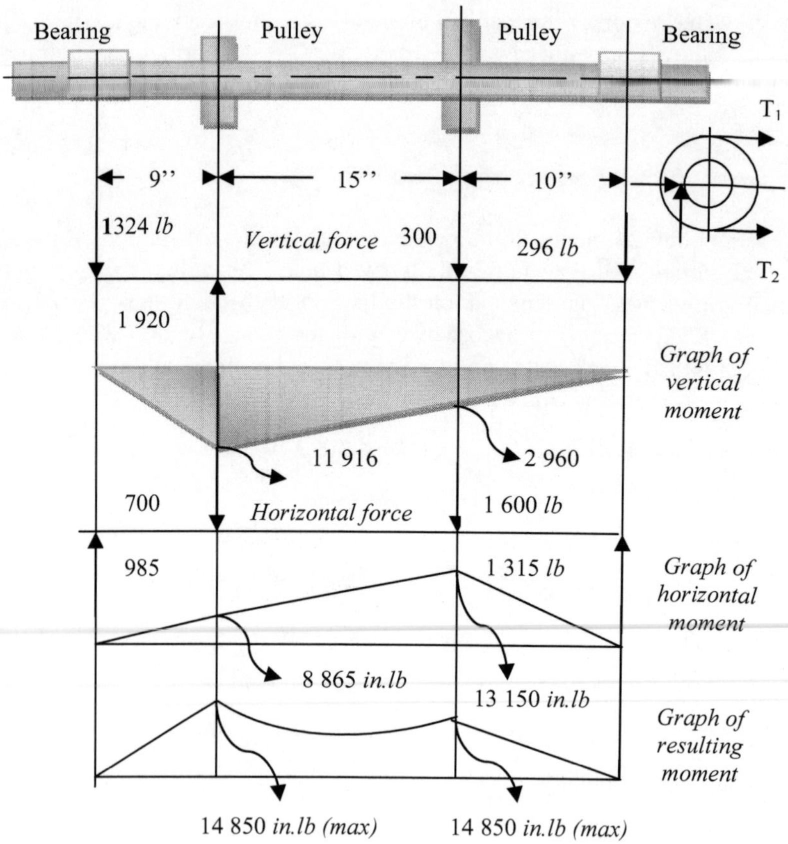

Figure 5.30. *Graph of calculations of the moments acting on the shaft. For a color version of this figure, see www.iste.co.uk/grous/design.zip*

$$\begin{cases} M_v = F_1 \times L_1 = 1324\ lbf \times 9\ in = 1.192 \times 10^4\ in.lbf = 1.346 \times 10^6\ N.mm \\ M_h = F_2 \times L_1 = 9851\ lbf \times 9\ in = 8.865 \times 10^3\ in.lbf = 1.002 \times 10^6\ N.mm \end{cases}$$

2) Determination of the *maximum bending and torsion moment* acting on the shaft:

$$\begin{cases} M_{f_{max}} = \sqrt{M_v^2 + M_h^2} = \sqrt{\left(1.192\times10^4\right)^2 + \left(8.865\times10^3\right)^2} = 1.678\times10^6\ Nmm \\ M_h = \left(T_1 - T_2\right) R_p = (1200-400)12 = 9.6\times10^3\ lbfin = 1.0085\times10^6\ Nmm \end{cases}$$

3) Determination of the shaft diameter in accordance with the ASME calculation code:

$$d = \sqrt{\frac{16}{\pi\times\sigma_s}\times\sqrt{\left(K_f M_{f\,max}\right)^2 + \left(K_\tau M_{\tau\,max}\right)^2}} = 3.037\ inch = 77.148\ mm$$

In accordance with the ASME code, we choose a diameter of 3 *in* and thus calculate the torsion angle:

$$\theta = 584\frac{M_{f\,max} L_1}{G\left(D_{ASMECode}\right)^4} = 0.052° \rightarrow \text{Proof } \theta = 584\times\frac{9.6\times10^3\times9}{12\times10^6\times(3)^4} = 0.052°$$

CONCLUSION.–

The diameter of the shaft conforms to the requirements of this calculation code. Hence, these dimensions are suitable to proceed to design and manufacture.

5.8. Sizing and design of couplings

The design of the mechanical elements – specifically, the shafts, beams and other columns – is at the heart of mechanical design and civil engineering. As is the case with shafts, the design of beams and columns is subject to precise calculation codes attached to the standards in force. As the geometric forms are not usual, the problem of which formulae to use for the calculations for such structures (sometimes complex) poses the question of the stressed geometry.

Case study, solved problem: consider a stiff steel coupling. The two parts of the coupling are held together by six bolts, subject to shearing. We can overlook the fit of the bolts, because it is a coupling which conveys power with the least possible friction. However, friction does, of course, occur, and it is important to take account of it in all mechanical designs.

Grade of the material: SAE 1045 annealed steel. Limit traction stress σ_p = 8.557×10^4 psi (600 MPa); failure threshold under traction (tension) $\sigma_p = 4.5\times10^4$ psi (315 MPa).

Diameter of the transmission shaft, d_{shaft} = 2 in = 50.8 mm.
Grade of the material SAE 1045
Heat treatment: annealing.
Number of bolts in the mechanism: N_b = 6 bolts
Rotation speed (RPM)
RPM = 150 rpm.
Power delivered by the pulley, h_p 22 HP.
Limit traction stress
σ_τ = ? and moment = (torque).

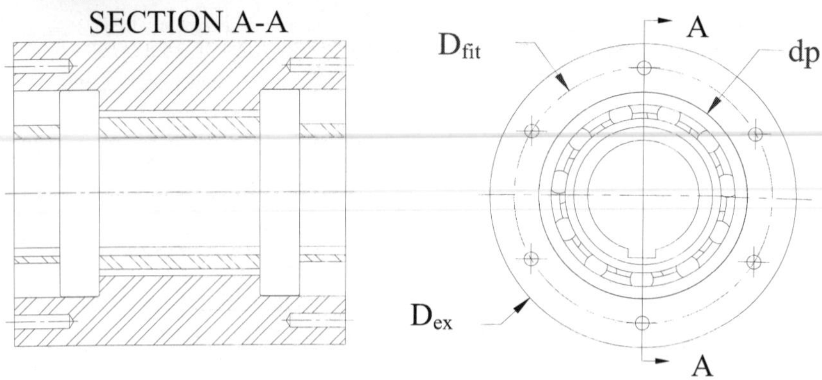

Figure 5.31. *Drawing defining a stiff coupling. For a color version of this figure, see www.iste.co.uk/grous/design.zip*

DOCUMENTED SOLUTION.–

1) Calculation of the contact radius (friction):

$$R_{friction} = \frac{2}{3}\left(\frac{R_0^3 - R_i^3}{R_0^2 - R_i^2}\right) = \frac{2}{3}\left(\frac{4^3 - 3.5^3}{4^2 - 3.5^2}\right) = 3.756\ in = 95.391\ mm$$

2) Calculation of the moment M_τ:

$$M_\tau = F_{axial}\mu R_{frict} = 30000\times0.15\times3.756 = 1.69\times10^4\ lbfin = 1.909\times10^6\ Nmm$$

3) Calculation of the power on the shaft:

$$P_{\tau} = M_{\tau}\omega_{coup}/Coeff. = 1.69\times10^4 \times 300/63000 = 81.476\ hp = 59.95\ KW$$

4) Calculation of the torque on the shaft:

$$\begin{cases} \sigma_{sys} = 30\%\sigma_{threshold} = 1.371\times10^4\ psi\ and\ \sigma_{sys} = 18\%\sigma_{\lim} = 1.54\times10^4\ psi \\ T_{shaft} = 0.75\pi\sigma_{sys}\left(d^3_{shaft}\right)/16 = 0.75\pi1.371\times10^4 = 1.54\times10^4\ lbf.in \end{cases}$$

5) Calculation of the power on the shaft:

$$P_{shaft} = T_{shaft}\omega_{coup}/Coeff = 1.59\times10^4 \times 300/63000 = 75.685\ hp = 56.44\ KW$$

5.8.1. *Design of a universal coupling, known as a Hooke coupling*

In this problem, our aim is to design and size Hooke joints as shown in Figure 5.32. These universal joints are so called because they are able to link at least two shafts which need not necessarily run along the same line. The angular velocity (ω_o) of the output shaft is different to that developed by the input shaft (ω_i). Consider a universal Hooke coupling, made of steel. The torque applied to the drive shaft (A_1) is T_i = 360 *in.lb*. Calculate:

1) the torque on the output shaft T_o (driven shaft, Figure 5.32);

2) the diameter of the cotter pin of the crosspieces, subject to bending. The bending stress applied to the bearings is of the order of 2000 *psi*. The admissible stress is 2×10^4 *psi* and the shear stress is 10^4 *psi*;

3) the maximum shear stress along the section E-E (2 *in*) along the Y-Y axis.

where:

– the grade of the material SAE 1045	annealed;
– the number of bolts in the mechanism	n = ?;
– the rotation speed (RPM)	RPM = 150 rpm;
– the power delivered by the pulley	h_p 22 HP;
– the grade of the material: SAE 1045	annealed;
– the limit traction stress	$\sigma_p = 8.557\times10^4$ psi (600 MPa);
– the failure threshold under traction (tension)	$\sigma_p = 4.5\times10^4$ psi (315 MPa).

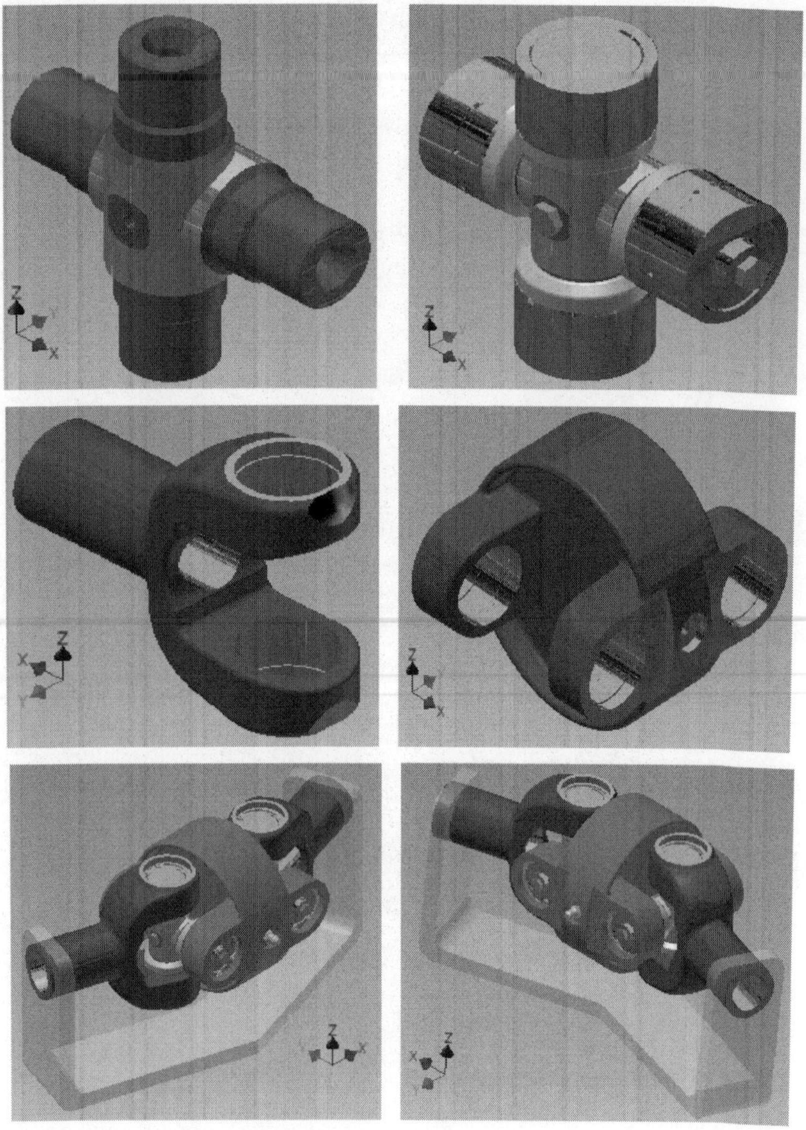

Figure 5.32. *Drawing of definition of a Hooke coupling. For a color version of this figure, see www.iste.co.uk/grous/design.zip*

SOLUTION.–

The transmission ratio is expressed by the following relation:

$$i = \omega_s / \omega_e = \left(\cos(\theta) / \left(1 - \cos^2(\alpha)\sin^2(\theta)\right)\right) \qquad [5.28]$$

where:

– θ is the angle formed between the axes of the shafts;

– α is the angle formed by the drive shaft, measured from the cotter pin;

– (ω_s) is the angular velocity of the output shaft;

– (ω_e) is the angular velocity of the input shaft.

Calculation of the contact radius (friction):

$$\frac{T_{s1} \times \omega_{s1}}{63024} = \frac{T_{s2} \times \omega_{s2}}{63024};\; T_{s1} \times \omega_{s1} = T_{s2}\omega_{s1}\frac{\cos(\theta)}{1 - \cos^2(\theta) \times \sin^2(\theta)} \qquad [5.29]$$

INITIAL DATA.–

Cardan joints (named after *Jérôme Cardan*) or Hooke joints:

– angle of rotation formed by the axes of the two shafts: θ = 12 deg; 15 deg … 90 deg;

– angle formed by the axes of the two shafts: α = 12 deg; 15 deg … 90 deg;

– angular velocity at input of the motion: ωS1;

– angular velocity at output of the motion: ωS2;

– power transmission ratio: R = ω_{S2}/ω_{S2} and n = 1.

$$R_{tr} = \frac{\omega_{s2}}{\omega_{s2}} \textit{ or indeed } R_{tr}(\alpha, \theta) = \frac{\cos(\alpha)}{1 - \sin^2(\alpha)\sin^2(\theta)} \qquad [5.30]$$

$R_{tr}(\alpha,\theta) =$

	0
0	0.980
1	0.969
2	0.955
3	0.939
4	0.920
5	0.899
6	0.875
7	0.850
8	0.821
9	0.791
10	...

etc. ...

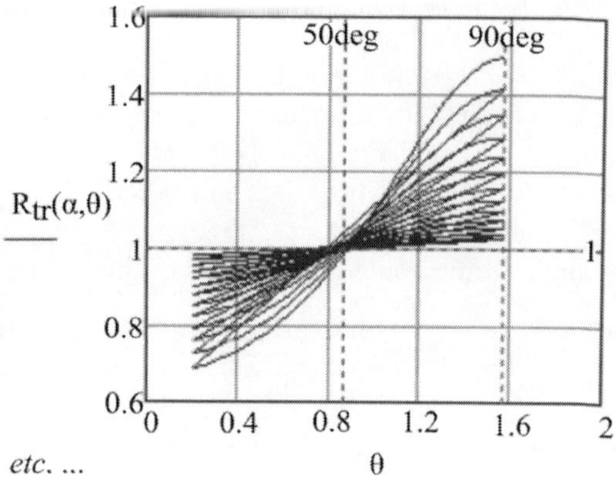

Figure 5.33. *Fluctuation of the velocity ω_{S2} as a function of the angle of rotation (θ). For a color version of this figure, see www.iste.co.uk/grous/design.zip*

5.9. Design of beams and columns

The design of mechanical elements – specifically, shafts, beams and other columns – is at the heart of mechanical design and civil engineering. The design of shafts is subject to precise calculation codes attached to the standards employed by the designers. As the geometric forms are not usual, the problem of which formulae to use for the calculations for such structures (sometimes complex) raises the question of the stressed geometry.

SOLVED PROBLEM 1.–

Consider a shaft of diameter ∅ = 1.98 inch (50 mm), made of steel, subject to a maximum bending moment M_b = 5600 lbf.in (M_b = (632715 N.mm). The torque applied is of the order of T = 2400 lbf.in ((T = (271164 N.mm).

1) Does this shaft conform to the safety requirements if the maximum admissible stress for the material is $\sigma_{adm} = 9600$ lbf/in^2 ($\sigma_{adm} = 66$ MPa)?

2) What is the ideal torque (in *psi* and MPa) for this shaft to operate safely?

We have:

– the maximum moment applied to the shaft: $M._b$ = 5600 lbf.in = 632715 N.mm;

– the rotating moment (torque) applied to the shaft: T = 2400 lbf.in = 271164 N.mm;

– the diameter of the shaft $\varnothing = d = 1.98$ inch = 50 mm;

– the maximum admissible stress of the material: $\sigma_{adm} = 9600$ lbf/in^2 = 66 MPa.

SOLUTION WITH DISCUSSION.–

1) Let us first calculate the equivalent bending moment of the shaft:

$$M_{equivalent} = \frac{M_b + \sqrt{M_b^2 + T^2}}{2} = \frac{5600 + \sqrt{5600^2 + 2400^2}}{2} = 5846\ lbfin = 660544\ Nmm$$

2) Calculate the stress applied to the shaft: $\sigma_{applied} = M \times c/J$:

c = (d/2) is the lever arm. It is the distance from the axis of the shaft (neutral fiber) to the generator of the shaft (r). $J = \pi d^4/64$ mm^4, for a homogenous, round, solid cross-section. By applying the above relation, we are able to write:

$$\sigma_{applied} = Mc/J = 5846(1.98/2)/\pi\left(1.98^4/64\right) = 7672\ psi = 53\ MPa$$

We deduce that the stress applied to the shaft (53 MPa) is less than the maximum admissible stress (66 MPa). The shaft is therefore safe. In addition, by calculating the equivalent maximum bending moment, we can write:

$$\begin{cases} M_{equivalent-Max} = Mc/J = \dfrac{7672\pi 1.98^4/64}{1.98/2} = 660544.25\ Nmm = 5846.309\ lbfin \\ M_{equivalent-Max} - M_{équivalent} = (5846.309) - (5846.309) = 0\ lbf.in = 0\ Nmm \end{cases}$$

Thus, we declared the safety of the shaft to be proven. Hence, the ideal torque would be:

$$T_{ideal} = M + \sqrt{M^2 + T^2} = 5600 + \sqrt{5600^2 + 2400^2} = 11693\ lbfin = 1321\ 089\ Nmm$$

CONCLUSION AND COMMENTS.–

We are within a safe range of operation even towards 11,693 *lbf.in*. The initial torque (given) is underestimated (T = 2400 lbf.in), because according to our SOM calculations, it is possible to apply an *ideal* torque of the order of 11,693 lbf.in, and still remain within the safe zone. Therefore, we need to *conceptualize* the choice of material for the shaft in order to preserve ideal optimization between the material and the stress to which it is subjected. This calculation procedure is valid for any shaft (subject to similar stresses), with a homogeneous cross-section, made of steel, aluminum, brass, bronze, etc.

5.9.1. *Solved case study: bending and torsion of a shaft*

Consider a steel shaft with the diameter = 1.98 inch (50 mm), subject to a maximum bending moment M_b = 5600 lbf.in = 632715 N.mm. The torque (moment) applied is T = 2400 lbf.in (T = 271164 N.mm). Does the shaft illustrated below conform to the safety requirements if the maximum admissible stress for the material is σ_{adm} = 9600 psi = 66 MPa?

where:

– length of the shaft, L = 1.75 ft = 4.8 m;

– relative weight (kg/m) of the shaft, Ws= 94.5 lb/in = 140.63 kg/m;

– RPM of the shaft, RPM = 500 rpm;

– radius of the pulley, r_p = 20 in = 508 mm;

– weight of the pulley, W_p = 150 lb = 68.04 kg;

– radius of the wheel (pinion), r_g = 10 in = 254 mm;

– weight of the wheel, W_g = 48 lb = 21.77 kg;

– power applied to the gear system (wheel) h = 75 HP = 55927.49 Watts;

– admissible stress for the material [σ] = 8200 lbf/in^2 = 56.54 MPa.

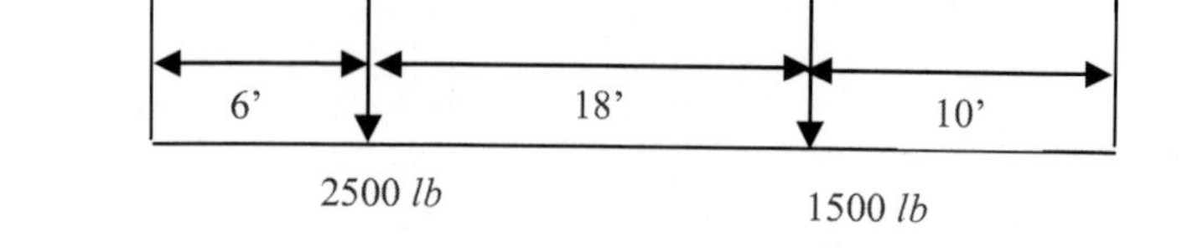

Figure 5.34. *Solid shaft subject to bending*

Calculations of the respective loads on the pulley and the gear system

From the below relation, we calculate the moment applied to the wheel of the pulley:

$$T_{applied} = h/2\pi R = 9454\ lbf.in = 1068136\ N.mm$$

We admit that the maximum tension (tight side) on the belt is twice that of the loose side. Therefore, let us set $R_p = 3T/r_p = 1418$ lbf = 6308 N. The total concentrated load on the pulley is then calculated as follows:

$$T_{total-pulley} = R_p + W_p \times g = 1568.07\ lbf = 6975.13\ N$$

Total concentrated load (gear system) $F_g = T/r_g = F_{tot-wheel} = 945.38\ lbf =$ $4205.26\ N$.

The total concentrated load on the gear and the pulley is thus calculated depending on the application of the forces: $F_g + W_g \times g = F_{total} = 993.38\ lbf =$ $4418.78\ N$.

Calculations of the reactions on the horizontal axis (x-x) RA and RB

Lever arm (8+10) = 28 in and 10 in →

$$L \times R_B - (F_{total.p} \times 28) - W_g \times g(L \times (L/2)) - (F_{total.g} \times 10) = 0.$$

Isolate R_B and R_A by solving the following system of equations with two unknowns:

$$\begin{cases} R_B = \left(48F_{total.p} + W_g g \, L^2/2 + 20F_{total.g}\right)/2L = 3764.35\ lbf = 167744.65\ N \\ R_A = \left(12F_{total.p} + W_g g L^2 + 48F_{total.g}\right)/2L = 2855\ lbf = 12701.00\ N \end{cases}$$

In summary: $R_B + R_A = 6619.62\ lbf = 29445.52\ N$

This increases the forces to: $F_{total.g} + F_{total.p} + W_g \times g \times L = 4049.83\ lbf = 18014.52.$

Calculations of the reactions on the vertical axis (y-y) VA and VB

Lever arm (108+10) = 28 in and 10 in. Similarly to the method applied above, we calculate the vertical reactions. The load of the pulley being 6 (*ft*) from the left side: $V_{6B} = R_B = 3764.35\ lbf = 16744.65\ N.$

On the right-hand side of the pulley, we set:

$$V_B = R_B - \left(6 \times W_g \times g\right) = 3291.85\ lbf = 14642.86\ N.$$

On the right side, we would have: $V_{6B} = V_{6A} - F_{total.p} = 1723.78\ lbf = 7667.74\ N.$

The gear system is loaded from the lever arm (24 in) with L = 15.75 in.

$$V_{24B} = V_{6A} - \left[L - (6' + 10')\right] \times W_g \times g = 1463.9\ lbf = 6511.75\ N.$$

On the right-hand side of the load, we have:

$$V_{24B} = V_{24A} - F_{total.g} = 470.52\ lbf = 2092.98\ N.$$

On the left-hand side of the load, we have:

$$V_{34A} = V_{24B} - 10' \times W_g \times g = -285.48\ lbf = -1269.88\ N.$$

On the right-hand side of the load (shaft), we have:

$$V_{34B} = V_{34A} + R_B = 3000.00\ lbf = 11430.99\ N.$$

Calculations of the *maximum bending moment* of the shaft: *Pp* is on the positive side of the equilibrium condition. V_{6A} in that shaft and *x* the lever arm (imagined) from V_{6A} where the shear is null (= 0). Reading from the graph of the maximum bending moment, let us set: $Pp - x\left(W_g g\right) = 0.$

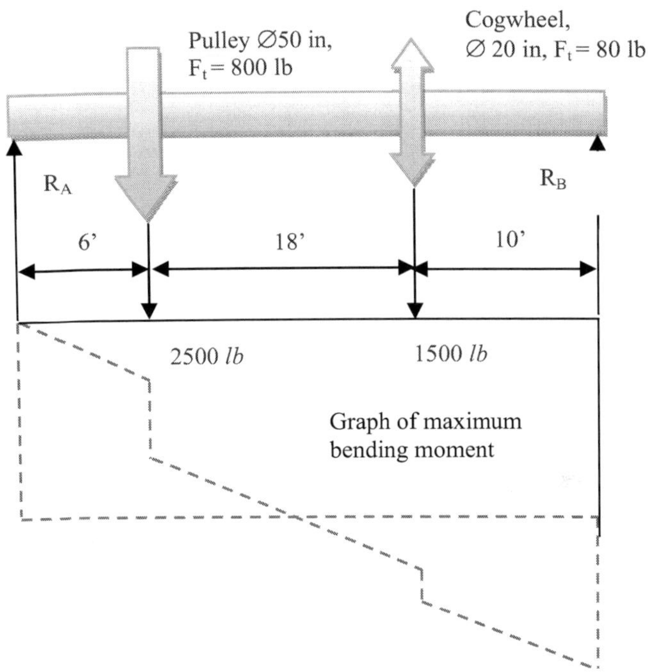

Figure 5.35. *Graph of bending moment of the main shaft*

$Pp = V_{6A}$; *at* $V_z = 6\,ft$, we would therefore have $x = Pp/W_g g = 18.24\ ft = 5.56\ m$

Thus, the expression of the maximum bending moment is written as:

$$M_{bending.\max} = V_A F_{total-p} x - W_g g\left(V_z + x\right)\frac{V_z + x}{2} = 34883\ lbf.ft = 47295\ Nm$$

Required diameter of the shaft

By virtue of the relation of the equivalent moment, in conditions of shear and bending, we set:

$$M_{equivalent} = \sqrt{M_{bending}^2 + \tau_{torsion}^2} = 34891.90\ lbf.ft = 47307.06\ N.m$$

The required diameter is: $d_{required} = 1.720 \times \sqrt{M_{equivalent} / [\sigma]} = 6.4\ in \cong 162.08\ mm$

→ We consider a diameter of 6.5 in or 165 mm.

COMMENT.– This approach can be used for any homogeneous solid with a uniform cross-section (steel, lightweight alloys, etc.). Upon reading the graph, we see that the dangerous section (0) is around the middle, where the maximum shear (torsion) moment applies. The results clearly confirm the intuitive hypothesis made. The choice of material, in design, is extremely important, as it affects the value of the admissible stress [σ].

– Choosing a higher stress is tantamount to a needless investment, because the value used in our calculations is chosen from established SOM tables used by manufacturers (metallurgy).

– Choosing a lowing admissible stress would leave our shaft at risk of premature failure, which is to be avoided.

It is not always necessary to build a prototype, as that would drive up the cost of the design. Robust finite-element simulation software packages are sufficiently reliable to confirm or invalidate our calculations.

5.9.2. *Case study 3: equivalent bending moment and ideal moment on a shaft*

Initial data	Imperial unit	SI unit
– d, diameter of steel shaft	2 in	50 mm;
– [σ], admissible stress for the material	9000 *lbf*/in^2	62 MPa;
– T, maximum moment (torque) on the shaft	2000 lbf.in	225970 N.mm;
– M_{fl}, maximum bending moment on the shaft	5600 lbf.in	M_f = 632715.

QUESTIONS.–

Calculate the "ideal" moment of the shaft to perform the required task.

1) The equivalent bending moment, applied to a solid shaft, is:

$$M_{equivalent} = 0.5\left[M_{bending} + \sqrt{M_{bending}^2 + T^2}\right] = 5773.2\ lbf.in = 652285{,}6\ \text{N}.mm$$

2) Calculation of the stress applied in psi and MPa. J is the quadratic moment.

$$\begin{cases} \sigma = M.c/J = M(d/2)/J\text{, round solid section, } J = \pi d^4/64 = 1\ in^4 = 31\ cm^4 \\ \sigma = M(d/2)/J = 7576\,lbf/in^2\ (psi) = 52\,N/mm^2 \prec 62\ [MPa] \end{cases}$$

3) The calculated stress is less than the admissible stress (52 MPa < 62 MPa), so the shaft is reliable and will work safely. Calculate the equivalent maximum moment $M_{maxi\text{-}equiv}$. $M_{\max} = 652285.6\ Nmm \rightarrow M_{\max} - M_{equiv} = 0.00\ lbfin = 0.0\ Nmm$

The shaft is therefore reliable and the state is safe.

4) The ideal moment on the solid shaft is $M_{fl} = 5600\ lbf.in = 632715\ Nmm$

$$T_{ideal} = M_{bend} + \sqrt{M_{bend}^2 + T^2} = 11546\ lbf.in = 1304571\ Nmm$$

This is valid for any shaft with a round, solid cross-section, made of steel or lightweight alloys.

5.9.3. *Case studies: maximum performance of pre-stressed bi-materials*

Consider a compound of material (steel and aluminum 6061). A steel stem passing through a steel tube of the same length attaches the two parts together at both ends. The attachment is made with adjustable nuts, holding the stem and tube together. The steel is subject to the initial tension, and the aluminum to initial compression. The initial loading is such that both parts reach their admissible tensions at the same time. A and E are respectively the area of the cross-section and the elasticity modulus.

Parameter	Imperial unit in^2	SI unit mm^2
A, area of the cross-section of the steel bearing	$A_{steel} = 2$	$A_{steel} = 1.29\times10^3$
E, elasticity modulus of the steel bearing	$\mathbf{E}_{steel} = 25\times10^6$ lbf/in²	1.724×10^5 MPa
[σ] stress admissible of the steel bearing	$[\sigma]_{steel} = 25\times10^3$ lbf/in²	124.106 MPa
A, area of the cross-section of the aluminum stem	$A_{alu} = 2.5\ in^2$	$1.613\times10^3\ mm^2$
E, elasticity modulus of the aluminum stem	$\mathbf{E}_{alu} = 12\times10^6$ lbf/in²	8.274×10^4 MPa
[σ], stress admissible of the 6061 stem	$[\sigma]_{alu} = 15\times10^3$ lbf/in²	103.421 MPa

Table 5.5. *Results of calculations of the design parameters, both in imperial and SI units*

When $[\sigma]_{adm}$ is reached, the force **P** in the aluminum 6061 and the steel is written as:

$$\begin{cases} P_{steel} = A_{steel} \times \left[\sigma\right]_{steel} = 3.6\times10^{4}\ lbf = 1.601\times10^{5}\ N \\ P_{alum} = A_{alum} \times \left[\sigma\right]_{alum} = 3.75\times10^{4}\ lbf = 1.668\times10^{5}\ N \end{cases}$$

Total load on the components: $P_{total} = P_{steel} + P_{alum} = 3.269\times10^{5}\ N.$

Calculate the initial tension or compression in the components and consider the tension (+) and the compression (-). The initial tension on the steel bearing is:

$$P_{initial/steel} = P_{steel} - P_{total}\left(\frac{A_{steel} \times E_{steel}}{A_{steel}E_{steel} + A_{alum}E_{alum}}\right) = -\ 9.937\times10^{3}\ lbf = -\ 4.42\times10^{4}\ N$$

The initial compression on the 6061 aluminum tube is:

$$P_{initial/alum} = P_{alum} - P_{total}\left(\frac{A_{alum}E_{alum}}{A_{steel}E_{steel} + A_{alum}E_{alum}}\right) = -\ 9.937\times10^{3}\ lbf = -\ 4.42\times10^{4}\ N$$

If the bi-material were not pre-stressed, the stress in the steel would still be just three times that in the aluminum, because it would withstand the same strain. Its elasticity modulus is three times as high.

5.9.4. *Case study: deflection and buckling of elements of machines*

PROBLEM.–

Calculate the horizontal deflection of the point A/*OX*, if diameter of a steel piece Dex = 2 in with an internal radius r_{in} = 4 in, as shown by the data in Figure 5.36. At any angle φ, the bending moment, with respect to the axis of the center of gravity of the given form (CG) is calculated below. E_M is the energy due to the bending moment M_{bend}.

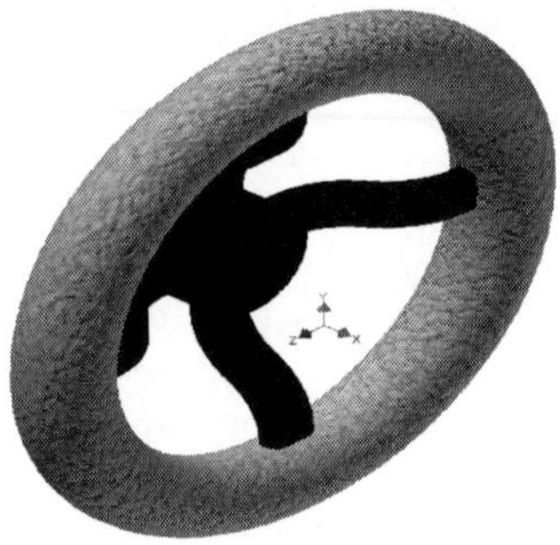

- $P = 185 lbf = 822.921$ N
- $A_{portionJ} = \pi$
- $r_{in} = 3.56$ in $= 90.424$ mm
- rex $= 5.56$ in $= 141.224$ mm
- Dex $= 2.25$ in $= 57.15$ mm
- $R_{mean} = 4.56$ in $= 115.824$ mm
- $\varepsilon =$ in and mm, to be calculated
- $E_{steel} = 28\times10^6$ psi $= 1.931\times10^6$ MPa

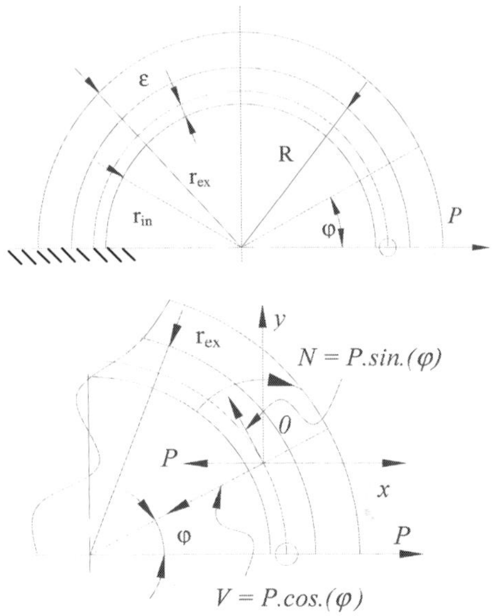

Figure 5.36. *Diagram and 3D drawing of calculations. For a color version of this figure, see www.iste.co.uk/grous/design.zip*

SOLUTION.–

Let us symbolically evaluate the expression of the energy E_M created by the bending moment $M_{bending}$ $M_{bend} = P\times 5\sin(\varphi)$ and deduce E_m:

$$\rightarrow E_M = \int_0^{\pi} \frac{M_{bending}^2}{2A\varepsilon E} d\varphi \ \ knowing\ that\ \int_0^{\pi} \frac{P\times 5\sin(\varphi)}{2A\varepsilon E} d\varphi \rightarrow \frac{25\pi P^2}{4A\varepsilon E} \qquad [5.31]$$

For a round, solid cross-section, the excentricity (ε) is the distance from the axis of the centroid. $\varepsilon = R_{mean} - \left(\sqrt{r_{ex}} + \sqrt{r_{in}}\right)^2 / 4 = 0.055500\ in = 1.409701\ mm.$

Knowing that the differential of the energy with respect to P gives the expression of the deflection (f, mm), let us set:

$$\partial E_M / \partial P = 25\pi P / 2 A_{portionJ} \varepsilon E_{steel} = 1.488\times10^{-3}\ in = 0.037798\ mm \qquad [5.32]$$

We can see that the deflection found by this method of calculation is controllable in terms of mechanical design (around 4/100 of a mm or 1/1000 of an inch).

5.10. Case studies using the Castigliano method

STATEMENT OF PROBLEM 1.–

Consider the frame illustrated below (Figure 5.37). Our goal is to calculate the horizontal displacement over the entire time the load is applied, considering that the form is identical on the three sides, so the moments of inertia in each section are the same. The resulting deflection is related exclusively to the bending moment.

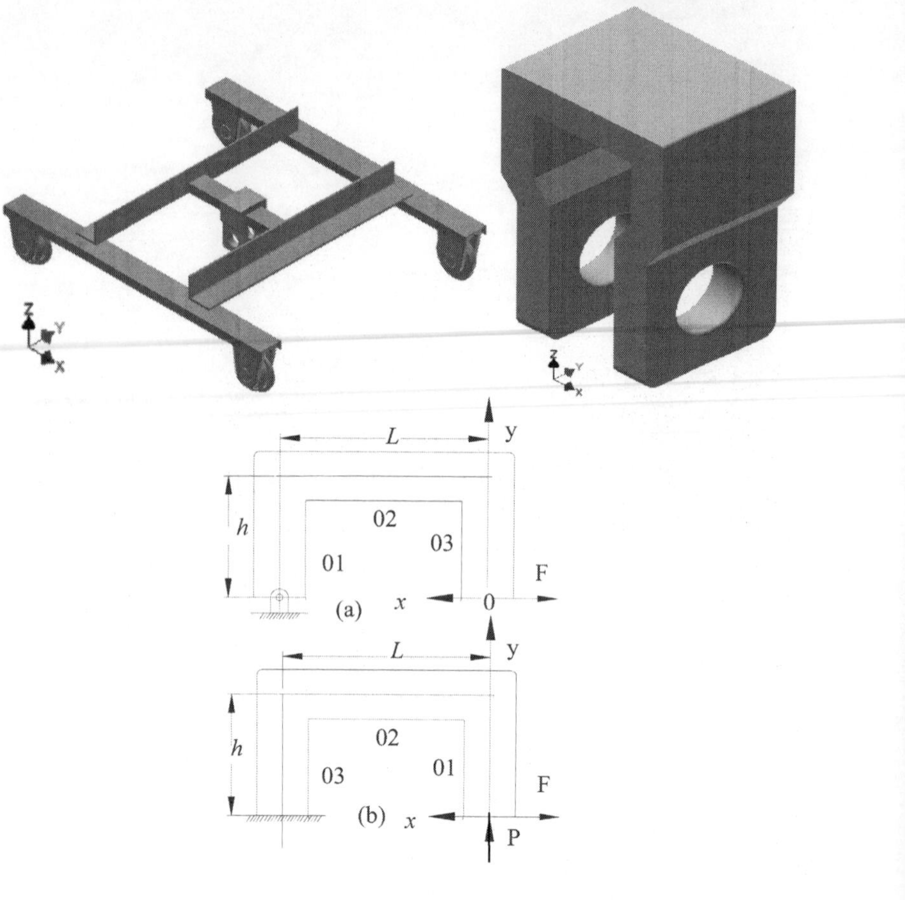

Figure 5.37. *Diagrams and 3D drawings of the frame to be calculated. For a color version of this figure, see www.iste.co.uk/grous/design.zip*

SOLUTION.–

Using Castigliano's theorem, the energy of strain of the frame corresponds to the sum of the energies of the three component parts of the structure.

The energies of members (1, 2, 3) E, E2 and E3, and the total energy on the frame. The sum of the energies is therefore written:

$$\begin{cases} E_{\{1-3\}}^{vert} = 2\int_0^h \frac{1}{2}\frac{M^2}{EJ}dy = 2\int_0^h \frac{1}{2}\frac{(F\times y)^2}{EJ}dy = \frac{F^2h^3}{3EJ} \\ E_{\{2\}}^{hori} = \int_0^L \frac{1}{2}\frac{M^2}{EJ}dx = \int_0^L \frac{1}{2}\frac{(F\times h)^2}{EJ}dx = \frac{F^2h^2L}{2EJ} \end{cases} \quad [5.33]$$

The total energy would be: $E_{total} = \left\{2\frac{F^2y^3}{6EJ} + \frac{F^2h^2L}{2EJ}\right\} \Rightarrow \Delta L = ?$ [5.34]

The horizontal displacement, in the direction of the force F, is:

$$\Delta f = \frac{\partial E_{total}}{\partial F} = \left\{\frac{2}{3}\times\frac{Fy^3}{EJ} + \frac{Fh^2L}{EJ}\right\} \text{ in the direction of } \vec{F} \quad [5.35]$$

APPLICATION.–

F = 156 lbf = 693.923 N; E = 3×10^7 psi = 2.068×10^5 MPa; b = 0.5 in = 12.7 mm; h = 2 in = 50.08 mm; $J_{xx}= bh^3/12 = 0.33$ in^4 = 1.387×10^5 mm^4; L = h = 2 in = 50.08 mm.

Calculation of the total energy in Joules

$$\begin{cases} E_{\{1-3\}}^{vert} = 2\int_0^h \frac{1}{2}\frac{(Fy)^2}{EJ}dy \rightarrow \frac{F^2h^3}{3EJ} = 7.332\times10^{-4}\ J \\ E_{\{2\}}^{hori} = \int_0^L \frac{1}{2}\frac{(Fh)^2}{EJ}dx \rightarrow \frac{F^2h^2L}{2EJ} = 1.100\times10^{-3}\ J \end{cases} \rightarrow E_{total} = E_{\{1-3\}}^{v} + E_{\{2\}}^{h} = 1.833\times10^{-3}\ J$$

Calculation of the deflection with respect to OX

$$\Delta f = \frac{2F^2h^3}{3EJ} + \frac{Fh^2L}{EJ} = 5.283\times10^{-3}\ mm$$

Considering the horizontal and vertical deflections at the level of the operated layer, as shown in Figure 5.37b:

P = 120 lbf = 533.787 N; a = 0.5 in = 12.7 mm, and therefore A = a_2 = 0.25 in^2 = 161.29 mm^2.

Energies of members (1, 2 and 3) E, E2 and E3, and the total energy on the frame: the energy distributed across each component member of the frame (b) is:

$$\left\{ E_{\{1\}} = \int_0^h \frac{1}{2}\frac{(Fy)^2}{EJ}dy + \int_0^h \frac{1}{2}\frac{P^2}{AE}dy \rightarrow \frac{F^2h^3}{6EJ} + \frac{P^2h}{2EA} = 5.835\times10^{-4}\ J \right.$$

$$\begin{cases} E_{\{2\}} = \int_0^L \frac{1}{2}\frac{(Px+Fh)^2}{EJ}dx + \int_0^L \frac{1}{2}\frac{F^2}{AE}dx \rightarrow \left\{ \frac{P^2L^3}{6EJ} + \frac{PL^2Fh}{2EJ} + \frac{F^2h^2L}{2EJ} + \frac{F^2L}{2AE} \right\} = \\ = 2.529\times10^{-3} J \ or\ indeed\ E_{\{2\}}\frac{F^2L}{2EA} + \frac{L\left[3F^2h^2 + 3FLPh + L^2P^2\right]}{6EJ} = 2.529\times10^{-3}\ J \end{cases}$$

$$\begin{cases} E_{\{3\}} = \int_0^h \frac{1}{2}\frac{(PL\times Fy)^2}{EJ}dy + \int_0^h \frac{1}{2}\frac{P^2}{AE}dy \rightarrow \left\{ \frac{P^2L^2h}{2EJ} + \frac{PLFh^2}{2EJ} + \frac{F^2h^3}{6EJ} + \frac{P^2h}{2AE} \right\} = \\ = 2.08\times10^{-3} J \ or\ indeed\ E_{\{3\}} = \frac{h\left(F^2h^2 + 3FLPh + 3L^2P^2\right)}{6EJ} + \frac{P^2h}{2AE} = 2.08\times10^{-3}\ J \end{cases}$$

$$E_{total} = E_{\{1\}} + E_{\{2\}} + E_{\{3\}} = \left\{ \begin{array}{l} \int_0^h \frac{1}{2}\frac{(Fy)^2}{EJ}dy + \int_0^h \frac{1}{2}\frac{P^2}{AE}dy + \\ + \int_0^L \frac{1}{2}\frac{(Px+Fh)^2}{EJ}dx + \int_0^L \frac{1}{2}\frac{F^2}{AE}dx \\ + \int_0^h \frac{1}{2}\frac{(PL\times Fy)^2}{EJ}dy + \int_0^h \frac{1}{2}\frac{P^2}{AE}dy \end{array} \right\} = 5.193\times10^{-3}\ J$$

Written differently, these equations still produce the same result:

$$E_{total} = \left\{ \begin{array}{l} \left\{ \frac{F^2h^3}{6EJ} + \frac{P^2h}{6AE} \right\} + \left\{ \frac{P^2L^3}{6EJ} + \frac{PL^2Fh}{2EJ} + \frac{F^2h^2L}{2EJ} + \frac{F^2L}{2AE} \right\} + \\ \quad \ldots + \left\{ \frac{P^2L^2h}{2EJ} + \frac{PLFh^2}{2EJ} + \frac{F^2h^3}{6EJ} + \frac{P^2h}{2AE} \right\} \end{array} \right\} = 5.193 \times 10^{-3}\ J$$

The horizontal deflection is determined by the partial differential of E_{total} with respect to F. The vertical deflection is determined by the partial differential of E_{total} with respect to P, as follows:

$$\begin{cases} \Delta f_x = \dfrac{\partial E_{total}}{\partial F} = \left\{ \dfrac{Fh^2}{EJ}\left(\dfrac{2h}{3} + L \right) + \dfrac{FL}{AE} \right\} = 2.496 \times 10^{-4}\ in = 6.34 \times 10^{-3}\ mm \\ \Delta f_y = \dfrac{\partial E_{total}}{\partial P} = \left\{ \dfrac{FLh}{2EJ}(h+L) \right\} = 1.248 \times 10^{-4}\ in = 3.17 \times 10^{-3}\ mm \end{cases}$$

5.11. Conclusion

In this chapter, we have presented case studies focusing specifically on machine elements and civil engineering. The dilemma which constantly arises is between the choices of materials [HER 89] to respect the admissible stresses which must not be exceeded, and the imposed geometry. Designers often need to make choices based on the needs of the technical specifications. Note that a design which is valid at one point in time may not necessarily be so in the future, in the same conditions, owing to the evolution of the characteristics of the materials and manufacturing processes.

6

Noise and Vibration in Machine Parts

6.1. Noise and vibration in mechanical systems

In theorizing about vibrations, there is a tendency to treat them as a deterministic process, as opposed to random vibrations (which are, in fact, a closer representation of reality). It is obvious that the calculations are most meaningful in mechanical systems involving dynamic processes. It is therefore of crucial importance to determine the random displacement factors in order to extract information on a system's reliability using its statistical properties [GRO 13]. Elastic suspensions are used to study noises in solids, such as soil structures. They are also of interest in suspensions for machines, and for this reason, we also measure vibration isolation, expressed in decibels, and referred to as gain. In metrology, vibration isolation is expressed [GRO 11] by the gain (*db*) = 100/[100-ε] of non-aerial and solid substances. In the area of machines, we speak of the shock (Q), which represents a variation of the applied force, also known as excitation over time (τ). The maximum force (F_M) is inversely proportional to the displacement (course). Thus, shock causes a displacement (U or stiffness) in terms of variation of velocity ΔV, whose differential gives rise to an acceleration (γ) of the mass M subject to the shock. In design, we closely study the energy absorbed by the shock absorber to limit its intensity by decreasing the acceleration. Let us express the energy E (work) in equation form:

$$E = \frac{1}{2} Q_u U^2 = \frac{1}{2} F_M U \text{ with the stiffness} : U = \sqrt{2E/Q_u} \qquad [6.1]$$

Shocks give rise to *rebounds*. For designers, choosing elastic suspension means the machine will move more smoothly. In this case, we use rebound travel *limiters* which would not *generate* too much normal oscillation in a given suspension system. The case studies presented here discuss examples of models with one or more degrees of freedom (*dof*).

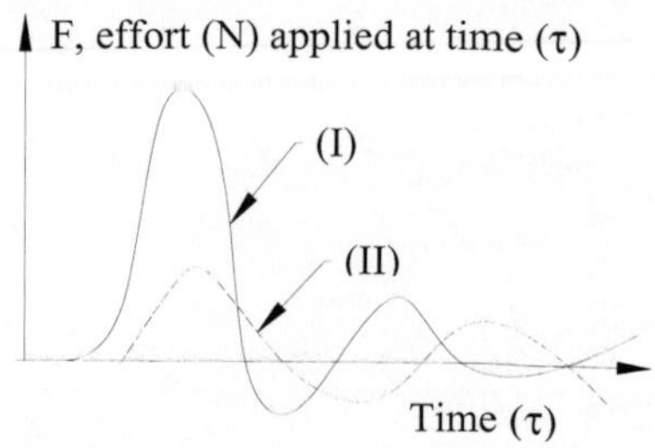

The acceleration (γ) resulting from the shock is harmful, and can even be destructive, and vibrations inevitably cause the eventual failure of machine parts. Designers need to choose the right damping mechanism to use in order to reduce rebounds and their successive amplitudes. In dynamic processes, we study possible influencing factors; sometimes the nature of such factors is less than obvious, such as unexpected, uncontrolled (i.e. random) external disturbances. For example, we could cite the state of the road surface or runway surfaces, which give rise to vibrations in the machines driving on those surfaces. We could also point to the numerous examples of vibrations in the transmissions even of stationary installations.

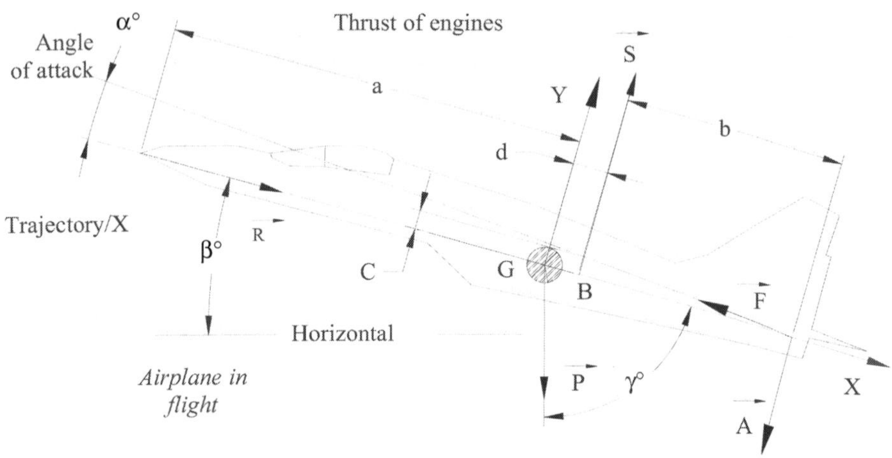

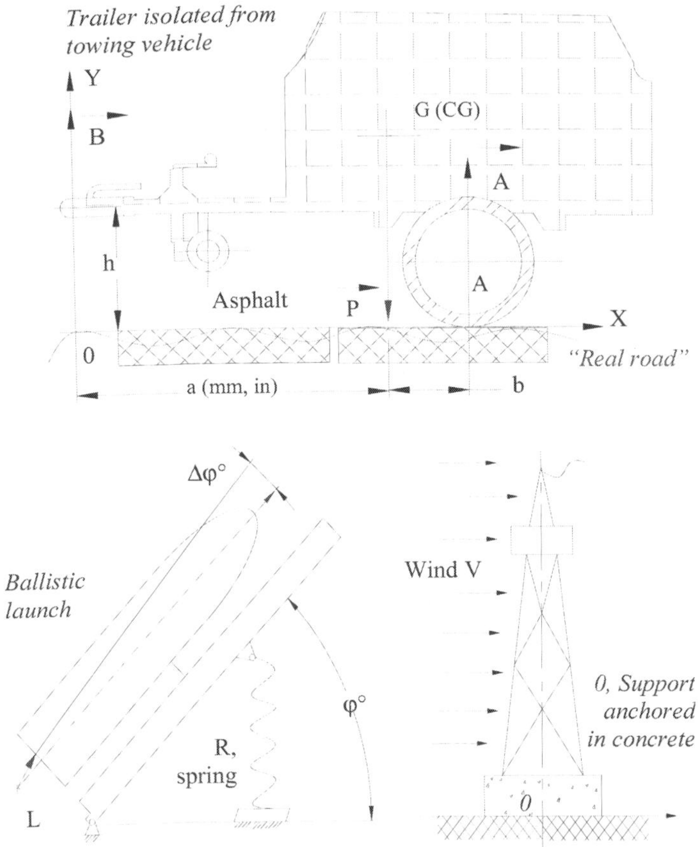

Figure 6.1. *Examples of sources of random, "uncontrolled" vibrations. For a color version of this figure, see www.iste.co.uk/grous/design.zip*

In many cases, there is a need to calculate the probability that the random function will surpass a given level, as is the case of movement of vehicles along a *road* whose surface exhibits random irregularities. It is common for the suspension to suddenly *snap*. This results in shocks.

For example, if we consider a vertical relative displacement, $y(\tau)$, as a random function, the point affecting the suspension is proportional to the probability of surpassing $y_0(\tau)$, so $y(\tau) \geq y_0(\tau)$, which represents free motion of the suspension: exceeding the given levels. Beginning with the vehicle's stationary oscillations, we set the relation between the spectral density $y_0(\tau)$ and the relative displacement $D_h(\omega)$. This expression includes $|W_y|$, which is a modulus of the transfer function between the input $H(\tau)$ and the relative displacement corresponding to the suspension. The state of the road surface can be simply formulated as:

$$D_y(\omega) = \left|W_y\right|^2 \times D_h(\omega) \; [m] \leftarrow vehicle \qquad [6.2]$$

If, for example, the distribution of y_0 were Gaussian (Weibull or log-normal, etc.), the probability that the relative displacement y_0 will exceed the value $y_0(\tau)$ – i.e. the probability that $y(\tau) \geq y_0(\tau)$ – is written as follows:

$$\begin{cases} P\{y(\tau) \geq y_0\} = \dfrac{1}{2\pi} \times \displaystyle\int_{\tau_1}^{\infty} Exp^{-\tau^2/2} \, d\tau = \Phi(\infty) - \Phi(\tau_1), \text{ where } \lrcorner \\ \sigma_{y_0} = \text{standard deviation and reduced centered value } \tau_1 = \left(y_0 - m_{y(\tau)}\right) / \sigma_{y_0} \end{cases} \qquad [6.3]$$

In this book, we shall not discuss the random vibrations to which mechanical systems are subject. Instead, we limit ourselves to the aspects pertaining purely to the design of such systems.

6.1.1. *Aerodynamism of moving mechanical bodies*

It is tricky to measure the parameters which characterize a body's air resistance, in view of random factors such as wind and pressure. In supposedly "*still*" air, objects flying at a velocity (v) encounter an opposing force (F). This is known as air resistance in motion. In aviation, ballistics and vehicle manufacture, the aerodynamism of objects plays a crucially important role, as the form has a considerable influence on the force to which the object is subject. Experience has demonstrated the advantage of so-called aerodynamic forms, which offer less resistance to their *air penetration*. These forms help reduce the amount of power needed to maintain their *displacements*. Here, we shall quite deliberately limit the discussion to basic applications, where $P/T = Cz/Cx$ is the fineness of the plane in terms of breadth and depth:

$$\begin{cases} R = C_x \rho A v^2/2 \text{ or } R = \kappa A v^2 \; [N \text{ or } lbf] \leftarrow vehicles \\ P = C_z \rho A v^2/2; \; T = C_x \rho A v^2/2 \; [N \text{ or } lbf] \leftarrow aviation \end{cases} \qquad [6.4]$$

The fineness depends solely on the angle of incidence, for a plan with a given breadth and depth.

Incidence in degrees	C_z	C_x	Fineness
(– 2.50)°	10.50×10^{-2}	1.01×10^{-2}	10.3
(+ 0.60)°	32.20×10^{-2}	1.46×10^{-2}	22.0
(+ 3.60)°	54.40×10^{-2}	2.37×10^{-2}	22.9

Table 6.1. *Fineness of the plane of breadth and depth*

– R is the air resistance in N;

– C_x is the body's drag coefficient (air penetration), dependent on the form;

– ρ is the air density in the specific temperature and pressure conditions, kg/m^3;

– A is the area of the cross-section (frontal area) in m^2;

– v is the velocity in m/s;

– κ is a coefficient which depends on the form of the body and the density of the air. Therefore, in aviation – or in ballistics – κ depends on the altitude, temperature and water content of the air;

– T is the drag (N) and P the lift (N). C_x is the drag coefficient of the wing in question.

Design is a compromise between choice and calculations, which is often difficult to strike. In aviation, for example, in the design of machines propelled by enormous forces, the structure must be as lightweight as possible, but at the same time stiff, so as to preserve the aerodynamics of the wings. *Safety*, strength and sometimes even cost are of little importance in relation to previous records. The transverse longeron supports the wings. The longitudinal longeron supports the tail. Both longerons are subject to bending stress. Torsion (twisting) is less than the bending stress. Latest-generation airplanes use epoxide–carbon-fiber composites, shaped by casting.

Once again, in aviation just as in other areas, the choice of material is of critical importance for the next stages of modeling. Thus, we attempt to find an optimal compromise between the material and the geometry, with a view to minimizing weight whilst retaining stiffness and bending capability which will withstand the hazards the machine will encounter. This balance is formulated thus:

$$i\left\{\begin{matrix} airplanes \\ performance \end{matrix}\right\} = \frac{\sqrt{\Psi\left\{\begin{matrix} \text{exp.} \\ stiffness \end{matrix}\right\} E}}{\rho} \quad and \quad i\left\{\begin{matrix} airplanes \\ performance \end{matrix}\right\} = \frac{\sqrt[3]{\Psi\left\{\begin{matrix} \text{exp.} \\ resistance \end{matrix}\right\} E}}{\rho} \qquad [6.5]$$

Woods (such as balsa wood and spruce) are excellent candidates in terms of performance index for reduced-scale models.

6.2. Case study 1

6.2.1. *Lightweight vehicles and trucks*

Consider a vehicle moving at a variable velocity v = between 50 and 120 km/h. The air resistance coefficient is C_x = ½ and the air density is ρ = 1.225 kg/m^3.

The frontal area is A = 3 m^2. Using R(v) = [$C_x.r.A.v^2$]/2, in (N), let us calculate the air resistance for v = 50 kph, 60 kph, …, 120 kph.

$$R(v) = \begin{pmatrix} 127.013 \\ 182.899 \\ 248.946 \\ 325.154 \\ 411.523 \\ 508.054 \\ 614.745 \\ 731.597 \end{pmatrix} N$$

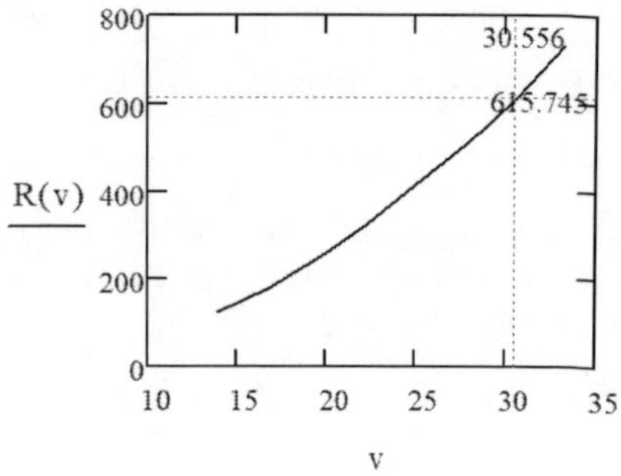

Figure 6.2. *Evolution of the resistance (power consumed) as a function of the velocity. For a color version of this figure, see www.iste.co.uk/grous/design.zip*

The velocity of displacement has an exponential influence on the resistance, and therefore considerable energy consumption. Every second at the velocity v = 30 m/s, the resistance force R produces resistance work which absorbs the powers:

$$R(v) = C_x \times \rho \times A \times v^2/2 \ \ and \ \ P(v) = R(v) \times V \ \ [W] \qquad [6.6]$$

In airplanes flying at very great speeds, the fuselage and the rest of the projecting parts must have a profile which is adapted to offer the least resistance during flight which will therefore save on fuel. In the solved problem below, we examine a simple case of aviation.

APPLICATION.– Consider a vehicle moving at a variable velocity v = (225–625) km/h. The air resistance coefficient is C_x = ½ and the density of the air ρ = 1.225 kg/m^3. The frontal area is A = 3 m^2.

R(v) =

	0
0	$3.589 \cdot 10^3$
1	$3.75 \cdot 10^3$
2	$3.915 \cdot 10^3$
3	$4.083 \cdot 10^3$
4	$4.255 \cdot 10^3$
5	...

N

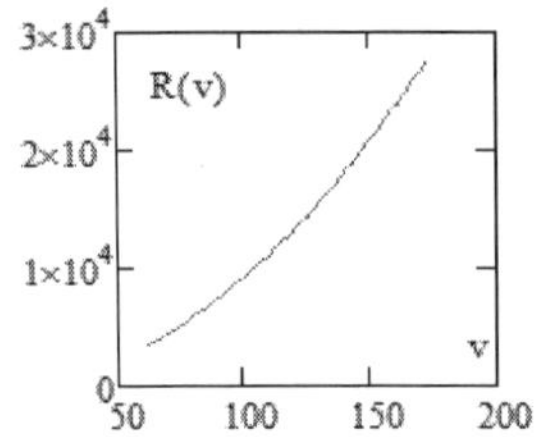

Figure 6.3. *Evolution of the resistance for an airplane (Inventor Pro 2017). For a color version of this figure, see www.iste.co.uk/grous/design.zip*

6.2.2. *Case study 1*

Consider an airplane during its ascent at a constant velocity.

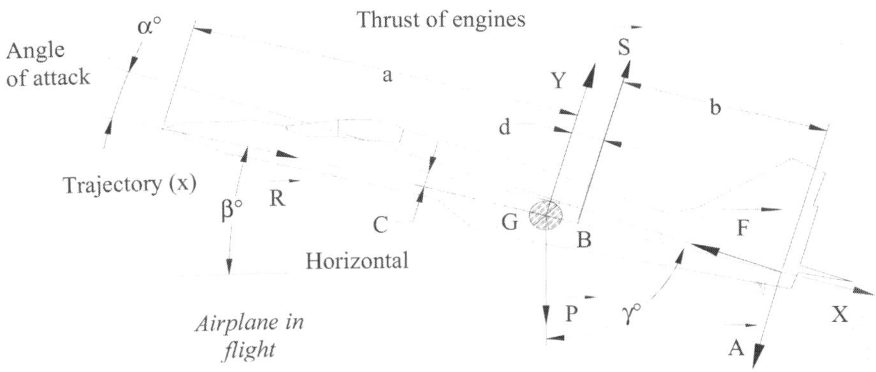

Figure 6.4. *Approximate simulation of stress zones. For a color version of this figure, see www.iste.co.uk/grous/design.zip*

The thrust P_{reac} of the reactors at [φ], where the polygonal representation of forces in action gives rise to the action of the air resistance $\vec{R}_{air}$ of the order of 5.54×10^4 N = 1.145×10^4 *lbf*. This is the air resistance (including Cx) of the whole structure. That structure is irregular, because of the required aerodynamics, which is why sweep diagrams are used. At the wings, $\vec{\omega}$ is the resultant of the forces of lift, while $\vec{S}$ is the resultant of the stabilizing forces.

Here:

– $\varphi = 5.54°$ is the angle of the airplane's flight;

– $\beta = 73° = (\pi/2 - 17°)$ is the complementary angle of the airplane in flight;

– C_x is the air penetration coefficient (*aerodynamism*);

– P_{reac} is the thrust of the reactors of the hydroplane in N or *lbf*;

– R_{air} is the air resistance across the whole structure = 5.54×10^5 N = 1.245×10^4 *lbf*;

– ω_{sust} is the resultant of the forces of lift on the wings of the hydroplane;

– A_{stab} is the resultant of the forces of stabilization on the hydroplane's rear aileron;

– W is the weight of the hydroplane when loaded = 2.54×10^5 N = 5.71×10^4 *lbf*;

– a = 1 m = 39.37 in; b = 5.52 m = 206.693 in and c = 10 m = 393.701 in.

The diagram of the vectors represents the respective forces: $\left\{\vec{\omega}_{sust}, \vec{A}_{stab}, \vec{P}_{reac}\right\}$. In this context, let us present the solution approach using the classic graphic-analytical method (vectors) known as the Culmann method, which is also based on the polygonal representation of the vectors to find the unknown forces, in the (XYZ) frame of reference. Looking at the diagram below, we see that the hydroplane is subject to five separate external forces:

$$\left\{\vec{W}_{weight}, \vec{R}_{air}, \vec{\omega}_{sust}, \vec{A}_{stab}, \vec{P}_{reac}\right\}$$

For ease of calculation, let us set out the main equations governing the condition of equilibrium of the system in flight, or the total equilibrium equation:

$$\begin{cases} \left\{\vec{W}_{weight}, \vec{R}_{air}, \vec{\omega}_{sust}, \vec{A}_{stab}, \vec{P}_{reac}\right\} = \vec{0} \downarrow \\ W_{weight} + R_{air} + \omega_{sust} + A_{stab} + P_{reac} = 0 \end{cases} \rightarrow \qquad [6.7]$$

Calculation method in comparison to Culmann graphical-analytical method:

$$\begin{cases} \textit{Condition of equilibrium in flight}: W_{weight} + R_{air} + \omega_{sust} + A_{stab} + P_{reac} = 0 & [00] \\ \textit{With respect to } 0x \textit{ axis}: W_{weight}\cos(\beta) + R_{air} + 0 - P_{reac}\cos(\varphi) + 0 = 0 & [01] \\ \textit{With respect to } 0y \textit{ axis}: -W_{weight}\sin(\beta) + 0 + \omega_{sust} + P_{reac}\sin(\varphi) - A_{stab} = 0 & [02] \\ /G\ (CG): M_{CG}\vec{W}_{weight} + M_{CG}\vec{R}_{air} + M_{CG}\vec{\omega}_{sust} + M_{CG}\vec{P}_{reac} + M_{CG}\vec{A}_{stab} = 0 & [03] \end{cases} \quad [6.8]$$

From equation [6.8-01], with respect to the (OX) axis, we set:

$\mathrm{W_{weight}}\cos(\beta) + \mathrm{R_{air}} + 0 + \mathrm{P_{reac}}\cos(\varphi) + 0 = 0 \rightarrow$ let us deduce the value of $\mathrm{P_{reac}}$:

$$W_{weight}\cos(\beta) + R_{air} + P_{reac}\cos(\varphi) = 12966N - 1.0P_{reac}\cos(5.54\deg)) = 0$$

Hence: $P_{reac} = 129662.41299N/\cos(5.54\deg)) = 1.303\times10^5\ [N] = 2.929\times10^4\ [lbf]$

From equation [6.8-02], with respect to the (OY) axis, we set:

$$W_{weight}\sin(\beta) - P_{reac}\sin(\varphi) = 2.303\times10^5\ N = 5.178\times10^4\ lbf$$

Knowing that:

$$(\omega_{sust} - A_{stab}) = W_{weight}\sin(\beta) - P_{reac}\sin(\varphi) = 2.126\times10^5\ N = 4.779\times10^4\ lbf$$

or from equation [6.8-03], we set: $a\omega_{sust} + (a+b)P_{reac}\sin(\varphi) - (a+b)A_{stab} = 0$

$$\left(aW_{weight}\sin(\beta) - P_{reac}\sin(\varphi) + A_{stab}\right) + (a+b)P_{reac}\sin(\varphi) - (a+b)A_{stab} = 0 \rightarrow \textit{simplify}$$

$$\rightarrow 242901.40801461101292\ Nm - 5.25\ mA_{stab} + 683922.23972868075\ Nm\sin(5.54\deg) = 0$$

$$A_{stab} = \left[242901\ Nm + 683922\ Nm\sin(5.54\deg)\right]/5.25\ m = 5.884\times10^4\ N = 1.323\times10^4\ lbf$$

In addition: $\omega_{sust} = W_{weight}\sin(\beta) - P_{reac}\sin(\varphi) + A_{stab}\textit{factor} \rightarrow$

$$\omega_{sust} = \frac{301744780083780612917\ \mathrm{N}}{1000000000000000} - \frac{130270902805463\ \mathrm{N}\times\sin(277\deg/50)}{1000000000} \swarrow$$

$\omega_{sust} = 2.892 \times 10^5\ N = 6.501 \times 10^4\ lbf$, or using the following manual calculation:

$$\omega_{sust} = W_{weight} \sin(\beta) - P_{reac} \sin(\varphi) + A_{stab} = 2.892 \times 10^5\ N = 6.501 \times 10^4\ lbf \ \rightarrow QED$$

On a simulated example, we have calculated the resultant of the stabilizing forces on the hydroplane's rear aileron: A_{stab}; the resultant of the forces of lift on the wings: ω_{sust}; and the thrust of the reactors in N and *lbf*: P_{reac}:

$$\text{Overview:} \begin{pmatrix} A_{stab} = 5.884 \times 10^4\ N = 1.323 \times 10^4\ lbf \\ \omega_{sust} = 2.892 \times 10^5\ N = 6.501 \times 10^4\ lbf \\ P_{reac} = 1.303 \times 10^5\ N = 2.929 \times 10^4\ lbf \end{pmatrix}$$

6.2.3. *Case study of the rotor blade of a fire brigade helicopter*

STATEMENT OF THE PROBLEM.– Consider a helicopter whose rotor blade spins at an RPM of 1200 (rpm; *revolutions* = *rotations* = *1*); $R_1 = 0\beta = 0\alpha = 1.25$ m and $0\delta = R_1/2 = 0.625$ m. We wish to find the respective velocities V_α, V_β and V_δ and to analyze the movements of the rotor blade, in order to steer the helicopter with greater efficiency.

SOLUTION WITH DISCUSSION.–

$$1) \begin{cases} \textit{Angular velocity}: \ \omega = 2\pi\ RPM = \pi \times 1200/30 = 125.664\ rad/s \\ \textit{Linear velocities}: \ V_\alpha = V_\beta = \omega R_1 = 157.08\ m/s \ \textit{and}\ V_\delta = \omega(R_1/2) = 78.54\ m/s \end{cases}$$

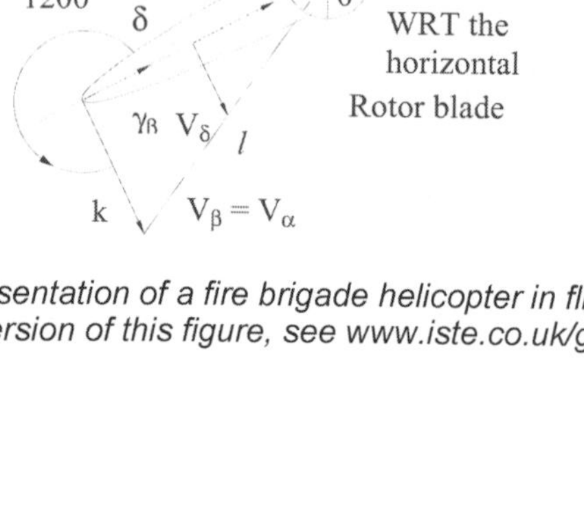

Figure 6.5. *Representation of a fire brigade helicopter in flight (lab simulation). For a color version of this figure, see www.iste.co.uk/grous/design.zip*

For ω = constant = 0, thus φ = 0, so the acceleration is written as: $\gamma_\tau = 0$:

$$\begin{cases} \gamma_\tau = \varphi \times R_1 = 0 \; and \; \gamma_N = \omega^2 R_1 = 1.974 \times 10^4 \left[m/s^2 \right], \rightarrow \\ As \; \vec{\gamma}_{N\beta} = \vec{\gamma} = \vec{\gamma}_N \rightarrow \vec{\gamma}_{N\delta} = \omega^2 \; R_1/2 = 9.87 \times 10^3 \left[m/s^2 \right] \end{cases}$$

$$2) \begin{cases} \Phi = 2\pi \times 60 = (\rho - \rho_0); \; \omega_0 = 0 \text{ equation of motion} \rightarrow \\ \omega^2 = \omega_0^2 + 2\varphi_1 (\rho - \rho_0) = 0 \rightarrow \varphi_1 = -\left(\omega^2 - \omega_0^2\right) / \left[2(\rho - \rho_0) \right] = -20.944\left(1/s^2\right) \end{cases}$$

The deceleration is: $\gamma_\tau = \varphi_1 \times R_1 = -26.18 \; m/s^2$. The equation of motion is written thus:

$$\gamma_M = (\gamma_\tau/2)\tau^2 + \omega\tau \; and \; \omega_M = \gamma_\tau \tau_M + \omega, \; where \; \omega_M = 0 \rightarrow \tau_M = ?$$

$$\gamma_\tau \tau_M + \omega = 0, \; simplify \rightarrow 125.66s - 26.18.m.\tau.M / s^2 \rightarrow \tau_M = 125.66.s/26.18 = 4{,}8s$$

COMMENTS.– During flight, the velocity V_α and acceleration $a_{N\alpha}$ do not vary as a function of the deceleration. Therefore, we can write: $0 \le V_\alpha \le 157.08$ and $0 \le \gamma_\alpha \le 1.974 \times 10^3$. If the helicopter's rotor blade were spinning at 60 rpm to begin with and then climbed to 1200 rpm, the results would simply be reversed – i.e. φ = 20.9445 rad/s² and $a_\tau = 26.18$ m/s², $\tau_\delta = 4.8$ s.

6.3. Vibration of machines in mechanical design

There are numerous published works on vibrations in machines [SVE 76]. In this book, we shall limit our discussion to case studies with their solutions, to present the main points that are inherent to mechanical design. Vibrations in machines are practically unavoidable, especially in tooled machines, such as transporter-conveyor belts, to cite only a few examples. The design of their parts, in terms of elasticity, must take account of all the forces generated by the system itself in operation. It is not at all problematic to sketch out a machine part, but once those parts have been designed, assembled and are in motion, the real problem manifests itself. If the effects of these forces are not taken into account, this is a simple technical drawing, rather than actual design, which is our focus in this book. The vibrations in machines, undesirable as they are, have to be damped, or at least attenuated for each of the functions. In mechanical design, vibration analysis requires exact conditions:

– assessment of the masses and understanding of the elastic properties of the parts;

– estimation (i.e. calculation) of the effects of the induced friction (tribology);

– idealized (diagramatically represented) design using a mechanic device, representing it as an equivalent system of masses, springs, dampers, guide pins, etc.;

– differential equation of the motion of the "idealized system", taken to be homogeneous;

– solving of the differential equations and correct interpretation of the results.

We shall now set out the theoretical basis of machine vibrations, beginning with the differential equation of the mechanical system's displacement over time:

$$\varphi(\tau) = m\ddot{x} + c\dot{x} + kx = m\frac{\partial^2 X(\tau)}{\partial \tau^2} + c\frac{\partial X(\tau)}{\partial \tau} + kx \qquad [6.9]$$

where:

– x is the displacement measured from the position of static equilibrium;

– m is the mass in kg;

– k is the constant of the spring, or its damping stiffness in lb/in or kg/mm;

– c is the damping constant (coefficient of friction between the materials);

– $\ddot{x}$ is the second differential or the acceleration with respect to time, and is measured from the position of static equilibrium;

– $\dot{x}$ is the velocity/time measured from the position of static equilibrium;

– (x) is the displacement over time, and is measured from the position of static equilibrium;

– $\omega.\tau$ is the angle of rotation of the non-balanced mass about the horizontal axis;

– $\varphi(\tau)$ is the external force as a function of time;

– m_e, c_e and k_e are, respectively, the equivalent mass, the damping constant and the stiffness of the spring. The displacement $x(\tau)$ may be either linear or angular;

– ω is the rotation frequency of the engine in rad/min or rad/s;

– ω_n is the natural frequency in rad/min or rad/s.

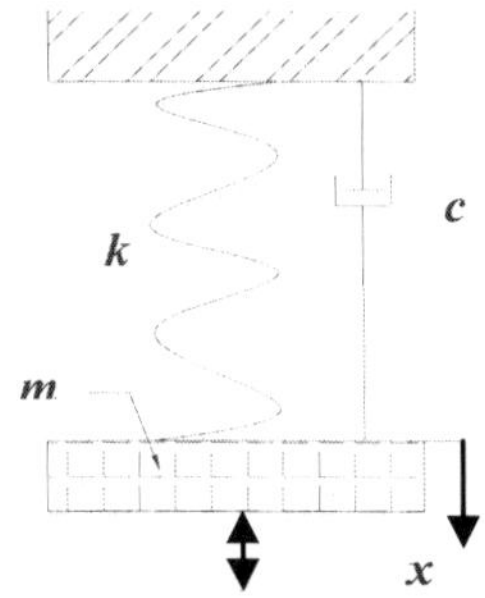

Figure 6.6. *Idealization of the vibrational system with one degree of freedom*

Any system with one degree of freedom is represented in equation form as follows:

$$\varphi(\tau) = m_e\ddot{x} + c_e\dot{x} + k_e x = m_e \frac{\partial^2 X(\tau)}{\partial \tau^2} + c_e \frac{\partial X(\tau)}{\partial \tau} + k_e x \qquad [6.10]$$

The function of *forced* oscillation (vibration) $\varphi(\tau)$ may take any shape (*linear or sinusoidal*). Therefore, we posit: $\varphi(\tau) = \varphi_0 \times sin(\omega\tau)$. Here, φ_0 is the amplitude of the external force applied and ω is its frequency. Free oscillation takes place after an initial excitation with no external force – i.e. $\varphi(\tau) = 0$. Let us set that at the start time $\tau = 0 \rightarrow m_e\ddot{x} + c_e\dot{x} + k_e x = 0$, which has the following as a solution:

$$x = A_1 \times \exp^{\lambda_1 \times \tau} + A_2 \times \exp^{\lambda_2 \times \tau} \rightarrow \left\{ \begin{array}{l} where\ \lambda_1 = -\dfrac{c_e}{2m_e} + \sqrt{\left(\dfrac{c_e}{2m_e}\right)^2 - \left(\dfrac{k_e}{m_e}\right)} \\ and\ \lambda_2 = -\dfrac{c_e}{2m_e} - \sqrt{\left(\dfrac{c_e}{2m_e}\right)^2 - \left(\dfrac{k_e}{m_e}\right)} \end{array} \right\} \qquad [6.11]$$

A_1 and A_2 are constants determined by the initial conditions. In special cases – i.e. with critical damping – we set: $(c_e/2m_e) \equiv (k_e/m_e)$; $A_1 = A_2 = \lambda$ and thc solution emerges: $x = (A_1 + B\tau) E\text{xp}\{-\lambda\tau\}$. Thus, critical damping for a free vibration is presented as:

$$x = \exp^{-\alpha \times \tau} \times X sin(\omega_a \tau + \phi), \text{ where } \alpha = \left(\frac{c_e}{2m_e}\right), \text{ and } \omega_a = \sqrt{\left(\frac{k_e}{m_e}\right) - \left(\frac{c_e}{2m_e}\right)^2} \qquad [6.12]$$

The expression of the damping frequency of the equivalent system is: $\omega_a = k_e/m_e$ (natural frequency). X and ψ are determined by the initial conditions. For forced oscillations, the solution to the differential equation is written thus:

$$x=\left\{\exp^{-\alpha\times\tau}\right\}\left\{X\sin\left[\omega_a\tau+\phi\right]\right\} \text{ where } \alpha=\frac{c_e}{2m_e} \text{ and } \omega_a=\sqrt{\frac{k_e}{m_e}-\left(\frac{c_e}{2m_e}\right)^2} \qquad [6.13]$$

We know that dampers reduce high-amplitude so-called *transient* vibrations and as well as permanent ones. Part *I* represents the transient vibration. Such is the case with aircraft propellers receiving a vibration signal from the propeller system, which attenuates (completely dies out) over time. The second part (*Y*) corresponds to the stable state of the system. It is this part which is of interest to us in mechanical design, given the influence it has. The *amplitude* of the stable state (*Y*) is written thus:

$$\begin{cases} Y=\varphi_0\Big/\sqrt{\left(k_e-m_e\times\omega^2\right)^2+\left(c_e\times\omega\right)^2}=\left(\varphi_0/k\right)\Big/\sqrt{\left(1-r^2\right)^2+\left(2\times\rho\times r\right)^2} \\ r=\omega/\omega_n \text{ frequency ratio, } \rho=\left(c_e/c_e\right)_c \text{ damping ratio} \end{cases} \qquad [6.14]$$

The *magnification factor* Φ_a is written as:

$$\Phi_a=\frac{Y}{\varphi_0/k}=1\Big/\sqrt{\left(1-r^2\right)^2+\left(2\rho r\right)^2} \qquad [6.15]$$

In the so-called stable-state phase, the magnification factor Φ_a is a ratio between the amplitude of displacement and the displacement caused by the static force φ_0. The phase angle is expressed by tan(ψ) and is written:

$$\tan(\psi)=\frac{c_e\omega}{k_e-m_e\omega^2} \text{ or indeed } \sin(\psi)\frac{c_e\omega}{\sqrt{\left(k_e-m_e\omega^2\right)^2+\left(c_e\omega\right)^2}} \qquad [6.16]$$

The *force transmitted* to the base of the mechanical system would be: $k_e x + c_e\dot{x}$:

$$F_{transmission}=F_T=\varphi_0\sqrt{k_e^2+\left(c_e\omega\right)^2}\Big/\sqrt{\left(k_e-m_e\omega^2\right)^2+\left(c_e\omega\right)^2} \qquad [6.17]$$

By virtue of the above, the ratio of the force transmission to the base, particularly with bolted assemblies – e.g. setups with no intermediary springs, dampers or guide pins – is written as:

$$R_{transmission} = R_T = \frac{F_T}{\varphi_0} = \frac{\sqrt{k_e^2 + (c_e\omega)^2}}{\sqrt{\left(k_e - m_e\omega^2\right)^2 + (c_e\omega)^2}} = \frac{\sqrt{1+(2\rho r)^2}}{\sqrt{\left(1-r^2\right)^2 + (2\rho r)^2}} \quad [6.18]$$

In case of forced *oscillations* such as in Figure 6.7 below:

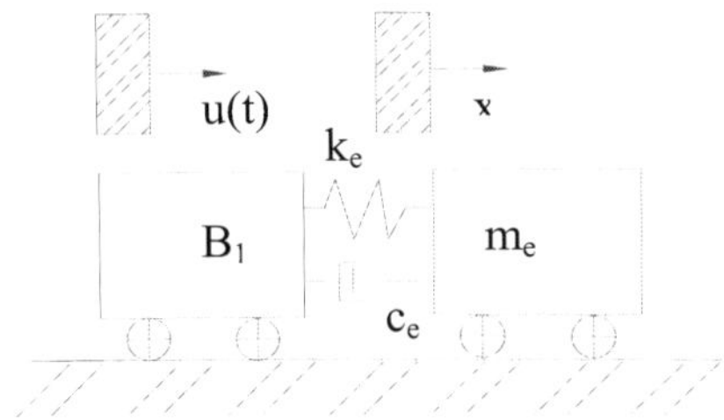

Figure 6.7. *Vibration system with periodic force applied*

The periodic motion is induced by the masses. This is idealized by a spring and/or a given damper. The mass is then reduced by the amplitude of the motion. Considering the sinusoidal motion $U(\tau)$, let us set: $U(\tau) = u.sin(\omega\tau)$. Similarly to the above, the differential equation of the motion of mass is:

$$\begin{cases} m_e\ddot{x} + c_e\dot{x} + k_e x = u\sqrt{k_e^2 + (c_e\omega)^2}\sin(\omega\tau - \psi), \ \psi = phase\ angle \\ \cos(\psi) = k_e \Big/ \sqrt{k_e^2 + (c_e\omega)^2} \ \ and \ \sin(\psi) = -c_e\omega \Big/ \sqrt{k_e^2 + (c_e\omega)^2} \end{cases} \quad [6.19]$$

From the academic literature on vibrations of systems, the amplitude of the stable state of mass is presented, similarly to above, as follows:

$$\left\{ Y = \frac{u\sqrt{k_e^2 + (c_e\omega)^2}}{\sqrt{\left(k_e - m_e\omega^2\right)^2 + (c_e\omega)^2}} \ \ and\ \mathrm{R}_T = \frac{Y}{u} = \frac{\sqrt{k_e^2 + (c_e\omega)^2}}{\sqrt{\left(k_e - m_e\omega^2\right)^2 + (c_e\omega)^2}} \right\} \quad [6.20]$$

These systems cannot be solved with the I-order differential equation, as developed above. The solution requires the use of the II-order differential equations with (n) degrees of freedom (*dof*) in the system. Systems with more than one *dof* exhibit two modes of vibrations. Finally, mechanical systems with more than one *dof* arc represented thus:

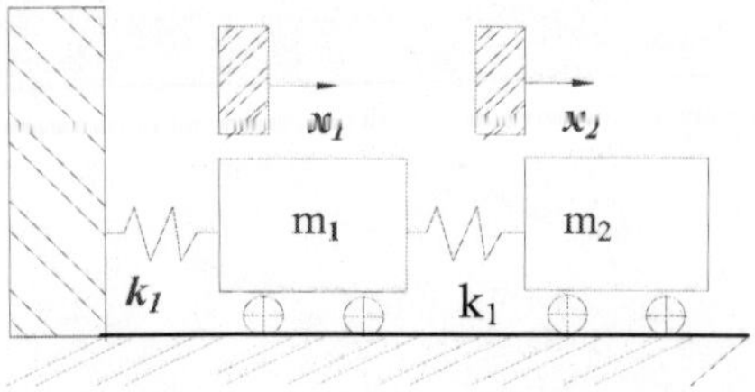

Figure 6.8. *Idealization of the vibrational system with more than one dof*

In vibrational mode I, the masses constitute the primary phase, reaching the maximum displacement in the same direction. In mode II, the two masses are in an opposing phase, reaching the maximum displacements in the opposite direction. In addition, the energy method for calculating the vibrations is worthy of interest. We shall present the salient points of this method later on. We ignore the friction and set the expression of the energy balance of the system, represented by the kinetic and potential energy values. X_1 and X_2 are, respectively, the displacements of the respective masses m_1 and m_2. Suppose that the system's motion is sinusoidal, with the frequency (ω), and set:

$$\text{Energy balance} = \begin{cases} E_c^{\max} = \frac{1}{2} m_1 X_1^2 \omega^2 + \frac{1}{2} m_2 X_2^2 \omega^2 \\ E_p^{\max} = \frac{1}{2} k_1 X_1^2 + \frac{1}{2} k_2 \times (X_2 - X_1)^2 \end{cases} \quad [6.21]$$

Without taking the friction into account, the kinetic energy would be equivalent to the potential energy:

$$E_c^{\max} = E_p^{\max} \rightarrow \left\{ \omega = \sqrt{\frac{k_1 X_1^2 + k_2 (X_2 - X_1)^2}{m_1 X_1^2 + m_2 X_2^2}} \quad or \quad \omega = \sqrt{\frac{k_1 + k_2\, X_2 / (X_2 - X_1)^2}{m_1 + m_2 (X_2 / X_1)^2}} \right\} \quad [6.22]$$

If we know the ratio between the amplitudes (X_2/X_1), the above equation for (ω) gives us the natural frequency of vibration. When there is resonance, (ω_n) is presented thus: $\omega_n = \sqrt{k_e / m_e}$. In the case of an idealized viscous system, the forced damping is written: $\varphi_0 \times \sin(\omega \tau)$ with the following maximum frequency:

$$\omega_{\max} Y = \omega_n \sqrt{1 - 2\rho^2} \quad Hence \quad \omega_a = \omega_n \sqrt{1 - \rho^2} \quad [6.23]$$

6.4. Case studies with a numerical solution

Consider a moving machine of mass M, attached to the wall by a spring which exerts a recall force in the sense of Hooke's law F(x) = |*k.x*| in a viscous medium with a damping force $F(\tau)=-c\times(\partial x/\partial \tau)$.

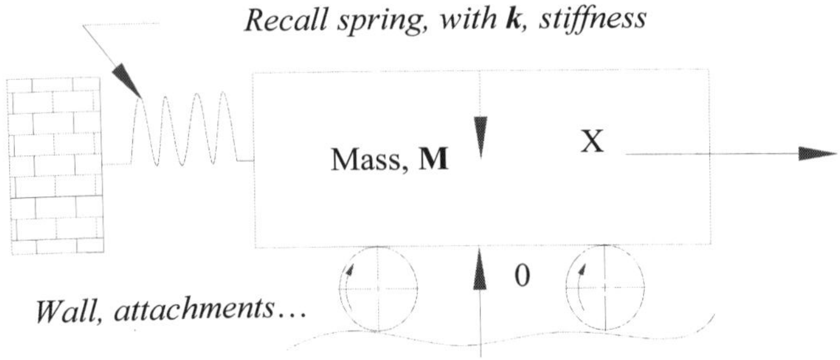

Figure 6.9. *Model of a free-oscillating mechanical system*

The system is drawn in the equilibrium position, meaning that X = 0. Let us set the differential equation of the mechanical system as follows:

$$\begin{cases} M\times\dfrac{\partial^2}{\partial\tau^2}X(\tau)+c\times\dfrac{\partial}{\partial\tau}X(\tau)+k\times X(\tau)=0 \\ For\ X(0)=X_0 \rightarrow \dfrac{\partial}{\partial X}X(\tau)=0 \leftarrow at\ \tau=0 \end{cases} \quad [6.24]$$

6.4.1. *Case study: input parameters: M = 1; k = 1; φ₀ = 1 and c = 2.25*

By transforming the system into standard form, we can write: $a=\sqrt{k/M}$ *and* $b=\sqrt{c/2M}$

– Set up solve block $t_e=7$ *and* $N_t=14$;

– t_e = 7 is the rightmost endpoint criterion, and N_t = 14 *Number of time steps*:

$$For\ X(0)=X_0\ and\ X_0'(0)\rightarrow\frac{\partial^2 X(\tau)}{\partial\tau^2}+2b\frac{\partial X(\tau)}{\partial\tau}+a^2X(\tau)=0 \quad [6.25]$$

– *ODEsolve – Ordinary Differential Equation;*

– $disp=Odesolve\{\tau,T_e\}, exact\ solution\ for\ m_1=-b+\sqrt{b^2-a^2}\ and\ m_2=-b-\sqrt{b^2-a^2}$

$$X(\tau) = if\left\{b = a,\ X_0 \times Exp^{-a\tau}(1 + a\tau)\ ,\ \frac{X_0}{m_1 - m_2}\left(m_1 Exp^{-m_2\tau} - m_2 Exp^{-m_1\tau}\right)\right\} \quad [6.26]$$

For $n = 0, 0.3 \ldots T_e$, let us plot the exact solution and the approximate solution of the mechanical system. We can see that there is a near perfect match between the two solutions.

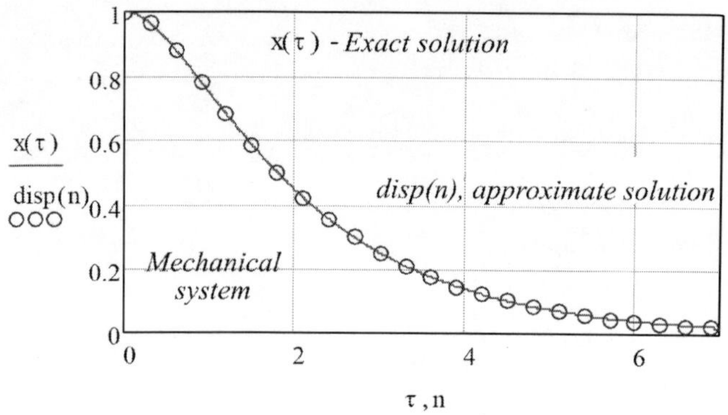

Figure 6.10. *Model of a free-oscillating mechanical system. For a color version of this figure, see www.iste.co.uk/grous/design.zip*

6.4.2. *Case study: system with free vibrations*

QUESTIONS.– Write the differential equation of the free vibrations in the system. The force generated by the spring is equal to [– (k.x)]. The same principle applies to the damping force, written as [-c.(dx/dτ)]. By virtue of Newton's second law, we can write:

$$\begin{cases} m\ddot{x} = -c\dot{x} - kx \Rightarrow m\ddot{x} + c\dot{x} + kx = 0 \\ m\left(\partial^2/\partial\tau^2\right)\times\varphi(x) + c\left(\partial/\partial\tau\right)\times\varphi(x) + kx = 0 \end{cases} \quad [6.27]$$

Fictitious data: $\tau = 5$; $k = k_e = 1.25$; $m_e = 20$; $c_e = 10$; $A_1 = 1$; $A_2 = 1$ and $B = 1$

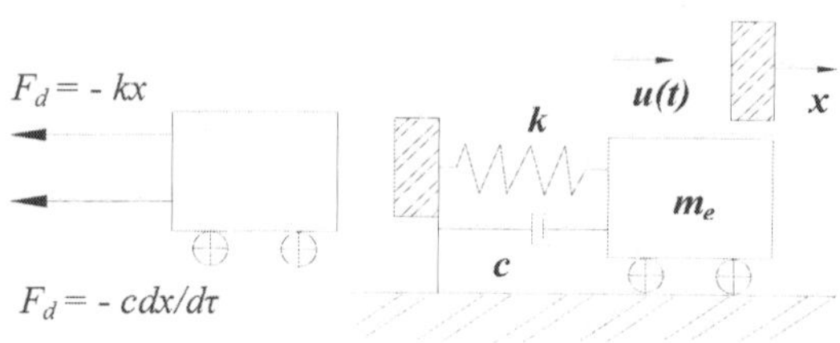

Figure 6.11. *Ideal representation of a freely-vibrating system*

Let us prove that: $(k_e/m_e) = (c_e/2m_e)^2 \rightarrow 1.25/20 = (10/2\times 20)^2 = 0.063 \leftarrow QED$

Let us first prove that:

$$\begin{cases} s_1 = \left(\dfrac{c_e}{2m_e}\right) + \sqrt{\left(\dfrac{c_e}{2m_e}\right)^2 - \left(\dfrac{k_e}{m_e}\right)} = -0.25 \; and \\ s_2 = -\left(\dfrac{c_e}{2m_e}\right) - \sqrt{\left(\dfrac{c_e}{2m_e}\right)^2 - \left(\dfrac{k_e}{m_e}\right)} = -0.25 \end{cases} \quad [6.28]$$

$$\begin{cases} \alpha = \left(\dfrac{c_e}{2m_e}\right) = 0.25; \; \omega_{damping} = 0 \rightarrow \omega_{damping} = \sqrt{\left(\dfrac{k_e}{m_e}\right) - \left(\dfrac{c_e}{2m_e}\right)^2} = 0 \\ \omega_{natural} = 0.25 \rightarrow \sqrt{k_e/m_e} = 0.25 \; QED \; and \; c_e = 10 \rightarrow 2\sqrt{k_e m_e} = 10 \; QED \end{cases} \quad [6.29]$$

$$\begin{cases} For \; s_1 = s_1 = s = -0.25 \; set \rightarrow x(\tau) = (A + B\times\tau)\,Exp^{-s\tau} \; because \checkmark \\ x(\tau) = \left\{A_1 Exp^{-s_1\tau} + A_2 Exp^{-s_2\tau}\right\} \rightarrow at \; time \; \tau = 5 \rightarrow x(\tau) = 6.981 \\ x(\tau_{var}) = \left\{A_1 Exp^{-s_1\tau_{var}} + A_2 Exp^{-s_2\tau_{var}}\right\}, at \; times \; \tau = 0 \; to \; 10 \rightarrow x(\tau) \end{cases}$$

$x(\tau_V) =$

	0
0	2.000
1	2.568
2	3.297
3	4.234
4	5.437
5	...

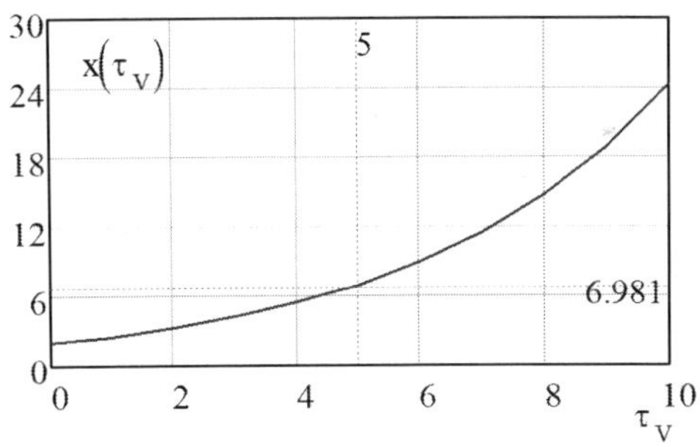

Figure 6.12. *Representation of the function x (τ v) at time (τ v). For a color version of this figure, see www.iste.co.uk/grous/design.zip*

For $F_0 = 1$, the unit force of initial excitation and w varying between 0.01, 0.02 and 1 gives us the expression of Y(ω).

From $Y(\omega) = F_0 / \sqrt{k_e - (m_e\omega^2)^2 + (m_e\omega)^2}$, we obtain the following graph on the state of the vibrational amplitudes as a function of the velocities (ω):

$Y(\omega) =$

	0
0	0.88
1	0.842
2	0.788
3	0.728
4	0.667
5	0.61
6	...

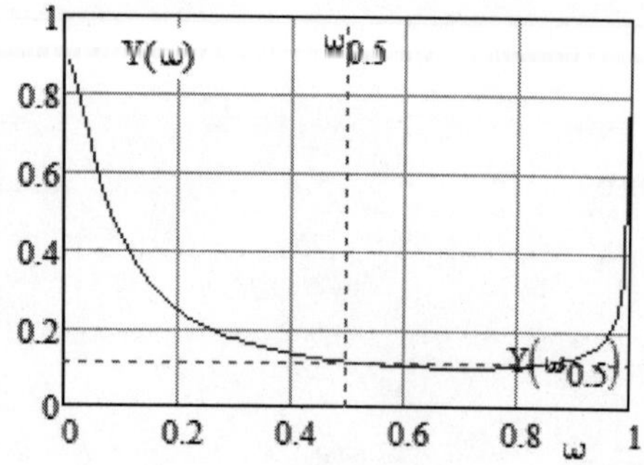

Figure 6.13. *Representation of amplitudes as a function of the variation of speed (ω). For a color version of this figure, see www.iste.co.uk/grous/design.zip*

At the angular velocity $(\omega_{0.5}) = 0.5$, the transmissibility ratio of the forces R_{TF} would be:

$$R_{TF} = \frac{F_{tramissible}}{F_0} = R_{TF} = \frac{\sqrt{(k_e)^2 + (c_e\omega_{0.5})^2}}{\sqrt{k_e - \left(m_e\omega_{0.5}{}^2\right)^2 + (c_e\omega_{0.5})^2}} = 4.61 \quad [6.30]$$

$$\begin{cases} F_{tramissible} = \dfrac{F_0\sqrt{(k_e)^2 + (c_e\omega_{0.5})^2}}{\sqrt{k_e - \left(m_e\omega_{0.5}{}^2\right)^2 + (c_e\omega_{0.5})^2}} = 4.61 \\ Y(\omega_{0.5}) = F_0 \Big/ \sqrt{k_e - \left(m_e\omega_{0.5}{}^2\right)^2 + (c_e\omega_{0.5})^2} = 0.115 \end{cases} \quad [6.31]$$

6.4.3. *Case study: problem with solution and discussion*

In the model of the design sketch below (Figure 6.14), we wish to write the differential equation of the vibrational system, without taking account of the lever's own mass (L). We also wish to determine:

– ω_n is the natural frequency, and ω_A the damping frequency;

– the critical value of the damping factor (c).

The moment of the forces expressed above is fairly small in relation to the angular displacement (α°). The moment created by the forces is written: $M_f = L\times\sin(\alpha) \equiv L\times(\alpha)$. As the angle, i.e. the displacement, is small, we treat the argument and the function as being identical, and set: M = Lx(α). The moment of inertia is $J_\omega = M\times L^2$.

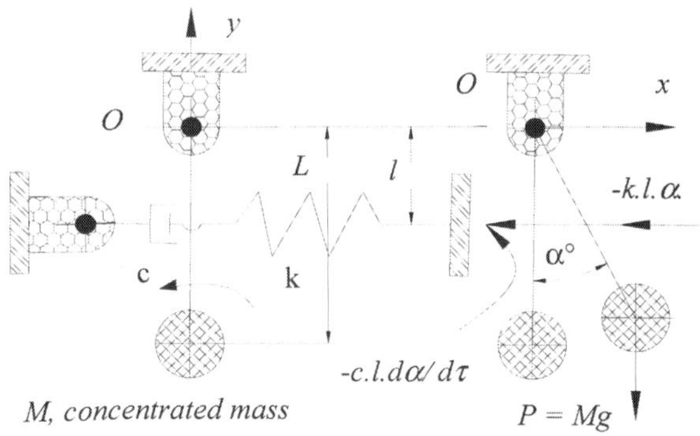

Figure 6.14. *Sketch of the calculation model with angular vibration. For a color version of this figure, see www.iste.co.uk/grous/design.zip*

The inertia system for the equilibrium is then presented as:

$$\begin{cases} ML^2\left(\partial^2\alpha/\partial\tau^2\right) = -\left\{cl\times(\partial\alpha/\partial\tau)\right\}l - (kl\alpha)l - Mg(l\alpha) \\ 0 = ML^2\left(\partial^2\alpha/\partial\tau^2\right) + c\times l^2\times(\partial\alpha/\partial\tau) + \left(kl^2 + MgL\right)\alpha \end{cases} \quad [6.32]$$

Parameters and equations used when designing vibration with one *dof*:

$$\begin{cases} x \equiv \alpha\ ;\ \dot{x} = \dfrac{\partial\alpha}{\partial\tau} = \dot{\alpha}\ ;= \ddot{x} = \dfrac{\partial^2\alpha}{\partial\tau^2} = \ddot{\alpha}\ ;\ M_e = ML^2;\ c_e = cl^2;\ k_e = kl^2 + MgL \\ \omega_n = \sqrt{\dfrac{k_e}{M_e}} = \sqrt{\dfrac{kl^2 + MgL}{ML^2}}\ ;\ \omega_a = \sqrt{\dfrac{k_e}{M_e} - \left(\dfrac{c_e}{2M_e}\right)^2} = \sqrt{\dfrac{kl^2 + MgL}{ML^2} - \dfrac{cl^2}{2ML^2}} \\ [c_e]_c = [c_c]l^2 = 2k_eM_e = 2\sqrt{\left(kl^2 + MgL\right)ML^2};[c_c] = \dfrac{2}{l^2}\sqrt{\left(kl^2 + MgL\right)ML^2} \end{cases} \quad [6.33]$$

6.4.4. *Case study: problem 3 with solution*

The parameters M_e, c_e and k_e are, respectively, the equivalent mass, the damping constant and the stiffness of the spring. The displacement (x) as a function of time may be linear or angular. Consider a platform subjected to stress at a frequency of 7 cycles/s at an amplitude of 0.039 in.

– $M = M_e$ is the equivalent mass of the support platform;

– P is the weight of the platform = 50 kg = 110.213 lb or M = P/32.2 = 3.423 lb;

– k is the constant of the spring – stiffness – in lb/in or kg/mm;

– c is the damping constant. In other words, it is the friction coefficient;

– ω is the frequency of rotation of the engine in rad/min or rad/s;

– ω_n is the natural frequency in rad/min or rad/s;

– f is the frequency in cycles per second = 7 cycles/s;

– A = 0.039 in = 1 mm is the target amplitude for the system.

Evaluating the vibration φ0 (ωvar) in conditions of resonance

Evaluation of the real vibration: P = Mg → hence M = P/g = 50/9.81 = 5.09 kg or else M = P/32.2 = 3.423 lb = 50/9.81 = 5.09 kg. The frequency is 7 cycles/s. We calculate: ω = 2πf = 43.982 rad/s $k = M\omega^2/2 = 276.054\ lb/in = 4.93\times10^3\ kg/m$. The system is idealized by two springs: $k_e = 2k = 552.109\ lb/in = 9.86\times10^3\ kg/m$.

We calculate c: $\begin{cases} c_e = c = 0.05c_c \rightarrow 2\sqrt{k_e M} = 448.341\dfrac{kg.s}{m} = 988.424\dfrac{lb.s}{m} \\ c_{effective}\ will\ therefore\ be\ c_e = 0.05c_c = 22.417\dfrac{kg.s}{m} = 49.421\dfrac{lb.s}{m} \end{cases}$

On condition of resonance, the amplitude **A** is calculated as a solenoid:

$$A = \varphi_0 / c\omega\ \ Hence\ \rightarrow\ \varphi_0 = c\times\omega\times A = 2.174 lb = 0.986\times10^3 kg$$

To clearly see the evolution of φ_0 (ω_{var}), let us vary the frequency ω between 0 and 50 and construct the semi-log plot. The results φ_0 (ω_{var}) are shown below.

$\varphi_0(\omega_{var}) =$

	0
0	0.000
1	0.112
2	0.224
3	0.336
4	0.448
5	0.560
6	0.673
7	...

kg·s

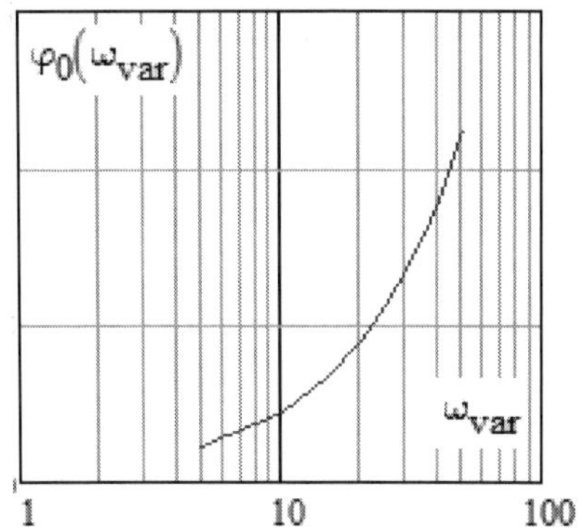

Figure 6.15. *Influence of frequency on the vibration of a mechanical system. For a color version of this figure, see www.iste.co.uk/grous/design.zip*

This expression is known as the natural frequency:

$$\omega_n = \sqrt{k_e/M_e} \qquad [6.34]$$

By damping, we are able to keep the frequency at the natural value – that is:

$\omega_a = \sqrt{k_e/M_e} = 43.982[1/s]$ For c = c_e, let us calculate the system's frequency:

$$\omega_a^{systeme} = \sqrt{(k_e/M_e)-(c_e/2M_e)^2} = 43.927[1/s] \qquad [6.35]$$

We can see that there is practically no difference between the frequency of the system and the damping frequency for a system with one *dof*.

6.4.5. *Case study: problem 2. Engine represented on two springs*

– M = M_e is the equivalent mass of the engine;

– m_e represents the disequilibrium of the rotor (mass of the rotor not at equilibrium);

– k is the constant of the spring – stiffness – in lb/in or kg/mm;

– c is the damping constant. In other words, it is the friction coefficient;

– ω is the frequency of rotation of the engine in rad/min or rad/s;

– (ω.τ) is the angle of rotation of the non-equilibrated mass (horizontal plane);

– f is the frequency in cycles per second = 7 cycles/s;

– x is the (vertical) displacement of the engine measured from the position when the system is at static equilibrium.

Evaluating the vibration φ0 (ωvar) in conditions of resonance

Consider an engine mounted on two springs. The rotor causes vibrations when the engine is in operation, because it is not equilibrated. We wish to perform a conceptual analysis of the system in order to choose the appropriate characteristics for the spring, considering only the vertical plane, as illustrated below.

DESIGN SOLUTION WITH DISCUSSION.– The engine has an acceleration ($\ddot{x}$). The non-equilibrated mass M has a vertical acceleration expressed by [-e.ω^2sin(ωτ)]. The external forces applied are the forces of the spring (-kx) and the damping force $c\dot{x}$ in addition to the weight in Mg. In conditions of equilibrium, therefore, we can write:

$$\begin{cases} -c\dot{x} - kx - Mg + Mg = M\ddot{x} - Me\omega^2 \sin(\omega\tau) \\ Condition\ of\ equilibrium\ M\ddot{x} + c\dot{x} + kx = Me\omega^2 \sin(\omega\tau) \end{cases} \quad [6.36]$$

In the state of equilibrium: $x = Y\sin(\omega\tau - \phi) \rightarrow \begin{cases} \dot{x} = Y\omega\cos(\omega\tau - \phi) \\ \ddot{x} = -Y\omega^2 \sin(\omega\tau - \phi) \end{cases}$ [6.37]

After substitution into the differential equation, we have:

$$M\left[-Y\omega^2 \sin(\omega\tau - \phi)\right] + \left[cY\omega\cos(\omega\tau - \phi)\right] + \left[kY\sin(\omega\tau - \phi)\right] = me\omega^2 \sin(\omega\tau) \quad [6.38]$$

This expression can also be written as:

$$\begin{cases} -MY\omega^2 \times \left[\sin(\omega\tau)\times\cos(\phi) - \cos(\omega\tau)\times\sin(\phi)\right] + \\ + ck\omega \times \left[\cos(\omega\tau)\times\cos(\phi) + \sin(\omega\tau)\times\sin(\phi)\right] + \\ + kY \times \left[\sin(\omega\tau)\times\cos(\phi) - \cos(\omega\tau)\times\sin(\phi)\right] = me\omega^2 \sin(\omega\tau) \end{cases} \quad [6.39]$$

We equalize the coefficients of sin (ωτ) and set:

$$\begin{cases} -MY\omega^2 \times \cos(\phi) + cY\omega \times \sin(\phi) + kY \times \cos(\phi) = me\omega^2 \\ -MY\omega^2 \times \sin(\phi) + cY\omega \times \cos(\phi) - kY \times \sin(\phi) = 0 \end{cases} \quad [6.40]$$

When we solve the above system of equations, we can write:

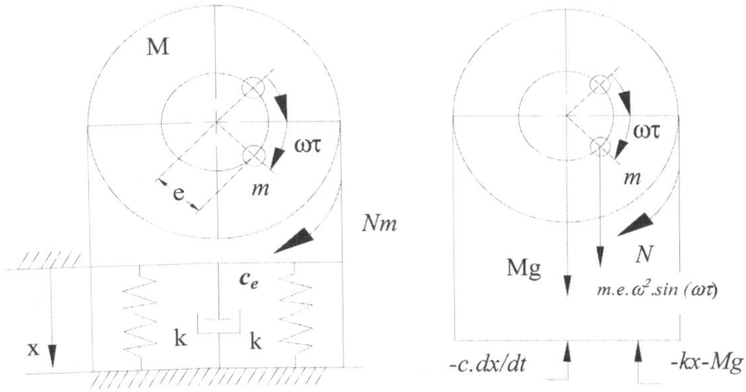

$$Y = \frac{me\omega^2}{\sqrt{\left(k - M\omega^2\right)^2 + (c\omega)^2}} \text{ and } \begin{cases} \cos(\phi) = \dfrac{\left(k - M\omega^2\right)}{\sqrt{\left(k - M\omega^2\right)^2 + (c\omega)^2}} \\ \sin(\phi) = \dfrac{(c\omega)}{\sqrt{\left(k - M\omega^2\right)^2 + (c\omega)^2}} \end{cases} \quad [6.41]$$

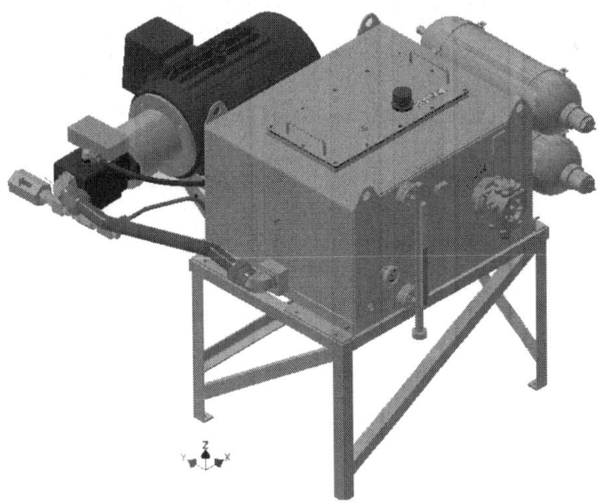

Figure 6.16. *Platform subject to vibrations and engine mounted on two springs. For a color version of this figure, see www.iste.co.uk/grous/design.zip*

On the condition of equilibrium, the function of state of the differential equation is written:

$$x = \left\{ \left(me\omega^2\right) \Big/ \sqrt{\left(k - M\omega^2\right)^2 + (c\omega)^2} \right\} \times \sin(\omega\tau - \phi) \qquad [6.42]$$

The forces transmitted to the base come from the spring and the damping forces:

$$\begin{cases} c\dot{x} + kx \text{ or indeed } kY\sin(\omega\tau - \phi) + cY\omega\cos(\omega\tau - \phi) = \ldots \\ \ldots = Y\sqrt{k^2 + (c\omega)^2}\,\sin(\omega\tau - \phi + \alpha),\ (\phi + \alpha) = Phase \end{cases} \qquad [6.43]$$

The angle [-ϕ + α] is the phase between the excitation force $M_e\omega^2 \sin(\omega\tau)$ and the force transmitted F_{TR} – i.e.:

$$F_{TR} = Y\sqrt{k^2 + (c\omega)^2} = M_e\omega^2\sqrt{k^2 + (c\omega)^2} \Big/ \sqrt{k - \left(M\omega^2\right)^2 + (c\omega)^2} \qquad [6.44]$$

Remember that: ω_n is the natural frequency, and c_c the critical damping factor:

$$f = \frac{\omega}{\omega_n} = \frac{\omega}{\sqrt{k/M}} \text{ and } \lambda = \frac{c}{c_c} = \frac{c}{2\sqrt{kM}}, \text{ and hence } \frac{F_{TR}}{M_e\omega^2} = \frac{\sqrt{1 + (2\lambda f)^2}}{\sqrt{\left(1 - f^2\right)^2 + (2\lambda f)^2}} \qquad [6.45]$$

The ratio of the amplitudes of the transmission force and the applied force is expressed by the transmissibility ratio, written:

$$R_{TR} = \sqrt{1 + (2\lambda f)^2} \Big/ \sqrt{\left(1 - f^2\right)^2 + (2\lambda f)^2} \qquad [6.46]$$

NUMERICAL APPLICATION.– For $f = 2.75$ and $\lambda = 0.035$, $n = 4$ springs; P = 35 lb = 15.876 kg; RPM = 960(1/s) and $R_{FT} = 0.16$; therefore, let us prove that:

$$R_{TR} = \sqrt{1 + (2\lambda f)^2} \Big/ \sqrt{\left(1 - f^2\right)^2 + (2\lambda f)^2}$$

$$f = \omega/\omega_n = 2.75\ ;\ \lambda = 0.035;\ P = Mg = 35\ lb = 15.876\ kg \rightarrow M = P/g = 1.619\ kg.s^2/m$$

$$RPM = 960\ rad/s \rightarrow \omega = 2\pi\, RPM/60 = 100.531\ rad/s\ \ From\ \sqrt{k/M} = \omega/f$$

$$\rightarrow k = M\omega^2/f^2 = 1.454\times10^3\ lb/ft = 121.148\ lb/in = 2.163\ kg/mm$$

For four springs, we have: $k_{n\_springs} = k/n = k/4 = 30.287\ lb/ft = 0.541\ kg/mm$

$$For\ \lambda = 0.035 \rightarrow c = \lambda\times\left(2\sqrt{kM}\right) = 4.143[kg.s/m] = 0.232[lb.s/in]$$

$$R_{FT}^{\text{Projected}} = 0.16 \rightarrow R_{FT}^{\text{Calculated}} = \sqrt{\left(1+(2f\lambda)^2\right)\Big/\left(\left(1-f^2\right)^2+(2f\lambda)^2\right)} = 0.155$$

6.4.5.1. *Design of elastic supports*

The support illustrated below matches the platform in Figure 6.16, above, for which the slice A-A shows the following dimensions.

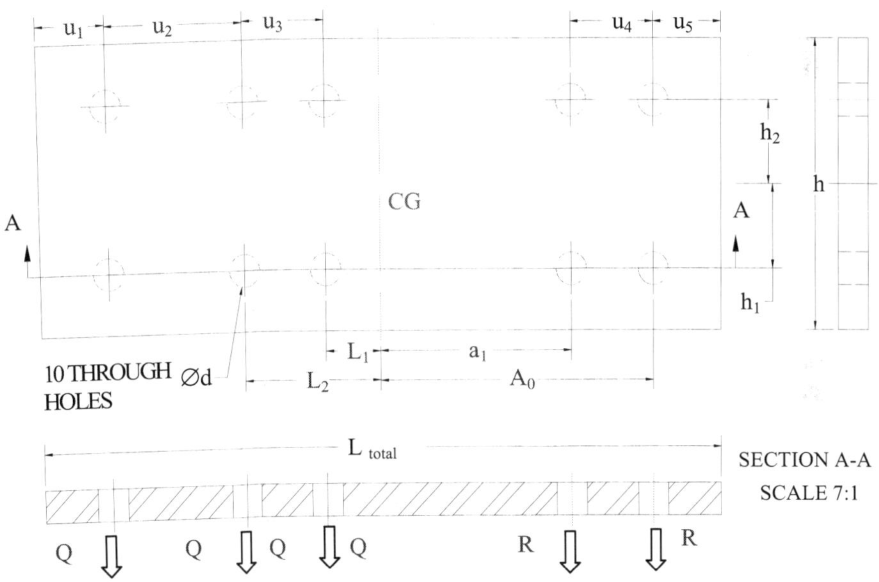

Figure 6.17. *"Ideal" representation of loading of the platform subject to vibrations. For a color version of this figure, see www.iste.co.uk/grous/design.zip*

Data

$a_1 = 0.5$ m $= 1.64$ ft
$L_1 = 0.35$ m $= 1.148$ ft
$L_2 = 1$ m $= 3.281$ ft
$P = 62\times10^3$ N $= 1.394\times10^4$ lbf

$$Q = \frac{\sum_{i=1}^{p} a_i}{2n\sum_{i=1}^{p} a_i + 2p\sum_{i=1}^{n} L_i} P \ \ and\ \ R = \frac{\sum_{i=1}^{n} L_i}{2n\sum_{i=1}^{p} a_i + 2p\sum_{i=1}^{n} L_i} P$$

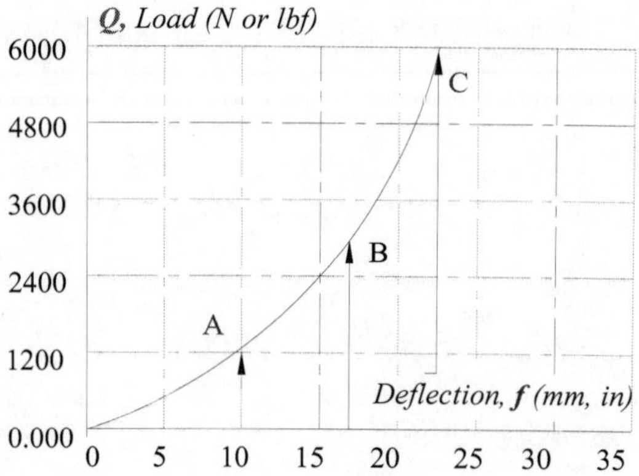

Figure 6.18. *Representation of loading as a function of the deflection. For a color version of this figure, see www.iste.co.uk/grous/design.zip*

$$\text{Solution:} \rightarrow \begin{cases} Q = \dfrac{a_1}{4a_1 + 2(L_1 + L_2)} \times P = 6.596 \times 10^3 \ N = 1.483 \times 10^3 \ lbf \\ R = \dfrac{L_1 + L_2}{4a_1 + 2(L_1 + L_2)} \times P = 1.781 \times 10^4 \ N = 4.004 \times 10^3 \ lbf \end{cases}$$

The deflection, under the imposed load, exhibits an almost linear slight curve, in the zones of static loading (*0n*), which corresponds to the static load stated by the manufacturer. (*AB*) represents the zone of dynamic loading, where there are repeated shocks at a normal cadence. (BC) represents the zone of exceptional or harmful shocks, where the Hooke stiffness increases with the deflection. To attenuate undesired vibrations, we opt for the compromise of an elastic suspension.

6.4.6. *Case study based on a concrete problem with solution*

Appropriate choice of springs and damping of the vibrations of the drum in a clothes dryer:

Data:

- total weight of the dryer drum — P_{total} = 45 lb = 20.412 kg;
- speed of rotation of the dryer drum, RPM — 560 rpm;
- placement angle (α) between the springs — α = 33°;
- supposed acceptable moment of disequilibrium (including excentricity) of the dryer drum — P_e = 18 lin.in = 0.207 m.kg;
- amplitude of the vibration of the dryer drum, in all directions, before resonance — Y = 0.5 in = 12.7 mm.

SOLUTION WITH DISCUSSION.– Consider a frame of reference (X0Y), with a slight deflection (x) with respect to the center of the dryer drum. As we can see in Figure 6.19 for the design, spring (1) is stretched and spring (3) remains compressed. Spring (2) exhibits no notable change, so its length remains constant. The forces applied to the springs shown in Figure (a), along the OX axis, would be:

$$F_{springs} = -2\cos(\alpha)\times k\times x\times\cos(\alpha) = -1.5k\times x \quad [6.47]$$

The *effective work of the springs* corresponds, along the *0X* axis, to *1.5k*. Similarly, we can make the same observation along the *OY* axis. By strict analysis of the damping forces, in the Cartesian space XOY, we can find the expression of the effective damping factor: *1.5c*:

– x is the displacement in mm or in;

– the first differential of the displacement with respect to time (τ), V = (dx/dτ);

– the second differential of the displacement with respect to time (τ), $\gamma = (d^2x/d\tau^2)$.

Differential equation of the system of vibrations $m_e\omega^2\sin(\omega\tau) = M\ddot{x} + 1.5c\dot{x} = -1.5kx$

The amplitude of the displacement is written:

$$Y = m_e\omega^2 \Big/ \sqrt{\left(1.5k - M\omega^2\right)^2 + 1.5c\omega} \quad [6.48]$$

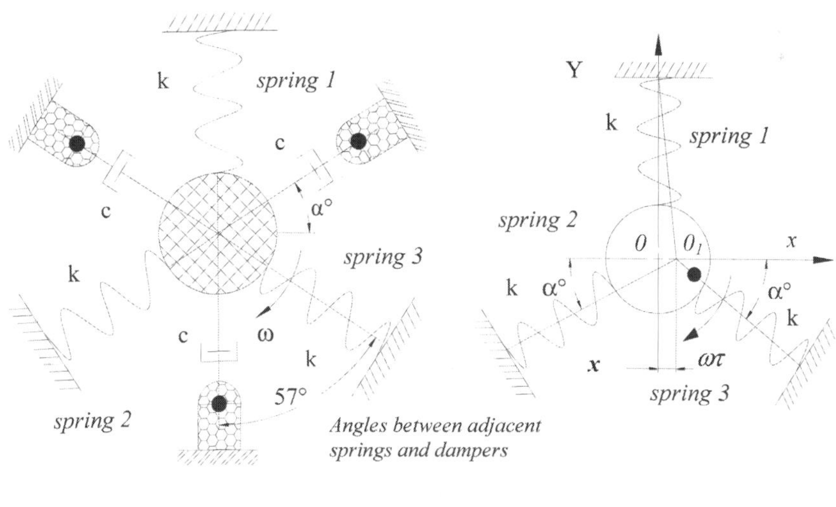

Figures 6.19. *Ideal representation of the springs under stress. For a color version of this figure, see www.iste.co.uk/grous/design.zip*

$$\left|F_{spring1}\right| = kx\cos(\alpha)\,;\ \left|F_{spring2}\right| = 0 \text{ and } \left|F_{spring3}\right| = kx\cos(\alpha)$$

The transmission function (factor):

$$F_{TR} = \frac{m_e\omega^2\sqrt{(1.5k)^2 + 1.5c\omega}}{\sqrt{\left(1.5k - M\omega^2\right)^2 + (1.5c\omega)^2}} \qquad [6.49]$$

In design, we suggest the following: $(\omega/\omega_n) = 3$ and therefore:

$$\omega_n = \sqrt{1.5(k/M)} \qquad [6.50]$$

where: $\omega = 2\pi\left(N_{rpm}/60\right) = 58.643[rad/s]$ *and* $\omega/\omega_n = 3 \rightarrow \omega_n = \omega/3 = 19.548[rad/s]$

$M = P_{total}/g = 5.589\left[s^2.lb/m\right] = 2.081\left[s^2.kg/m\right]$ and thus deduce the value of (k):

$$1.5k = \omega_n^2 \times M = \left(\frac{\omega}{\omega_n}\right)^2 \times M \rightarrow k = \left(\frac{\omega}{\omega_n}\right)^2 \times \frac{M}{1.5} = 29.691\left[\frac{lb}{in}\right] = 530.221\left[\frac{kg}{m}\right]$$

Damping factor c, to limit the amplitude of displacement Y, to (1/2) in conditions of resonance. Limitation imposed by the hypothesis Y = 0.5 in = 12.7 mm.

$$m_e = P_e/g = 0.047\left[s^2.lb\right] = 0.021\left[s^2.kg\right] \text{ and } c = \frac{m_e \times \omega_n}{1.5Y} = 1.215\left[\frac{s.lb}{in}\right] = 21.7\left[\frac{s.kg}{m}\right]$$

Assured stability and controlled vibration

$$Y^{calculated}_{verified} = m_e\omega^2 \Big/ \sqrt{\left(1.5k - M\omega^2\right)^2 + (1.5c\omega)^2} = 0.5\ in = 12.7\ mm \qquad [6.51]$$

Resonance: The peak of the force of the solenoid required is F_0:

$$c_c = 0.05\,c,\ Y = F_0/c\omega \rightarrow F_0 = Y \times c\omega = 0.808\ kg = 1.781\ lb \qquad [6.52]$$

Let us calculate the differential equation for the system: $m_e\omega^2\sin(\omega_n\tau) = M\ddot{x} + 1.5c\dot{x} = -1.5kx \rightarrow$

$m_e\omega^2 \sin(\omega_n\tau) = M\ddot{x} + 1.5c\dot{x} = -1.5kx = F(\tau)$. The transmissibility factor of the vibration forces is written:

$$F_{TR} = m_e\omega^2 \sqrt{(1.5k)^2 + (1.5c\omega)^2} \Big/ \sqrt{\left(1.5k - M\omega^2\right)^2 + (1.5c\omega)^2} = 49.91[lb] = 22.639[kg]$$

At times τ between 0 s, 0.5 s and 30 s, we have:

$F(\tau) =$

0	kg
-24.873	
46.747	
-62.982	
71.62	
-71.62	
62.982	
...	

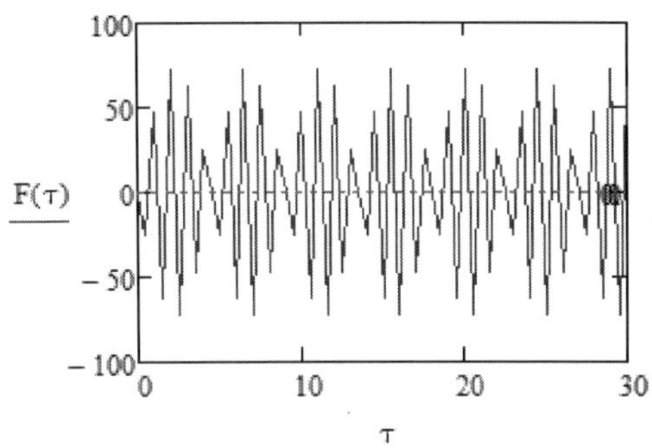

Figure 6.20. *Ideal representation of F(τ). For a color version of this figure, see www.iste.co.uk/grous/design.zip*

6.5. Critical speeds of shafts in mechanical systems

All rotating shafts, even without any loading, undergo deflection due to the influence of their own weight. The extent of the deflection depends on their stiffness. The notations used in talking about such matters are as follows:

– m is the mass in kg or lb;

– k is the constant of the spring of the shaft;

– f is the static deflection localized to the attached mass in mm or in;

– Y = A is desired amplitude in mm or in;

– ω is the frequency of rotation (angular velocity) in rad/s;

– ω_n is the so-called natural frequency, in rad/s;

– ω_c is the critical frequency, in rad/s;

– $P = mg$ is the force (the weight) in N or *lbf*;

– g is the gravitational constant, 9.81 m/s^2 (SI) or 32.2 ft/s^2 = 386 in/s^2.

CASE 1.– At the first critical velocity – i.e. at the natural frequency (ω_n) – the shaft simply flexes. At its second critical velocity, the shaft flexes again to the degree II, and so on. This is illustrated below.

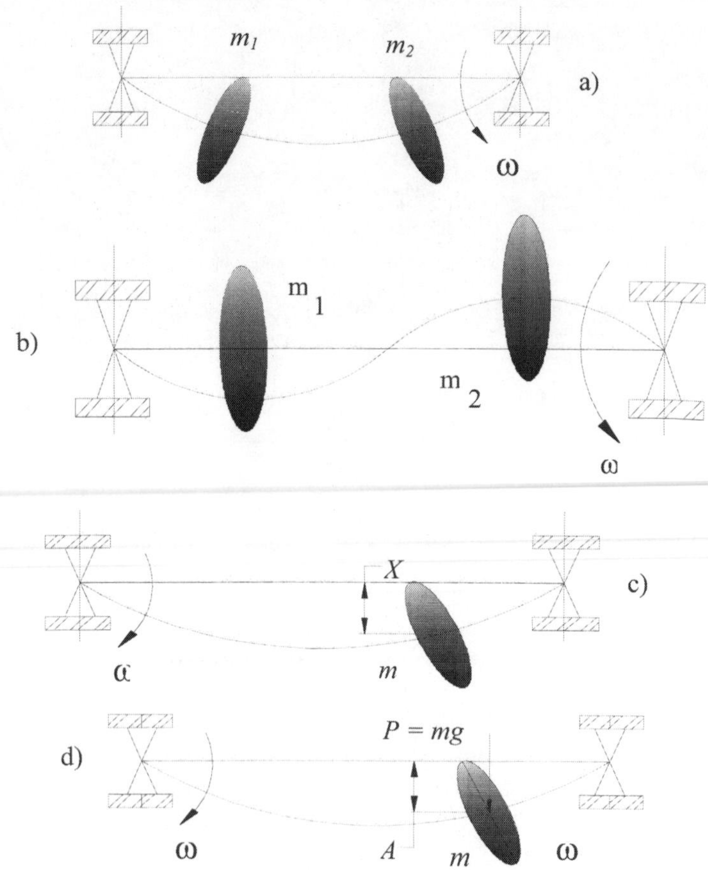

Figure 6.21. *Representation of masses on shafts rotating at critical velocities. For a color version of this figure, see www.iste.co.uk/grous/design.zip*

CASE 2.– When the shaft's own mass is reasonably small in comparison to the attached mass, the critical frequency is calculated thus:

$$\omega_{critical} = \sqrt{k/m}\ \left[rad/s\right] \qquad [6.53]$$

This expression holds true regardless of the orientation of the shaft, whether the inclination is vertical, horizontal or even between the two. The critical velocity is expressed thus:

$$\omega_{critical} = \sqrt{g / f_{stat}} \ \ [rad/s] \qquad [6.54]$$

The action of the masses is similar to the phenomenon of a gyroscope, which explains the very slight (negligible, even) difference between the two cases.

CASE 3.– When the shaft is subjected to loading only at the endpoints – i.e. without any additional masses – the critical velocity is written as follows:

$$\omega_{critical} = \sqrt{\frac{5}{4}\frac{g}{f_{static}^{\max}}} \ \ [rad/s] \qquad [6.55]$$

$f_{static}^{\max}$ is the maximum static deflection resulting from the uniform loading of the shaft (P = mg), when the mass of the shaft is *supposedly negligible* but it supports several other concentrated masses, as in the cases illustrated below.

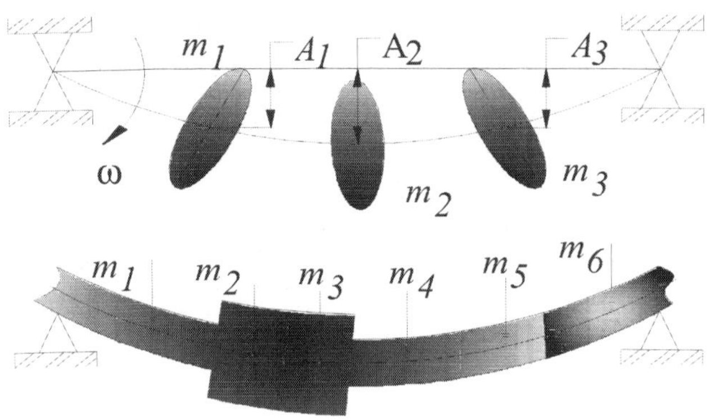

Figure 6.22. *Ideal representation of a mass attached to a rotating shaft. For a color version of this figure, see www.iste.co.uk/grous/design.zip*

These cases are handled by what is known as the Rayleigh–Ritz equation:

$$\omega_{critical} = \sqrt{g \times \sum_{i=1}^{j} P_n \times f_n \Big/ \sum_{i=1}^{j} P_n \times f_n^2} \ \ [rad/s] \qquad [6.56]$$

where:

– j represents the total number of masses forming the load;

– P_n represents the weight of the n masses forming the load;

– f_n represents the deflection of the n masses forming the load.

There are a variety of experimental and theoretical methods to distribute the masses around every part of the system. We shall not detail these methods here, but present the following example of the so-called Dunkerley equation:

$$\frac{1}{\omega_c^2} = \frac{1}{\omega_1^2} + \frac{1}{\omega_2^2} + \frac{1}{\omega_3^2} + \ldots\ldots\frac{1}{\omega_n^2} \quad [6.57]$$

where:

– ω_c is the first critical velocity of rotation, of the multi-system illustrated above;

– ω_1 is the first critical velocity of rotation, linked to mass 1, etc., ω_n.

Note simply that the equations used here (Rayleigh–Ritz and Dunkerley) are approximations of the natural frequency of vibrations. At the highest critical velocity of the shaft, note the loading of two masses:

$$\frac{1}{\omega^4} - \frac{1}{\omega^2}\left[a_{11}m_1 + a_{22}m_2\right] + m_1 m_2\left[a_{11}a_{22} - a_{12}a_{21}\right] = 0 \quad [6.58]$$

This is a quadratic form ($1/\omega^4$) of positive roots with ω_1 and ω_2 respectively representing the first and second critical velocity – i.e. the natural critical velocities of the vibration. Linked to the respective masses, we present the formulation as a matrix, where a_1 and a_2 are the constants (coefficients). In this matrix, a_{12} is the deflection located on the mass m_1, caused by the localized mass m_2; a_{11} is the deflection 1 caused by the mass m_1, etc. The reciprocal work theorem is known as Maxwell's theorem in states $a_{12} = a_{21}$. For a multi-mass system, the equation of the frequency is the result of the matrix calculation with a null determinant ($\Delta = 0$):

$$\begin{bmatrix} \left(a_{11}m_1 - \frac{1}{\omega^2}\right) & a_{12}m_2 & a_{13}m_3 & \ldots \\ a_{21}m_1 & \left(a_{22}m_2 - \frac{1}{\omega^2}\right) & a_{23}m_3 & \ldots \\ a_{31}m_1 & a_{22}m_2 & \left(a_{33}m_3 - \frac{1}{\omega^2}\right) & \ldots \\ \ldots & \ldots & \ldots & \ldots \end{bmatrix} \quad [6.59]$$

6.5.1. *Case study with solution and discussion*

Consider a rotating shaft where two gears are attached to it at the masses m_1 and m_2 (see Figure 6.23). The pinion represents $m_1 = 35\ lb = 15.876$ kg, and the gear wheel represents $m_2 = 75\ lb = 34.019$ kg. The respective deflections are $f_1 = 0.0010$ in $= 0.025$ mm and $f_2 = 0.0002$ in $= 0.026$ mm. We wish to find the appropriate value of d and test the shaft by calculating its critical velocity of rotation.

$$m_1 = 35\ lb = 15.876\ kg;\ f_1 = 0.001\ in = 0.025\ mm \rightarrow P_1 = m_1 f_1 = 155.688\ N = 35\ lbf$$

$$m_2 = 75\ lb = 34.019\ kg;\ f_2 = 0.00025\ in = 6.35\times10^{-3}\ mm \rightarrow P_2 = m_2 f_2 = 333.617\ N = 75\ lbf$$

Calculating the moments

$$M_1 = \sum\{P_1 f_1 + P_2 f_2\} = \sum(35\times0.001 + 75\times0.00025) = 6.073\ N.mm = 0.054\ lbf.in$$

$$M_2 = \sum P_1 f_1^2 + P_2 f_2^2 = \sum 35\times0.001^2 + 75\times0.00025^2 = 0.114\ N.mm^2 = 3.969\times10^{-5}\ lbf.in^2$$

Calculating the critical velocity

$$\omega_{critical} = \sqrt{g\frac{M_1}{M_2}} = \sqrt{g\frac{\sum\{P_1\times f_1 + P_2\times f_2\}}{\sum\{P_1\times f_1^2 + P_2\times f_2^2\}}} = 723.029\left[\frac{rad}{s}\right] = 6904\ [rpm]$$

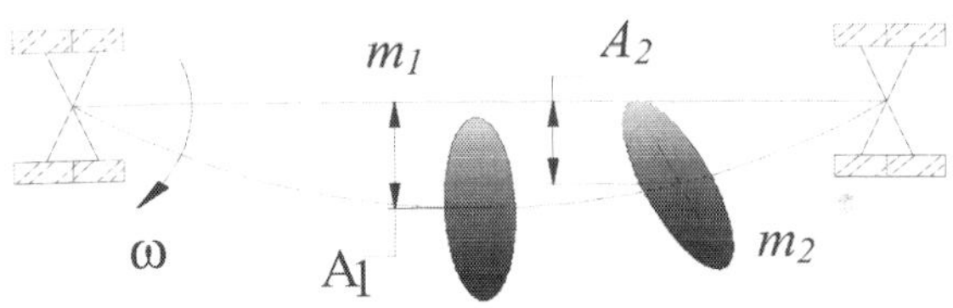

Figure 6.23. *Masses attached to a rotating shaft. For a color version of this figure, see www.iste.co.uk/grous/design.zip*

The shaft with the two masses (wheel–pinion) is capable of spinning at a critical frequency of around 6904 rpm (rad/s). This design helps achieve equilibrium in the system to prevent:

– pronounced deflection, and therefore an abnormal excentricity (ε);

– loss of energy, noise and vibration, lack of safety, premature wearing out of the bearings.

As ε is the excentricity resulting from the above, let us set:

– $k.x$ is the force idealized by the spring (Hooke vibrations);

– $(X+\varepsilon).\omega^2$ is the acceleration of the center of gravity of the mass.

$$kX = m(X+\varepsilon)\omega^2 \quad [6.60]$$

Solving this equation gives us X, which would be the deflection of the shaft of mass (m):

$$kX = m(X+\varepsilon)\omega^2 \rightarrow X\left(k - m\omega^2\right) = m\varepsilon\omega^2 / \left(k - m\omega^2\right) \quad [6.61]$$

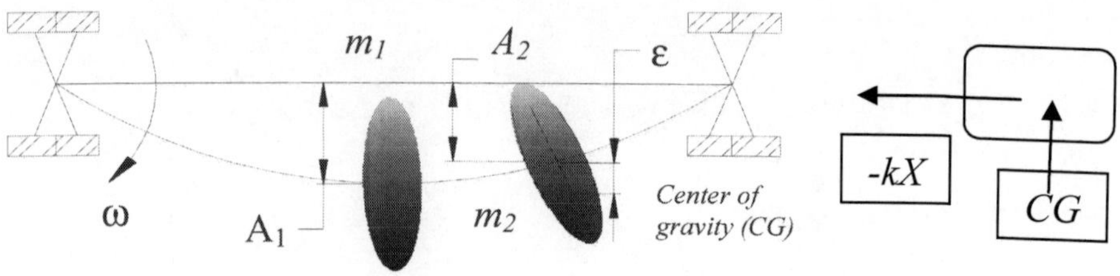

Figure 6.24. *Masses attached to a rotating shaft. For a color version of this figure, see www.iste.co.uk/grous/design.zip*

The deflection then becomes important, because it is quite significant when $k = m\omega^2_{critical}$. The critical velocity is written: $\omega_{critical} = \sqrt{k/m}$. On the basis of $m = P/g$, we can state $k/m = k(g/P) \equiv g/f$. The static deflection f is caused by the force P, so we set P/k, and hence: $\omega_{critical} = \sqrt{g/f}$.

$\omega_c(f) =$ (units: $\frac{1}{s}$)

	0
0	$1.243 \cdot 10^3$
1	$1.134 \cdot 10^3$
2	$1.05 \cdot 10^3$
3	982.344
4	926.163
5	878.635
6	837.746
7	802.081
8	...

$f := 0.00025\text{in}, 0.00030\text{in} .. 0.001\text{in}$

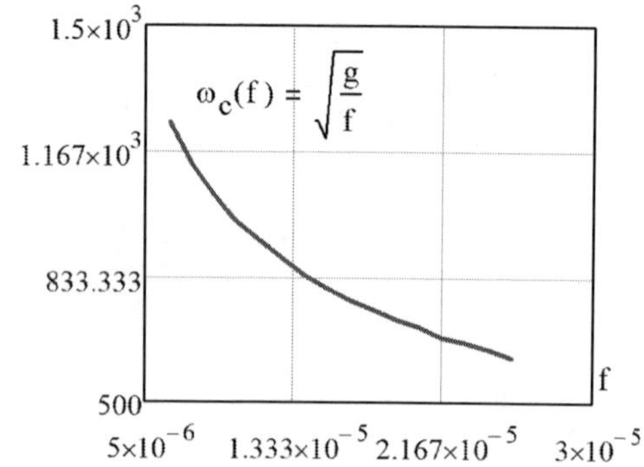

Figure 6.25. *Evolution of the critical velocity with respect to the deflection. For a color version of this figure, see www.iste.co.uk/grous/design.zip*

If we were to calculate the critical velocity at all the simulated deflections, we would clearly see how the deflection decreases in a manner inversely proportional to the critical velocity of rotation. Therefore, we can see that the critical velocity of the shaft is dependent upon the ratio of the gravitation to the deflection. It is also possible to use the energy method to validate the expression of the critical velocity:

$$E_{total} = E_{kinetic}^{\max} + E_{potential}^{\max} \rightarrow \begin{cases} E_{kinetic}^{Max} = \frac{1}{2}m_1V_1^2 + \frac{1}{2}m_2V_2^2 + \ldots = \frac{1}{2}m_nV_n^2 \\ E_{potential}^{Max} = \frac{1}{2}k_1X_1^2 + \frac{1}{2}k_2X_2^2 + \ldots = \frac{1}{2}k_nX_n^2 \end{cases} \quad [6.62]$$

Each (k) represents the stiffness (Hooke's constant) of a spring. The forces applied are therefore F_1, F_2, ... F_n, giving rise to the respective deflections X_1, X_2, ... X_n. The motion of mass is considered to be sinusoidal in form. The maximum velocity of mass for a displacement of amplitude X_n takes the form $\omega.X_n$:

$$E_{kinetic}^{Max} = \frac{1}{2}m_1\omega_1^2 + \frac{1}{2}m_2\omega_2^2 + \ldots = \frac{1}{2}\omega_n^2 \times \sum_{i=1}^{n} X_n^2 m_n \quad [6.63]$$

The work supplied by each force applied, with the slope (k), is represented below.

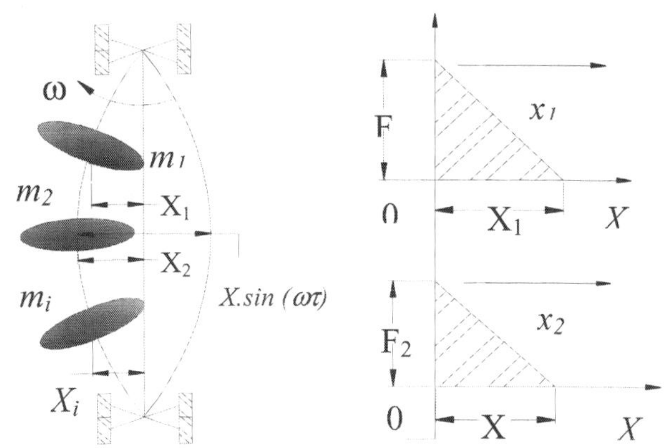

Figure 6.26. *Evolution of the critical velocity with respect to the deflection. For a color version of this figure, see www.iste.co.uk/grous/design.zip*

The equilibrium condition equalizes the kinetic energy E_c and the potential energy E_p to give us the angular velocity of the system. The form of each vibrating shaft is supposed to be identical in static deflection of the curvature, i.e. $X_1 = cA_1$; $X_2 = cA_2$; $X_n = cA_n$. This approximation gives rise to the equation of the angular velocity (ω). A_1, A2, …, A_n are the damping values in the system. Its own weight $P_n = m_n g = k_n A_n$, and therefore $m_n = P_n / g$:

$$\omega = \sqrt{\frac{\sum_{i=1}^{n} k_n X_n^2}{\sum_{i=1}^{n} m_n X_n^2}} = \sqrt{\frac{\sum_{i=1}^{n} k_n A_n^2}{\sum_{i=1}^{n} m_n A_n^2}} = \sqrt{\frac{g \times \sum_{i=1}^{n} P_n A_n}{\sum_{i=1}^{n} P_n A_n^2}} \qquad [6.64]$$

When the natural frequency of rotation (ω) is equal to the critical frequency (ω_c), we write:

$$\omega_{critical} = \sqrt{g \sum PA / \sum PA^2} \qquad [6.65]$$

NUMERICAL APPLICATIONS.– The masses m_1 and m_2 represent the pinion and the gear wheel mounted to the shaft. The respective weights are P_1 and P_2. In keeping with the deflections noted during the motions of masses, we have written the influence of the coefficients of the shaft as follows:

$$\begin{cases} P_{wheel} = 135\ lbf = 61.235 kg \\ P_{pinion} = 60\ lbf = 25.401 kg \end{cases} \rightarrow \begin{Bmatrix} a_{11} = 1.5\times10^{-6} \dfrac{in}{lb} = 8.4\times10^{-5}\ mm/kg \\ a_{22} = 10\times10^{-6} \dfrac{in}{lb} = 5.6\times10^{-4}\ mm/kg \\ a_{12} = a_{21} = 3.25\times10^{-6} \dfrac{in}{lb} = 1.82\times10^{-4} \end{Bmatrix}$$

(a_{11}) corresponds to the deflection of shaft 1, caused by the respective force. To determine the critical frequency (velocity), we shall use the approximations of the Rayleigh–Ritz equations, the Dunkerley equations, and the approximations derived from the Dunkerley method, discussed below.

6.5.2. *Method of approximation using the Dunkerley equations*

According to the approximation derived from the Dunkerley equation, we can set:

$$\begin{cases} \omega_{wheel} = \sqrt{\dfrac{g}{A_{11}}} = \sqrt{\dfrac{g}{P_1 a_{11}}} = 1.381\times10^3 \left[\dfrac{rad}{s}\right] \\ \omega_{pinion} = \sqrt{\dfrac{g}{A_{22}}} = \sqrt{\dfrac{g}{P_2 a_{22}}} = 830.328 \left[\dfrac{rad}{s}\right] \end{cases}$$

The frequency of rotation derived from Dunkerley's approximation is:

$$\left\{\frac{1}{\omega_c^2} = \left(\frac{1}{\omega_{wheel}^2} + \frac{1}{\omega_{pinion}^2}\right) = X = 1.975\times10^{-6} \rightarrow \omega_{c\text{Dunkerley}} = \omega_{cD} = \sqrt{\frac{1}{X}} = 711.58 \ \frac{1}{s} \right. \quad [6.66]$$

6.5.3. *Method of approximation using the Rayleigh–Ritz equation*

According to the approximation derived from the Rayleigh–Ritz equations, we can set the expression of the deflections (f) as follows:

$$\begin{cases} f_{wheel} = P_{wheel}a_{11} + P_{pinion}a_{12} = 3.845\times10^{-4}\ [in] = 9.766\times10^{-3}\ [mm] \\ f_{pinion} = P_{pinion}a_{22} + P_{wheel}a_{21} = 9.987\times10^{-4}\ [in] = 0.025\ [mm] \end{cases}$$

$$\begin{cases} Balance\ of\ the\ [wheel]\ and\ [pinion] \rightarrow P\times f^2 = \ \downarrow \\ P_{wheel}\times f_{wheel}^2 = 1.996\times10^{-5}\left[lb.in^2\right] = 5.841\times10^{-5}\left[kg.mm^2\right] \\ P_{pinion}\times f_{wheel}^2 = 5.586\times10^{-5}\left[lb.in^2\right] = 0.016\left[kg.mm^2\right] \end{cases}$$

$$\sum\nolimits_{(1)wheel/pinion} = P_r f_r^2 + P_p f_r^2 = 7.581852\times10^{-5}\ lb.in^2 = 0.022\ kg.mm^2$$

$$\begin{cases} Balance\ of\ the\ [wheel]\ and\ the\ [pinion] \rightarrow P\times g = \ \downarrow \\ P_{wheel}\times f_{wheel} = 0.052\ [lb.in] = 0.598\ [kg.mm] \\ P_{pinion}\times f_{pinion} = 0.056\ [lb.in] = 0.644\ [kg.mm] \end{cases}$$

$$\sum\nolimits_{(2)wheel/pinion} = P_r f_r + P_p f_p = 0.107867[lb.in] = 1.242[kg.mm]$$

The frequency of rotation stemming from the RR (Rayleigh–Ritz) approximation is:

$$\omega_{cRayleigh-Ritz} = \omega_{cRR} = \sqrt{g\sum\nolimits_{(1)}Pf \Big/ \sum\nolimits_{(2)}Pf^2} = 741.038\ [rad/s]$$

6.5.4. *Method of approximation using the equations of the rotation frequencies*

$$\left\{\begin{array}{l} a_{11}m_{wheel}+a_{22}m_{pinion}=a_{11}\left(P_{wheel}/g\right)+a_{22}\left(P_{pinion}/g\right)=1.975\times10^{-6}\left[s^{2}\right] \\ \left(a_{11}a_{22}-a_{12}a_{21}\right)m_{p}m_{r}=\left(a_{11}\left(P_{wheel}/g\right)+a_{22}\left(P_{pinion}/g\right)\right)P_{wheel}P_{pinion}= \\ \quad =\left(a_{11}a_{22}-a_{12}a_{21}\right)P_{wheel}\times P_{pinion}/g^{2}=2.251\times10^{-13}\left[s^{4}\right] \end{array}\right\}$$

$$\begin{cases} F\left(\omega_c\right)=\left(1/\omega_c^4\right)-1.975\times10^{-6}\left(1/\omega_c^2\right)+2.251\times10^{-13}\ rad/s \\ \ldots\ and\ \text{let us prove}\ for\ which\ values\ of\ \left(\omega_c\right)\rightarrow F\left(\omega_c\right)=0 \end{cases}$$

6.5.5. *Method for solving the function F(ω$_c$): roots → (r$_0$ and r$_1$)*

We enter a hypothetical value for the solution, and modify it gradually until we obtain: *F(ω$_c$) = 0, which is the condition of equilibrium, in our design.* The solver that was used then demonstrates good convergence of the solution. We can represent, on a graph, the function to find a value which is reasonably close to the root corresponding to the solution adopted in the initial hypothesis $F(\omega_c) = 0$, and then present the proof. For example, for $\omega_c = 711.58$ rad/s, we find the roots of $F(\omega_c) = 0$:

$$\left\{\begin{array}{ll} \left[r_0=root\left(F\left(\omega_c\right),\omega_c\right)\right]\{\ and & \left\{r_1=root\left[F\left(q\right),q.-750,-700\right]\right. \\ r_0=734.5087 & and\quad r_1=-734.509 \end{array}\right\} \qquad [6.67]$$

We have just proved that r_0 and r_1 are the roots which return a zero value of $F(\omega_c) = 0$. $r_0 = 734.5087$ rad/s and $r_1 = 734.509$ 1/s are the precise values of the critical velocity $\omega_{cequat\text{-}Frequencies}$, found by the frequency method. We can deduce that the approximation using the Dunkerley equation $\omega_{cD} = 711.580$ 1/s is better than the Rayleigh–Ritz approximation $\omega_{cRR} = 741.038$ 1/s.

In the case of *complex roots*, only the real part (**r$_0$**) is shown in the f(z) graph below, and encapsulates the root r_0.

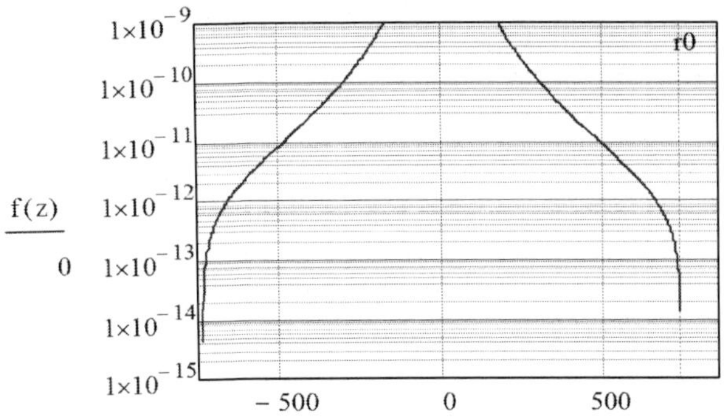

Proof

f (r0) = 0

f (r1) = 0

Cqfd

f (−734.509) = 0 f (734.5087) = 0

Figure 6.27. *Graphical solution of the roots of the function F(ωc) = 0. For a color version of this figure, see www.iste.co.uk/grous/design.zip*

Comparison between the two approximations

$$\left\{\begin{array}{l}\text{Dunkerley}\downarrow \quad \text{and} \quad \text{Rayleigh-Ritz}\downarrow \\ \\ \omega_{cD} = 711.58[rad/s] \ and \ \omega_{cRR} = 741.038[rad/s]\end{array}\right\} \rightarrow \frac{\text{Rayleigh-Ritz}}{\text{Dunkerley}} = \frac{741.038}{711.580} = 1.041$$

Note that the Dunkerley equation underestimates the approximation. The Rayleigh–Ritz equation overestimates it. There is an interval of (1.048 rad/s) between the two.

COMMENTS ON THE DESIGN.– The *TRUE* value lies partway between the two approximations. It can quite easily be shown that the design without preliminary calculation is doomed to fail if the pre-project phase relies on it. Design is not a matter of "sketching", but rather of careful mathematical calculation.

6.6. Conclusion

Of the various methods available to provide *long-lasting* solutions to the problems of vibrations, noise and shocks, the principle of conservation of energy is a potent tool to help better understand the vibrations occurring in complex systems. Systems which have slight damping generate errors due to the calculations of the natural frequencies. The natural frequency tends to be treated as being zero, but in our view, this is a mistake. The *Rayleigh* method, though, is readily applicable to formed pieces and other shafts with associated masses. This method is based on the observation that the systems (shaft-pulleys – masses – shaft-bearings, etc.) reach their maximum deflections simultaneously at the lowest natural frequency (ω_n).

7

Principles of Calculations for Fatigue and Failure

7.1. Mechanical elements of failure through fatigue

Here we present the concept of safety coefficients. For a given structure, such as a vehicle, hydroplane or even a tooling machine, to safely fulfill its *mission* (purpose) in light of the loads to which it is subjected, it needs to be able to withstand those loads for the set periods of time. Thus, a reassuring-sounding factor known as the safety coefficient represents that ratio of strengths, written as:

$$S_{\text{structure}} = \left\{ \frac{\text{Admissible loads}}{\text{Applied loads}} \right\} = \left\{ \frac{\text{Real strength}}{\text{Necessary strength}} \right\} \qquad [7.1]$$

For fragile materials such as concrete, wood and composites, it is recommended to consider the failure limit as equivalent to the material's elastic limit:

$$S_{structure} = \left\{ \frac{R_{elastic}}{R_{tolerated}^{in\ practice}} \right\} = \left\{ \frac{R_{failure}}{R_{tolerated}^{in\ practice}} \right\} \qquad [7.2]$$

The value, often written as $[S_s]$, is dependent upon numerous phenomena acting on the structure, such as shocks, fatigue, repeated loads, stress intensifications, the morphology (nuance) of the material, the environment, the surface state, etc. When choosing a safety coefficient, designers often find themselves facing needlessly high costs. It would be very wise to have multidisciplinary experts choose that coefficient. In addition, it is codified, and only legislation can govern the choice. For example, a low (and therefore cheap) safety coefficient would expose the designer and manufacturer to the danger of failure – often prematurely.

Conversely, a high, and therefore expensive, safety coefficient could mean overloading the structure beyond that allowed for in the technical specifications.

Many researchers prefer to make statements on the basis of reliability, or the likelihood of destruction in relation to a safety zone, also probabilistic. This is what appears to be favored by fracture mechanics. In many areas of mechanics, we also use the load factor, defined thus:

$$S_s = \left\{ \frac{[F]_{admissible}}{F_{usual}} \right\} = \left\{ \frac{\text{Maximal admissible load}}{\text{Usual load during operation}} \right\} \qquad [7.3]$$

In civil engineering, aviation, offshore structures, fatigue fracture mechanics [GRO 94, GRO 13], it is common to allow a *safety margin*:

$$\text{Safety margin} = M_s = \{S_s - 1\} \qquad [7.4]$$

This author's own previous work has focused in great depth on the above [GRO 94, GRO 13]. This book is dedicated primarily to design, so we shall limit ourselves, here, to presenting a few values as illustrative examples of typical safety coefficients. However, it is the responsibility of the designer to ensure a suitable value of [S] is chosen.

[S] coefficient	Loads exerted on the structure	Stresses in the structure	Behavior of the material
$1 < S < 2$	Known and regular	Known	Known after test
$2 < S < 3$	Regular or well known	Fairly well known	Average knowledge even after a test
$3 < S < 4$	Averages known, or uncertain	Averages known, or uncertain	Known without a test

Table 7.1. *Indicative values of the safety coefficient*

This table is merely indicative. In the real world, it is important to refer to the specialized tables attached to standards and/or legislation. In the case studies in all chapters of this book, we refer to a "homogeneous and isotropic" material. At this juncture, it is worthwhile recapping the definition of these two important concepts.

A material is referred to as homogeneous when all of its component crystals are identical (surface state, grains, internal structure, etc.). A material is said to be *isotropic* when, within its own structure, all points (atoms) exhibit the same mechanical characteristics along the three axes (*xyz*). For example, we made no mention of this concept when presenting the case studies on the wooden platform, as wood is not isotropic. It exhibits better resistance in the direction of the fibers and poorer resistance perpendicularly to it. The starting assumption for many discussions is that metals are homogeneous and isotropic. In our view, this hypothesis should be treated *with a degree of caution*. Indeed, it is the typical trap into which certain designers cannot avoid falling.

7.2. Analysis of materials and sizing in applied design

Mechanical design must, unavoidably, be based on the materials used for the product. The calculations of stresses in a material are based on the six components along (x, y, z). By direct application, we shall find the value of the stresses along the three axes of the frame of reference. The *axial stresses* under longitudinal tension, vertical and lateral compression, are as follows:

$$Traction \rightarrow \{\sigma_x \ \sigma_y \ \sigma_z\} = \{44.816 - 17.237 \ 51.711\} \ MPa = \{6500 - 2500 \ 7500\} \ psi$$

$$Shear \rightarrow \{\tau_{xy} \ \tau_{xz} \ \tau_{yz}\} = \{-1000 \ 4000 \ \text{-}2000\} \ psi = \{-6.895 \ 27.579 \ \text{-}13.79\} \ MPa$$

Note that the (-) signs correspond to anticlockwise directions. The three stresses $\{\sigma_0, \sigma_1, \sigma_2\}$ are written thus:

$$\begin{pmatrix} C_2 \\ C_1 \\ C_0 \end{pmatrix} = \begin{pmatrix} \sigma_x + \sigma_y + \sigma_z \\ \sigma_x.\sigma_y + \sigma_y.\sigma_z + \sigma_x.\sigma_z - \tau_{xy}^2 - \tau_{xz}^2 - \tau_{yz}^2 \\ \sigma_x.\sigma_y.\sigma_z + 2.\tau_{xy}.\tau_{xz}.\tau_{yz} - \sigma_x.\tau_{yz}^2 - \sigma_y.\tau_{xz}^2 - \sigma_z.\tau_{xy}^2 \end{pmatrix} \qquad [7.5]$$

After solving the above equation, we find:

$$\begin{cases} C_2 \left(\frac{C_1}{MPa}\right) \left(\frac{C_0}{MPa^2}\right) = \left\{79.29 - 344.648 - 3.257\times10^4\right\} MPa, \; or \swarrow \\ C_2 \left(\frac{C_1}{psi}\right) \left(\frac{C_0}{psi^2}\right) = \left\{1.15\times10^4 - 7.250\times10^6 - 9.937\times10^{10}\right\} psi \end{cases} \quad [7.6]$$

The main stresses are the result of the function f(σ), written as:

$$f(\sigma) = \sigma^3 - C_2 \times \sigma^2 + C_1 \times \sigma - C_0 \quad [7.7]$$

Given the range of variables, which may take negative values, we assign (σ) a range ω, from 1 to 27, and set: σ_i = i(*psi*) or i(*MPa*). The fitting of the main root is obtained by shifting (σ) to see the negative regions of the function. Thus, at the scale 700, we have:

$$\begin{cases} Shift_i = \left\{\sigma_i - 10\times\left(lbf/in^2\right)\right\} \; at\ a\ scale = 700, we\ set \lrcorner \\ fl_i = (scale\times shift_i)^3 - C_2(scale\times shift_i)^2 + C_1(scale\times shift_i) - C_0 \end{cases} \quad [7.8]$$

The graph of the function fl_i with respect to (0) shows the variation of stresses (+σ) and (–σ). After fitting the σ axis, we are able to state the appropriate values of the stresses on the three axes:

$$\{\sigma_1 \ \sigma_2 \ \sigma_3\} = (12000 \ 2400 - 4600) \ psi = (82.737 \ 16.547 - 31.716) \ MPa$$

$$\begin{cases} \sigma = \left\{ root(f(\sigma_1),\sigma_1) \ root(f(\sigma_2),\sigma_2) \ root(f(\sigma_3),\sigma_3)\right\} \\ \sigma = \left(1.137\times10^4 \ 3.023\times10^3 - 2.892\times10^3\right) psi = (78.385 \ 20.842 - 19.937) \ MPa \end{cases} \quad [7.9]$$

The precise value of the primary longitudinal stress is then written as:

$$\sigma = \left\{1.137\times10^4 \ 3.023\times10^3 - 2.892\times10^3\right\} psi = \{78.385 \ \ 20.842 - 19.937\} \ MPa$$

After a second adjustment (fitting) using the following equation, we set:

$$f(\sigma^T)_0 \ f(\sigma^T)_1 \ f(\sigma^T)_2 = \left\{-7.678\times10^{-5} \ \ 0 \ \ 0\right\} psi^3 = \left\{-2.517\times10^{-11} \ \ 0 \ \ 0\right\} MPa^3$$

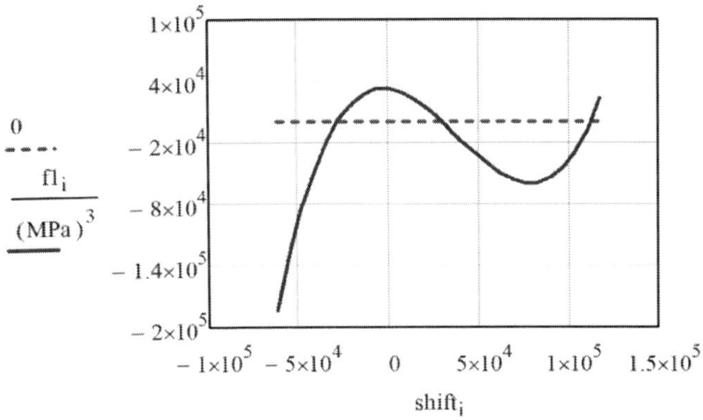

Figure 7.1. *Function fli with respect to (0). Variation of stresses (+σ) and (-σ). For a color version of this figure, see www.iste.co.uk/grous/design.zip*

Therefore, the maximum stress is written:

$$\{\max\sigma = if\left(\left|\max\sigma\right|\right) \succ \left|\min(\sigma)\right|, \left|\max\sigma, \min(\sigma)\right| = 1.113\times 10^4\ psi = 78.385\ MPa$$

With the three primary stresses σ, let us find the maximum stress C:

$$\begin{cases} C=\sigma^T \rightarrow \begin{pmatrix} C_0 & \sigma_y & \tau_{xz} \\ \sigma_x & \tau_{xy} & \tau_{yz} \end{pmatrix} = \begin{pmatrix} 1.137\times10^4 & -2.5\times10^3 & 4\times10^3 \\ 6.5\times10^3 & -1\times10^3 & -2\times10^3 \end{pmatrix} psi \\ \sigma = C^T \rightarrow \begin{pmatrix} C_0 & \sigma_y & \tau_{xz} \\ \sigma_x & \tau_{xy} & \tau_{yz} \end{pmatrix} = \begin{pmatrix} 14200.psi & \sigma_y & \tau_{xz} \\ \sigma_x & \tau_{xy} & \tau_{yz} \end{pmatrix} \end{cases} \quad [7.10]$$

$$\begin{cases} C=\sigma^T \rightarrow \begin{pmatrix} C_0 & \sigma_y & \tau_{xz} \\ \sigma_x & \tau_{xy} & \tau_{yz} \end{pmatrix} = \begin{pmatrix} 78.385 & -17.237 & 27.579 \\ 44.816 & -6.895 & -13.79 \end{pmatrix} MPa \\ \sigma = C^T \rightarrow \begin{pmatrix} C_0 & \sigma_y & \tau_{xz} \\ \sigma_x & \tau_{xy} & \tau_{yz} \end{pmatrix} = \begin{pmatrix} 97.906\ MPa & \sigma_y & \tau_{xz} \\ \sigma_x & \tau_{xy} & \tau_{yz} \end{pmatrix} MPa \end{cases} \quad [7.11]$$

The direction cosines enable us to write $\begin{pmatrix} I_{1x} & I_{1y} & I_{1z} \end{pmatrix} = \begin{pmatrix} 0.5 & 0.5 & 0.5 \end{pmatrix}$. Given that:

$$\begin{cases} \left[\sigma_x - \left(\sigma^T\right)_0\right]\times I_{1x} + \tau_{xy}\times I_{1y} + \tau_{xz}\times I_{1z} - 0\ psi \\ \tau_{xy}\times I_{1x} + \left[\sigma_y - \left(\sigma^T\right)_0\right]\times I_{1y} + \tau_{yz}\times I_{1z} = 0\ psi \end{cases} \quad [7.12]$$

$$l_{1x}^2 + l_{1y}^2 + l_{1z}^2 = 1 \rightarrow Find\left(l_{1x}, l_{1y}, l_{1z}\right) = \begin{pmatrix} +0.646 \\ -0.154 \\ +0.748 \end{pmatrix} \quad [7.13]$$

Finally, let us write the direction cosines along the three primary directions and the maximum shear stress at the greatest magnitude of stress ω:

$$\begin{cases} \omega = \{(\sigma_1 - \sigma_2) \quad (\sigma_1 - \sigma_3) \quad (\sigma_2 - \sigma_3)\}/2 \\ \omega = \{4.8\times10^3 \quad 8.3\times10^3 \quad 3.5\times10^3\}\ psi = \{33.095 \quad 57.226 \quad 24.132\}\ MPa \end{cases} \quad [7.14]$$

The maximum shear stress is:

$$Max(\tau) = if\left|\max(\omega)\right| \succ \left|\min(\omega)\right|, \max(\omega), \min(\omega) = 8.3\times10^3\ psi = 57.226\ MPa \quad [7.15]$$

7.3. Sizing of pivot joints with bearings

The performance of a bearing is based on its lifetime, which is dependent upon the number of rotations the bearing carries out before showing signs of wear. “Jamming” may also occur due to an insufficient supply of lubrication. The bearing has contact between rolling bodies and rings, which are the site of specific types of stresses. In this book, we shall not discuss the phenomena of wear and tear. Let us simply remind readers that it is important to take account of these phenomena during a live project. Little by little, designers (particularly in the field of aeronautics) are coming to use quite advanced calculations to optimize the performances of the mechanical systems they create. For instance, in industries where the weight-to-power ratio needs to be as low as possible, in the design calculations, CAD software offers calculation tools in line with the drawings made. Evidently, for a designer, a bearing is one of the most frequently used mechanical components in connections between rotating sub-systems. The catalogues from bearing manufacturers (such as SKF, among others), supported by ISO standards, help in choosing and installing the right components for the mechanism at hand.

Methods for sizing of bearings are not all covered by ISO 281: 2007 [ISO 07], which applies only to constructions where the rolling parts transfer power to the shaft, not to the bore, unless the surface area is equivalent to the inner ring. We then calculate the clearance fit, as illustrated by Stribeck’s work [STR 07], where the link with Hertz’s theory (contact between the ball bearing and the track) applies. The work of Lundberg and Palmgren [LUN 07] constitutes the basis for Norm 76 [NOR 76]. By studying the fatigue behavior of bearings, it is possible to put forward a sizing criterion.

7.3.1. *Basic formulae for calculating lifetime*

The below formulae are inspired by manufacturers' documentation. The lifetime of a bearing can be calculated in a number of different ways depending on the operating conditions. The simple method based on ISO 281: 2007 can be used to calculate the lifetime achieved by 90 % of bearings working under dynamic loading. It is linked to the phenomenon of fatigue of the material as a cause of failure. To determine the lifetime in accordance with ISO 281: 2007, we calculate the following.

1) The equivalent dynamic radial load P

$$P = X \times F_r + Y \times F_a, \qquad [7.16]$$

where:

– X and Y are load factors defined in a table provided by the manufacturer;

– F_a and F_r are the axial and radial forces applied. The nominal lifetime L_{10} is calculated thus:

$$L_{10} = \left(\frac{C}{P}\right)^n \times 10^6;\ \text{turns or indeed } L_{10} = \left(\frac{C}{P}\right)^n \times \frac{10^6}{60N}\ \text{hours} \qquad [7.17]$$

for: P = C; L10 = 106 rotations, which is the dynamic load capacity (in N). $n = 3$ constant for ball bearings and n = 10/3 for roller bearings.

2) Basic dynamic load of the bearing

The basic dynamic load of the bearing is calculated using ISO 281: 2007:

$$\begin{cases} \textit{Ball bearing},\ d_{balls} \prec 25.4\ mm, C = f_c\left(i \times \cos(\alpha)^{7/10} \times Z^{2/3} \times D_w^{1.8}\right) \\ \textit{Ball bearing},\ d_{balls} \prec 25.4\ mm\ \alpha\ =\ 90°, C = f_c\left(\cos(\alpha)^{2/3} \times D_w^{1.8}\right) \\ \textit{Roller bearing}, C = f_c\left(i \times l \times \cos(\alpha)^{7/9} \times Z^{3/4} \times D_w^{29/27}\right) \end{cases} \qquad [7.18]$$

– D_w is the diameter of the rolling part, which is big;

– Z is the number of rolling parts.

3) Load capacity of double bearings

Double bearings are ones with two sets of rolling parts ($i = 2$) or indeed sets made up of two identical bearings. The capacity (C_e) of the whole system is that of one set (C), multiplied by the number of sets of balls $\{2^{7/10} \equiv 1.625$ and $2^{7/9} \equiv 1.715\}$.

According to documentation published by SNR and SKF, doubling a bearing improves the platform's load capacity by around 70 %, depending on the type used:

– *axial load factor Y*: the axial load factor Y, which depends on the contact angle, is calculated differently depending on the type of bearing:

- *radial-contact ball bearings*: the contact angle (α) is zero, with a single radial load. The axial load creates local deformations at the point of contact between the balls and the bearing tracks. This leads to a relative axial displacement of the two rings. Thus, (α) increases as a function of the axial force applied. A ratio F_a/C_0 is used to calculate Y. It takes account of the modification of (α) due to the axial force;

- *angular-contact bearings*: the contact angle (α) varies little depending on the combined loads. The axial load factor Y for a given (α) is roughly constant. Oblique-contact ball bearings, with an identical value of (α) for all the bearings, are calculated with the same (Y). Conical roller bearings have a Y which varies depending on series and size.

4) Static capacity

The dimensions of the bearing are chosen on the basis of the static load when the bearing is at rest, or undergoing slight oscillations, and subject to continuous or intermittent loads. Owing to the stresses in the contacts of the rolling bodies with the tracks, the application of a static load causes permanent local deformations. When rotating, these deformations are damaging to the proper function of the bearing. A maximum admissible radial load is imposed. The stress resulting from that load can be tolerated for the majority of applications. The basic static capacity C_0 is also its maximum admissible load. The basis static capacity of a bearing $\mathbf{C_0}$ is defined by ISO 76: 2006 [ISO 06]. The same is true for the radial load (or axial load in the case of stops) which creates a Hertz pressure in the contact under greatest load. Manufacturers suggest:

– 4000 MPa for roller bearings and roller stops (all types of bearings);

– 4200 MPa for ball bearings and ball stops (except roller hinges on balls);

– 4600 MPa for roller hinges on balls.

5) Equivalent static load P_0 and safety factor

When the bearing is subject to combined static loads of which F_r is the radial component and F_a the axial component, an equivalent static load is calculated and compared to the bearing's static capacity. The safety factor f_s is important in the case of bearings operating in harsh conditions:

$$F_{safety} = C_0 / P_0 \text{, dimensionless} \qquad [7.19]$$

Here are a few minimal values, in principle, for the safety factor F_s:

– 1.5–3 for severe requirements;

– 1.0–1.5 for normal conditions;

– 0.5–1 for operations without requirements in terms of noise or accuracy.

6) Equivalent static load

The equivalent static load is the larger of the two values:

$$P_0 = \{X_0 \times F_r + Y_0 \times F_a\},\ F_a = 0 \rightarrow for\{P_0 = F_r\}\ or\ negligible \qquad [7.20]$$

Coefficients X_0 *and* Y_0 *from SKF or SRN tables.* F_r *and* F_a *are the static forces applied*

7) Variable loads or speeds

When a bearing operates under variable loads or at variable speeds, an equivalent load and speed are determined in order to calculate the lifetime.

8) Constant load and variable speed of rotation

$$\text{Equivalent speed: } N_e = \{\tau_1 N_1 + \tau_2 N_2 + \ldots + \tau_k N_k\}; \sum_{i=1}^{k} \tau_k = 1 \qquad [7.21]$$

9) Constant load and speed of rotation

$$\text{Equivalent load: } P_e = \left(\tau_1 P_1^n + \tau_1 P_1^n + \ldots + \tau_k P_k^n\right)^{1/n} \ ; \sum_{i=1}^{k} \tau_k = 1 \qquad [7.22]$$

10) Periodic load and constant speed of rotation

– *Equivalent* sinusoidal load: $P_e = \{(0.32 \times P_{min}) + (0.68 \times P_{max})\}$ [7.23]

– *Equivalent* linear load: $P_e = \{P_{min}/3 + 2P_{max}\}$ [7.24]

If the speed and load are variable, we calculate the lifetime for each rate of use, and then the weighted lifetime for the whole of the cycle.

11) Variable load and speed of rotation

$$\text{Weighted lifetime: } L_{\text{Weighted}} = \left(\frac{\tau_1}{L_1} + \frac{\tau_2}{L_2} + \ldots + \frac{\tau_k}{L_k}\right)^{-1} \ ; \sum_{i=1}^{k} \tau_k = 1 \qquad [7.25]$$

where:

– τ_k: usage rate;

– N_k: rotation speed for usage rate τ_k;

– P_k: load for usage rate τ_k;

– L_k: lifetime for usage rate τ_k;

– $n = 3$: bearings and ball stops;

– $n = 10/3$: bearings and roller bearings.

12) Axial equilibrium of the shaft

As the paths of angular-contact bearings are inclined, the radial loads F_{r1} and F_{r2} produce an axial reaction force, known as the induced axial force. RxQ_{a1} and RxQ_{a2} are the axial loads applied to the bearings. If the bearing is that whose induced axial force is in the direction of the external axial force A, the equilibrium of the shaft is calculated as follows:

$$\textit{Case of loading:}\quad A+\left\{\frac{F_{r1}}{2Y_1}\succ\frac{F_{r2}}{2Y_2}\right\} \qquad [7.26]$$

Bearing 1 operating with CLEARANCE FIT	Bearing 1	Bearing 2
Applied axial load	$R\times Q_{a1}=\{F_{r1}/2Y_1\}$	$R\times Q_{a2}=A+\{F_{r1}/2Y_1\}$
Axial load used in calculation of the equivalent dynamic load	$F_{a1}=\{0\}$	$F_{a2}=\{R\times Q_{a2}\}$

Table 7.2. *Axial equilibrium*

$$\textit{Case of loading:}\quad A+\left\{\left(\frac{F_{r1}}{2Y_1}\right)\prec\left(\frac{F_{r2}}{2Y_2}\right)\right\} \qquad [7.27]$$

Bearing 2 operating with CLEARANCE FIT	Bearing 1	Bearing 2
Applied axial load	$R\times Q_{a1}=\{F_{r2}/2Y_2\}-A$	$R\times Q_{a2}=\{F_{r2}/2Y_2\}$
Axial load used in calculation of the equivalent dynamic load	$F_{a1}=\{R\times Q_{a1}\}$	$F_{a2}=\{0\}$

Table 7.3. *Axial equilibrium. Case of loading*

The required lifetime for the bearing is set by the manufacturer of the machine.

13) Corrected nominal lifetime and reliability of bearings

The basic nominal lifetime L_{10} is a satisfactory estimate of the performances of a bearing. It represents 90 % reliability, in normal operating conditions. Beyond a certain level of loading C_u, a quality bearing could achieve a very long lifetime, in favorable lubrication conditions. ISO 281: 2007 introduces a lifetime correction factor, a_{ISO}, which can be used to calculate a nominal lifetime as follows:

$$L_{nm} = a_1 \times a_{ISO} \times L_{10} \qquad [7.28]$$

The coefficient L_{nm} represents the estimated influence of the lubrication and contamination of the bearing. It takes account of the fatigue limit of the materials, with which the bearing may begin to exhibit deterioration at a random time. The reliability value F for an operational lifetime L is expressed in mathematical form as a function of the reference lifetime L_{10}:

$$F = \exp\left\{\left(\ln\frac{9}{10}\times\frac{L}{L_{10}}\right)^{\beta}\right\} \; and \; a_1 = \left(\frac{L}{L_{10}}\right) = \left(\frac{\ln F}{\ln 0.9}\right)^{1/\beta} \qquad [7.29]$$

The correction coefficient a_1 is calculated with a slope of the Weibull line; $\beta = 1.5$ (mean value for all bearings and stops). These reliability values demonstrate the great dispersion characteristic of the lifetime of bearings:

– 30 % of bearings in a given batch achieve a lifetime that is five times the nominal lifetime L_{10};

– 10 % have a lifetime equal to eight times the nominal lifetime L_{10}. Bearing performance can only be analysed by means of multiple identical tests. Based on the statistical results, it is possible to draw meaningful conclusions.

Unlike what is expressed by the above formulae (employed in ISO 281: 2007 to calculate the coefficient a_1), there is a certain value of operating time below which the bearings exhibit no risk of failure (reliability 100 %). This value is roughly equal to 5 % of the lifetime ($\alpha \times L_{10}$). In reliability calculations for short periods of operation, we use the above formula, but with a correcting factor $\alpha = 0.05$:

$$F = \exp\left\{\left(\ln\frac{9}{10}\left(\frac{L}{L_{10}}\right) - \alpha\right)^{\beta} \times (1-\alpha)^{-\beta}\right\} \; and \; D = \{1 - F\} \qquad [7.30]$$

For every value of reliability F (or $\Re$), there is a corresponding likelihood of failure D = [1-F]. This is illustrated on a Weibull plot (log-log scale) by a line having the slope ß. The values of (a_1) and of reliability for a chosen duration are read as follows:

Reliability	L_{nm}	a_1	Reliability	L_{nm}	a_1	Reliability	L_{nm}	a_1
90	$L10m$	1	99	$L1m$	0.25	99.9	$L0.1m$	0.093
95	$L5m$	0.64	99.2	$L0.8m$	0.22	99.92	$L0.08m$	0.087
96	$L4m$	0.55	99.4	$L0.6m$	0.19	99.94	$L0.06m$	0.080
97	$L3m$	0.47	99.6	$L0.4m$	0.16	99.95	$L0.05m$	0.077
98	$L2m$	0.37	99.8	$L0.2m$	0.12			

Table 7.4. *Reliability 100 %, L_{nm}, a_1. Sourced from manufacturer SNR*

According to the theory of compound probability, the reliability of a set of bearings is the product of the reliability values for its individual components: $F = F_1 \times F_2 \times \ldots \times F_n$. We can deduce the lifetime of a set on the basis of the lifetime L_{10} of each of the bearings:

$$L_e = \left((1/L_1)^{1.5} + (1/L_2)^{1.5} + \ldots + (1/L_n)^{1.5} \right)^{-1/1.5} \qquad [7.31]$$

The probability of failure of a set is, roughly speaking, the sum of the probabilities of failure of each bearing (for small failures): $D = D_1 + D_2 + \ldots + D_n$. The reliability of a mechanical system will be greater, in terms of its bearings, when the individual bearings have longer lifetimes. The lubricant separates the active metal surfaces in the bearing by maintaining a film of oil between the rolling parts and their tracks. This helps prevent wear and tear, limits abnormal stresses and limits the buildup of heat. The lubrication of the bearing is characterized by the ratio of the thickness h of the film of oil to the equivalent roughness of the surfaces in contact: $R_a = \sqrt{R_{a1}^2 + R_{a2}^2}$. Here, R_{a1} is the mean roughness of the bearing tracks and R_{a2} is the mean roughness of the rolling bodies. The efficiency of the lubrication defined by the ratio h/Ra has a great influence on the effective lifetime of the bearings.

7.3.2. *Determination of the minimum viscosity necessary*

The oils used are often mineral oils with a viscosity index close to 90. Oil is characterized by its nominal viscosity (in *centistokes*) at the nominal temperature of 40°C. From this, we deduce the viscosity at the operating temperature. The normal operating temperature of the bearing is between –20°C and +120°C. Beyond these operational limits, it has an impact on:

– the characteristics of the steel;

– the internal operational clearance fit;

– the toughness of the joints and cages made of synthetic material.

The operational conditions of bearings outside of “normal” temperature ranges are presented thus:

<table>
<tr><td>Operational T°C</td><td colspan="4">-40 -20 0 40 80</td><td colspan="2">120 160 200 240</td></tr>
<tr><td>Steel alloyed with Cr (100 Cr6)</td><td colspan="4">Standard</td><td colspan="2">Fatigue resistance. Thermal treatment</td></tr>
<tr><td>CLEARANCE FIT</td><td colspan="4">Normal operation</td><td colspan="2">Increased CLEARANCE FIT</td></tr>
<tr><td>Grease</td><td colspan="2">Low T° special</td><td>Standard</td><td>Loss of performance</td><td>High T° special</td><td>Dry lubrication</td></tr>
<tr><td>Joint</td><td colspan="2"></td><td colspan="2">Standard, acrylic nitrite</td><td colspan="2"></td></tr>
<tr><td rowspan="3">Cage (bore) Casing</td><td></td><td colspan="4">Special (fluoride elastomer)</td><td></td></tr>
<tr><td colspan="2"></td><td colspan="2">Polyamide 6/6</td><td colspan="2"></td></tr>
<tr><td colspan="6">Metal</td></tr>
</table>

Table 7.5. *Continuous operational temperature in °C. Source SNR*

7.4. Faults of form and position of ranges on the operating clearance fit

The basic dynamic load of a bearing is defined supposing that the operational radial clearance fit (clearance fit of the bearing after construction) is zero; that means that half of the rolling parts bear a load. In practice, the operational clearance fit is *never zero*. Faults of alignment on stiff bearings (without ball joints) are expressed by an angle between the axis of the inner ring and that of the outer ring. The bearing is a precision part, and in calculating its fatigue resistance, we assume a homogeneous and steady distribution of the load amongst the rolling parts. It is necessary to calculate the stresses using the finite element method when the distribution is non-homogeneous. Faults may stem from a lack of:

– concentricity between the two ranges of the shaft or casings;

– alignment between the axis of the shaft and that of the casing of a given bearing;

– linearity of the shaft and perpendicularity between the shoulders and ranges.

The maximum value of the admissible alignment faults, without significantly impacting the lifetime for a normal operating clearance fit, would be as follows

Calculation formulae	$(F_a/F_r) \prec e$	$(F_a/F_r) \succ e$
Bearing with one set of balls	0.17°	0.09°
Stiff bearing with two sets of balls Bearing with cylindrical or conical rollers	0.06°	0.06°

Table 7.6. *Alignment faults. Source SKF*

7.5. Friction and speed of bearings

The friction of a bearing and its increase in temperature depend on numerous parameters such as the applied load, the friction of the cage, the lubrication, etc. For most applications below the limit speed and without an excessive amount of lubrication, the friction in the bearings is calculated thus:

$$M_R = (\mu/2) \times F \times D_m \ [N.mm] \ and \ P_R = M_R \omega / 9550 \ [W] \qquad [7.32]$$

where:

– M_R, resistant moment (N.mm);

– P_R, absorbed power (W);

– F, radial load for bearings and axial load for stops (N);

– D_m, mean diameter of bearing. $D_m = (d + D) / 2$ (mm);

– ω, rotation speed (min^{-1});

– μ, friction coefficient.

Friction coefficient	μ	Friction coefficient	μ
Radial contact ball bearing	0.0015	Ball stop	0.0013
Ball joint bearing on balls	0.0010	Cylindrical roller bearing	0.0050
Oblique contact ball bearing		Conical roller bearing	0.0018
One set of balls	0.0020	Ball joint bearing on rollers	0.0018
Two sets of balls	0.0024		

Table 7.7. *Bearings without watertight joints. Source SKF*

ISO 15312 introduces the concepts regarding the velocities of bearings such as the reference thermal velocity, the maximum admissible velocity and the limiting velocity. The reference thermal velocity is the speed of rotation of the inner ring where thermal equilibrium is achieved between the heat produced by friction in the bearing (N_r) and the heat flow emitted through the site (shaft and casing) of the bearing (Φr). It operates in reference conditions, such that the heat generated by friction (N_r = Φr) in a bearing at the reference thermal velocity in the reference conditions is written thus:

$$N_r = \pi \times n_{\theta r} / \left(30 \times 10^3\right) \times N_{0r} \times N_{1r} \qquad [7.33]$$

where:

– M_{0r}: moment of friction independent of the load;

– M_{1r}: moment of friction dependent on the load.

$$N_r = \pi \times n_{\theta r} / 30 \times 10^3 \times \left\{10^{-7} f_{0r} \sqrt[3]{v_r n_{\theta r}} \times d_m^3 + f_{1r} P_{1r} d_m\right\} \qquad [7.34]$$

where:

– P_{1r} is the reference load;

– f_{0r} is a correction factor for the moment of friction independent of the load but dependent on the velocity in the reference conditions;

– d_m is the mean diameter of the bearing $d_m = (D + d)/2$;

– f_{1r} is a correction factor for the moment of friction dependent on the load.

According to ISO 15312, a bearing in operation may attain a maximum admissible thermal velocity, which depends on the reference thermal velocity.

7.6. Sizing of bearing pivot joints and lifetime

The sizing of bearings is assumed to be known in manufacturers' mechanical specifications. In mechanical design, the bearings are chosen so as to conform to requirements rclating to the conditions of bulk, speed, precision, metrology, setup, etc. The same is true when calculating the tolerances (clearance fits). The range of available bearings is clearly set out in manufacturers' manuals, which offer tables and exact grids to reliably calculate lifetimes. The dynamic criteria are dependent upon phenomena of fatigue due to the cyclic stresses to which the material is subjected. ISO 281: 2007 is in force in this field. Crack propagation leads to the appearance of chips in the "bearing paths". These phenomena are detected by noise analysis (see Chapter 6), by analyzing the temperature or the increase in the resistant moment on the shaft. According to ISO 281: 2007, the bearing shows signs of fatigue after a series of cycles, of greater or lesser length. In design, we cater to the desire for reliability: the probability of reaching, or even surpassing, a certain lifetime (L in hours of rotation or L_{10} in 10^6 rotations) at a load P (N). Two bearings, even if they are identical, may behave differently when mounted in the designed mechanism. To resolve the question of reliability, we use statistical methods. Taken from the manufacturer SKF, below are a few values of lifetimes based on experience.

Lifetime	L in hours
Velocity reducers	20,000–50,000
Tooling machines	20,000–80,000
Textile machines	50,000–65,000
Automobiles	900–10,000
Machines such as tractors	500–16,000

Table 7.8. *Indicative examples of lifetimes. Source SKF*

Data for calculation of the reliability of a bearing using Weibull's law:

– L is the lifetime of the bearing in hours of work;

– L_{10} is the lifetime of the bearing in millions of rotations (10^6 rot.);

– F is the reliability of the bearing according to Weibull's law;

– D = 1-F is the probability of failure (D %);

– α is a position parameter, $\alpha = 0.025$;

– β is a form parameter, $\beta = 1.5$;

– η is a scale parameter, $\eta = 1$.

The reliability F in % of the bearing is expressed thus:

$$\begin{cases} F = exp\left(\ln\frac{9}{10}\right)\times\left(\frac{L}{L_{10}}\right)^{\beta} \; ; \beta = slope \text{ and } L = \left(\frac{\ln F}{\ln 0.9}\right)^{\frac{1}{\beta}} \times L_{10} \\ With\ (\alpha \text{ and } \beta) \text{ we set} \rightarrow F = exp\left(\ln\frac{9}{10}\right)\times\left(\frac{(L/L_{10} - \alpha)}{1-\alpha}\right)^{\beta} \end{cases} \qquad [7.35]$$

The probability of failure D in % is expressed thus:

$$D = 1 - F = 1 - exp\left(\ln\frac{9}{10}\right)\cdot\left(\frac{(L/L_{10} - \alpha)}{1-\alpha}\right)^{\beta}, \ Low\ D \approx 0 \rightarrow \text{for } L \prec \alpha L_{10} \qquad [7.36]$$

7.7. Case study: statement of the problem

The results below confirm those presented in the technical literature published by SKF and SNR, where the reliability during the warranty period is $F_g = 99.75$ % for a reliability of 95 % with D = 5 %, $\alpha = 0.025$ and $\beta = 1.5$. L = 24 hours, 48 hours up to 24,000 hours.

Probability of failure D% or cumulated % of deteriorated bearing

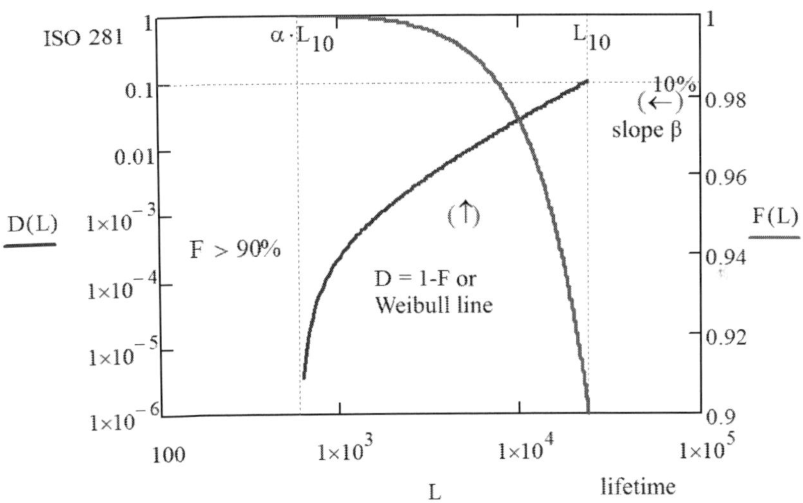

Figure 7.2. *Calculation of the lifetime in accordance with ISO 281: 2007. For a color version of this figure, see www.iste.co.uk/grous/design.zip*

For $L = 0.62 \times L_{10} = 1.488 \times 10^4\ h$ and $\alpha \times L_{10} = 0.025 \times 24000 = 600\ h$, we can see that at the 401st iteration of our calculations, F = 97.50 %.

7.7.1. *Internal clearance fit of bearings*

Bearing manufacturers publish tables guiding the choice of fits of the rings in typical conditions. It is wise to closely study the influence of pressing on the internal clearance fit of the bearings.

Load	Shaft		Bore	
	C/P >5	C/P <5	C/P >5	C/P <5
OR rot./load	g6	J6	M7-N7	N7-P7
IR rot./load	J6-k6	M6-p6	J7	H7 or Rollers: M7-P4

Table 7.9. *Representation of shaft/bore clearance fits. Source SNR*

ISO 5753 describes the radial clearance fit J_0 for the different types of bearings. For that purpose, the manufacturer SNR has published the table below.

Type of bearing, R	J_{rmean} (μm)	J_a/Jr
Stiff ball bearing	$10^{-3}\sqrt{d}$	$\sqrt{(D-d)/20}$
Oblique contact ball bearing	–	$\cong 0.83$
Cylindrical roller bearing	$4\times10^{-3}\sqrt{d}$	–
Ball joint bearing, balls	$2\times10^{-3}\sqrt{d}$	$Y_0/0.44$
Ball joint bearing, rollers	$5\times10^{-3}\sqrt{d}$	$Y_0/0.44$
Conical roller bearing	–	$Y_0/0.8$

Table 7.10. *Representation of a number of shaft/bore clearance fits. Source SNR*

Measured in a static position with no load, the internal clearance fit in bearings represents the dimension of displacement of the ring with respect to the other. This gives us J_a and Jr, and therefore the initial mean radial clearance fit (J_{rmean}) and the residual mean radial clearance fit (J_{rrmo}).

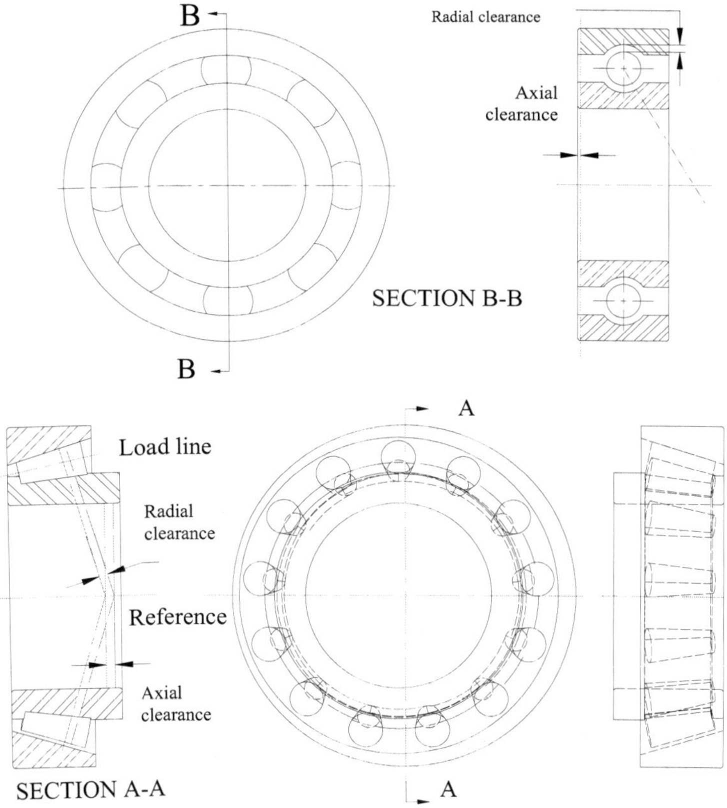

Figure 7.3. *Representations of clearance fits on bearings. Figure (a): bearing with radial and axial clearance fit (radial contact) and figure (b): Axial clearance fit for oblique contact. For a color version of this figure, see www.iste.co.uk/grous/design.zip*

Shaft d = 25k5			Bore = 62 N6		
+2 µm		+11 µm	–33 µm		+14 µm
–2 µm	–11.5 µm	–21 µm	– 1 µm	–17 µm	–33 µm
–5 µm		–18 µm	–5.5 µm		–28.5 µm

Table 7.11. *Representation of gaps, clearance fits and press fits. Source: SNR*

Elements of bearing	Range	Δτ, Repercussion rate
IR, Inner ring	Solid shaft	$\Delta\tau i = 8/10$
	Hollow shaft	$\Delta\tau i = 6/10$
OR, Outer ring	Steel/Cast iron	$\Delta\tau o = 7/10$
	Lightweight alloys	$\Delta\tau o = 5/10$

Table 7.12. *Repercussion rate. $\Delta\tau$,. Source SNR*

Class		C2		C0		C3		C4		C5	
J_{0min}	J_{0min}	1	11	5	20	13	28	23	41	30	53
J_{0mean} [μm]		6		12.5		20.5		32		41.5	

Table 7.13. *Classes of clearance fit for the shaft (d) of 24–30 mm. $\Delta\tau$,. Source SNR*

7.8. Biaxial stresses combined with shear for ductile materials in concrete application

Consider:

– stress in direction (*ox*), $\sigma_x = 220$ psi = 1.517 MPa;

– stress in direction (*oy*), $\sigma_y = 120$ psi = 0.827 MPa;

– shear stress, $\tau_{xy} = 200$ psi = 1.379 MPa;

– variation of the angle (θ) = 0°, 5° … 180°. Target angle: (θ_p) = 15°.

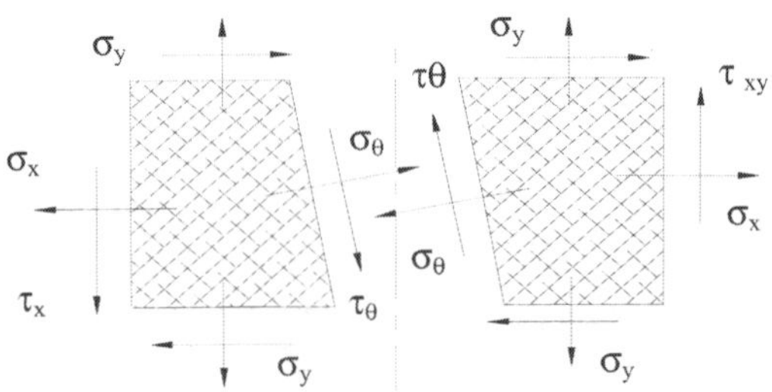

Figure 7.4. *Representation of stresses in the frame of reference*

QUESTIONS.– Calculate the normal and shear stresses and, on two different plots, show the evolution of the stresses as a function of the variation of the angle (θ). The calculations must be shown both in SI and imperial units.

– *Respective calculations of the normal stress $\sigma(\theta)$ and shear stress $\tau(\theta)$*

$$\begin{cases} \sigma(\theta)=\frac{1}{2}(\sigma_x+\sigma_y)+\frac{1}{2}(\sigma_x-\sigma_y)\cos(2\theta)+\tau_{xy}\sin(2\theta)=313.301\ psi=2.16\ MPa \\ \tau(\theta)=\frac{1}{2}(\sigma_x-\sigma_y)\sin(2\theta)-\tau_{xy}\cos(2\theta)=-148.205\ psi=-1.022\ MPa \end{cases} \quad [7.37]$$

– *Expressions of maximum normal and shear stresses*

$$\begin{cases} For\ \ \theta=\arctan\left(2\tau_{xy}/(\sigma_x-\sigma_y)\right)/2=37.982°,\ then\ \sigma_1=?... \\ \sigma_1=(\sigma_x+\sigma_y)/2+\sqrt{\left((\sigma_x-\sigma_y)^2/4\right)+\tau_{xy}^2}=376.155\ psi=2.593\ MPa \end{cases}$$

$$\begin{cases} For\ \theta=\left\{\arctan\left(\dfrac{2\tau_{xy}}{\sigma_x-\sigma_y}\right)+\dfrac{\pi}{2}\right\}\Big/2=127.982°,\ then\ \sigma_2\rightarrow \\ \sigma_2=(\sigma_x+\sigma_y)/2-\sqrt{(\sigma_x-\sigma_y)^2/4+\tau_{xy}^2}=-36.155\ psi=-\ 0.249\ MPa \end{cases}$$

$$\begin{cases} In\ the\ same\ conditions\ but\ with\ shear,\ \tau_{\max}\rightarrow \\ \tau_{\max}=\sqrt{(\sigma_x-\sigma_y)^2/4+\tau_{xy}^2}=206.155\ psi=1.421\ MPa \end{cases}$$

– *Graph plots of the stresses*

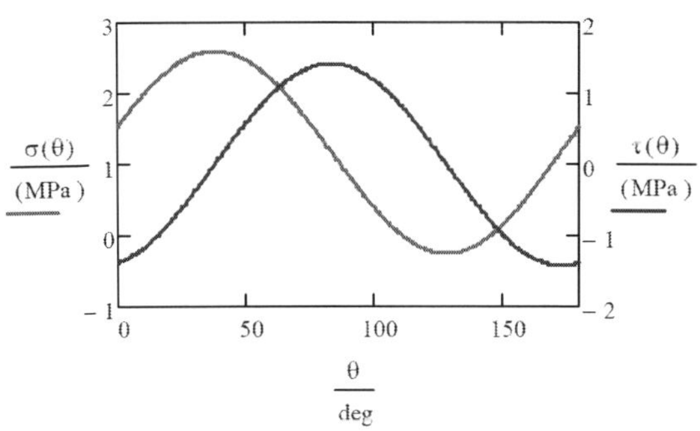

Figures 7.5. *Representation of stresses as a function of the angle (θ). For a color version of this figure, see www.iste.co.uk/grous/design.zip*

7.9. Fundaments of sizing in mechanical design. Soderberg equations in fatique of ductile materials

Numerous factors come into play in the sizing and specifications of elements of machines. They are very often connected to the movements, the forces, the geometry and the materials. Solutions may be reached in many case studies without the equations introduced being absolutely exact. Particular attention must be paid to the formulation of the problem at hand, in terms of stresses, to calculate the safety factor [SOD 30, SOD 35]. We set:

$$\begin{cases} \dfrac{DE}{OC-OD}=\dfrac{OB}{OA} \text{ in termes of stresses } \dfrac{K_f\sigma_r}{\sigma_u/f_s-\sigma_{av}}=\dfrac{\sigma_e}{\sigma_y}\downarrow \\ \text{Hence } \dfrac{1}{f_s}=\left(\dfrac{K_f\sigma_r}{\sigma_e}+\dfrac{\sigma_{av}}{\sigma_y}\right)\rightarrow f_s=\dfrac{\sigma_y}{\sigma_{av}+\left(\sigma_y/\sigma_e\right)K_f\sigma_r} \end{cases} \quad [7.38]$$

The term in the denominator is simply the side OC.

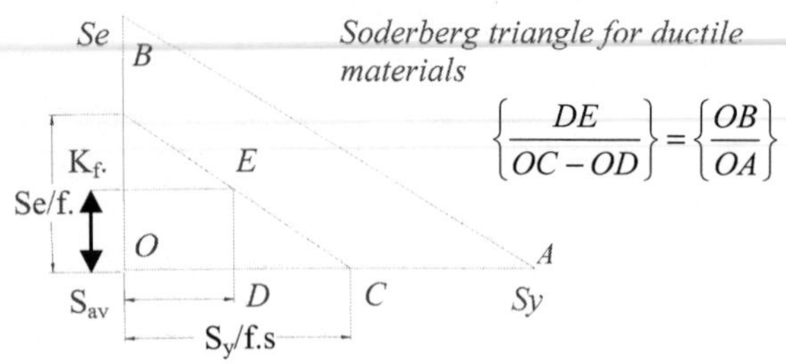

Figure 7.6. *Representation of Soderberg triangle for ductile materials*

7.9.1. *Application of Soderberg equations*

Axial load: $\sigma = F/A$ where F is the load in N and A the area of the cross-section in mm^2.

$$fs=\frac{A\sigma_y}{F_{av}+\left(\sigma_y/\sigma_e\right)K_fF_r}\rightarrow A=\frac{F_{av}+\left(\sigma_y/\sigma_e\right)K_fF_r}{\sigma_y/fs}\left[mm^2 \text{ or } in^2\right] \quad [7.39]$$

Pure bending: $\sigma = M\times c/J$, *in psi or MPa* [7.40]

$$fs=\frac{(J/c)\sigma_y}{M_{av}+\left(\sigma_y/\sigma_e\right)K_fM_r}\rightarrow J/c=\frac{M_{av}+\left(\sigma_y/\sigma_e\right)K_fM_r}{\sigma_y/fs} \quad [7.41]$$

where:

– c is the radius to the furthest fiber in mm or in;

– M is the bending moment;

– J is the polar quadratic moment of inertia in mm^4 or in^4;

– $W = J/c$ is the resistant moment (or resistance modulus).

Case of a shaft subject to simple shear $\tau = T \times l / J$ *psi or MPa* [7.42]

$$fs = \frac{(J/c)\sigma_{sy}}{T_{av} + (\sigma_{sy}/\sigma_{se})K_{fs}T_r} \rightarrow J/l = \frac{T_{av} + (\sigma_{sy}/\sigma_{se})K_{fs}T_r}{\sigma_{sy}/fs} \quad [7.43]$$

where:

– T is the torsion moment (or torque) in *N.mm* or *lbf.in*;

– J is the polar moment of inertia of the section of the area A in mm^4 or in^4;

– l is the length of the radius to the furthest fiber in *mm* or *in*;

– K_{fs} is the torsion-fatigue *notch factor*.

Shafts and pulleys are often subjected to combined forces of bending and torsion. For cases where the material is a ductile steel σ_s and σ_e, we set:

$$fs = \frac{(J/c)\times(\sigma_s/2)}{T_{av} + (\sigma_y/\sigma_e)K_{fs}T_r} \rightarrow J/l = \frac{T_{av} + (\sigma_y/\sigma_e)K_{fs}T_r}{(\sigma_y/2)fs} \quad [7.44]$$

7.9.2. *Stress intensification factors (SIFs)*

The term "stress intensification" refers to the result of changes in geometric form, of shoulders, bores, notches, grooves, etc. The nominal stress is sometimes calculated for the minimum cross-section. This is not always the case, because the designer needs to take account of the exact geometry.

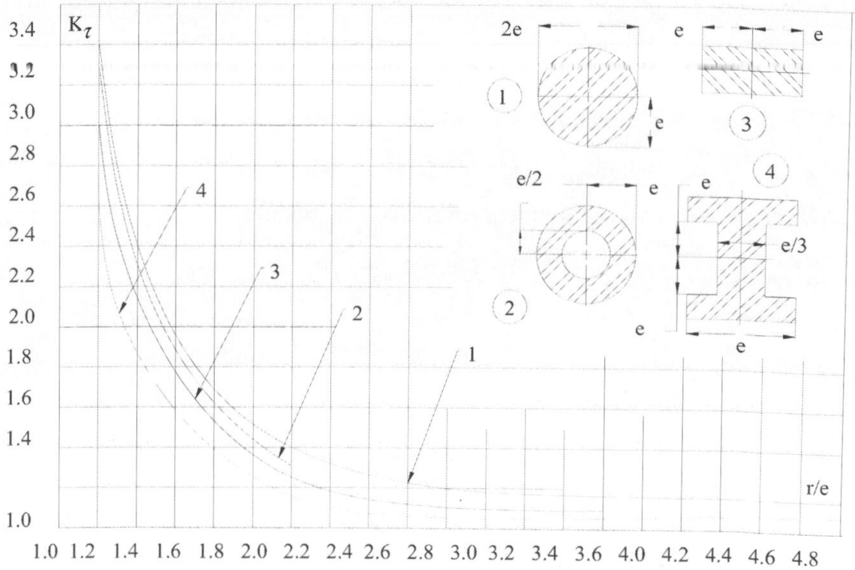

Figure 7.7. *Theoretical stress intensification factor for a bending bar. For a color version of this figure, see www.iste.co.uk/grous/design.zip*

The maximum stress (σ_{max}) is ordinarily calculated on the basis of the geometry:

$$\sigma_{\max} = \left[(2a/b)+1\right]\times\sigma \rightarrow K_\tau = \frac{\text{Applied stress}}{\text{Nominal stress}} = \frac{\sigma_{\max}}{\sigma} = \left(2\frac{a}{b}+1\right) \qquad [7.45]$$

For plates (sheet metal) [TIM 51], when the geometric ratio (a/b) is large, the ellipse moves closer to the transverse crack, and the SIF then becomes fairly high. Conversely, when the geometric ratio shrinks, the ellipse approaches the longitudinal notch, and the decrease in stress is slight. For "circular" holes, the geometric ratio (a/b) = 1, giving rise to a maximum stress three times as great ($3\times\sigma_{max}$).

7.9.3. *Case study*

Consider a centrifugal fan, spinning at 560 rpm. The main shaft is made from machined SAE 1018 steel. The drive belt is subjected to a moment M_{app} = 2.560 *lbf.in* = 289.241 *N.mm*, given that the force F = 620 *lbf* = 2758 *N*. As illustrated below from the preliminary stages of the design project, the ratio of the diameters D/d ≈ 1.2, and the threads are presented as: r = (D – d) / 2 ≈ 0.1×d.

QUESTION.– The designer must clearly specify the sizing [d, D and r] in inches and mm for a safety factor f_s supposed to be no greater than 2.75.

Here:

– σ_{scc} is the endurance limit *without* stress intensification;

– σ_{acc} is the endurance limit *with* stress intensification;

– $K_G = 1.62$ is the correction factor for the torsion-fatigue notch;

– $F_s \leq 2.75$ is the safety coefficient;

– R is the radius of the shaft (*D/2*, lever arm of moment) = 1.969 in = 50.013 mm;

– F is the applied force = 620 lbf = 2758 N;

– M_{app} is the applied moment of force (given torque) = 2.5600 lbf.in = 289.241 N.mm;

– $\varphi = M_{cal} x K_\tau$ is the correction of the torque at the level of the cotter pin (see diagram);

– $C_d = 0$ is the radius of the external fiber furthest from the axis of the shaft;

– σ_y = 54,000 psi = 0.372 MPa is the traction resistance stress;

– σ_e = 25,000 psi = 0.172 MPa is the elasticity stress;

– σ_{ult} = 75,000 psi = 0.517 MPa is the ultimate traction stress.

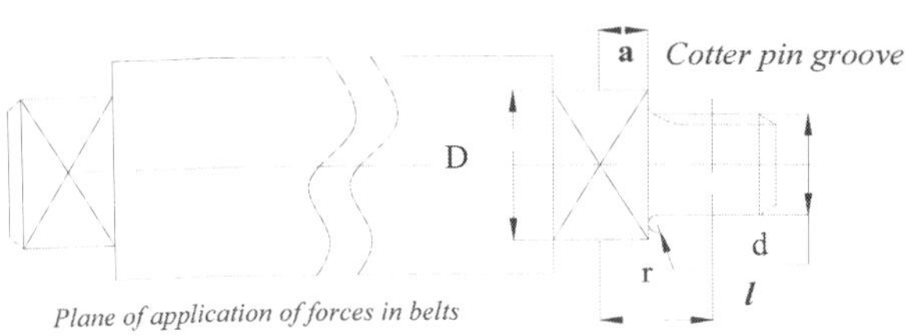

Figure 7.8. *Main shaft of fan. For a color version of this figure, see www.iste.co.uk/grous/design.zip*

SOLUTIONS WITH DISCUSSION.– Calculation of the moment applied to the shaft, M_{cal}:

$$M_{calculated} = F \times R = 620 \times 1.969 = 1221\ lbf.in = 1{,}379 \times 10^5\ N.mm \qquad [7.46]$$

The correction factor for the geometry of the notch is given:

$$K_G = \frac{endurance\ limit\ without\ stress\ intensification}{endurance\ limit\ with\ stress\ intensification} = \frac{\sigma_{SLL}}{\sigma_{acc}} = 1.62 \quad [7.47]$$

Estimation of correction coefficient for the calculated moment, $\varphi = (M_{cal} x K_\tau)$:

$$\varphi = K_G M_{app} = 2.5600 \times 1.62 = 4.147\ lbf.in = 468.571\ N.mm \quad [7.48]$$

In the knowledge that $C_d = 0$, let us set $C_d = d_i/d = (\pi d^3/16)(1-C_d^4) = 0$ and calculate the resistant moment ($W_{yy'}$), also known as the resistance modulus of the cross-section in question:

$$W_{yy'} = \frac{f_s}{\sigma_y\left(1-C_d^4\right)}\sqrt{\left(\frac{\sigma_y}{\sigma_e}\varphi(M_{cal})\right)^2 + M_{app}} = 0.474\ in^3 = 7775\ mm^3 \quad [7.49]$$

In addition, for $d = 1.969\ in = 50.013\ mm$, we calculate the resistance modulus (resistant moment) $W_{yy'}$. The resistance modulus W_{yy} is written with respect to the *xx'* (or *yy'*) axis. We read the expressions of these moduli as a function of the geometry of the cross-section. If $d_{sketched}$=1.969 in = 50.013 mm, then:

$$W_{yy'} = \pi d_{sketched}^3/32 = 0.749\ in^3 = 12281\ mm^3 \quad [7.50]$$

The verified radius (r) and (D) are written as:

$$\begin{cases} r_{verified} = 0.1 d_{sketched} = 0.1969\ in = 5.0013\ mm \\ D_{verified} = 1.2 d_{sketched} = 2.3628\ in = 60.0151\ mm \end{cases} \quad [7.51]$$

From $W_{yy'} = \pi d_{sketched}^3/32$, we draw $d_{verified} = \sqrt[3]{32 W_{yy'}/\pi} = 1.6907\ in = 42.9436\ mm$. As (*r*) is now defined by the standard calculations used in strength of materials, it is possible to verify K_f using the notch sensitivity factor $q = 0.93$, as follows:

$$For\ q = 0.93\ and\ K_\tau = 1.62 \rightarrow K_{sensitivity} = 1 + q(K_\tau - 1) = 1.5766 \quad [7.52]$$

RECAP.– If necessary, in the interests of exact design, it is helpful to recalculate (d) using the calculated value of $K_{fsensitivity}$. We suggest keeping the value of $K_\tau = 1.62$ instead of $K_{fsensitivity}$ and choosing the appropriate diameters for the design, i.e.:

$$\begin{cases} d_{project}^{design} = \left(d_{sketched}\right) = 1.9690\ in = 50.1260\ mm \\ r_{project}^{design} = 0.1\left(d_{verified}\right) = 0.0197\ in = 0.50013\ mm \\ D_{project}^{design} = 1.2\left(d_{sketched}\right) = 2.3628\ in = 60.0151\ mm \end{cases} \qquad [7.53]$$

The safety factor is kept at around 2.75, by the optimization of the design based on the principles of stringent calculations.

7.10. Welding and fatigue

Welding always creates internal tensions, which are often quite close to the elastic limit. They are usually dealt with thermally by relaxation. Designers try, wherever possible, to minimize costs by avoiding the critical deformations that occur because of welds. Welded joints must be rigorous and governed by very clear norms. In mechanical design, welded structures are built in accordance with established codes and standards, such as ASW, ASME, AISC (USA), the Canadian Welding Bureau (CWB) – an example is CSA W 47.1-2009.

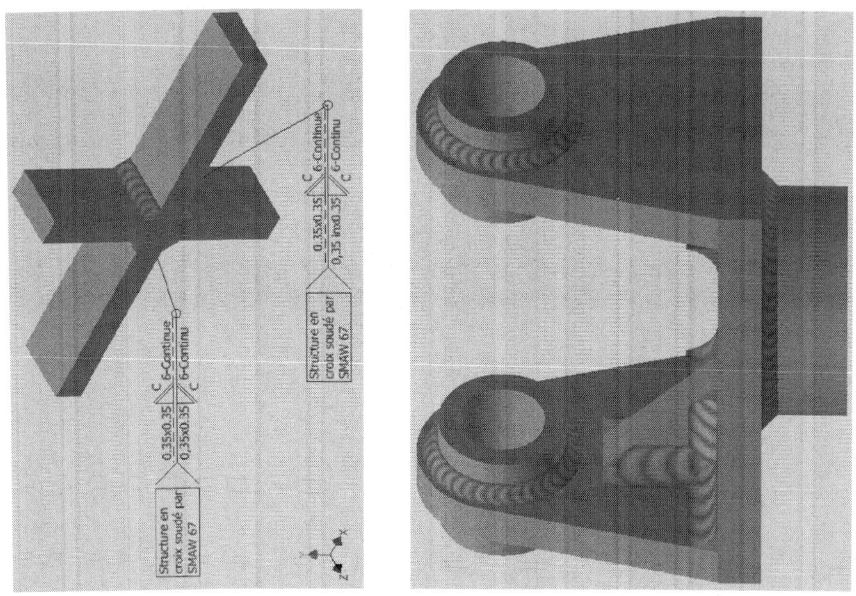

Figure 7.9. *Elements of welded cross-shaped structure [GRO 94, GRO 13]. For a color version of this figure, see www.iste.co.uk/grous/design.zip*

7.10.1. *Case study: calculation of resistance of weld joints in design*

Calculate the length (L) required for a weld which conveys a load Q = 12000 lb = 5443 kg, if the thickness of the metal to be welded is e = 0.5 in = 12.7 mm (parallel weld) as illustrated below. The stress intensification factor (SIF) K_f = 2.7 [JEN 36]. Calculate the area A of the welded joint (in^2 and mm^2) and prove that the design calculations account for the load *P* at the preliminary phase of the project.

SOLUTION.– Based on the literature, the expression of the loading is written as follows:

$$Q = A\times(\sigma_a/K_f),\ [N\ or\ lbf] \tag{7.54}$$

where:

– Q = 12000 lb = 5443 kg;

– σ_a = 5000 psi = 34.474 MPa, from the reference [JEN 36];

– K_f = 2.7 (tabulated) of 7 from [JEN 36].

Let us calculate the area of the double weld – hence the factor (2) in the equation of the area:

$$A = 2\times(0.707)\times(0.5)\times L_{weld},\ \left[in^2\ and\ mm^2\right]$$

Initial approximation for the transverse weld:

$$Q = (\sigma_a/K_f)0.707\,L \rightarrow 12000 = (5000/2.7)0.707\,L \rightarrow$$

$$L = 9.165\ in = 0.233\ m = 6480/707$$

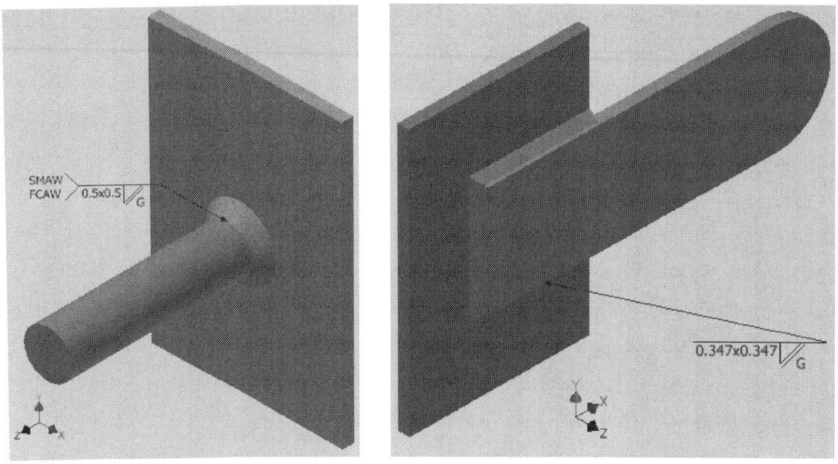

Figure 7.10. *Elements of welded structure. For a color version of this figure, see www.iste.co.uk/grous/design.zip*

From relation [7.54] and where $K_f = 1.5$, let us deduce the length of the weld required to transmit the load Q as follows:

$$Q = (\sigma_a / K_f) 0.707 L \rightarrow L = Q \times K_f / 0.707 \sigma_a = 5.092\ in = 0.129\ m,$$

or indeed:

$$Q = (\sigma_a / K_f) 0.707 L \rightarrow simplify \rightarrow L = 5.0919377652050919378\ in = 5.092\ in = 0.129\ m$$

The area A will therefore be:

$$\begin{cases} A = 2 \times (0.707) \times (0.5) \times L_{weld} \rightarrow simplify,\ A = \ ? \\ A = 3.5999999999999999733\ in^2 = 2.323 \times 10^{-3}\ m^2 \end{cases}$$

or indeed: $A = 2(0.707)(0.5) L_{weld} 3.600000\ in^2 = 2.323 \times 10^{-3} m^2 \leftarrow QED$

Verification of load Q (lb and kg) which is transferred by the length L (in and m):

$$Q = 0.707 L (\sigma_a / K_f) \rightarrow simplify \rightarrow 11999.9999\ lb \rightarrow 5.443 \times 10^3\ kg \leftarrow QED$$

7.10.2. *Real-world case study: welded cross-shaped structure*

Consider a welded cross-shaped structure [GRO 94, GRO 13] for which we need to sketch the design and calculate the parameters below.

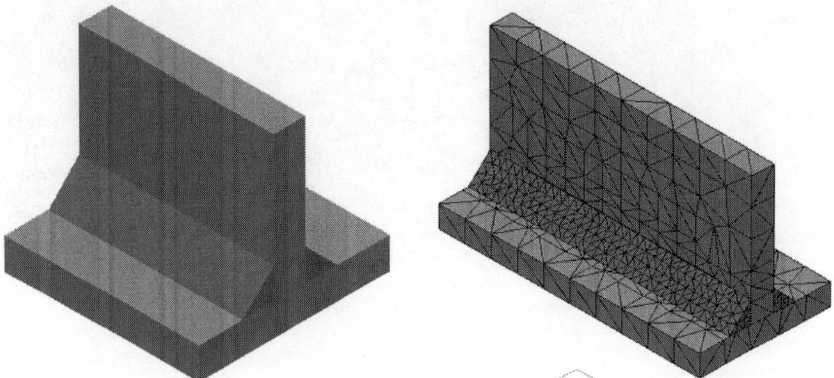

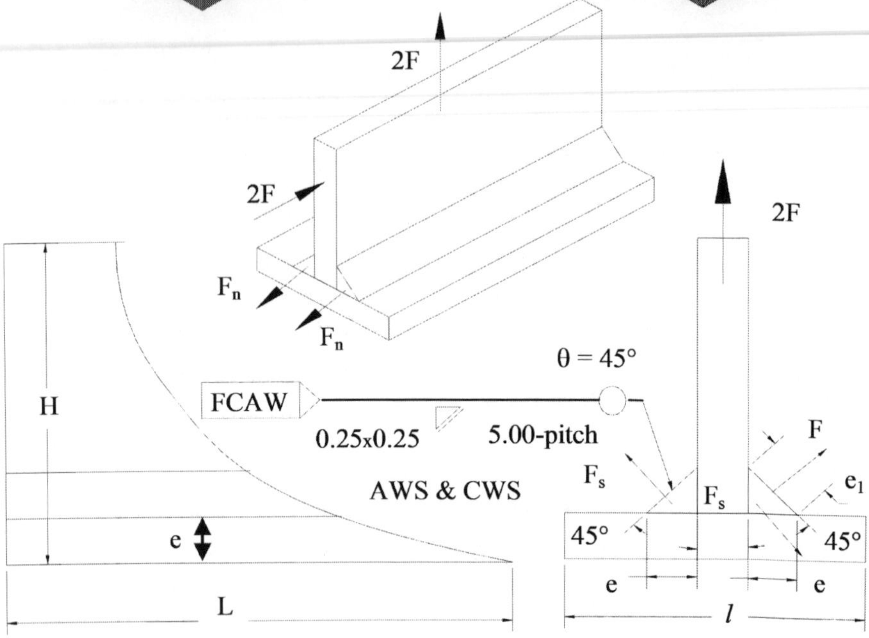

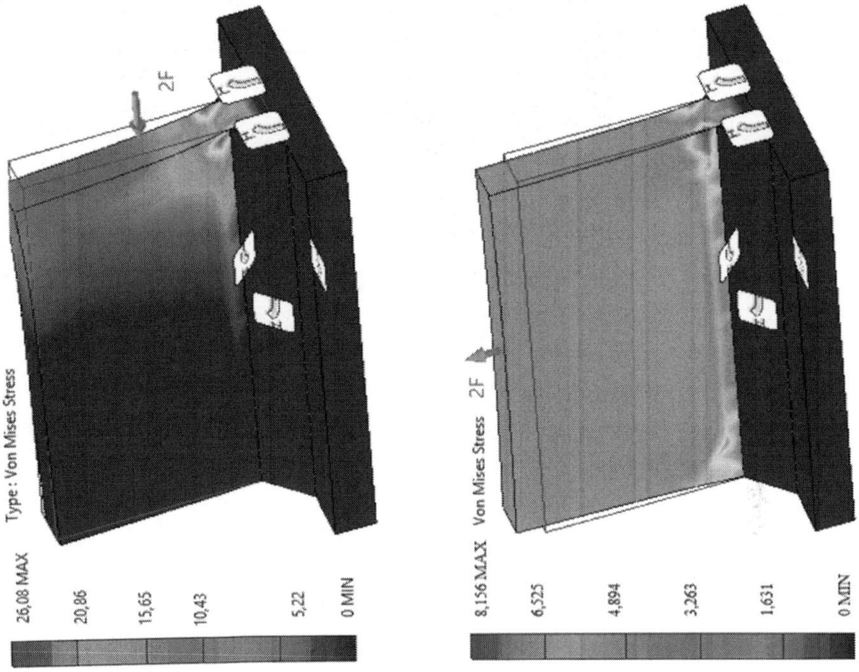

Figure 7.11. *Stress analysis by the finite element method. Welded cross-shaped structure. For a color version of this figure, see www.iste.co.uk/grous/design.zip*

Estimate the admissible force on the weld line for the following data:

– σ_{shear} = Shear stress in psi and MPa;

– F = force applied in N or lbf (2F applied along y-y – i.e. horizontally);

– e = side (length of foot of the weld line) at 45°; e_1 = (throat);

– A = area of weld in mm^2 or in^2;

– θ = 45 deg angle of weld;

– L = total length of weld (mm or in);

– $F_{[adm]}$ = admissible force per inch (or mm) of the weld.

EXERCISE 1.– Calculate the admissible force on the cross-shaped weld line.

SOLUTION.– $A = e_1 \times L_{weld} = 967.74\ mm^2 = 1.5\ in^2;\ \sin(\theta) = \cos(\theta)$

$$\begin{cases} F = \dfrac{\tau_{shear} \times L_{weld} \times e_1}{\sin(\theta) + \cos(\theta)} = 5.898 \times 10^4 \, N = 132\ 260\ lbf;\ and\ \swarrow \\ \dfrac{\partial \tau_{shear}}{\partial \theta} = \dfrac{F}{L_{weld} e_1}\left[\cos(\theta) - \sin(\theta)\right] = 0 \rightarrow \tau_{shear} = \dfrac{F}{L_{weld} e_1} = 60.0942\ MPa \end{cases}$$

Verification: $F_{[adm]} = \tau_{shear1} L_{weld} \times e / \sqrt{2} = 5.898 \times 10^4 \, N = 13\ 260\ lbf \leftarrow QED$

EXERCISE 2.– Declaration of variables (2F applied along x-x – i.e. vertically):

where:

– σ_{shear} = shear stress 12500 psi = 86.184 MPa;

– F = applied force N or lbf;

– e = side (length of foot of weld line) at 65°; e_1= (throat);

– A = area of weld in mm^2 or in^2;

– θ_1 = 65° weld angle;

– L = total length of weld (mm or in);

– $F_{[adm]}$ = admissible force per inch (or mm) of the weld.

SOLUTION.– The two welds share the same load (exercises 1 and 2); only the angle changes.

$$\begin{cases} Equilibrium: 2F - 2F_s - \sin(\theta_1) - 2F_n \cos(\theta_1) = 0,\ and \\ \tau_{shear} = \dfrac{F}{L_{weld} e_1} \rightarrow F = \tau_{shear} \times L_{weld} e_1 = 13\ 260\ lbf = 58\ 980\ N \end{cases}$$

If we suppose that the resultant of F_s and F_n is vertical, the horizontal components are identical and have opposite signs. Therefore, we set:

$$2F - 2F_s - \sin(\theta_1) - \frac{2F_s \cos(\theta_1) - \cos(\theta_1)}{\sin(\theta_1)} = 0 \rightarrow F_s = F_1 \sin(\theta_1) = 12020\ lbf = 53450\ N$$

The maximum shear stress on the weld at 65° would therefore be:

$$\begin{cases} \tau_{cis} = \dfrac{F}{eL}\left[\sin(\theta_1) \times \cos(\theta_1) - \sin(\theta_1)\right] + \ldots \\ \ldots\ + \left[\cos(\theta_1) + \sin(\theta_1) \times \sin(\theta_1) \times \cos(\theta_1)\right] = 1089\ psi = 7511\ MPa \end{cases}$$

We can clearly see that the weld angle has an important role to play with regard to the strength. These two examples are simply enlightening.

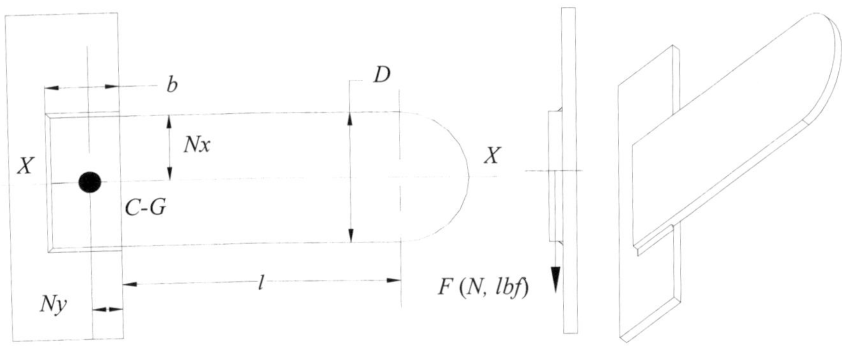

F_v= vertical force = 72000 lb

L = length 7.5 in

$L+N_x$ = distance to diameter = Lω 10.25 in

N_x distance from symmetry = 5.25 in

b = distance between vertical wall and horizontal axis of CG of the weld joint = 5.2 5 in

A = surface of weld = ? = 20.75 in

C = linear distance between point A (or B) and the CG (center of gravity) = 6.25 in

$F_{[adm]}$ = admissible force = 10025 lb

C_h = horizontal distance between point A (or B) and the vertical axis of the CG = 4 in

C_v = vertical distance between point A (or B) and the vertical axis of the CG = 5.25 in

D = d = diameter = 12 in

r = distance between the CG of the weld and the point of application of the vertical load (F) = 24.5 in

Plates welded together

Figure 7.12. *Assembly welded together. For a color version of this figure, see www.iste.co.uk/grous/design.zip*

SOLUTION.–

Center of gravity Ny: $Ny = b/(2b+d) = 1.225\ in = 31.115\ mm.$

The torsion moment: $M = r \times F_V = 1.764 \times 10^6\ lb.in = 20320\ kg.m.$

The force applied to the weld joint: $F_s = F_V / A = 3470\ lb/in = 61960\ kg/m.$

Resistant moment of the structure due to torsion:

$$J_\omega = \left(\left(2b+d\right)^3/12\right) - \left(b^2\left(b+d\right)^2/\left(2b+d\right)\right) = 584.705\ in^3 = 9.582 \times 10^6\ mm^3$$

The maximum load due to torsion (e) $F_s = M \times C / J\omega = 18860\ lb/in = 336700\ kg/m$

Vertical and horizontal direction:

$$\begin{cases} F_v = \dfrac{Ch}{\sqrt{Cv^2 + Ch^2}} f = 2.041 \times 10^5\ kg/m = 1.143 \times 10^4\ lb/in \\ F_h = \dfrac{Cv}{\sqrt{Cv^2 + Ch^2}} f = 2.678 \times 10^5\ kg/m = 1.5 \times 10^4\ lb/in \end{cases}$$

The components in the horizontal and vertical directions give us:

$$F_{combined} = \sqrt{F_h{}^2 + (F_v + F_w)^2} = 3.775 \times 10^5\ kg/m = 2.114 \times 10^4\ lb/in$$

Size of weld (factual = F_{Comb}) $W = Factual / F_{[adm]} = 2.109\ in^{-1} = 0.083\ mm^{-1}$

The shear force per (in or mm) is therefore:

$W = \dfrac{2}{3} \dfrac{F_v}{L\omega} = 1.054 \times 10^4 \dfrac{lb}{in} = 1.882 \times 10^5 \dfrac{kg}{m}$, so the size of the real weld would be:

$$W = f_{factual} / F_{[adm]} = 1.881\ in^{-1} = 0.074\ mm^{-1};\ where\ fs = f_{factual} \quad [7.55]$$

7.10.3. *Case study: fracture mechanics and stresses*

STATEMENT OF THE PROBLEM.– Calculate the thickness (e) of a circular bar of 38 mm = 1.496 in in diameter. The plate withstands uniformly distributed pressure P = 1.5 MPa = 217.557 psi. The plate is made of gray cast iron; the ultimate stress of the material σ_u = 350 MPa = 50760 psi. Also, ν is Poisson's ratio, ν = 0.3.

SOLUTION.– The applied stress is calculated as follows: σ_{app} = 0.15 σ_{ult} = 52.5 MPa = 7614 psi → From the specialized strength-of-materials literature, we consider the maximum stress σ_{max} = 52 MPa, and set the expression:

$$\sigma_{\max} = \left(3W_p / 8\pi \times e^2\right) \times (3+\nu) = 52\ MPa \quad [7.56]$$

The maximum load applied to the plate is: $W_p = \pi p d/2 = 1701\ N = 382.439\ lbf$

Thickness (e) of the plate: $e = \sqrt{\dfrac{3P}{8\pi \times \sigma_{\max}}(3+\nu)} = 3.59\ mm = 0.141\ in$ [7.57]

Let us choose a safety factor f_s = 6.75. Thus, the design of the plate means it will only fracture after surpassing the following: $f_s x W_p$ = 11347 N = 2551 lbf. Now let us apply the above in terms of safety to the design. For fragile materials such as gray cast iron, the concept of calculating plasticity moments does not apply.

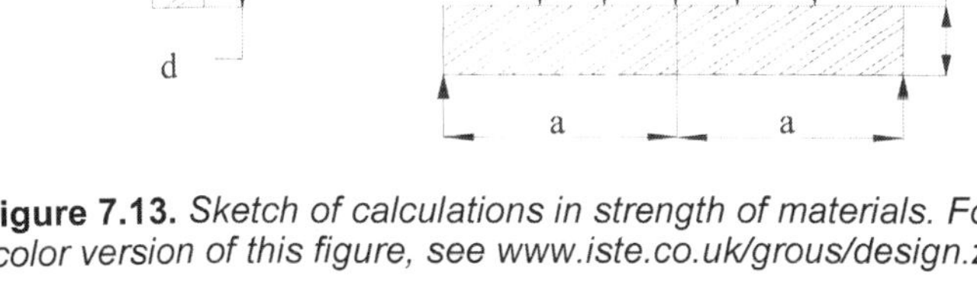

Figure 7.13. *Sketch of calculations in strength of materials. For a color version of this figure, see www.iste.co.uk/grous/design.zip*

As the failure factor is an experimental result, we consult the following table:

Form and type of loading of the plate	K_{fail} as a function of the material	Ratio of σ_{max} and the failure modulus K_{fail} under bending and torsion
Rectangular piece with support at ends. Central load = l/d = 8 and +	Cast iron and steel, $K_{fail} = 1.70$	Cast iron and carbon steel = 1
Circular plate, support at ends with uniform load	$K_{fail} = 1.9$	Cast iron and carbon steel = 1.07
Concentric circular load. Circular plate supported at ends. The area is localized at $x = (r_0/A)^{1/6}$	$\kappa_{fail} = 2.4 - x/2$	Cast iron and carbon steel [1.40 – $x/3$]

Table 7.14. *Values of failure factors for fragile materials*

In the specific case of interest to us here, the failure factor corresponds to gray cast iron. Therefore, we can recalculate the applied stress:

$\sigma_{app} = 1.9\sigma_{ult} = 665\ MPa = 96\ 450\ psi\ (\kappa_{fail} = 1.9)$. Using the above SOM formulae, we set:

$$\sigma_{max} = \frac{3f_s W_p}{8\pi e_p^{\ 2}}(3+\nu) \text{ we isolate } e_p = \sqrt{\frac{3f_s W_p}{8\pi\sigma_{max}}(3+\nu)} = 2.593 = 0.102\ in \quad [7.58]$$

The solution, under maximum stress, depends essentially on the accuracy of the failure factor found experimentally. In many cases, we employ the failure modulus:

$$+\ failure\ modulus = \kappa_{failure} \times \sigma_u = 665\ MPa = 96\ 450\ psi \quad [7.59]$$

7.10.4. *Case study in fatigue fracture mechanics*

In terms of multi-stresses, the design of rods for high-performance engines is a crucially important mechanical element because, should a failure occur, it could be very damaging.

To optimize the calculations of the forces of inertia and the Hertz pressures (contact), we design lightweight rods. Thus, we are guided toward choosing materials which are both strong and light. The rods examined up until now, made of inexpensive materials such as cast iron or certain ordinary grades of steel, are chosen for economic reasons.

In terms of modeling, it tends to be modern materials, which are capable of withstanding stress beyond the fatigue limits and prevent elastic buckling, which are chosen. With that goal in mind, the appropriate geometric characteristics need to be selected. The figure below shows a rod which has been deliberately represented as a "rectangle", with area $A = a \times b$ [mm^2].

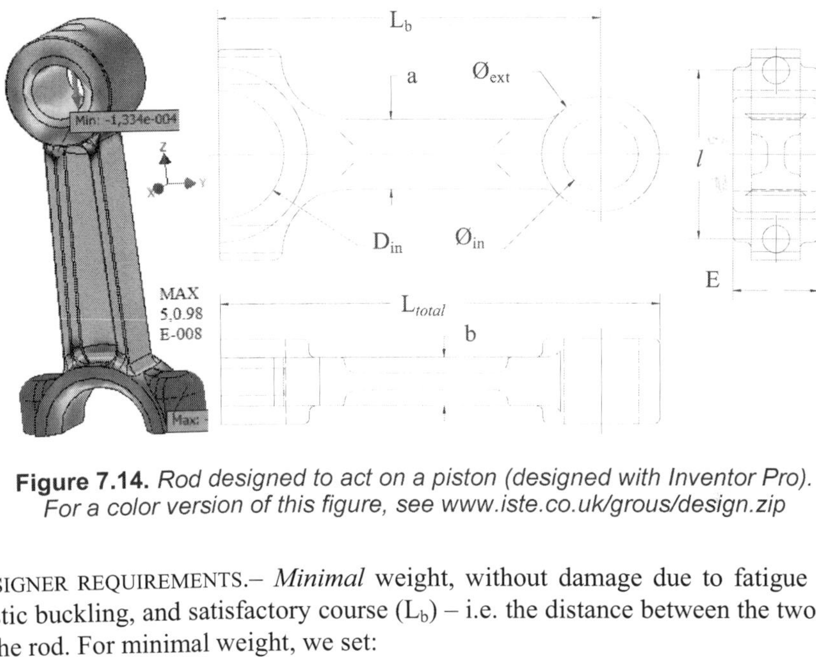

Figure 7.14. *Rod designed to act on a piston (designed with Inventor Pro). For a color version of this figure, see www.iste.co.uk/grous/design.zip*

DESIGNER REQUIREMENTS.– *Minimal* weight, without damage due to fatigue or to elastic buckling, and satisfactory course (L_b) – i.e. the distance between the two axes of the rod. For minimal weight, we set:

$$\begin{cases} \sigma\left\{\begin{matrix} endurance \\ fatigue \end{matrix}\right\} = F_{app} / A_{\sec} \ with \ m_b = \alpha A L_b \rho, \ [kg] \\ Hence \ m\left\{\begin{matrix} fatigue \\ satisfaction \end{matrix}\right\} = \alpha F_{app} L_b \dfrac{\rho}{\sigma\left\{\begin{matrix} endurance \\ fatigue \end{matrix}\right\}}, \ [kg] \end{cases} \quad [7.60]$$

where:

– m_{rod} is the mass of the rod in kg (or mg) or lb;

– A is the area of the section in mm^2;

– L_{rod} is the interaxial distance of the rod in mm;

– ρ is the density of the material from which the rod is made in kg/cm³ or lb/in³;

α is a coefficient reflecting the bores of platforms;

– λ is a coefficient dependent upon the proportions of the transverse section;

– σ_e is the limit stress or endurance limit of the material (MPa or psi).

It is plain to see that the performance index of the rod is dependent upon the endurance limit (MPa) and the density of the material (kg/cm³). It is therefore possible to write an Eulerian equation to estimate the maximum buckling load under compression, as that is the main form of stress to which the rod is subjected:

$$\begin{cases} F\left\{{\max \atop \text{Euler}}\right\} \leq \pi^2 EJ/L_b^2 \ \ in\ N\ or\ lbf,\ with\ J = ab^3/12,\ cm^4 \\ b = \lambda \times a \rightarrow m\left\{{\text{Performance} \atop \text{rod}}\right\} = \dfrac{\alpha}{\pi} \times \left\{L_{rod}^2 \dfrac{\rho}{\sqrt{E}}\right\} \times \left\{\sqrt{\dfrac{12F}{\lambda}}\right\} \end{cases} \quad [7.61]$$

We can see that the material's performance index depends on the elasticity modulus and the density of the material. The existing body of literature suggests the following:

Nuance USA	E, MPa	Density ρ in mg/m³	σ_e, MPa endurance limit	$\mathbf{m_{fatigue}}$	$\mathbf{m_{buckling}}$	$\mathbf{m_b = \alpha AL_b\rho}$
FGS	1.78×10^5	7150	250	2.21	0.13	0.21
4140	2.10×10^5	7850	590	0.10	0.13	0.13
Al-Si-C	2.10×10^5	2880	230	**0.09**	**0.07**	0.09
Ti-6-4	1.15×10^5	4400	530	0.06	0.10	0.10

Table 7.15. *Choice of materials for the rod*

CALCULATED PROOF.– For: $L_b = 132$ mm; $E_{Al\text{-}Si\text{-}C} = 1.10\times10^5$ MPa; $F_{app} = 1.5\times10^5$ N; $\lambda = 0.5$ and $\rho_{Al\text{-}Si\text{-}C} = 2880$ mg/m³; and $\alpha = [0.75, 0.8, \ldots 1]$.

$m_b(\alpha) =$

	0
0	0.069
1	0.073
2	0.078
3	0.082
4	0.087
5	0.091

$\cdot mg$

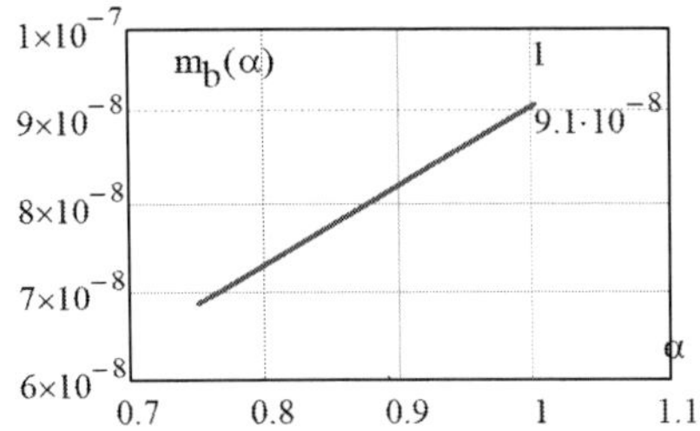

Figure 7.15. *Mass of the material of the rod made of Al-Si-C m = f(L and α). For a color version of this figure, see www.iste.co.uk/grous/design.zip*

Consider a rod machined and straightened by SAM, with the ultimate stress, σ_{ult} = 689.476 MPa (100,000 psi). The elastic limit of the material is σ_y = 372.317 MPa (54,000 psi). It has a stress intensification factor (SIF) K_f = 1.5 from [JEN 36]. The loads are respectively listed below:

– from 68.948 MPa to 413.685 MPa for 0.5 % of cycle time;

– from 137.895 MPa to 344.738 MPa for 9.5 % of cycle time;

– from 137.895 MPa to 310.264 MPa for 20 % of cycle time;

– from 206.843 MPa to 275.790 MPa for 30 % of cycle time;

– from 206.843 MPa to 241.317 MPa for 40 % of cycle time.

QUESTION.– Calculate the lifetime (N_{cycles}) of that rod.

SOLUTION.– First let us calculate the endurance limit of the material for an ultimate stress of σ_{ult} = 689.476 MPa. Reading the grid below, we find the value of the endurance limit σ_e = 45000 psi = 310.264 MPa for a machined, straightened fluted bar.

For σ_{ult} = 689.476 MPa = 100,000 psi, we read: σ_e = 45000 psi = 310.264 MPa. For each cycle (N_{cycle}), the equivalent bending stress gives us:

$$\sigma_{v,i} = \left(\sigma_e/\sigma_y\right)_{mean} + K_f\left(difference/2\right) \text{ where } K_f = 1.5 \quad [7.62]$$

1) $\sigma_{v,1} = \left[\sigma_e/\sigma_y\right] \times (234.422\ MPa) + K_f(165.474\ MPa) = 443.563\ MPa = 64\ 333\ psi$

2) $\sigma_{v,2} = \left[\sigma_e/\sigma_y\right] \times (234.422\ MPa) + K_f(96.527\ MPa) = 340.142\ MPa = 49\ 333\ psi$

3) $\sigma_{v,3} = \left[\sigma_e/\sigma_y\right] \times (206.843\ MPa) + K_f(68.948\ MPa) = 275.791\ MPa = 40\ 000\ psi$

4) $\sigma_{v,4} = \left[\sigma_e/\sigma_y\right] \times (234.422\ MPa) + K_f (33.095\ MPa) = 244.994\ MPa = 35\ 533\ psi$

5) $\sigma_{v,5} = \left[\sigma_e/\sigma_y\right] \times (206.843\ MPa) + K_f (155.132\ MPa) = 405.067\ MPa = 58\ 750\ psi$

The fatigue lifetime, for the initial cycle, is $\sigma_{v,i} = \sigma_{v,1}$. According to the literature published by *Product Engineering*, the lifetime corresponding to that cycle would be:

$$N_{cycle}^{1} = 10^3 \left(\sigma_{ult}/\sigma_{v,i}\right)^{3 / \log\left(\frac{\sigma_{ult}}{\sigma_e}\right)} = 45414\ cycles \quad [7.63]$$

σ_{vi} for 1 psi = 6.895×10^{-3} MPa	σ_{v1}/σ_{vi}	σvi /σv1 7.2406	*i*	σvi /σv1 7.2406
60,830	1.0000	1.00000	0.005	0.00500
45,830	1.3273	0.12873	0.095	0.01223
40,420	1.5051	0.05179	0.200	0.01036
30,830	1.9730	0.007296	0.300	0.00219
25,420	2.3934	0.001802	0.400	0.00072
Fatigue life cycle analysis values				Mean = 0.03050

Table 7.16. *Fatigue life cycle analysis values*

For an exponential law, the fatigue lifetime would be:

$$N\left\{\begin{matrix} Exp\ law \\ cycle\ life \end{matrix}\right\} = 2.55/\log\left(\sigma_{ult}/\sigma_e\right) = 7.3532 \quad [7.64]$$

The fatigue lifetime is generally calculated by the following equation:

$$\frac{N\left(\begin{smallmatrix}Exp\ law\\ cycle\ life\end{smallmatrix}\right)_1}{\eta_1}+\eta_2\left(1\Big/\frac{\sigma_{v,1}}{\sigma_{v,2}}\right)^{2.55\Big/\log\left(\frac{\sigma_{ult}}{\sigma_e}\right)}+ \\ \ldots\ \eta_3\left(1\Big/\frac{\sigma_{v,1}}{\sigma_{v,2}}\right)^{2.55\Big/\log\left(\frac{\sigma_{ult}}{\sigma_e}\right)}+\ldots+\eta_i\left(1\Big/\frac{\sigma_{v,1}}{\sigma_{v,2}}\right)^{2.55\Big/\log\left(\frac{\sigma_{ult}}{\sigma_e}\right)} \qquad [7.65]$$

For each application of a bending load, this equation is updated with η_1, η_2, etc., which are the percentages of the lifetimes under the various loads. For example, looking at Table 7.16, we could write the following:

$$N^1_{cycle\ life}=N_1/0.03050=1\ 488.977\ life\ cycles \qquad [7.66]$$

The data reproduced here are taken from the tables published in Marks' Standard Handbook for Mechanical Engineers [AVA 06]. The equations we are using here take account of simple and compound loading.

7.11. Limits of performance and of strength in the elastic domain

Beams, columns and even shafts often have standard revolving or prismatic sections. Whether solid or hollow, these sections are of varying states, of which certain performances are expected. There are limits to the performances. We can cite two such limits: the elastic limit and the strength limit.

Here are two case studies: two materials – wood and steel – are used in design, where A = the area of the section in m^2 and $\varphi_{material}$ is the form factor drawn from the specialized literature [GER 85, ASH 91]. Let us vary the areas between $A = \{0.25\times10^2, 0.5\times10^2, \ldots\ 5\times10^2\}\ m^2$ for two simulated materials. By way of example, we shall use $\varphi_{wood} = 4$ and $\varphi_{steel} = 62$.

CASE 1.– *Elastic performance limit (J) presented thus:*

$$J^{wood}_{inertia}(A)=2\log(A)+\log\left(\frac{\varphi_1^{wood}}{4\pi}\right)\ \text{and}\ J^{steel}_{inertia}(A)=2\log(A)+\log\left(\frac{\varphi_1^{steel}}{4\pi}\right) \qquad [7.67]$$

$\log(J_{wood}(A))$	$\log(J_{steel}(A))$
0.361	0.543
0.69	0.785
0.739	0.824
0.765	0.846
0.783	0.861
0.797	0.872
0.808	0.881
0.816	0.889
...	...

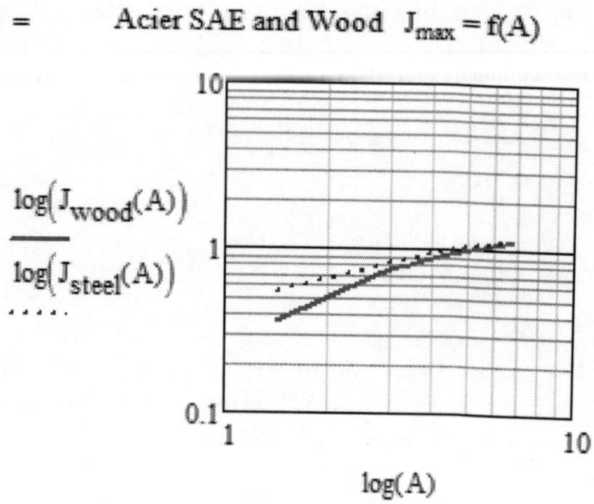

Figure 7.16. *Performance indices of wood and steel in terms of elasticity (J). For a color version of this figure, see www.iste.co.uk/grous/design.zip*

CASE 2.– *Strength performance limit (W) presented thus:*

$$W_{inertia}^{wood}(A)=\frac{3}{2}\log(A)+\log\left(\frac{\varphi_1^{wood}}{4\sqrt{\pi}}\right) \text{ and } W_{inertia}^{steel}(A)=\frac{3}{2}\log(A)+\log\left(\frac{\varphi_1^{steel}}{4\sqrt{\pi}}\right) \quad [7.68]$$

$\log(W_{wood}(A))$	$\log(W_{steel}(A))$
0.267	0.483
0.58	0.698
0.627	0.734
0.653	0.755
0.67	0.769
0.683	0.779
0.694	0.787
0.702	0.794
...	...

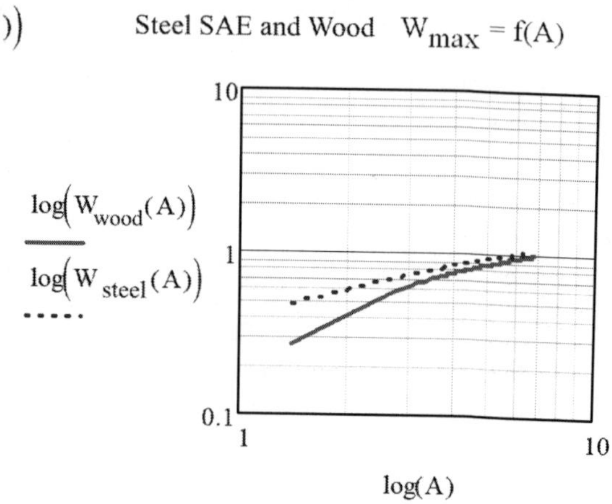

Figure 7.17. *Strength performance of wood and steel (W). For a color version of this figure, see www.iste.co.uk/grous/design.zip*

It is clear, from reading the two simulated graphs for wood and steel that the performance depends on the materials. It is precisely there that the informed designer must make the right choice of materials and their geometric features. The objective is to minimize the wear of the materials [BEE 14, ROA 11].

7.12. Proposed project: outboard motor for a small boat

NOTE.– Subsidiary proposal

As part of our projects as an initiation to mechanical design, we redesigned an outboard boat motor.

In order to do that, we obtained a real motor (from a junkyard). Having taken it apart completely, we measured and drew each component (both in 2D and 3D). The assembly with the software (Inventor Pro) gave rise to what is represented in Figure 7.18.

In addition, we exploited the true dimensions and surface states of the materials in the real motor to run the calculations for the machine elements. The case studies presented here represent examples which readers could employ in their own laboratories.

This is an example of a project which is useful from a pedagogical standpoint, because it includes all of the components of calculations in the training programs for design and calculating strength.

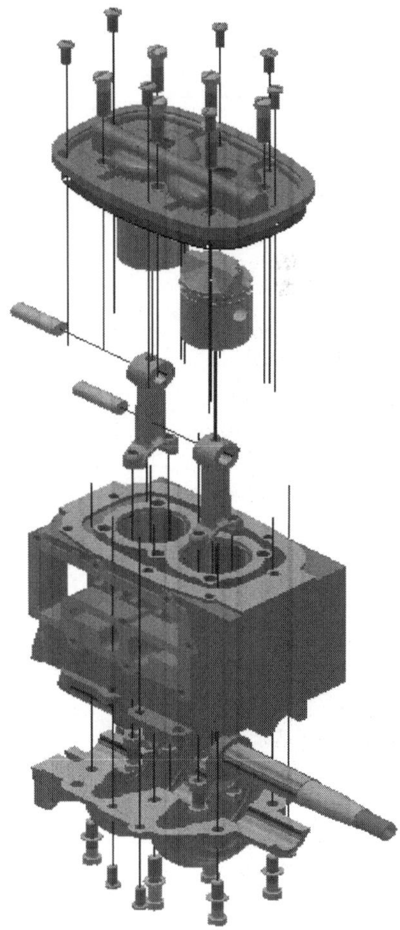

Figure 7.18. *Example of a project for an outboard motor. For a color version of this figure, see www.iste.co.uk/grous/design.zip*

7.13. Conclusion

This chapter essentially discusses a number of aspects of machine elements assembled by welding, and fatigue. The aim here was not to present the theory of these topics, but a variety of case studies relevant to the calculations. The design of such topics, in and of itself, requires whole books, examining the matter in depth. Therefore, we chose to present some very specific, relevant case studies here. Besides the considerations of price, esthetics and the environment, good designers must be conscious of the questions of:

– the extreme limits of efficiency of the form factors of the materials ($\varphi_{material}$);

– clearly establishing the dependence of that link ($\varphi_{material}$) between the materials;

– finding the best possible compromise, in terms of performance, between the geometry and the material.

Readers able to succinctly answer the above three questions prove themselves to be well-qualified designers. Indeed, that knowledge is the primary goal of this book.

8

Friction, Brakes and Gear Systems

8.1. Friction, materials and design of assembled systems

In this chapter, we present the principles of calculations of design of braking systems and other gear systems. In addition, a sizeable portion of the chapter focuses on the materials used in braking and gear systems.

Friction on an inclined plane

The term "friction" is used to denote the force of resistance to the motion and displacement of one solid with respect to another, if they are pressed together. In terms of slip, the following is the simple tribological representation:

$$F = \mu \times N \left[N\right] \textit{ where } \mu = \tan(\rho) \qquad [8.1]$$

Theoretically, the coefficient (angle) of friction (μ) depends on the materials (grades SAE 1018, ANSI 4012), the surface states (Ra, roughness in mm or min), the specific pressure (Ps, in MPa) and the rate of displacement (V in m/s). N is the normal reaction to the displacement in (N):

$$\left\{\begin{matrix} \textit{If } \alpha \succ \rho \rightarrow P = S\tan(\rho) \textit{ Hence } F = S\mu = S.\tan(\rho), \rho \succ \mu \textit{ and } \alpha \succ 0 \\ \textit{If } \alpha \prec \rho \rightarrow P \prec F \; \alpha \prec 0; \textit{ If } \alpha = \rho \rightarrow P = F \textit{ and } V = Const. \textit{ and } \alpha = 0 \end{matrix}\right\} \qquad [8.2]$$

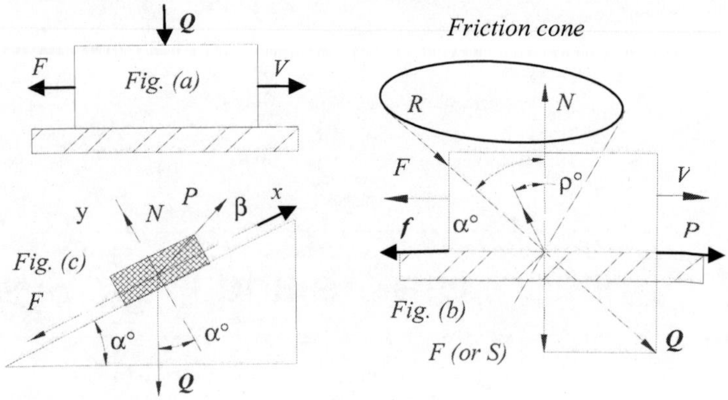

Figure 8.1. *Graphical illustrations of the components of tribology calculations. For a color version of this figure, see www.iste.co.uk/grous/design.zip*

For a uniform motion, as illustrated by Figure 8.1c, let us set the equilibrium equations:

$$\begin{cases} \sum F_x = 0 \; ; P\cos(\beta) - F - Q\sin(\alpha) \quad (1) \\ \sum F_y = 0 \; ; N + P\sin(\alpha) - Q\cos(\alpha) \quad (2) \end{cases} \qquad [8.3]$$

From (2), we can derive the expression of N. as follows:

$$N = g.\cos(\alpha) - P.\sin(\beta), \text{ hence } F = \mu N = \mu g\cos(\alpha) - \mu P\sin(\beta) \qquad [8.4]$$

We substitute the expression of F into (1) and deduce the expression of P:

$$P\cos(\beta) - \mu g\cos(\alpha) + \mu P\sin(\beta) - Q\sin(\alpha) = 0 \rightarrow$$

$$P\left[\cos(\beta) + \mu\sin(\beta)\right] = Q\left[\mu\cos(\alpha) + \sin(\alpha)\right] \textit{Hence} \qquad [8.5]$$

$$P = Q\frac{\mu\cos(\alpha) + \sin(\alpha)}{\cos(\beta) + \mu\sin(\beta)} \text{ in the knowledge that } \mu = \tan(\beta)$$

After trigonometric transformation, we obtain the expression of P as follows:

$$P = Q\frac{Cos(\alpha)\dfrac{Sin(\rho)}{Cos(\rho)} + Sin(\alpha)\times Cos(\rho)}{Cos(\beta) + \dfrac{Sin(\rho)}{Cos(\rho)}Sin(\beta)\times Cos(\rho)} = Q\frac{Sin(\alpha+\rho)}{Cos(\beta-\rho)} \qquad [8.6]$$

Three distinct scenarios then arise, as follows:

– *case 1*: on the horizontal plane, a = 0, we have the expression of *P* as follows:

$$P = Q\frac{\mu}{Cos(\beta)+\mu Sin(\beta)} = Q\frac{Sin(\rho)}{Cos(\beta-\rho)} \quad [8.7]$$

– *case 2*: on the horizontal plane, b = 0, we have the expression of *P* as follows:

$$P = Q\left[\mu.Cos(\alpha)+Sin(\alpha)\right] = Q\frac{Sin(\alpha+\beta)}{Cos(\beta)} \quad [8.8]$$

– *case 3:* on the horizontal plane, b = – α we have the expression of *P* as follows:

$$P = Q\frac{\mu.Cos(\alpha)+Sin(\alpha)}{Cos(\alpha)} = Q\frac{\mu.Cos(\alpha)+Sin(\alpha)}{Cos(\beta)+\mu.Sin(\beta)} = Q\frac{Sin(\alpha+\rho)}{Cos(\beta-\rho)} \quad [8.9]$$

Numerical applications

With:

– friction coefficient (brakes) $\boldsymbol{\mu} = 0.4$;

– for a load varying between **Q** = [0 N and 25 N];

– *case 3*: $\alpha_3 = 45$ deg; $\beta_3 = -\alpha_3 = -45$ deg and $\rho_3 = 15$ deg (Figure 8.1(a));

– *case 2*: $\alpha_2 = 45$ deg; $\beta_2 = 0$ deg and $\rho_2 = 45$ deg (Figure 8.1(b)).

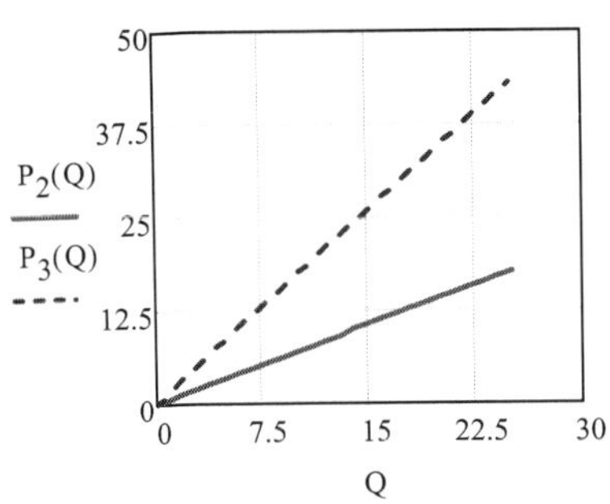

Influence of friction coefficient and setup angles (α , β and ρ) on the pressure exerted (N)

$P_2(Q)$ =	$P_3(Q)$ =
0	0.000
0.707	1.732
1.414	3.464
2.121	5.196
2.828	6.928
3.536	8.660
4.243	10.392
4.95	12.124
5.657	13.856
6.364	15.588
...	...

Figure 8.2. *Graphic illustration of the pressures, P = F {Q, μ, α, β and ρ}. For a color version of this figure, see www.iste.co.uk/grous/design.zip*

It is rare to deal with buttressing in sliding connections, and the phenomenon of stick-slip due to the jerky motion of the slide.

8.2. Buttressing of mechanical connections

There is a particular type of equilibrium, in cases of buttressing [RAB 95, POP 10], dependent on the forces applied (sliding connections) as illustrated below in Figure 8.3. The cross-section may be round or even prismatic. There is a guide at the points of contact (α and β) for which we calculate the location of the force F parallel to the guiding axis (x-x'). The guiding occurs over the length (b), $E = b/2\mu$, and we deduce $N_\alpha = N_\beta = F_{app}/2\mu$. We enforce that there is slippage if $E < b/2\mu$, and there is *buttressing* if $E > b/2\mu$. The system is then locked. Such is the case in "systems" of clamps, certain *jacks*, platforms of tooling machines, etc.

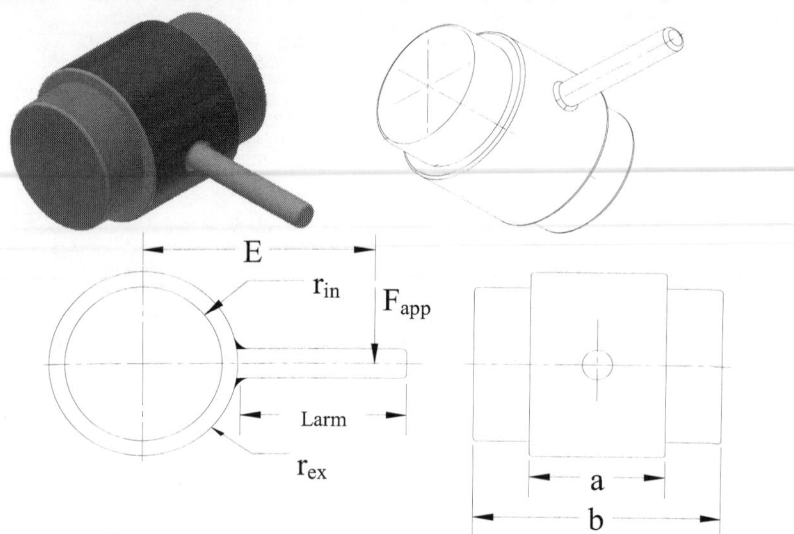

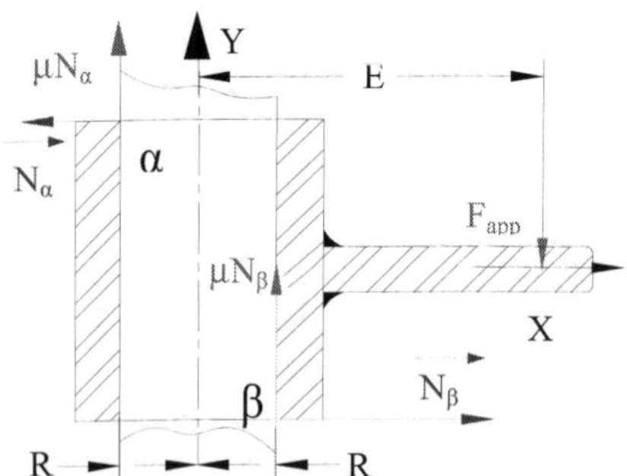

If μ is the friction coefficient of the materials in Hertzian contact, the slight horizontal disruption facilitates the following:

$$\begin{cases} -N_\alpha + N_\beta = 0 \\ \mu \times \left(N_\alpha + N_\beta\right) - F_{app} = 0 \\ -F_{app} \times E + L\left(N_\alpha + N_\beta\right)/2 = 0 \end{cases}$$

Figure 8.3. *Guiding of a slider in a sliding connection. Contact location. For a color version of this figure, see www.iste.co.uk/grous/design.zip*

$E \prec b/2\tan(\varphi) \rightarrow \{If\ \tan(\varphi) = 0.5 \rightarrow E \prec b\}$. If E is the guide length (mm); (μ) the adherence between the surfaces in contact, (d) is the distance between the direction of the force and the axis of the connection, the condition of non-buttressing as a function of the friction (μ) is $\delta \leq b/2\tan(\varphi) = b/2\mu$. If b = 540 mm = 21.26 in and μ = 0.12, buttressing can only be avoided if E < b/2μ = 2250 mm = 88.583 in.

The forces of friction are strong, as a function of the slip rate. Were we to add a slight stiffness to the mechanical components, we would see more of a jerky or stick-slip type of movement. This phenomenon takes place in dynamics, where friction comes into play. In connection with the velocity, when there is a significant variation in the friction, the poor rigidity accentuates the stick-slip motion of the slide. In terms of vibrational modeling, the parameters of stick-slip are written as follows:

$$\left\{\left\{X'_\alpha = X'_{\alpha 1}\right\} \rightarrow X_\alpha = X'_{\alpha 1}.\tau - X_{\alpha 0} \rightarrow Calibration: O\alpha = X_\alpha.X\ and\ O\beta = X.X\right\}$$

(k) is the stiffness of the spring; αβ = E_0 = L_0; length of the spring at rest X_{FT}, represents the tangential component of the friction, which depends on the velocity (X'):

$$X_{FT} = \left\{-\varphi X' + X_0\right\}; \qquad [8.10]$$

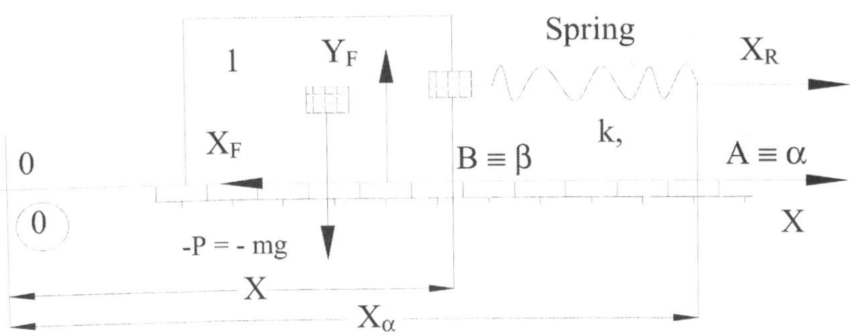

Figure 8.4. *Stick-slip model, represented as vibrational calibration. For a color version of this figure, see www.iste.co.uk/grous/design.zip*

X_0 corresponds to the tangential component at equilibrium. From the fundamental law of dynamics, (1/0) in projection with respect to X. Building on Chapter 6, where we discussed case studies relating to noise and vibration, let us set the equation:

$$m(\partial X'/\partial \tau) = mX'' = k \times (X_\alpha - X - L_0) - (\varphi X' + X_0) \qquad [8.11]$$

At the limit of the condition of static equilibrium, i.e. [X = 0, X' = 0 and X'' = 0], we set the equilibrium condition $k(X_{\alpha 0} - L_0) - (X_0) = 0$. We can then feed that equation back into equation [8.11], above, which would become:

$$\begin{cases} m\dfrac{\partial X'}{\partial \tau} = mX'' = k\left(X_\alpha - X_{\alpha 0} - X\right) - \varphi X', \textit{ where } X_\alpha = \left(X'_\alpha \tau - X_{\alpha 0}\right) \\ m\dfrac{\partial X'}{\partial \tau} = mX'' = k\left(X'_\alpha \tau - X\right) - \varphi X'; \textit{ thus } mX'' + \varphi X' + kX = kX'_\alpha \tau \end{cases} \quad [8.12]$$

The differential equation, for $[(\varphi^2/4) < k]$, is written thus:

$$\begin{cases} X = Exp^{A\tau}\left(\alpha cos\omega\tau + \beta \sin \omega\tau\right) + X'_\alpha\left(\gamma\tau + \phi\right);\ a = \dfrac{\varphi}{2\,m}\ ;\ \omega = \sqrt{\dfrac{4\ km - \varphi}{4\ m^2}} \\ X = Exp^{a\tau}\left\{A cos\left(\omega t\right) + B \sin\left(\omega t\right)\right\} + X'_\alpha\left(C\tau + D\right);\ C = 1 \rightarrow D = \varphi/k \end{cases} \quad [8.13]$$

We have simulated a case to demonstrate the graph of the slider exhibiting stick-slip behavior. We can see that the slider sticks at time t_1, and the motion only resumes (begins to slip once more) at time t_2, etc. In certain cases, this behavior may even be desired, such as in the case of clamps. A helpful example is that of a drawer which sticks and catches when we try to close it. For this reason, where possible, roller guides are used when constructing drawers. The action of the applied force (F_{app}) is off-center in relation to the axis of slip. As it tends to upset the carriage of the guiding path, the drawer cannot slide.

X(τ) = mm

6.571
8.717
10.677
12.421
13.93
15.188
...

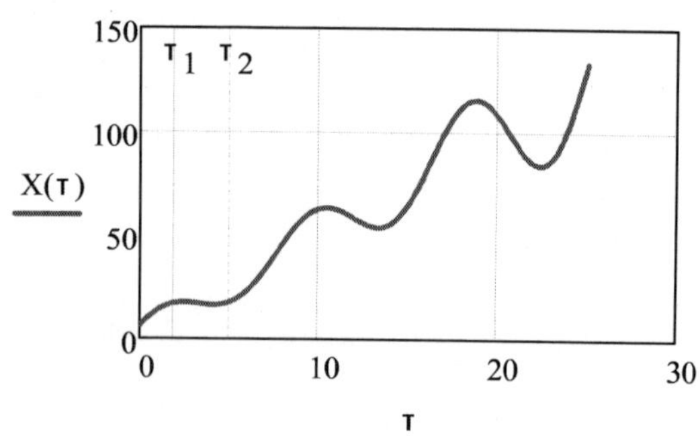

Figure 8.5. *Displacement of slider exhibiting stick-slip behavior. For a color version of this figure, see www.iste.co.uk/grous/design.zip*

To prevent buttressing, it is useful to choose a sufficient guide length (Figure 8.3), for a clearance (J = D-d→H7/f8 and/or H7/g6), defined, and an appropriate degree of adherence between the materials in contact, to avoid juddering (stick-slip). If (D) is the diameter of the bore, and (d) that of the shaft whose slip fit is H7/g6, we can show the general rule of the guiding ratio (k), in connection with the diameters or the angular deviation (φ°). The condition of non-buttressing is presented in the following form as a clearance:

$$clearance\ fit = D - d;\ \varphi = clearance\ fit/b \ \ and\ \ k = b/d \rightarrow condition\ \{1.5 \leq\ k \leq 4\}$$

The jack modeled below using the finite element method, on its own, could fill an entire chapter on friction. However, we shall limit ourselves to presenting the 3D drawings below.

$$clearance\ fit = D - d;\ \varphi = clearance\ fit/b \ \ and\ \ k = b/d \rightarrow condition\ \{1.5 \leq\ k \leq 4\}$$

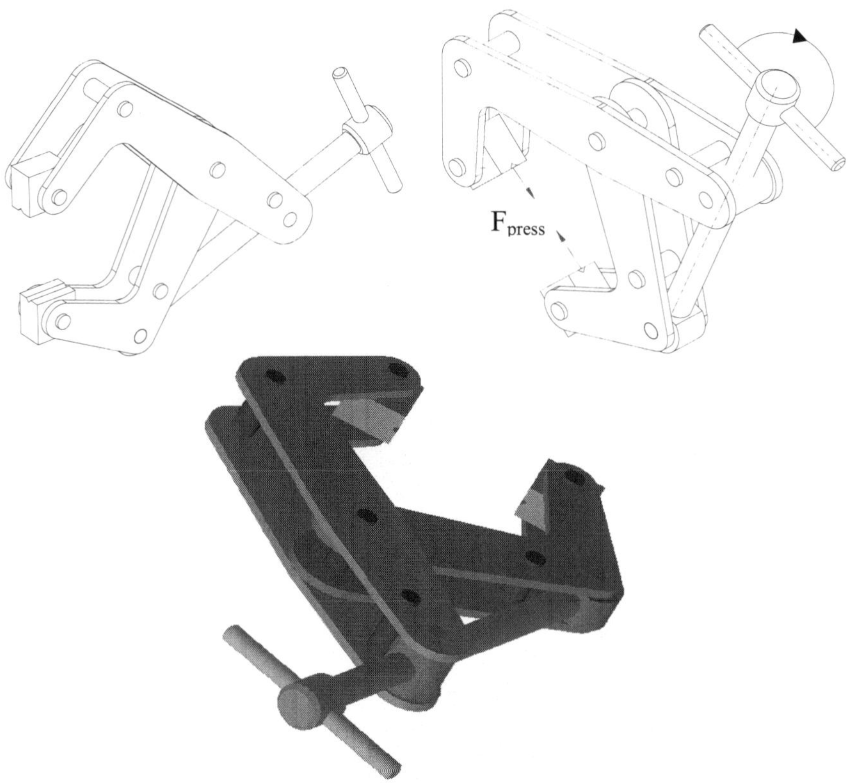

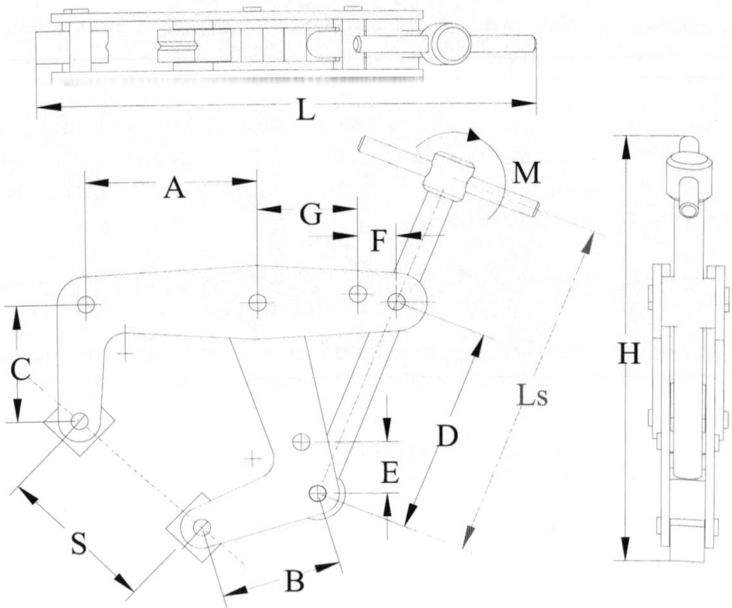
L
A
G
F
M
C
H
Ls
D
E
S
B

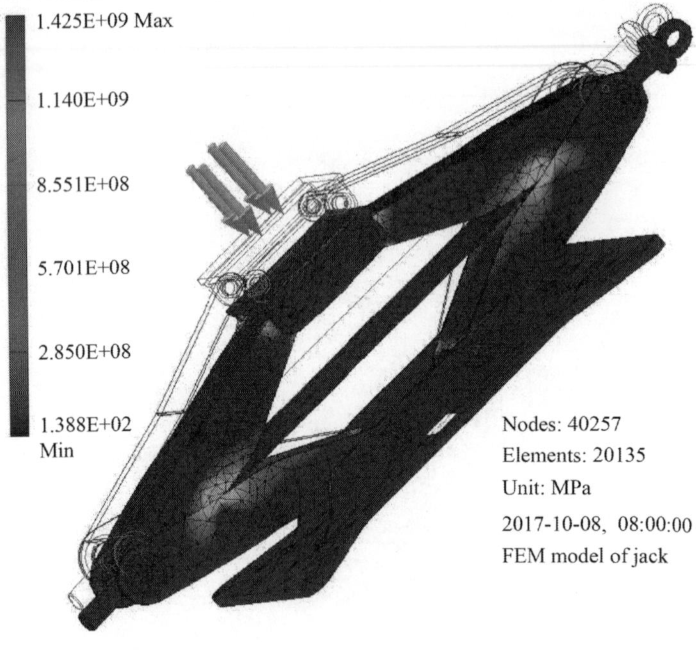
1.425E+09 Max
1.140E+09
8.551E+08
5.701E+08
2.850E+08
1.388E+02
Min
Nodes: 40257
Elements: 20135
Unit: MPa
2017-10-08, 08:00:00
FEM model of jack

Figure 8.6. *Design of shoes and jacks. For a color version of this figure, see www.iste.co.uk/grous/design.zip*

8.3. Case study: principles of calculations for brakes

Machine elements such as brakes and gear systems absorb both potential and kinetic energy. They can either slow or completely halt the machine's motion. Of course, the energy absorbed is dissipated as heat. Braking depends on the pressure exerted between the parts in contact with one another, such as the shoe and brake drum. The friction coefficient is a function of which materials are in contact. Below are some examples of case studies on braking systems.

Brake system with an external shoe, known as a block brake system

Design of a simple brake block by calculation: here, the design is based on the braking force and the moment applied to a brake block with a freely rotating shoe, as in Figure 8.7.

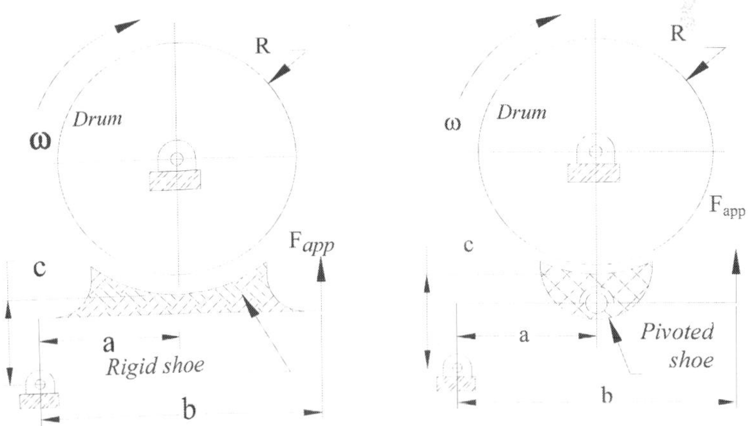

Figure 8.7. *Illustration of the elements involved in calculations relating to braking. For a color version of this figure, see www.iste.co.uk/grous/design.zip*

where:

– N is the normal force in (N);

– μN is the force of friction (N);

– μ is the friction coefficient;

– ρ is the contact angle formed by the drum on the shoe in degrees;

– R is the radius of the brake drum.

To avoid significant errors, let us consider the angle $\rho \geq 60°$. Thus, we can form the equilibrium equation around the pivot (O):

$$(N+Q)a-\mu Nc-Fb=0 \rightarrow F=\left[(N+Q)a-\mu Nc\right]/b \quad [8.14]$$

The partially automatic mechanism includes a physical component with at least one surface of the vehicle, and an actuator. Taking the friction coefficient into account, the braking block can also be designed in such a way that braking occurs when F = 0 N or indeed (F = negative). The previous equation would then be written as:

$$F=(Na-\mu Nc)/b \leq 0 \rightarrow \textit{Auto-braking} \text{ occurs for: } \{a/c\} \leq \mu \quad [8.15]$$

The braking moment (work) to prevent autobraking is formulated thus:

$$M=\mu \times N \times R \ \left[lbf.in\right] \ or \ \left[N.mm\right] \quad [8.16]$$

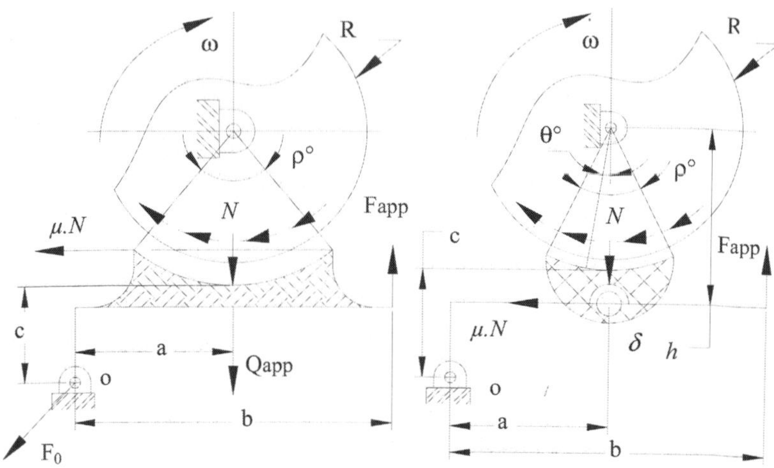

Figure 8.8. *Illustrated design of brake system components*

When the contact angle ρ becomes greater than 60°, errors can result from the friction coefficient μ and the acting forces (N and Q) at the median point (pivot O). A brief illustration shows that the force μN is drawn from the drum to the shoe at the point denoted by (δ), at a distance (*h*) from the pivot (O) (Figure 8.8). In mechanical design, the construction is finalized when the external shoes are "long". The braking moment (work) is then expressed thus:

$$M = \mu Nh = \mu N \frac{4R\sin(\rho/2)}{\rho + \sin(\rho/2)} \quad hence \Rightarrow h = \frac{4R\sin(\rho/2)}{\rho + \sin(\rho/2)} \qquad [8.17]$$

This expression assumes that the wear and tear taken in the direction of the force (N) is normal. Therefore, we are able to write the expression of the normal pressure P_n which varies as a function of *cosine* (θ) – i.e.:

$$P_n = C \times \cos(\theta) \; [N] \; or \; [lbf] \qquad [8.18]$$

$C \equiv \kappa_s$ is a constant equivalent to the specific pressure; l is the width of the brake shoe in (in or mm). In design, its expression is:

$$C = \frac{2N}{l \times R \times [\rho + \sin(\rho)]} \quad \left[\frac{N}{mm^2} = MPa \,\middle|\, ou\ psi\right] \qquad [8.19]$$

A good design integrates the height (*h*) as the proper location of the pivot. Thus, two main design conditions are met:

– the shoe is made up of two forces; the resultant of the normal force (*N*) and the force of friction inevitably passed through the pivot;

– the pressure is then normally distributed, as per the hypothesis.

If the pivot of the shoe were located at another distance different to (*h*), the resulting moment would take a value of zero (*M* = *0*). The ideal is to position the pivot (O) along the normal resultant, and the friction so that the pressure is distributed without affecting the design. The mean pressure suggested by the technical literature is written as:

$$P_{mean} = 2C \times \sin(\rho/2)/\rho \; [MPa] \; or\ indeed \; [psi] \qquad [8.20]$$

8.3.1. *Design of a double brake block by calculation*

In mechanical design, we invariably use double shoe brakes to decrease the loads on the shafts and bearings. To obtain a better yield and decrease the heat generated per unit surface area (mm^2):

$$M = (N_{left} + N_{right}) \times \mu R \qquad [8.21]$$

If the contact angle ρ of the shoe is greater than 60°, [$\rho \geq 60°$], the exact design can be derived from the following equation:

$$M = \mu\left(N_{left} + N_{right}\right) \times 4R\sin\left(\rho/2\right) / \left(\rho + \sin\left(\rho\right)\right) \qquad [8.22]$$

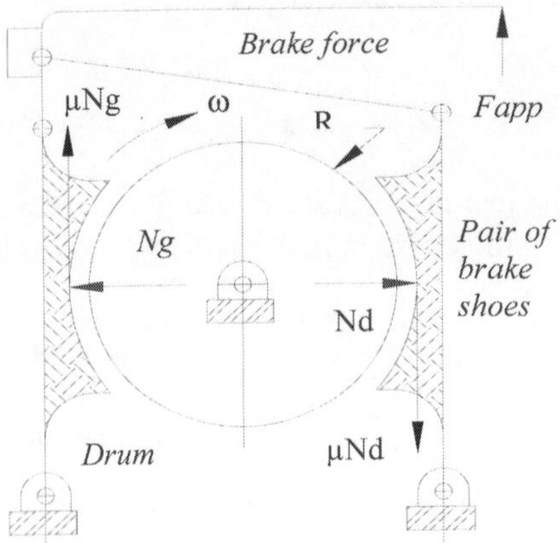

N_{Left} is the normal force in (N) on the left side.

N_{Right} is the normal force in (N) on the right side.

μ is the friction coefficient; ρ is the contact angle ≤ 60 °.

R is the radius of the brake drum.

Figure 8.9. *Graphic illustration of design of components of shoe braking system. For a color version of this figure, see www.iste.co.uk/grous/design.zip*

8.3.2. *Design of inner double-shoe block brake*

Block brakes are used for the reasons discussed above. However, their design differs, in accordance with the equations set out below. The moment M is calculated using the following relation:

$$M = \mu \omega R^2 \frac{\cos\left(\rho_1\right) - \cos\left(\rho_2\right)}{\sin\left(\rho_{\max}\right)}\left[P_{1\max} + P_{2\max}\right] \qquad [8.23]$$

where:

– μ is the friction coefficient;

– a is the distance from the center of the drum to the pivot of the shoe in mm or in in;

– R is the radius of the brake drum in mm or *in*;

– ρ_1 is the angle at the center, from the pivot of the shoe to the line of contact with the drum;

– ρ_2 is the angle at the center in (°), which runs from the pivot of the shoe to the line of contact with the tread of the drum;

– ρ_{max} is the angle at the center in (°), which runs from the pivot of the shoe to the maximum line where the pressure is applied, in MPa or psi;

– P_{1max} is the maximum pressure applied to the right shoe;

– P_{2max} is the maximum pressure applied to the left shoe.

$$P_{2\max} = C \times F \times P_{1\max} / \left(M_n + M_f\right) \qquad [8.24]$$

Suppose we have a *uniformly* distributed pressure, which is modeled thus:

$$P = P_{1\max} \times \sin(\rho) / \sin(\rho_{\max}) \qquad [8.25]$$

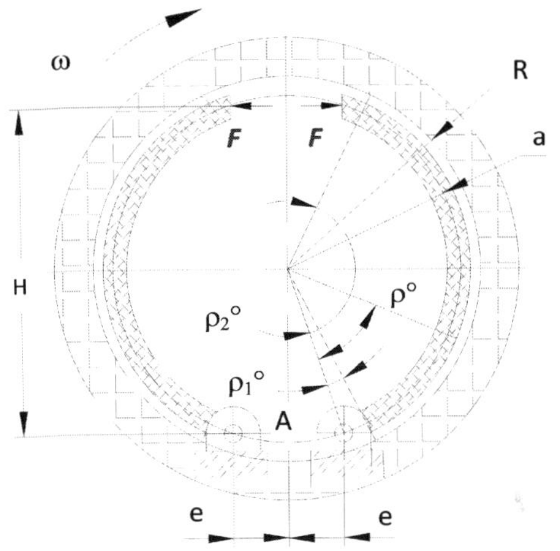

Figure 8.10. *Graphic design of components of internal shoe brakes*

As ρ_{max} is the angle at the center which goes from the pivot of the shoe to the maximum line where the pressure, in MPa (or psi), is applied, we exercise the following reasoning in design:

$$\left\{ If\ \rho_2 \succ 90^\circ \rightarrow \rho_{\max} = 90^\circ\ and\ if\ \rho_2 \prec 90^\circ \rightarrow \rho_{\max} = \rho_2 \right.$$

Because of the forces of friction on the shoe in contact with the drum, we respectively write the friction moments M_f and the normal moments M_n:

$$M_\mu = \frac{\mu P_{\max} \omega R}{\sin(\rho_{\max})} \int_{\rho_1}^{\rho_2} \sin(\rho)\left[R - a\cos(\rho)\right] d\rho \text{ and } M_n = \frac{a P_{\max} \omega R}{\sin(\rho_{\max})} \int_{\rho_1}^{\rho_2} \sin^2(\rho)\, d\rho \qquad [8.26]$$

The fitting force is the expression of the difference between the above moments in reference to the height (H is the lever arm in mm or in). Considering a clockwise rotation of the drum (ω in rad/sec), the right shoe self-activates by drive. Thus, we set:

$$Right\ shoe \rightarrow F = \left(M_n - M_\mu\right)/H\ and\ Left\ shoe \rightarrow F = \left(M_n^1 - M_\mu^1\right)/H \qquad [8.27]$$

The moments are also calculated by the respective expressions of the pressures:

$$M_n^1 = M_n P_{\max}^1 / P_{\max}\ and\ M_\mu^1 = M_\mu P_{\max}^1 / P_{\max} \qquad [8.28]$$

The above equations are based on three main hypotheses:

1) The shoe is rigidly mounted to the braking system.

2) The friction coefficient (μ) does not vary as a function of the pressure or of the velocity.

3) The normal pressure, at all points of contact of the shoe, is directly proportional to the vertical distances with respect to the pivot points.

The annexes contain a table of friction coefficients μ.

Materials	Max temperature 1°F = –17.222°C	μ, friction coefficient	Pmax 1 MPa = 145.038 psi
Metal on metal	315.5556°C	0.25	200 psi =1.379 MPa
Wood on metal	65.55556°C	0.25	70 psi = 0.483 MPa
Leather on metal	65.55556°C	0.35	25psi = 0.172 MPa
1°C = 33.8° Fahrenheit	1°C = 274.15 Kelvin		1°C = 493.47°Rankine

Table 8.1. *Friction coefficients as a function of the materials in contact*

8.3.3. *Design of a band brake block*

Band brakes are flexible. They consist of a band wrapped around the drum. These braking systems are usually driven by a pulley. The brake power is therefore a function of the angle of wrap, the friction coefficient and the tension on the band. The design of such systems is simple, using Euler's relation, which is written as:

$$\begin{cases} The\ tension\ ratio\ is\ written: T_1/T_2 = Exp^{(\mu\alpha)} \\ The\ braking\ moment\ is\ written: M_{br} = (T_1 - T_2)R \end{cases} \qquad [8.29]$$

These models of brake systems do not have auto-braking, with the exception of the diagram (Figure 8.11c) below.

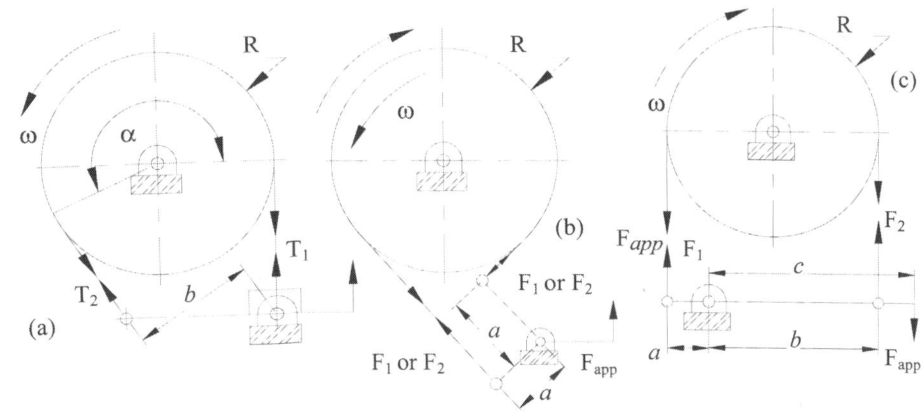

Figure 8.11. *Diagrams of band brakes*

where:

– T_1 is the tension on the tight side of the roller;

– T_2 is the tension on the loose side of the roller;

– μ is the friction coefficient, which is dimensionless;

– α is the angle of wrap in [°];

– M is the braking moment in N.mm or lb.in;

– a is the distance from the pivot to the tangent of the drum, left side, mm;

– b is the distance from the pivot to the tangent of the drum, right side, mm;

– c is the lever arm (from the pivot to the origin of the applied force, mm;

– R is the radius of the brake drum (R = a + b) in mm or in.

$$Tc - T_1 a - T_2 b = 0 \rightarrow T = \frac{(T_2 b - T_1 a)}{c}; T_1 = T_2 exp^{\mu\alpha} \text{ thus } T = \frac{T_2\left(b - Exp^{\mu\alpha} a\right)}{c} \quad [8.30]$$

In mechanical design, we often find ourselves dealing with sketches as follows. We observe that for $T = 0$ or *negative*, auto-braking occurs. In design, we need to estimate the expression of (b) or (b/a):

$$\left\{ b \leq a \times Exp^{(\mu \times \alpha)} \text{ or indeed } b/_a \leq Exp^{(\mu \times \alpha)} \right\} \quad [8.31]$$

Auto-brake systems accept unidirectional rotation. This prevents the reversal of the direction of rotation. Such is the case with distribution conveyors, winches and other hoists. The expression of the maximum pressure is written: $P_{max} = T_1/\omega R$. The mean pressure between the band and the drum is presented as follows:

$$P_{mean} = T_1/\omega R\mu\alpha \times \left(Exp^{(\mu\times\alpha)} - 1 \right) / Exp^{(\mu\times\alpha)} \qquad [8.32]$$

The heat generated by braking must be dissipated via a calorific transfer if we want to prevent the system from overheating. The ratio of generated heat, E_g, is:

$$E_g = P_{mean} \frac{A_c \mu V}{778} = \frac{E_{potential}^{absorbed} + E_{kinetic}^{absorbed}}{778} \left[\frac{Btu}{\min} \right] \qquad [8.33]$$

The generated heat is also expressed by means of the following energy balance:

$$\Delta E_g = E_{potential}^{absorbed} + E_{kinetic}^{absorbed} \; [J] \qquad [8.34]$$

where:

– 1Btu = 1055.056 Joules and 1 BTU/min = 17.584 watts;

– A_c is the contact area in (mm^2 or in^2);

– V is the linear velocity of the drum in m/min ft/min;

– μ is the friction coefficient;

– P_{mean} is the mean constant pressure in MPa or psi;

– Ep is the total potential energy absorbed in *ft-lb/min* or J;

– Ek is the total kinetic energy absorbed in *ft-lb/min* or J.

The dissipated heat is written as:

$$C_{dissipated} = C \times \Delta T \times A_{radiation} \; [Btu/\min] \qquad [8.35]$$

where:

– C_d is a specific heat transfer coefficient;

– ΔT is the temperature difference between the surface exposed to the heat (thermal radiation) and the ambient air in °C or K;

– A_r is the area of the part irradiated with heat in in^2 or mm^2.

It is worthwhile paying close attention to the expressions of the dissipated heat, which are approximate. A careful designer will conduct tests to better determine the thermal parameters. In mechanical design, we usually refer to the calculations of the brake forces, according to the design sketched in [hp/ld], which corresponds to the power (hp = *horse power*) in relation to the area (A_r) of the part irradiated with heat. In this ratio, (l) is the width of contact of the band (or shoe) in mm (in) and d the diameter of the drum in mm or in. With continuous loads, the product of the mean power P_{mean} and the velocity (V) must be limited to $P_{mean}.V \leq 28{,}000$.

8.3.4. *Examples of principles of calculations for brake design, with solutions*

The examples are offered as a guide to the calculations dedicated to the mechanical design of braking systems [ROA 75], found in numerous industries.

Example 1. Design hypothesis for a *single-shoe brake*

R = 350 mm = 13.78" is the radius for a single-shoe braking system. The moment M (force torque) is 2.5×10^5 N.mm = 2213 lbf.in. The system turns at a rotation frequency of 60 rad/s = 572.958 (rpm). Considering the materials in contact, we choose a friction coefficient (μ) = 0.35 (leather/metal). The design illustrated below has the following dimensions: a = 350 mm = 13.78 in; b = 900 mm = 35.433 in and c = 40 mm =1.575 in.

1) What is the normal force N (N) applied to the shoe, to be sketched in design?

2) What is the brake force required F (N) to be applied to the brake for a clockwise rotation?

3) What is the force F (N) required on the brake for an anticlockwise rotation?

4) What are the dimensions (C, in mm and in) required to allow the braking system to automatically lock if the measurements are as follows: a = 350 mm = 13.78 in; b = 900 mm = 35.433 in and c = 40 mm = 1.575 in?

5) What is the rate of heat generated (thermal transfer) E in BTU/min and in J?

SOLUTION WITH DISCUSSION.–

1) The expression of the moment M, depending on the design, can be employed to deduce N_n as follows:

$$\textit{From } M = \mu N_n R, \textit{ we deduce } N_n = M/\mu R = 2.5\times10^5/0.35\times350 = 2041\ N = 458.794\ lbf$$

2) The force of friction: $F_{friction} = \mu N_n = 0.35 \times 2.041 \times 10^3 = 714.286$ $N = 160.578$ lbf

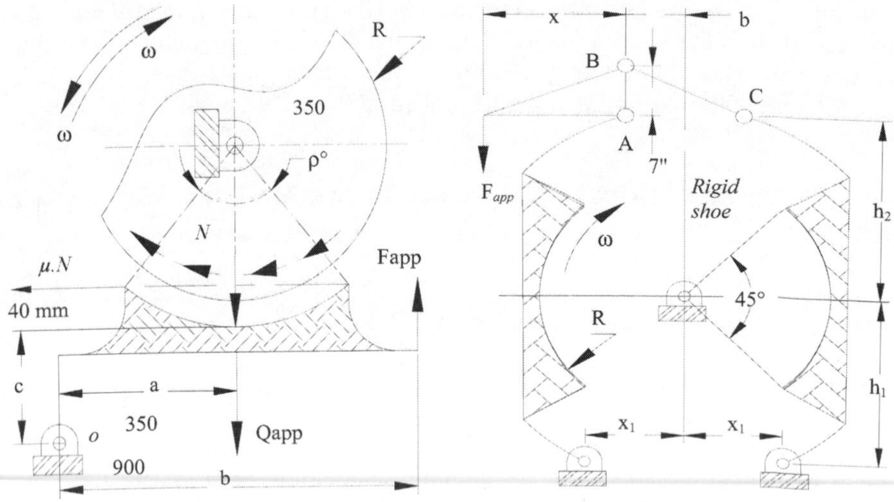

Figure 8.12. *a) Single shoe brake, b) double shoe brake*

3a) With clockwise rotation, the equilibrium expression is written as follows, along the OX axis. As Q = 0, we set:

$$\begin{cases} \left\{C\mu N_n + bF^{clockwise}_{equilibrium} - aN_n\right\} = 0 \rightarrow F^{clockwise}_{equilibrium} = \dfrac{aN_n - C\mu N_n}{b} = \\ = \left(350 \times 2.5 \times 10^5 - 40 \times 0.35 \times 2.5 \times 10^5\right) / 900 = 761.905 \ N = 171.283 \ lbf \end{cases}$$

3b) With anticlockwise rotation, the equilibrium condition is written thus:

$$\begin{cases} \left\{aN_n + c\mu N_n - bF^{anticlockwise}_{equilibrium}\right\} = 0 \rightarrow F^{anticlockwise}_{equilibrium} = \left(aN_n + c\mu N_n\right)/b \\ F^{clockwise}_{equilibrium} = \left(350 \times 2.5 \times 10^5 + 40 \times 0.35 \times 2.5 \times 10^5\right) / 900 = 825.395 \ N = 185.557 \ lbf \end{cases}$$

4) In auto-braking: $Condition \geq a/\mu$ $hence$ $C \geq (350/0.35) = 39.37$ $in = 1000$ mm

This condition does not exist with anticlockwise rotation, so will not be calculated.

5) The calculation of the generated thermal energy is written as follows:

$$E_{calorific} = \mu N\omega/778 \;\left[Btu/\min\right](\text{USA}) \rightarrow for\; N = 458.794\; lbf\; and\; \text{RPM} = 572.958$$

$$where\; \omega = \frac{\pi 28\times 572.958}{12} \rightarrow E_{calorific} = \frac{\mu N\omega}{778} = \frac{0.35\times 458.794\pi 28\times 572.958/12}{778}$$

$$866.87\left[\frac{Btu}{\min}\right] = 867.23\left[\frac{CBTU}{\min}\right] = 756.89\left[\frac{IBTU}{\min}\right] = 7.98\times 10^5\left[\frac{J}{\min}\right] = 1.91\times 10^5\left[\frac{cal}{\min}\right]$$

IBTU = according to international standard; CBTU = according to Canadian standard.

8.3.5. *Case study: hypothesis of the design of a double-shoe brake*

Consider a design of a double-shoe brake as illustrated below (Figure 8.13). We read the following essential dimensions:

h_1		**e_1**		**b_1**	
30 in	30 in	5 in	127 mm	20 in	508 mm
h_2		**$H = h_1 + h_1$**		**$B = 2\,b_1$ car $b_2 = b_1$**	
32 in	32 in	62 in	1575 mm	40 in	1575 mm
a		**F_{app}** (see diagram)		**μ,** friction coefficient	
7 in	177.8 mm	224.809 lbf	1000 N	0.35	Metal/ leather

Table 8.2. *Dimensions and coefficients*

QUESTIONS.– Calculate the braking moment (M = torque in N.m and lbf.ft) considering an anticlockwise motion of the system if the friction coefficient (μ) = 0.35. The sizing is presented in the design sketch (leather/metal). What are the normal forces F_N in N and lbf applied to the shoe on the right and left sides, and the total moment applied in N.m, lbf.ft applied to both sides of the shoe?

SOLUTION WITH DISCUSSION.– Consider the system to calculate the reactions on the horizontal and vertical planes to joints A and B, and set the equilibrium equations:

$$1)\begin{cases} Based\; on\; the\; equilibrium\; condition\; we\; set \rightarrow F_A^H = F_B^H\; and\; F_A^H - F_A^H - F_{app} = 0 \\ F_{app}h_1 = aF_B^H \rightarrow F_B^H = F_{app}h_1/a = (224.809\times 30/7) = 963.467\; lbf = 4286\; N \end{cases}$$

2) Consider the system at the joints of the lever BC and set the equilibrium equations on the vertical plane:

$$\begin{cases} \text{From the equilibrium condition: } F_B^H = F_C^H;\ F_B^V = F_C^V \text{ and } F_B^V B = F_B^H a \\ F_B^V = F_B^H a/B = (963.467 \times 7/40) = 168.607\ lbf = 750\ N \end{cases}$$

3) Consider the right and left sides of the system with free levers, and set the equilibrium equations of the normal forces in light of the equilibrium conditions:

3a) *Left side of shoe*

$$\begin{cases} N_G h_1 - \mu a N_G - F_B^H H - F_B^V e_1 = 0 \\ N_G = \dfrac{F_B^H H + F_B^V e_1}{h_1 - \mu a} = 9781\ lbf = 2199\ N \end{cases}$$

3b) *Left side of shoe*

$$\begin{cases} N_D h_1 + \mu a N_D - F_B^H H + F_B^V e_1 = 0 \\ N_D = \dfrac{F_B^H H + F_B^V e_1}{h_1 + \mu a} = 8073\ lbf = 1815\ N \end{cases}$$

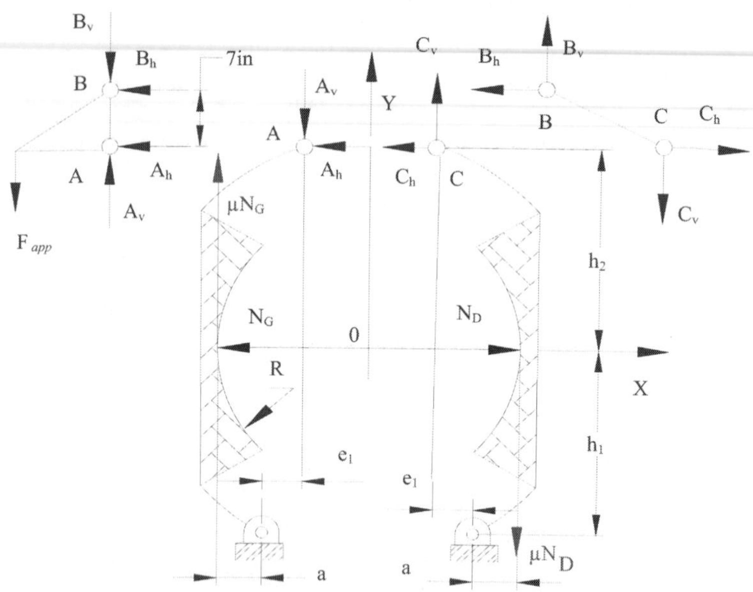

Figure 8.13. *Sketch of the proposed brake design (double shoe). For a color version of this figure, see www.iste.co.uk/grous/design.zip*

4) *Calculation of the moment*

$$M = \mu(N_G + N_D)(e_1/2) = 0.35\times(2199+1815)\times(5/2) = 292.665\ lbf.ft = 396.8\ Nm$$

8.3.6. *Case study: hypothesis of the band brake whose drum has a radius R (mm and in)*

We wish to calculate the tensions T_1 and T_2 applied to the band wrapped around a drum if the angle of wrap $\alpha = 260°$ at a distance a of 10 in (a = 254 mm) from the diameter of the drum. The moment applied is $M = 1345\ lbf.in\ (1.52\times10^5\ N.mm)$. The mechanism is driven at 1000 rpm. The materials in contact (cast iron in a humid environment) require a friction coefficient of μ = 0.2. The yield of the system η is 75 %.

SOLUTION WITH DISCUSSION.– By virtue of Euler's relation, we begin by writing:

$T_1/T_2 = Exp^{\mu\times\alpha}$ where $\alpha = \pi 280/180 = 280°$. The ratio of the forces is written thus:

$$T_1/T_2 = Exp^{\mu\alpha} = \Psi = e^{0,2\times280} = 2.657\ rad = 152.263°$$

The condition of equilibrium (inertia) enables us to set: $M = a(T_1 - T_2) = Exp^{\mu\times\alpha}$

From $T_1/T_2 = Exp^{\mu\times\alpha}$, we deduce T_1: $T_1 = \left(Exp^{\mu\times\alpha}\right)\times T_2$

We then substitute this expression into the expression of M. We obtain:

$$\begin{cases} M = a\left(Exp^{\mu\times\alpha}\times T_2 - T_2\right) = T_2 a\left(Exp^{\mu\alpha} - 1\right) \rightarrow T_2 = \dfrac{M}{a\left(Exp^{\mu\times\alpha} - 1\right)} = \dfrac{1345}{10(2.657-1)} \\ T_2 = 81.146\ lbf = 360.958\ N\ and\ T_1 = Exp^{\mu\alpha}T_2 = 2.657\times81.146 = 215.646\ lbf = 959.243\ N \end{cases}$$

Knowing the rate of rotation N, the angular velocity (ω in 1/S) would be: 2πN/60 = 10.966 [1/S]. From the power relation, we can set: $P = M\times\omega\times\eta\ [W]$, where η represents the yield = 75 %:

$$P = M\omega\eta = 1345\times10.966\times75\% = 71.078\ [Btu/\min] = 1.25\times10^3\ [W]$$

8.3.7. *Case study: differential brake using a roller pressed against a drum*

Consider the design of a double-shoe brake (also see Figure 8.14).

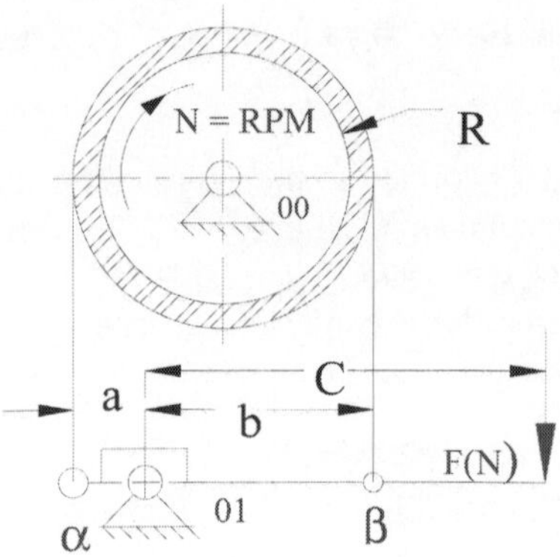

where:
– F_{app} = 56 *lbf* = 249.1 N
– μ = 0.35 Friction coefficient
– *M* = 5000 *lbf.in* =5.649×10^5 N.mm
– *a=1.5 in=38.1 mm; b=3.5 in=88.9 mm*
– *C=7.5 in=190.5 mm*
– *R=D/2=5.5 in=139.7 mm*

1. Determine the maximum and minimum forces.
2. Determine the maximum moment.
3. Determine the power in watts.

Figure 8.14. *Sketch of the design of a differential roller brake. For a color version of this figure, see www.iste.co.uk/grous/design.zip*

With clockwise rotation, let us first check whether the brake can automatically lock. (α) is the angle of wrap on the drum in degrees $\alpha = 180\times\pi/180 = 180°$. The condition justifying this is written:

$$\begin{cases}(b/a)\le\Psi,\ knowing\ \Psi = Exp^{\mu\times\alpha} = Exp^{0.35\times\pi} = 3.003\ and\ (b/a)\le 3.003\\ From\ \Psi = 3.003\ and\ from\ b/a = 2.43 \Rightarrow \Psi \le b/a\ because\ 2.43 \prec 3.003\end{cases}$$

The maximum and minimum forces are calculated on the basis of this system of equilibrium:

$$\begin{cases}M = R\times(F_1 - F_2) = 3600\ lbf\\ a\times F_1 + c\times F_{app} - b\times F_2 = 0\end{cases} \Rightarrow \begin{cases}R\times F_1 - R\times F_2 = 3600\ lbf\\ b\times F_2 - a\times F_1 = c\times F_{app}\end{cases}$$

Case 1: clockwise rotation (CW →-) unauthorized – non-self-locking brake:

By writing it in matrix form, we are able to solve the system with two unknowns:

$$\Delta = \begin{pmatrix} a & -b\\ R & -R\end{pmatrix},\ V = \begin{pmatrix}-c\times F_{app}\\ M\end{pmatrix} \Rightarrow \text{soln} = lsolve(\Delta, V) = \begin{pmatrix}1.185\times10^4\\ 6.029\times10^3\end{pmatrix} N = \begin{pmatrix}2.665\times10^3\\ 1.355\times10^3\end{pmatrix} lbf$$

$$F_{\min}^{\max} = \begin{pmatrix} F_{\max}^{H} = 1.185\times10^{4} \\ F_{\min}^{H} = 6.029\times10^{3} \end{pmatrix} [N] \text{ or indeed } \begin{pmatrix} F_{\max}^{H} = 2.665\times10^{3} \\ F_{\min}^{H} = 1.355\times10^{3} \end{pmatrix} [lbf]$$

Case 2: anticlockwise rotation (ACW → +) the brake is self-locking: by writing it in matrix form, we can easily solve the system with two unknowns. The maximum and minimum forces are calculated on the basis of this system of equilibrium:

$$\begin{cases} \Psi = \dfrac{F_1^{AH}}{F_2^{AH}} = Exp^{\mu\alpha} = 3.003 \\ bF_1^{AH} - cF_{app} - aF_2^{AH} = 0 \end{cases} \Rightarrow \begin{cases} F_1^{AH} - \Psi\times F_2^{AH} = 0 \\ bF_2 - aF_1 - cF_{app} = 0 \end{cases} \Rightarrow$$

$$\Delta = \begin{pmatrix} 1 & -\Psi \\ b & a \end{pmatrix} \text{ and } V = \begin{pmatrix} 0 \\ cF_{app} \end{pmatrix} \Rightarrow \text{soln} = lsolve(\Delta, V) = \begin{pmatrix} 589.866 \\ 196.426 \end{pmatrix} N = \begin{pmatrix} 132.607 \\ 44.158 \end{pmatrix} lbf$$

$$F_{\min}^{\max} = \begin{pmatrix} F_{\max}^{AH} = 589.8661 \\ F_{\min}^{AH} = 196.4266 \end{pmatrix} N \text{ or indeed } \begin{pmatrix} F_{\max}^{AH} = 132.6072 \\ F_{\min}^{AH} = 44.158 \end{pmatrix} lbf$$

Results from substitution:

$$\begin{cases} F_2^{AH} = \dfrac{c\times F_{app}}{\Psi\times b - a} = 44.158\ lbf = 196.426\ N \\ F_1^{AH} = \Psi\times F_2^{AH} = 132.607\ lbf = 589.866\ N \end{cases} \leftarrow QED$$

Calculation of the force moment (torque) in N.mm and in lbf.in

$$M_{\sup} = R\left(F_{1\max}^{AH} - F_{2\min}^{AH}\right) = \frac{5.5}{2}(132.607 - 44.158) = 243.235\ lbf.in = 2.748\times10^{4}\ N.mm$$

Calculation of the power (in W and Btu/min) necessary for the applied moment: for N = 960 RPM, we calculate the frequency of rotation, ω = 2πN/60 = 10.528 [1/S]. The power in watts and Btu/min, if the yield (η) of the system is 70 %, is:

$$P_{syst} = M_{\sup}\times\omega\times\eta = 243.235\times10.528\times0.7 = 202.522\ [Watt] = 11.517\ [Btu/\min]$$

8.3.8. *Symmetrical shoe brake pressed against a drum with radius R*

Consider the design of a symmetrical external shoe brake (Figure 8.15). We wish to determine the resultant of the normal forces, the force of friction and the precise location of the applied forces. We determine the moment with respect to the center of the drum, taking account of the phenomenon of wear and tear. The wear is considered to be uniform. In other words, the diameter of the drum must remain circular (we impose a tolerance interval on the circularity [IT] circularity, O).

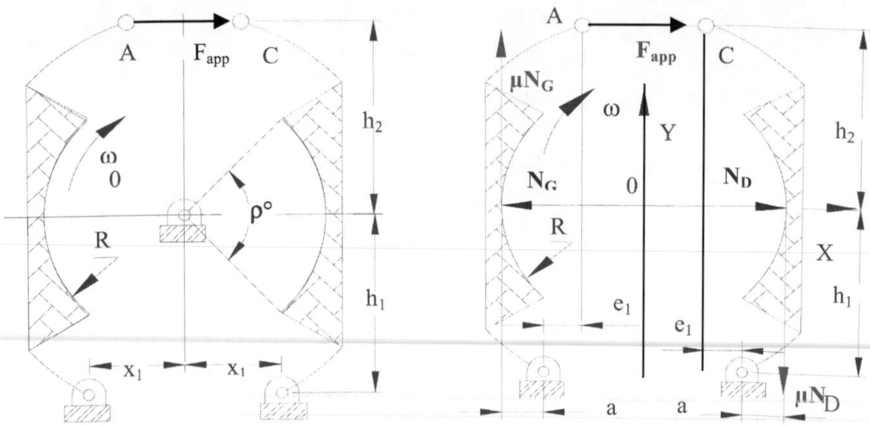

Figure 8.15. *Sketches of the proposed brake design (double symmetrical shoe). For a color version of this figure, see www.iste.co.uk/grous/design.zip*

The radial wear and tear is: $F_{radial} = \kappa_s \times (P_n \times V)$ [8.36]

After the motion and the radial wear and tear, the point δ moves to δ_1. The purpose of this is to maintain contact between the rigidly mounted shoe and the drum. The horizontal displacement (Δ_h) of the point (δ) is expressed thus:

$$\Delta_{horizontal} = \kappa_s \times P_n \times V / \cos(\varphi) \quad [8.37]$$

where:

– κ_s is a coefficient of the material, equivalent to the P_{max} when the angle is zero ($\rho = 0°$);

– P_n is the normal pressure and V is the linear velocity;

– Δ_h is a displacement (equal shift), constant at every point.

Normal pressure → $P_n = \kappa_s \times \cos(\varphi)$, where $\kappa_s = \Delta / \kappa V$ [8.38]

The shoe is symmetrical, and therefore the normal pressure is also symmetrical. We conclude that the outcome of the forces is equivalent to the sum of the horizontal components. The same is true for the outcome of the vertical forces of the differential of the forces of friction, as illustrated below (Figure 8.16).

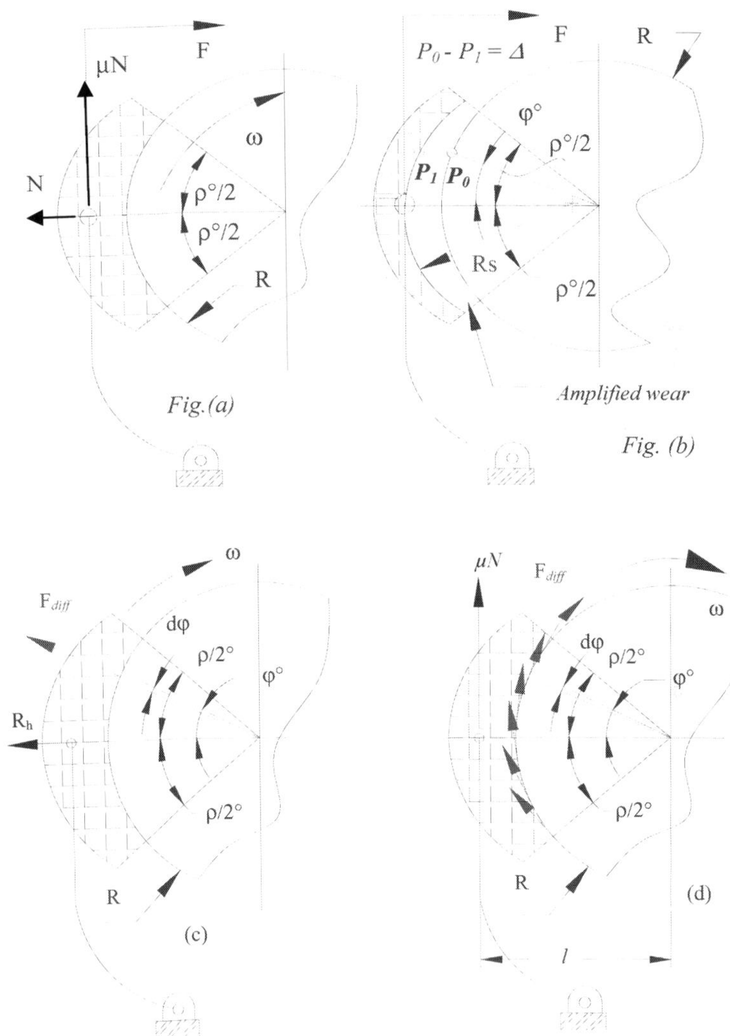

Figure 8.16. *Illustrations of the proposed brake design (symmetrical double shoe). For a color version of this figure, see www.iste.co.uk/grous/design.zip*

$$\begin{cases} R_F^h = N = 2\int_0^{\rho/2} \Delta_{horiz}\cos(\varphi)\,\kappa_s\cos(\varphi)R\,d\varphi = \dfrac{\kappa_s R\Delta_{horiz}}{2}\cos(2\rho) \\ R_F^V = 2\int_0^{\rho/2} \mu\kappa_s\cos(\varphi)R\cos(\varphi)\,d\varphi = \mu N \end{cases} \quad [8.39]$$

The position of the resultant force is found in relation to the load (Figure 8.16a and b). N passes through the center of the rod of the shoe along the horizontal axis of symmetry (Figure 8.16c). The position of the vertical force represented by (μN) is calculated by the principle of equilibrium of the moment applied and the forces resulting from the forces of friction. The center of the drum is considered the reference point. (**h**) is the distance from the center of the drum to the point of application of the force of friction (**μN**). Hence:

$$\mu Nh = 2\int_0^{\rho/2}\left[\mu\kappa_s\cos(\varphi)\right]R\Delta_{horiz}\,d\varphi = 2\mu R^2\Delta_{horiz}\sin\left(\rho/2\right) \quad [8.40]$$

The maximum pressure K_s, for $\rho = 0°$: $\kappa_s = \dfrac{2N}{R\Delta_{horiz}\sin(\rho)+\rho}$ [8.41]

The length would then be: $l = 4R\sin(\rho/2)/(\sin(\rho)+\rho)$ [8.42]

The moment of the resultant of the friction forces (μN) with respect to the center of the drum constitutes the brake force of the system and is written as follows:

$$M_{res} = \mu\times N\times h = \mu N\left[4R\sin(\rho/2)/(\sin(\rho)+\rho)\right] \quad [8.43]$$

$$Fig(c)\rightarrow\begin{cases} F_{differential} = F_{diff} = P_n\times R\times\omega\times d\varphi \\ N_{horizontal}^{resultant} = N_{hori}^{res} = \dfrac{\kappa_s R\omega}{2}\left[\sin(\rho)+\rho\right] \end{cases} \quad [8.44]$$

$$Fig(d)\rightarrow\begin{cases} F_{différential}^{friction} = F_{diff}^{fric} = \mu\times P_n\times R\times\omega\times d\varphi \\ F_{vertical}^{resultant} = F_{vert}^{res} = \mu N \text{ where } l = \dfrac{4R\sin(\rho/2)}{\sin(\rho)+\rho} \end{cases} \quad [8.45]$$

COMMENTS AND DESIGN ANALYSIS OF THE SYSTEM.– The rod is situated at the distance (h) from the center of the drum, as a free-moving body. It is represented by the normal force (N) passing through the center of the pin and the force of friction ($\mu.N$). It also passes through the center of the pin of the shoe in order to appropriately satisfy the pressure distribution $P_n = \kappa_s.\cos(\varphi)$. The analysis of the forces is conducted without reference to the pressures distributed on the shoe. If the pin were situated at a distance different to (h) as expressed above, the result of the forces would still have to pass through the center of the pin. The above assumption [$P_n = \kappa_s.\cos(\varphi)$] is no longer valid, and does not deliver equilibrium of the system. It is therefore useful to consider the components of the forces and (μN) to unfailingly act at the center of the pin. Remember that the pivot can be positioned along the resultant of the forces (N) and (μN) without affecting the distribution of pressures on the drum from the relative shoe. For example, if a shoe with a symmetrical pivot were positioned in such a way that it subtends a right angle (90 °) around a drum 304.8 mm = 25 in in diameter, what would be the distance (h) between the center of the drum and the pivot in order to avoid a moment of rotation due to the forces of friction, if we assume that the wear and tear is uniformly distributed? We assume that the pivot is thus situated along the normal force (N). The distance (h) from the pivot to the center of the drum is:

$$l = 4R\sin\left(\rho/2\right) / \left(\sin(\rho)+\rho\right) = 4\times 24\sin\left(\pi/4\right) / \left(\sin\left(\pi/2\right)+\pi/2\right) = 13.203\ in = 335.345\ mm$$

In the above figure, the radius has a linear influence on the height at which the forces are applied to the internal symmetrical shoes. We must take account of the uniformity of the wear and of the specific pressure κ_s. Without taking account of the above, the designed system would be rather unsatisfactory (in terms of safety, lifetime, etc.). For different diameters ranging from 12 to 26 in, for $R_{variables}$ (12 in, …, 26 in)/2 and for the same angle (ρ), let us plot the evolution of $h(R_{variables})$.

$h(R_{variable}) =$

	0
0	13.203
1	14.303
2	15.403
3	16.503
4	17.603
5	18.704
6	19.804
7	20.904
8	22.004
9	...

$$h(R_{variable}) = \frac{4 \cdot R_{variable} \cdot \sin\left(\frac{\rho}{2}\right)}{\sin(\rho) + \rho}$$

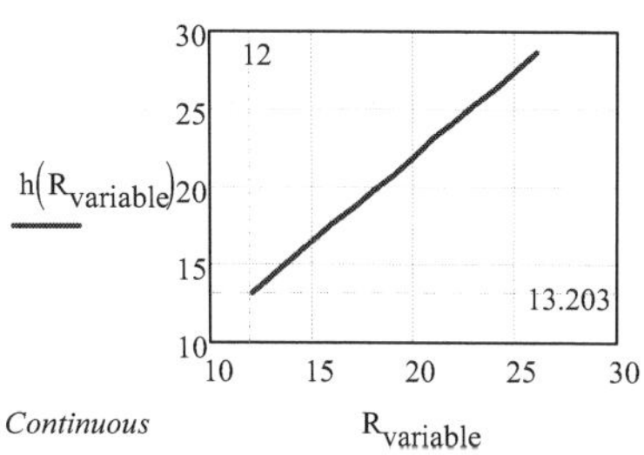

Figure 8.17. *Illustration of the proposed brake design (external symmetrical double shoe). For a color version of this figure, see www.iste.co.uk/grous/design.zip*

8.4. Principles of calculations of a gear system or gear disc

Gear systems, like couplings, are adherence power transmitters. Thus, they are temporary relays between a motor and a power receiver. The following are the main operational phases:

– *transient slip phase*: during gear shifting, power transmission is progressively engaged. The input and output shafts do not spin at the same speed. Slip occurs between the discs, giving rise to wear and tear, and therefore the dissipation of energy in the form of heat. Over a limited period of time, this phase gradually unites the engine with the gearbox;

– *engaged position*: the gear system transmits *all* of the power supplied to the bearing. This is the stable position with lack of command;

– *disengaged position* (dead point): the transmission is interrupted, the wheel is free and the vehicle stopped. The engine can continue to run without driving the wheels.

Consider an *internal* shoe brake, as illustrated below.

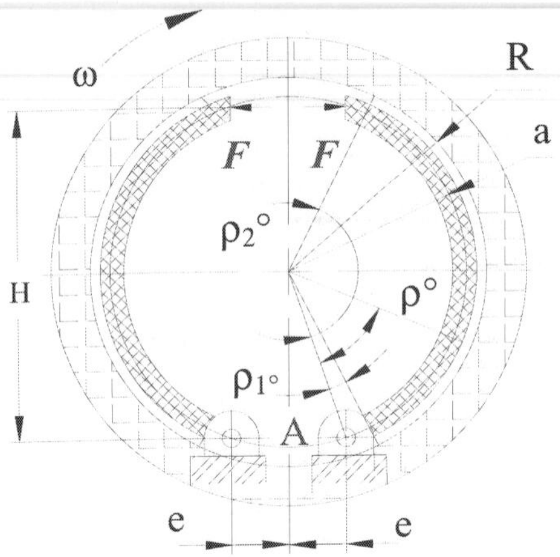

Hypothesis of calculations relative to the design:

– F = force applied (load);
– H = height;
– R = external radius;
– a = internal radius;
– e = excentricity;
– $\rho°$ = angle;
– $\rho_1°$ = small angle;
– $\rho_2°$ = winding angle;
– μ = friction coefficient;
– ω = rate of rotation.

Figure 8.18. *Internal symmetrical shoe brake. For a color version of this figure, see www.iste.co.uk/grous/design.zip*

The transmission torque delivered by a gear system depends on the material used for the brake linings, the number and dimensions of surfaces of friction between discs, and the forces exerted by the springs. The disc is the element associated with the output shaft. It is pinched between two elements linked to the engine shaft. It carries the friction *linings*, and is therefore subject to wear and tear. There is an even number of contact surfaces.

The contact surfaces may operate dry or be lubricated in an oil bath. The control may be mechanical, hydraulic or electric, with electronic command. The temperature increases as the discs spin. The performances are dependent on the temperature. Cooling is needed, because the linings are made of synthetic materials. The dimensions are dictated by the slip phase, which determines the maximum transmissible torque.

Multidisc gear systems (Figure 8.19b) work on the principle of a stack of discs. One out of every two discs is notched (grooved) around its circumference, meaning it can be connected in rotation to the *bell* of the gear system. The other, non-notched discs are connected to the *arm of the gear system.* The stack is kept under pressure by springs. The force is almost identical for each disc. This configuration, for the same value of the transmissible torque, is much more compact radially than that with a single disc. Multidisc gear systems are drawn in 3D.

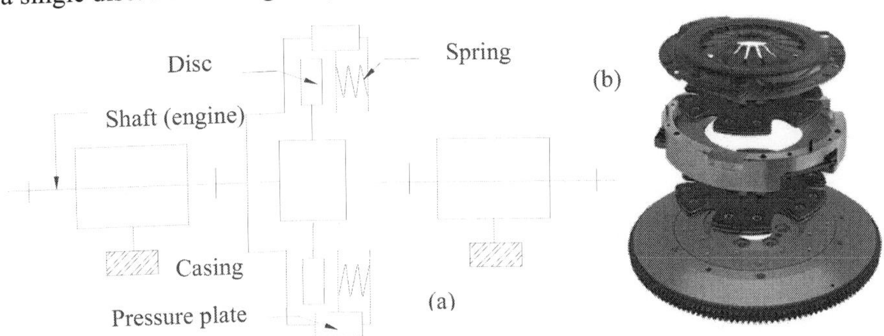

Figure 8.19. *Single-disc gear system (a). Multidisc gear system (b). For a color version of this figure, see www.iste.co.uk/grous/design.zip*

8.4.1. *Case study: principles of calculations for gear systems*

CAM GEAR SYSTEM.– Design a cam gear system to run a centrifugal pump. The gear system must transmit 117 hp at 1662 rpm to the pump, which starts up and shuts down 36 times an hour for 12 hours, 360 days per period. The pump's lifetime is estimated at 12 years.

With:

– power of the gear system, $P_{emb} = 117$ hp $= 8.725\times10^4$ W;

– frequency of rotation of the gear system, $\omega_{emb} = 1662$ rpm;

– number of cycles per hour, $Cy_h = 42$ Cycles/hr;

– number of hours per day, Hd = 1 h/d – 1 hr/day;

– number of days per year, Dy = 360 d/y = 360 day/yr.

Calculation of the maximum moment (torque) acting on the cam gear system:

$$M_{\max} = P_{emb}/\omega_{emb} = P_{emb}/2\pi N_{emb} = 501.291\ N.m = 4437\ lbf.in \quad [8.46]$$

ANALYSIS OF THE MOMENT (WORK) OF TORQUE ACTING ON THE GEAR SYSTEM.– For installations without shock loads (no-load installations) from start to finish, we write the maximum torsion moment which acts on the gear system. However, there may be a shock load at startup and cutoff. In other words, the shock torsion moment must be added to the torsion moment to determine the action of the total torsion moment. Calculate the shock torsion moment using the relation from step 1 and the true power (*hp*) and the velocity developed by the shock load.

Calculation of the total load (annual number of cycles):

$$N_{cycles} = C_{cy.year} \times H_j \times J_a = 1.512\times10^4\ [Cycles/year] \quad [8.47]$$

In 12 years, the gear system would perform this number of cycles before failing:

$$N_{cycles}^{12years} = \left(C_{cy.year} \times H_j \times J_a\right)\times 12 = 1.814\times10^5\ [Cycles] \quad [8.48]$$

8.4.2. *Analysis and choice of the dimensions of the cam gear system*

Using the grid in Figure 8.20, the maximum torsion moment, 365 *lbf.ft*, is on the left and the number of loading cycles, 1,728,000, is at the bottom. By horizontal and vertical projection to the point of intersection, we are able to choose the represented gear system by reading the next indicator above the curve. Thus, a type of gear system would be used for that particular load. (Note that the capacity of the gear system could be tabulated instead of read off a graph).

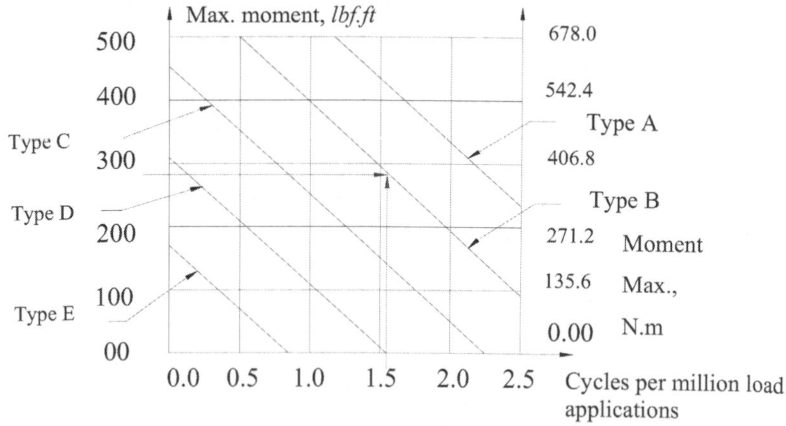

Figure 8.20. *Grid of choices for cam gear systems. For a color version of this figure, see www.iste.co.uk/grous/design.zip*

8.4.3. *Sizing of a cam gear system and case study*

As illustrated below, if the bore is too small, we must choose a larger bore and verify that the gear can fit in the available space. In order to do this, we usually use the general procedure for choosing cam gears (see the above grid) for compressors, transporters, cranes, ventilators, airplanes, 3D printers, pumps, punching machines, speed reducers, looms, etc.

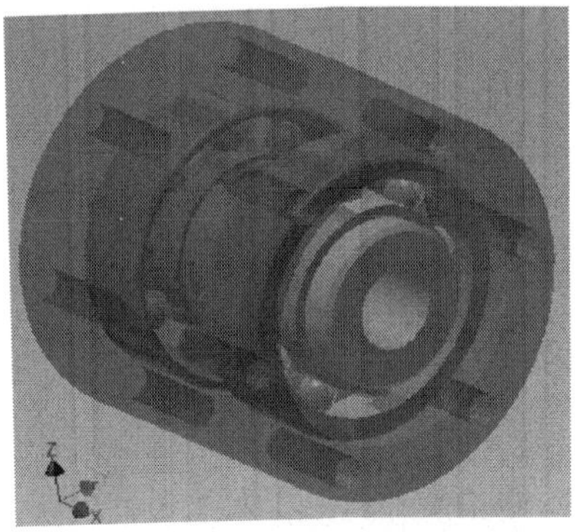

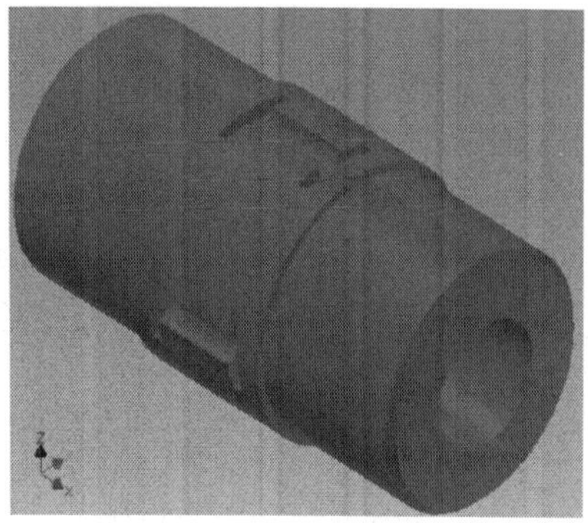

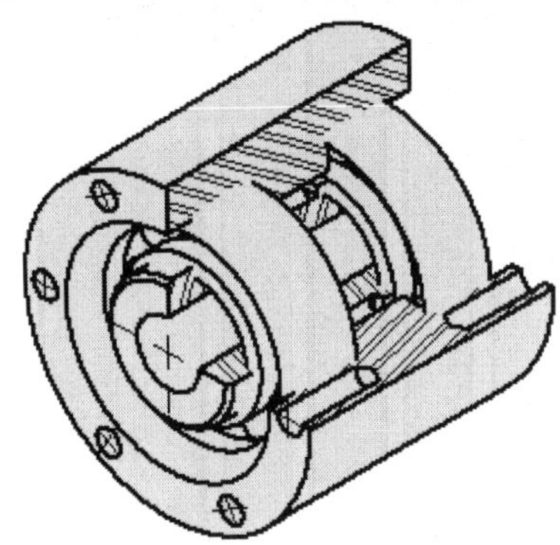

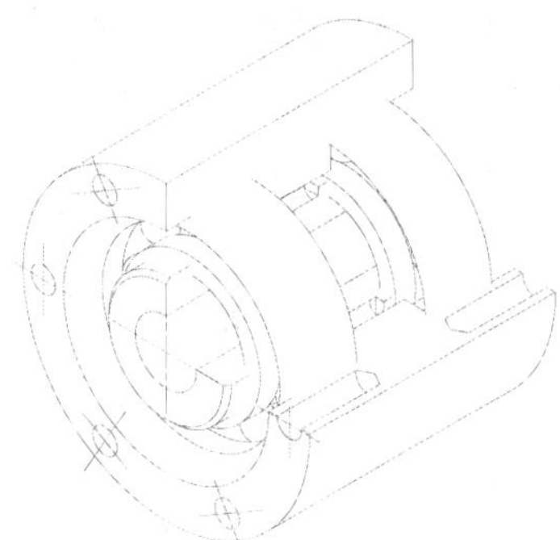

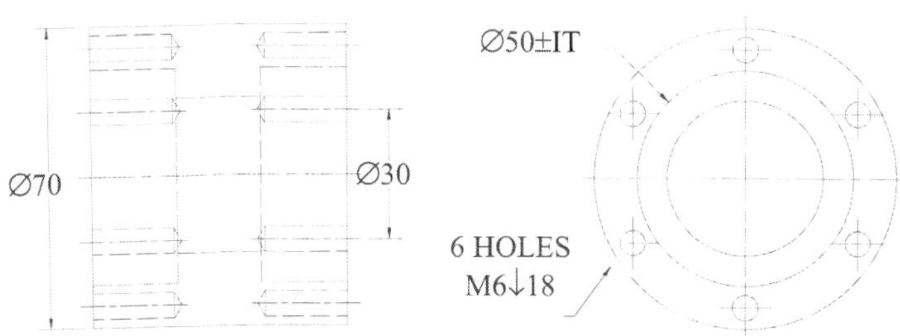

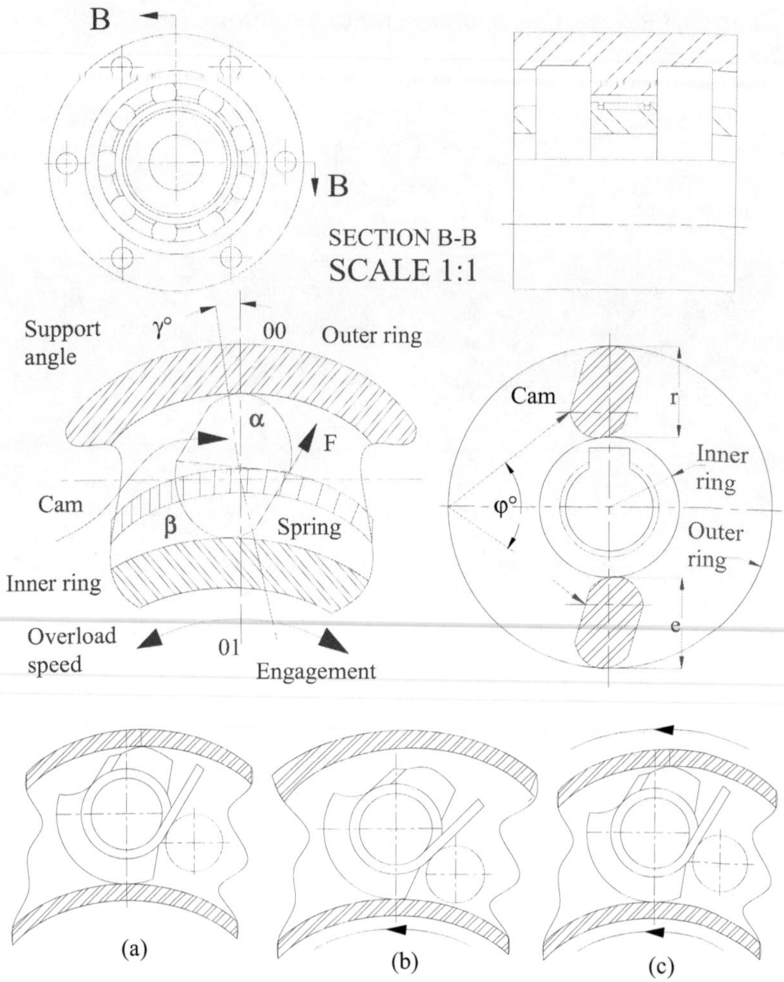

Figure 8.21. *Components of gear system: bearing, cam, springs and cages. (a) Cam stationary; (b) inner and outer rings locked, during rotation of the mechanism; and (c) only the inner ring is rotating. R = reduced shrinkage; e = reduced gap and φ is a slight inclination. For a color version of this figure, see ww.iste.co.uk/grous/design.zip*

Consider a mechanism with a cam gear system, made by the manufacturer Tsubaki [TSU 16], having a moment (M, work) required per service factor. The capacity of the moment of the gear system cam must be greater than the moment of the cam designed, in order to better withstand the overload in speed. This is formulated as follows:

– S = Clamping moment (%) is always > 100 %, and chosen from the manufacturers' catalogues;

– the service factor Fs = 1.5/day (numerous stops) and Fs = 2.0/day (sudden stops);

– μ is the friction coefficient of the bearing = 0.03;

– h is the total lift in m;

– l is the distance between the tail and the head of the pulley;

– l_0 is the distance corrected by the length (l). The manufacturer suggests $l_0 = 49$ m;

– Q is the maximum load in metric tons/h;

– F_s is the service factor (see manufacturer's coefficients);

– V is the velocity of the mechanism used – for instance, a conveyor – in m/min;

– W is the weight of the parts driven by the conveyor in a no-load situation (kg/m);

– N is the rate of rotation of the drive shaft of the gear system (rpm).

According to the manufacturer's documentation: $\eta_{\text{Imperial}} = 5252$ and $\eta_{\text{ISO}} = 9550$:

$$\begin{cases} M_{engine}^{torque} = P_{hp}^{engine} = \dfrac{P_{hp} \times \eta_{imperial}}{RPM_{shaft}} \times \dfrac{S}{100} \leq \left[M_{engine}^{Max\ torque} \right] \\ \text{or } M_{engine}^{torque} = P_{kW}^{engine} = \dfrac{P_{kW} \times \eta_{ISO}}{RPM_{shaft}} \times \dfrac{S}{100} \leq \left[M_{engine}^{Max\ torque} \right] \end{cases} \quad [8.49]$$

Belt width, mm	**400**	**450**	**500**	**600**	**750**	**900**
Estimated relative weight, kg/m	22.4	28	30	35.5	53	63
Belt width, mm	1050	1200	1400	1600	1800	2000
Estimated relative weight, kg/m	80	90	112	125	150	160

Table 8.3. *Estimated values of a conveyor belt. Source [TSU 16]*

Expressions of calculation of the powers, according to the manufacturer

Parasitic pinion and no-load belt:

$$P_I = P_{engine}^{unloaded\ belt} = (6/100)\times\mu\times W\times V\times((l+l_0)/367)\ [kW] \qquad [8.50]$$

Active pinion and belt loaded – horizontal motion:

$$P_{II} = P_{engine}^{horizontal} = \mu\times Q_\tau\times((l+l_0)/367)\ [kW] \qquad [8.51]$$

Active pinion and belt loaded – vertical motion:

$$P_{III} = P_{engine}^{vertical} = h\times Q_\tau\times(1/367)\ [kW] \qquad [8.52]$$

Backstop power:

$$P_{Backstop} = P_{III} - (7/10)\times(P_I + P_{II})\ [kW] \qquad [8.53]$$

BACKSTOP TORQUE (FORCE TORQUE).– It is important to select the appropriate gear cam which delivers the required backstop torque:

$$M = (9550/RPM)\times P_{Backstop}\times F_S\ [N.m] \qquad [8.54]$$

The principle behind a cam gear system is simple and astute. There are many different spring cams. Here is one example, inspired by Tsubaki [TSU 16].

8.4.4. *Case study: principles of calculations for gear systems in design*

The principles of calculations of gear systems are based on the force of friction between the materials in the mechanism, the exerted pressure and the geometry of the essential parts, such as their diameters. The figure below illustrates this.

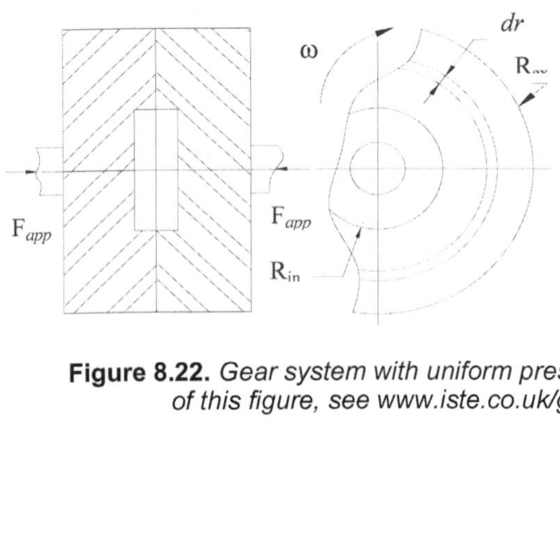

← *Hypothesis* of calculations relating to the design here illustrated:

- F_{app} = ?;
- R_{in} = 3.5 in = 88.9 mm;
- R_{ex} = 3.0 in = 76.2 mm;
- μ = 0.30;
- ω = 960 rad/s.

Figure 8.22. *Gear system with uniform pressure. For a color version of this figure, see www.iste.co.uk/grous/design.zip*

The expression of the total moment, in lbf.in or N.mm, is:

$$M_{total} = 2\pi\mu P \int_{R_{in}}^{R_{ex}} R^2 dR = \pi\mu P \frac{3}{2}\left(R_{ex}^3 - R_{in}^3\right) \qquad [8.55]$$

$$Based\ on\ F_{app} = \pi P\left(R_{ex}^2 - R_{in}^2\right) we\ write\ the\ pressure\ P = \frac{F_{app}}{\pi\left(R_{ex}^2 - R_{in}^2\right)} \qquad [8.56]$$

After substitution of the equation of pressure P into the equation of the moment, we find:

$$M_{total} = \frac{F_{app} 2\mu}{\pi\left(R_{ex}^2 - R_{in}^2\right)} \int_{R_{in}}^{R_{ex}} R^2 dR = \pi\mu P \frac{2}{3}\left(\frac{R_{ex}^3 - R_{in}^3}{R_{ex}^2 - R_{in}^2}\right) = \mu F_{app} \frac{2}{3}\left(\frac{R_{ex}^3 - R_{in}^3}{R_{ex}^2 - R_{in}^2}\right) = \mu F_{app} R_{\mu}$$

The friction radii are expressed by $R_{\mu p}$; under uniform pressure, we have:

$$R_{\mu p} = \frac{2}{3}\left(\frac{R_{ex}^3 - R_{in}^3}{R_{ex}^2 - R_{in}^2}\right) \ hence \Rightarrow M_{total} = \mu F_{app} R_{\mu p} \qquad [8.57]$$

The pressure is not always uniform when the gear system is new. The friction radii are also expressed by $R_{\mu U}$. With uniform wear and tear, we have:

$$Uniform\ wear \rightarrow R_{\mu u} = \left(R_{ex} - R_{in}\right)/2 \ hence \Rightarrow M_{total} = \mu F_{app} R_{\mu U} \qquad [8.58]$$

NUMERICAL APPLICATION.– For: R_{in} = 3.0 in = 76.2 mm and R_{ex} = 3.5 in = 88.9 mm:

$$\begin{cases} Uniform\ pressure \rightarrow R_{\mu p} = \frac{2}{3}\left(\frac{R_{ex}^3 - R_{in}^3}{R_{ex}^2 - R_{in}^2}\right) = \frac{2}{3}\left(\frac{3.5^3 - 3.0^3}{3.5^2 - 3.0^2}\right) = 82.713\ mm \\ Uniform\ wear \rightarrow R_{\mu u} = \left(\frac{R_{ex} - R_{in}}{2}\right) = \left(\frac{3.5 - 3.0}{2}\right) = 82.55\ mm \end{cases} \qquad [8.59]$$

The graphic plots of the respective forces of friction in relation to the respective external radii illustrate the behavior:

$$\left\{ f(R_{in}) = \frac{R_{\mu p}}{R_{ex}} = \frac{2}{3}\left(\frac{R_{ex}^3 - R_{in}^3}{R_{ex}^2 - R_{in}^2}\right) \Big/ R_{ex}\ and\ g(R_{in}) = \frac{R_{\mu u}}{R_{ex}} = \left(\frac{R_{ex} - R_{in}}{2}\right) \Big/ R_{ex} \right. \qquad [8.60]$$

By varying the respective internal radii R_{in} *from 0 in to 3.5 in and* R_{ex} *= 3.75 in,* we obtain the respective friction radii:

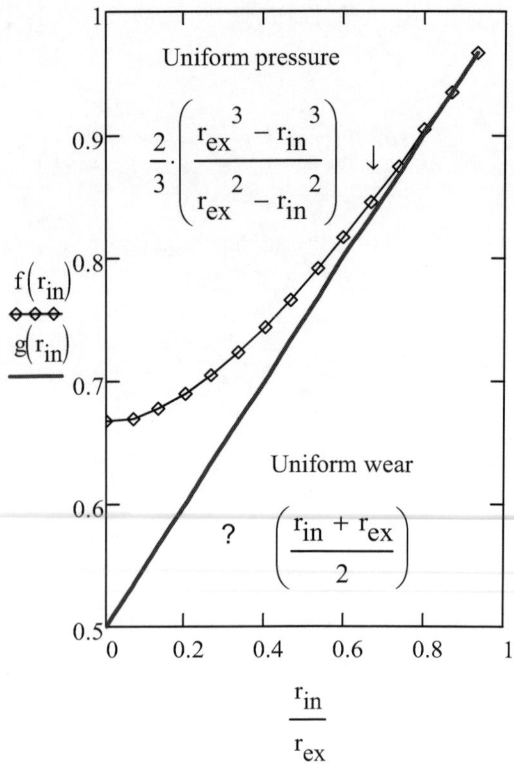

$f(r_{in}) =$

0.667
0.669
0.677
0.689
0.704
0.722
0.743
0.766
0.790
0.817
0.844
0.874
0.904
0.935
0.967

$g(r_{in}) =$

0.500
0.533
0.567
0.600
0.633
0.667
0.700
0.733
0.767
0.800
0.833
0.867
0.900
0.933
0.967

Uniform pressure ↓

$$R_{\mu p} = \frac{2}{3}\left(\frac{R_{ex}^3 - R_{in}^3}{R_{ex}^2 - R_{in}^2}\right)$$

Uniform wear ↓

$$R_{\mu u} = \left(\frac{R_{ex} - R_{in}}{2}\right)$$

$R_{up}(r_{in}) =$

	0
0	2.500
1	2.510
2	2.539
3	2.583
4	...

Uniform pressure

$R_{uu}(r_{in}) =$

	0
0	1.875
1	2.000
2	2.125
3	2.250
4	...

Uniform wear

Figure 8.23. *Radii of friction of a gear system with uniform pressure and uniform wear. For a color version of this figure, see www.iste.co.uk/grous/design.zip*

8.4.5. *Conical gear system*

In the case of conical gear systems, the problem of conicity is a serious one, unlike with cylindrical gear systems. The force of engagement is important in order to engage the two conjugal parts. This fact is attributable to the force required to rotate the cone and its conjugal part.

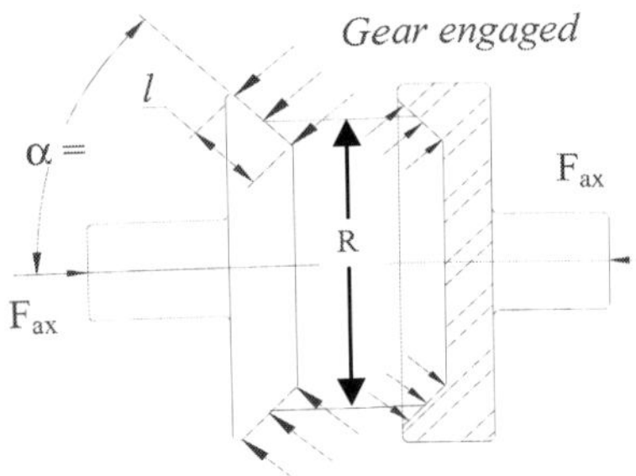

Uniform pressure ↓

$$M_{pu} = \frac{\mu F}{\sin(\alpha)}\left(\frac{2}{3}\frac{R_{ex}^3 - R_{in}^3}{R_{ex}^2 - R_{in}^2}\right)$$

Other formulation ↓

$$M_{pu} = \mu F\left(\frac{R_{ex}^3 - R_{in}^3}{3R_{mean} l \sin(\alpha)^2}\right) \quad [8.61]$$

Figure 8.24. *Principle of conical gear systems with uniform pressure. For a color version of this figure, see www.iste.co.uk/grous/design.zip*

where:

- M is the moment (torque) of forces in lb.in or N.m;
- F_{axe} is the axial force in lb or N;
- μ is the friction coefficient;
- R_{ex} is the external radius (large) in in or mm;
- R_{in} is the internal radius (small) in in or mm;
- R_{mean} is the mean radius in in or mm;
- *l* is the width in in or mm;
- (α) is the angle of the cone (in our case, α = 45°).

$$As\text{: } F_n = 2\pi R_{mean} \times P \times l \rightarrow M_{pu} = \mu F_n\left(\frac{2}{3}\frac{R_{ex}^3 - R_{in}^3}{R_{ex}^2 - R_{in}^2}\right) \quad [8.62]$$

If the wear and tear is uniform, the force torque of the conical gear system is written:

$$M_{pu} = \mu F_{axi} R_{mean} / \sin(\alpha) \text{ or indeed } M_{pu} = \mu F_n R_{mean} \quad [8.63]$$

Thus, the pressure variation, when the wear and tear is uniform, is written as:

$$P_{uu} = (F_{axi}/2\pi)\times(R_{ex} \quad R_{in})\times R \qquad [8.64]$$

The maximum pressure acts on R_{in} (small radius) and the minimum pressure on R_{ex} (large):

$$\left.\begin{cases} \text{Maximum pressure} \rightarrow P_{\max} = \dfrac{F_{ax}}{2\pi(R_{ex} - R_{in})R_{in}} \\ \text{Minimum pressure} \rightarrow P_{\min} = \dfrac{F_{ax}}{2\pi(R_{ex} - R_{in})R_{ex}} \end{cases}\right\} \rightarrow P_{mean} = \frac{F_{ax}}{\pi\left(R_{ex}^2 - R_{in}^2\right)} \qquad [8.65]$$

The analysis of design, in view of conicity, is tricky or complicated if using conventional means, given the orientation of the forces of friction, which are dependent upon the way in which the cones engage and interlock with one another. Considering the classic method whereby we mask the relative motion of rotation at the time of engagements, we introduce the axial force at its maximum value, which is modeled as follows:

$$F_{ax/\max} = F_n \times\left[\sin(\alpha) + \mu\cos(\alpha)\right] \qquad [8.66]$$

This force represents the maximum force required to obtain the normal force F_n which, whilst rotating, generates friction, which gives rise to the resistance moment. The axial force which supports and keeps the conical part engaged in its site varies between:

$$F = F_n \times \sin(\alpha) \ \textit{and / or} \ F = F_n \times\left[\mu\cos(\alpha) - \sin(\alpha)\right] \qquad [8.67]$$

This is due to the vibrations giving rise to the forces of friction: $F = F_n . \sin(\alpha)$. Thus, to disengage the two conical parts, we calculate the axial retraction force (disengagement) F_r as follows:

$$F_r = F_n\left[\mu\cos(\alpha) - \sin(\alpha)\right] \qquad [8.68]$$

It is not necessary to consider special forces for the retraction of the two parts of the gear system. It is even possible that $\mu\cos(\alpha) > \sin(\alpha)$, in which case the force F_r is sufficient and necessary to retract the two pieces out of contact with one another

$$F_r = F_n\left[\mu\cos(\alpha) - \sin(\alpha)\right] \qquad [8.69]$$

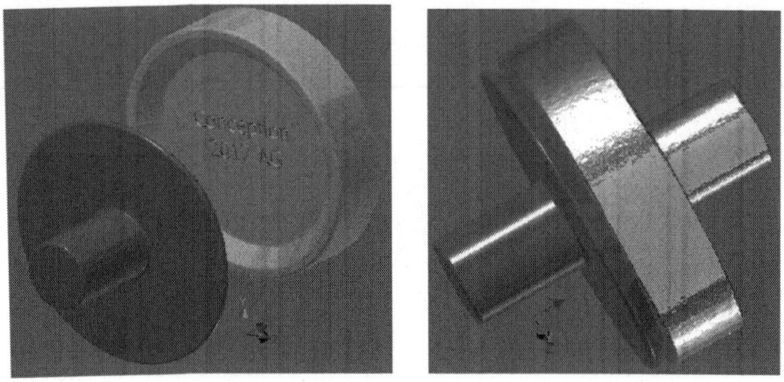

Figure 8.25. *Engagement and retraction of the gear system with uniform pressure and wear. For a color version of this figure, see www.iste.co.uk/grous/design.zip*

The power necessary for the conical gear system is expressed as follows:

$$\begin{cases} \textit{Uniform pressure},\ P = Hp = \dfrac{M \times N}{63000} = F_n\mu \times \dfrac{2}{3}\left(\dfrac{R_{ex}^3 - R_{in}^3}{R_{ex}^2 - R_{in}^2}\right)\dfrac{N}{63000} = \ldots \\ (N\ \textit{is the RPM}) \rightarrow \quad \ldots = \dfrac{N}{63000} \times \dfrac{F_n\mu}{\sin(\alpha)} \times \dfrac{2}{3}\left(\dfrac{R_{ex}^3 - R_{in}^3}{R_{ex}^2 - R_{in}^2}\right) \\ \textit{Uniform wear},\ P = Hp = \dfrac{M \times N}{63000} = \dfrac{NF_n\mu \dfrac{D_{mean}}{2}}{63000} = \dfrac{N}{63000} \times \dfrac{F_n\mu R_{mean}}{\sin(\alpha)} \end{cases} \quad [8.70]$$

The normal force F_n, once we know the mean pressure, is calculated as follows:

$$F_{nr} = p_{mean} \times 2\pi R_{mean} \times l \quad [8.71]$$

8.5. Flywheels and rims (discs and rims)

The weight of the rim is expressed by the following relation:

$$\begin{cases} p_j = \dfrac{K_1}{K_2} \times \dfrac{\gamma E}{V^2} = \dfrac{2K_1 \times \gamma E}{\left(V_1^2 - V_2^2\right)} = \dfrac{2K_1 \times \gamma E}{\left(\omega_1^2 - \omega_2^2\right)R^2} \quad [kg\ or\ lb] \\ K_2 = \dfrac{V_1 - V_2}{V_m} = \dfrac{\omega_1 - \omega_2}{\omega_m},\ V_m = \dfrac{V_1 - V_2}{2}\ and\ \omega_m = \dfrac{\omega_1 - \omega_2}{2} \end{cases} \quad [8.72]$$

Numerical applications of the principles of calculations for the designed rim

Where:

– E is the energy supplied to the quarter turn of the flywheel, E = 1500 ft.lb = 207.382 kg.m;

– N_1 and N_2 are the frequencies of rotation. N_1 is 200 rpm = 200 (1/s). If it decreases by $\eta = 13$ %, then N_2 becomes $N_2 = 139.200$ rpm = 139.2 (1/s);

– maximum radius, $R_{mean} = 35$ in = 0.889 m;

– acceleration due to gravity $\gamma = 32.2$ ft/s^2 = 9.81 m/s^2;

– K_1 is the empirical coefficient or the velocity ratio, $K_1 = 0.90$;

– ζ is the increase of the real weight of the rim $\zeta = 1.117$ ($\zeta = 1.15$);

– P_j is the weight of the rim in kg or lb;
– $K_1 \approx 0.90$ is an empirical coefficient due to the effect of the hub, the shaft, etc.;
– K_2 is the ratio (coefficient) of variations of velocity;
– γ is the acceleration in m/s^2;
– E is the storage energy of the flywheel in kg.m;
– V_1 is the maximum linear velocity in m/s at the point where R is average;
– V_2 is the minimum linear velocity in m/s at the point where R is average;
– V_m mean velocity in m/s at the point where the radius R is average ;
– ω_1 is the maximum angular velocity in rad/s;
– ω_2 is the minimum angular velocity in rad/s;
– ω_m is the mean angular velocity in rad/s.

Figure 8.26. *Car wheel rim drawn with Inventor 3D. For a color version of this figure, see www.iste.co.uk/grous/design.zip*

$$\begin{cases} p_{rim} = \dfrac{2K_1\gamma E}{\left(\omega_{max}^2 - \omega_{min}^2\right)R^2} = \dfrac{2\times 0.9\times 9.815\times 207.382}{\left(16.755^2 - 14.577^2\right)0.889^2} = 67.925\ kg = 149.749\ lb \\ \xi = 1.117 \rightarrow p_{rim}^{approximate} = p_{rim}\times\xi = 67.925\times 1.117 = 75.872\ kg = 167.270\ lb \end{cases}$$

For a mean velocity ω_{mean}, found here, we calculate the fluctuation of velocities, represented by the coefficient K_1, as follows:

$$\omega_{mean} = \frac{\omega_{max} + \omega_{min}}{2} = \frac{16.755 + 14.577}{2} = 16.666;\ K_2 = \frac{\omega_1 - \omega_2}{\omega_{mean}} = \frac{16.755 - 14.577}{16.666} = 0.139$$

8.5.1. *Flywheel for a solid disc*

The kinetic energy is written:

$$E_{kinetic} = \frac{1}{2} J_0 \left(\omega_{\max}^2 - \omega_{\min}^2\right) = \frac{1}{2} \frac{J_0}{R_{mean}^2} \left(R_{mean}^2 \omega_{\max}^2 - R_{mean}^2 \omega_{\min}^2\right) \quad [8.73]$$

$$p_d = \frac{2\gamma E}{V_{mean}^2 K_2} = \frac{4\gamma E}{\left(V_{1o}^2 - V_{2o}^2\right)} = \frac{4\gamma E}{\left(\omega_1^2 - \omega_2^2\right) R_o^2} \left[kg\right] \text{ and } K_2 = \frac{\left(V_{1o}^2 - V_{2o}^2\right)}{V_m} \quad [8.74]$$

where:
- R_0 is the external radius of the disc in mm or in;
- V_{1o} is the maximum linear velocity in m/s at the external point of the radius of the disc;
- V_{2o} is the minimum linear velocity in m/s at the external point of radius of the disc;
- V_m is the mean velocity in m/s at the point in relation to the external radius R;
- ω_1 is the maximum angular velocity in rad/s;
- ω_2 is the minimum angular velocity in rad/s;
- ω_m is the mean angular velocity in rad/s;
- ω is the angular velocity of the disc in rad/s;
- J_0 is the moment of inertia of the disc and the shaft combined in *lb.ft/s²* or *kg.m/s²*;
- R_{mean} is the mean radius.

Figure 8.27. *Flywheel drawn with Inventor 3D. For a color version of this figure, see www.iste.co.uk/grous/design.zip*

$$\begin{cases} E_{kinetic} = \dfrac{J_0}{2R_{mean}^2}\left(V_{\max}^2 - V_{\min}^2\right) = \dfrac{J_0}{2R_{mean}^2}\left(V_{\max} + V_{\min}\right)\left(V_{\max} - V_{\min}\right) \\ E_{kinetic} = \dfrac{J_0}{2R_{mean}^2} 2V \times K_2 V^2 \text{ where } K_2 = \dfrac{\left(V_{\max} - V_{\min}\right)}{V} \quad (\text{fluctuation}) \end{cases} \quad [8.75]$$

The moment of inertia J_0 is made up of two distinct parts (the disc and the shaft):

$$J_0 = \left(P_{rim} \times \rho_{rim}^2 / \gamma\right) + \left(P_{hub} \times \rho_{hub}^2 / \gamma\right) \quad [8.76]$$

where:

- ρ_j is the radius of gyration of the flywheel in m or ft;

– ρ_m is the radius of gyration of the hub in m or ft;

– γ is the acceleration (9.81) in m/s^2 or (32.2) ft/s^2.

If we consider that the radius of the rim is fairly small in comparison with the mean radius, ρ_j is assimilated to the mean radius (R_{mean}). Thus, by dividing equation [8.76] by R_{mean} and setting $(R_{mean})^2 \equiv (\rho_j)^2$, we find the expression of the strength modulus as follows:

$$\frac{J_0}{R_{mean}^2} = \frac{P_{rim}}{\gamma} + \frac{P_{hub} \times \rho_{hub}^2}{\gamma \times R_{mean}^2} \to E_{kinetic} = V^2 K_2 \frac{J_0}{R_{mean}^2} \qquad [8.77]$$

$$\begin{cases} E_{kinetic} = \left(\frac{P_{flywheel}}{\gamma}\right) + \left(\frac{P_{hub} \times \rho_{hub}^2}{\gamma R_{mean}^2}\right)\left(V^2 K_2\right) \\ \ldots \Rightarrow \frac{P_{flywheel}}{\gamma} = \left(\frac{E_c}{V^2 K_2}\right) - \left(\frac{P_{hub} \times \rho_{hub}^2}{\gamma R_{mean}^2}\right) \end{cases} \qquad [8.78]$$

The value of the coefficient of velocity fluctuation takes the following form:

$$\begin{cases} K_2 = 1 - \frac{P_{hub} \times \rho_{hub}^2}{\gamma R_{mean}^2 E} V^2 K_2 \text{ and } P_j \text{ will therefore be written} \downarrow \\ P_{flywheel} = \frac{\gamma E}{V^2 K_2}\left(1 - \frac{P_{hub} \times \rho_{hub}^2}{\gamma R_{mean}^2} \frac{V^2 K_2}{E}\right) = \frac{K_1 \gamma E}{V^2 K_2} \end{cases} \qquad [8.79]$$

where:

– ζ, adjustment of the real weight of the rim $\zeta = 1.15$.

We have already anticipated that the value of K_1 is essentially equivalent to 1 (0.90), and we often need to adjust the result by 10 %, supposing that the energy is distributed over 90 %. Experience tells us that the weight of the flywheel therefore needs to be adjusted with a coefficient of around 1.15 times – i.e. $(P_j x\ \zeta) = P_{approximate}$.

8.5.2. *Flywheel system with rim and discs (internal and external) made of cast iron*

Consider the following:

– radius of outer cylinder, R_{ex} = 29 in = 0.737 m;

– radius of inner cylinder, R_{in} = 25 in = 0.635 m;

– thickness of rim ε = 6.25 in = 0.159 m;

– width of rim of the corona: l = 8.25 in = 0.210 m;

– acceleration (zero gravity): γ = 32.2 ft/s² = 9.815 m/s²;

– density of cast iron: ρ = 0.255 lb/in³ = 7.0585 kb/m³.

QUESTIONS.– With the aim of optimizing the design, calculate the following:

1) The weight of the outer cylinder, P_{ex} (in kg and lb).

2) The weight of the inner cylinder, P_{in} (in kg and lb).

3) From the moment of inertia of the two cylinders, deduce an approximate moment and the resulting error.

4) The kinetic energy of the rim Ej in ft.lb or kg.m.

5) The coefficient of fluctuation of the angular velocities, if the adjustment is ζ.

6) The output energy per rotational cycle $M_\tau \times \theta$ in lb.ft or kg.m if K_1 = 0.95, which is the effect of the hub and the arms on the rim.

SOLUTION WITH DISCUSSION.–

$$\begin{cases} Weight\ of\ outer\ cylinder\ P_{ex} = \pi R_{in}^2 l \rho_{cast\ iron} = 2.521\times10^3\, kg = 5.558\times10^3\, lb \\ Weight\ of\ inner\ cylinder\ P_{in} == \pi R_{in}^2 l \rho_{cast\ iron} = 1.874\times10^3\, kg = 4.131\times10^3\, lb \end{cases}$$

Moments of inertia $J_{cylinders}$ of the two cylinders together in the constructed system:

$$J_{in}^{ex} = \left(\frac{1}{2}\frac{P_{ex}}{\gamma}R_{ex}^2\right) - \left(\frac{1}{2}\frac{P_{in}}{\gamma}R_{in}^2\right) = 31.201\times10^3\left[mkgs^2\right] = 225.675\left[lbfts^2\right]$$

The approximation of the moment of inertia (J_{in}^{ex}), on the basis of the initial hypothesis that it is a fine rim where the thickness of the corona is: ε = 6.25 in = 0.159 m, would be:

$$P = \left(P_{ex} - P_{in}\right) = 647.54\ kg = 1.43\times10^3\ lb;\ R_{mean} = \left(R_{ex} + R_{in}\right)/2 = 0.687\ m = 27\ in$$

Thus, we can calculate: $J^{inertia}_{approximate} = (P/\gamma)\times R^2_{mean} = 31.03\times 10^3\ mkgs^2 =$ 224.44 $lbfts^2$

The error in the approximation would be:

$$Error\ \% = \frac{J_{cylinders} - J^{inertia}_{approximate}}{J_{cylinders}} = 0.546\ \%$$

The kinetic energy of the cylinders is written: $E_{kinetic} = \frac{J_{cyl}}{2}\times\left(\omega_1^2 - \omega_2^2\right)$ [8.80]

The weight of the cylinders is: $P_{cyl} = 4\gamma E_{kin}\Big/ R^2_{ex}\left(\omega_1^2 - \omega_2^2\right)$ [8.81]

Knowing the expression of the moment of inertia: $J_{cyl} = \frac{P_{cyl/ex}}{2\gamma}R^2_{ex}$, let us calculate E_{kin} as follows:

$$E_{kin} = \frac{P_{cyl/ex}}{2\gamma}R^2_{ex}\left(\omega_1^2 - \omega_2^2\right),\ the\ total\ weight:\ P_{cyl} = \frac{4\gamma E_{kin}}{R^2_{ex}\left(\omega_1^2 - \omega_2^2\right)} \quad [8.82]$$

8.5.3. *Flywheel: numerical applications. Hypothesis II*

Consider:

– the kinetic energy of the rim is Ej in ft.lb or kg.m;

– the output energy per rotation cycle is $M_\tau\times\theta$ in lb.ft or kg.m.

SOLUTION WITH DISCUSSION.– Let K_1 be a coefficient of velocity fluctuation, K_1 = 90 % and $K_{adjustment}$ = 1.15. The kinetic energy supplied E_k = 1878 lb.ft = 259.643 kg.m. The RPM_1 is the rotational frequency = 250 rad/s at startup and is reduced by 10 %, i.e. by 25 rad/s, so that $RPM_2 = RPM_1 - 25$ rad/s = 225 rad/s.

The respective speeds of rotation are therefore ω_1 and ω_2, as follows:

$$\left\{\omega_1 = 2\pi\frac{RPM_1}{60} = 26.18\frac{rad}{s}\ and\ \omega_1 = 2\pi\frac{RPM_2}{60} = 23.562\frac{rad}{s}\right\}$$

Knowing that: $R_{mean} = (R_{ex} + R_{in})/2 = 0.686\ m$, the weight of the two cylinders is written as:

$$P_{cylinders} = 2K_1 \times \gamma \times E_{kin} \Big/ \left[R_{mean}^2 \left(\omega_1^2 - \omega_2^2 \right) \right] = 74.892\ kg$$

We hypothesize that the weight of the rim is 1.15 times less than that of the corona (cylinder) and we set $P_{adjusted} = P_{adjusted} \times K_{adjustment} = 86.125\ kg = 189.874\ lb.$ The velocity fluctuation coefficient K_2 is therefore:

$$\left\{ \omega_{mean} = (\omega_1 + \omega_2)/2 \ \ and\ \ K_2 = (\omega_1 + \omega_2)/\omega_{mean} = 0.105 \leftarrow QED \right\}$$

8.6. Conclusion

The machine elements for power transmission are numerous and varied. Their designs evolve with the progress of research, and manufacturers offer brake systems, gear systems and other connecting material that are both reliable and lightweight. The materials used are undergoing considerable development, and drawing inspiration from this for design studies is worthwhile. What we have presented is a condensed series of examples useful for mechanical design. However, whilst these calculations are helpful, they are no substitutes for a detailed, targeted study of each stage of the design process.

9

Sizing of Creations

9.1. Elastic machine elements and bolted assemblies

Elastic supports are used to damp the vibrations and shocks in machines placed on and bolted one another machine, as in the 3D illustration below. The illustration includes profile pieces and supports, where four compression springs are placed.

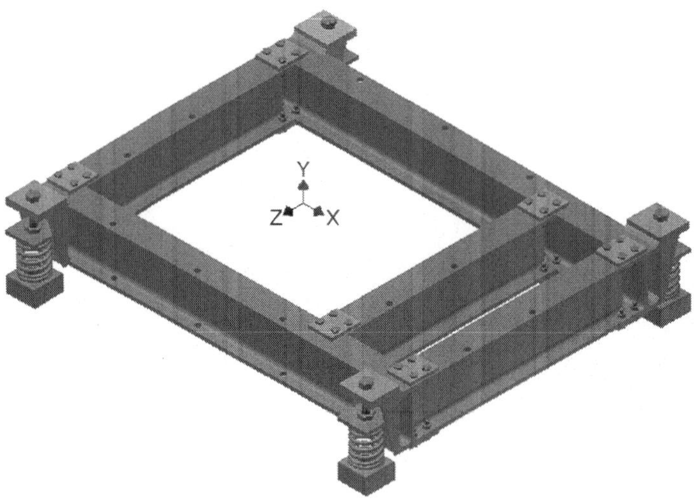

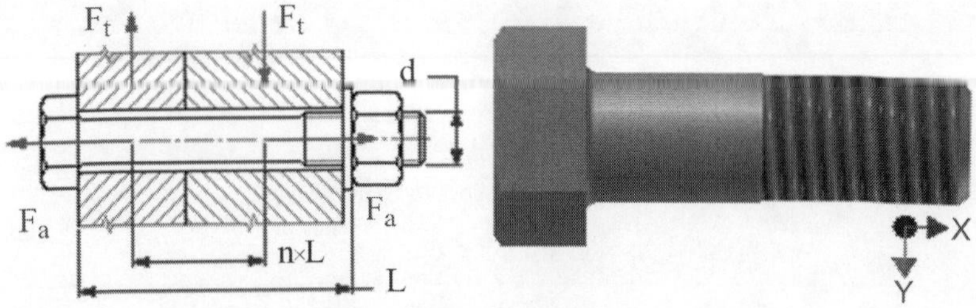

Figure 9.1. *Bolted assemblies and supports on compression springs. For a color version of this figure, see www.iste.co.uk/grous/design.zip*

The mathematical design of nut and bolt elements is governed by the ANSI standards, in the English-speaking world [ROA 75] and ISO standards using metric units. The parameters which come into play in the design are the following:

– Ψ, fitting factor (prestressed bolted assembly). Contact between the materials is not recommended because of the surface states;

– $k = (1+\Psi)$ where $\Psi = 0.5$ to 1.5 is a fitting factor, with a recommended value of 1.2;

– F_a is the maximum axial force imposed;

– *n* is the force input factor, which determines the shape of the end of the thread;

– F_t is the maximum tangential force imposed;

– z is the number of bolts;

– p is the step of the thread;

– d_s is the pitch diameter (∅) of the thread;

– d_{min} is the minimum diameter of the bolt thread;

– d is the nominal diameter of the thread in [mm or in];

– d_f is the pitch diameter of the thread in [mm or in];

– σ_y is the elastic limit;

– k_s is the safety factor;

– P_A is the acceptable (allowable) pressure on the nut thread;

– f is the friction coefficient of the assembly joint;

– f_1 is the friction coefficient of the threads on the bolt and nut (Table 9.1);

– f_2 is the friction coefficient between the bolt and the nut (Table 9.2);

– E_2 is the elasticity modulus of the materials from which the assembly is made;

– E_1 is the elasticity modulus of the bolt;

– F is the maximum axial force in [N or lb];

– M is the maximum moment in [N.m or lbf.ft];

– σ, S_e is the threshold of the elastic limit [MPa or psi];

– A_{min} is the minimum area in [mm^2 or in^2];

– H is the height of the screw vis in [mm or in];

– L is the total length subjected to the load on the thread in [mm or in];

– K_s is the safety factor [dimensionless];

– μ is the friction coefficient between the bolt and the metal plate.

Reduced length L_r; minimum section of the area S_{min} and the helix angle α:

$$\left\{L_{reduced} = n \times L;\ S_{\min} = \pi d_{\min}^2 / 4;\ \alpha = \arctan\left(p / \pi d_S\right)\right\} \qquad [9.1]$$

For a quarter turn (rapid rotation), the yield is written:

$$\begin{cases} \eta = \dfrac{\tan(\alpha)}{\tan\left[\alpha + \arctan(\mu_1)\right]} = \dfrac{1 - \mu_1 \tan(\alpha)}{1 + \mu_1 \mathrm{co}\tan(\alpha)},\ \eta \neq 0; if\ \alpha \prec \arctan(\mu_1) \downarrow \\ \ldots\ then \rightarrow \eta = \dfrac{\tan\left[\alpha - \mathrm{arc}\tan(\mu_1)\right]}{\tan(\alpha)} = \dfrac{1 - \mu_1 \mathrm{co}\tan(\alpha)}{1 + \mu_1 \tan(\alpha)}\ for\ \eta = 0 \end{cases} \qquad [9.2]$$

Force and maximum output moment:

$$F_{\max}^{axial} = 2M_{\max} 2\pi\eta / p \ \ and\ \ M_{\max}^{output} = F \times p / 2\pi\eta \qquad [9.3]$$

Slenderness coefficient: $\lambda_{slenderness} = 4L_{reduced} / d_S$ [9.4]

Pressure stress: $\sigma_{pressure} = P_t = F / A_{\min}$ [9.5]

Torsion stress: $\tau_{torsion} = 16M_{torsion} / \pi d_{\min}^3$ [9.6]

Reduced equivalent stress: $\sigma_{reduced} = \sqrt{\sigma^2_{pressure} + 3\tau^2_{torsion}}$ [9.7]

Rankine critical stress: $\sigma_{Rankine} = S_y \Big/ \left(1 + \frac{S_y \times \lambda^2}{\pi^2 \times E}\right)$ [9.8]

Euler critical stress: $\sigma_{Euler} = \lambda^2 \times E / \lambda^2$ [9.9]

Johnson critical stress: $\sigma_{Johnson} = S_y\left(1 - \frac{\lambda^2 \times S_y}{4\pi^2 E}\right)$ [9.10]

Pressure on the threads: $P_{threads} = (F/0.75) \times \pi \times d_s (d - d_s)(H/p)$ [9.11]

Materials	Transmission per screw of Pressure in psi	Transmission per screw of Pressure in Pa
Steel-cast iron	700-1000	5–7
Steel-bronze	700-2200	5–15
Steel-steel	1100-1450	7.5–10
$\mu = f_1$, Friction coefficient per thread		
ACME thread		$f_1 = \mu = 0.1–0.2$
Thread, power transmission per bearing		$f_1 = \mu = 0.01$

Table 9.1. *Power transmission factors*

Of the various methods of constructing an assembly, bolting is frequently used, because the structures can be dismantled when needed, but are also robust while standing. Below are a few examples of the friction coefficients.

Surface of threads in contact **Table 1**	Non-lubricated surface	Lubricated surface
Covered with black powder (matt, anti-UV) and phosphate plating	f_1 = 0.14-0.21	f_1 = 0.12-0.15
Zinc galvanization	f_1 = 0.13-0.18	f_1 = 0.12-0.17
Platinum cadmium plating	f_1 = 0.08-0.12	f_1 = 0.08-0.11
	Non-lubricated surface	Lubricated surface

Materials in contact **Table 2**	f_2 = friction coefficient of bolted assemblies	
Steel on steel	0.8	0.16
Steel on cast iron	0.4	0.21
Steel on brass	0.35	0.19
Cast iron on bronze	0.25	0.08
Bronze on bronze	0.25	0.10
Aluminum on aluminum	1.35	0.30
Copper on copper	1	0.08
Plexiglass on plexiglass	0.8	0.8
Steel on plexiglass	0.4-0.5	0.4-0.5

Table 9.2. *Table of the main friction coefficients for various materials*

9.2. Dimensions (sizing) of bolted assemblies

We use the parameters defined above to calculate the dimensions. It is helpful to refer to published works in mechanical construction and metrology [GRO 11, GRO 13] to find the formulae used for the threads:

$$\begin{cases} D_{\min}^{threads\ of\ nut} = D_1 = d - 1.082531 \times p \ \left[in\ mm\ or\ in\right] \\ d_{pitch}^{threads\ of\ nut} = d_2 = d - 0.649519 \times p \ \left[in\ mm\ or\ in\right] \end{cases} \qquad [9.12]$$

The calculations for the bolted assembly take account of the aforementioned parameters.

$$\begin{cases} F_{\max}^{assembly} = \dfrac{\psi}{Z}\left(F_a + \dfrac{F_t}{f}\right),\ F_0^{\Pr estressed} = F_{\max} - \dfrac{F_a}{Z}\left(\dfrac{C_2}{C_1 + C_2}\right) \\ where\ C_{10} = \dfrac{L + 0.8d}{E_1 \dfrac{\pi d_s^2}{4} + C_2};\ C_{20} = \dfrac{L}{\dfrac{\pi E_2}{4}\left(1.5d + \dfrac{L}{a}\right) - 1.05d^2};\ and \\ C_1 = C_{10} + C_{20}\left(1 - n\right)\ and\ C_2 = C_{20}n;\ a = coef.,\ f\left(materials\right) \end{cases} \qquad [9.13]$$

steels, a = 10; cast irons, a = 10; lightweight alloys, a = 6.

The pressing moment, after a prestress force, in light of the friction on the threads between the bolt and the nut, is written as follows:

$$M_u = F_0\left\{\frac{d_2}{2}\tan\left(\frac{p}{\pi d_2}+\frac{f_1}{\cos\left(30^{\circ}\right)}\right)+0.7d\times f_2\right\} \qquad [9.14]$$

Thus, the respective applied stresses of traction and torsion are expressed:

$$\sigma_{traction} = 4F_0/\pi d_{\min}^2 \text{ and } \tau_{torsion,k} = 16M_u/\pi d_{\min}^3 \text{ MPa or psi} \qquad [9.15]$$

The equivalent stress would therefore be:

$$\sigma_{equivalent} = \sqrt{\sigma_{traction}^2 + 3\tau_{k,torsion}^2} \text{ MPa or psi} \qquad [9.16]$$

Consider the following:

– F force applied in N or *lbf*;
– ψ press coefficient of the joint (bolt–nut);
– F_0 prestress force in N or in *lbf*;
– F_{max} pressing force applied to the bolt in N;
– ΔF_1 increase of the prestress force on the bolt;
– ΔF_2 reduction of pressing due to the force applied to the bolt;
– ΔL_1 degree of elongation of the screw or bolt;
– ΔL_2 degree of compression of the materials in contact, mm or in;
– C_1 constant (of the spring) of the bolt;
– L_F is the width of the part subjected to the force.

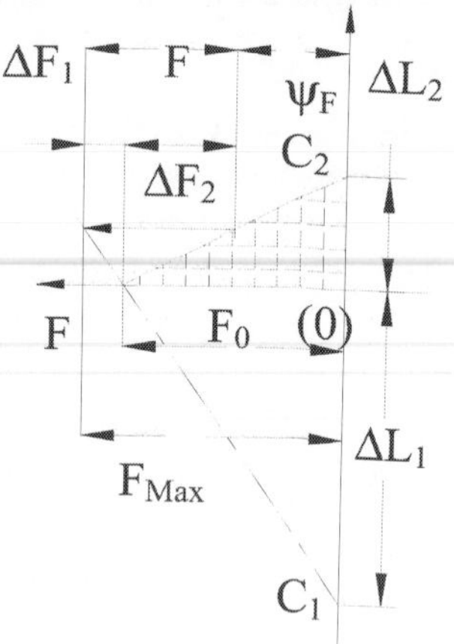

Assembly ↓	$\left\{\begin{matrix} n=1 \\ L_F = L \end{matrix}\right\}$	$\left\{\begin{matrix} n=0.75 \\ L_F = 0.75L \end{matrix}\right\}$	$\left\{\begin{matrix} n=0.5 \\ L_F = 0.5L \end{matrix}\right\}$	$\left\{\begin{matrix} n=0.25 \\ L_F = 0.25L \end{matrix}\right\}$
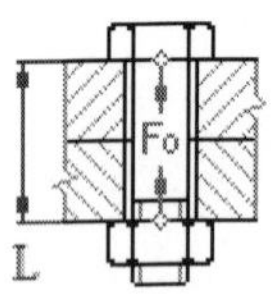		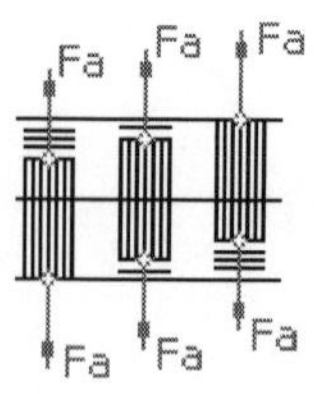	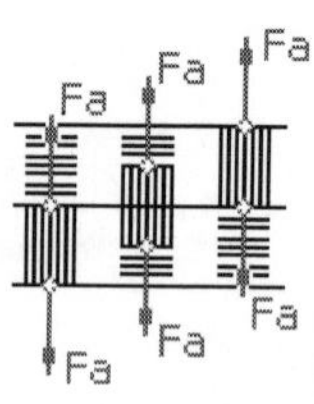	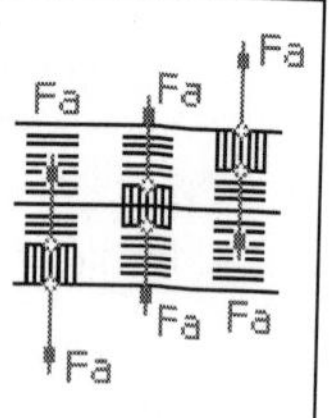

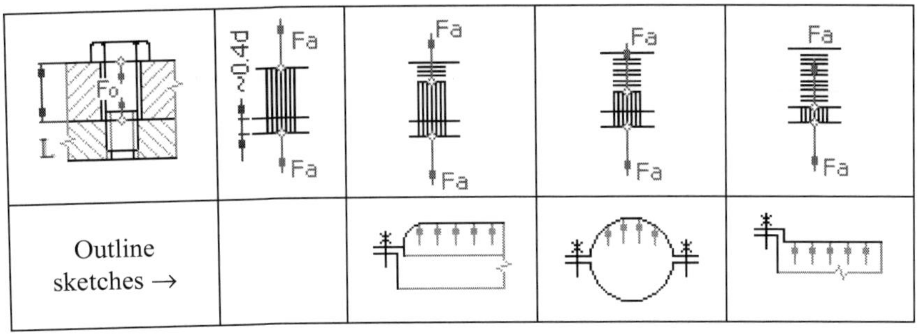

Figure 9.2. *Outline sketches of the forces-elongations modeled using Inventor. For a color version of this figure, see www.iste.co.uk/grous/design.zip*

The friction coefficients can be read from Table 9.2. The maximum force (F_{max}, load) applied to the bolt imposes a maximum stress and a pressure:

$$\sigma_{\max} = \frac{4F_{\max}}{\pi d_{\min}^2} \quad and \quad P_c = \frac{4F_{\max}}{\pi(0.8d/p).\left(d^2 - D_1^2\right)} \quad MPa \ or \ psi \qquad [9.17]$$

The calculation [ROA 75] to verifying the sizing of the assembly, in connection with the safety coefficient, enables us to write and plot the resulting diagram:

$$\left\{ p_c \leq p_A;\ \sigma_{\max} \leq S_y/k_s \text{ and } \sigma_{equivalent} \leq S_y/k_s \right\} \qquad [9.18]$$

By way of example, below are a few references regarding the grades of forces of the bolts as a function of the materials used, in Canada and the United States.

Material used for bolt ↓	Grades of forces of bolts for each material according to ISO and CSN standards (USA and Canada)									
	4A	4D	4S	5D	5S	6S	6G	8G, 8E	10K, 10G	12K
	3.6	4.6	4.8	5.6	5.8	6.8	6.9	8.8	10.9	12.9
	p.A [MPa] or [psi]									
Steel →	40	50	75	70	90	110	120	150	200	250K
Gray cast iron	25	30	45	40	55	70	80	90	125	150
Lightweight alloys	18	20	30	27	35	45	50	60	80	90

Table 9.3. *Grades of bolts for different categories of material*

9.3. Fatigue, shocks and endurance of bolted assemblies

Loading causes fatigue in bolted structures. There are also repeated blows (shocks) which affect the joints of bolted assemblys. Therefore, we need to take account of the endurance limits, the lifetimes per cycle of fatigue and the loading factors. We use the Wöhler curve (load/elongation, S-N). The mean values of the loads per cycle are typically calculated as follows:

$$\left\{F_m = (F_h + F_n)/2 \text{ and } F_a = (F_h - F_n)/2 \right\} \; N \; or \; lbf \tag{9.19}$$

We use dynamic shocks because of the working movements of the structure. Thus, we write the expression of the maximum loading as follows:

$$F_{\max} = F_m + \eta \times F_a \tag{9.20}$$

The mean stress (σ_m) and the stress corresponding to the upper cycle (σ_h) are calculated in reference to the mean force (F_m) and maximum force (F_{max}), in static conditions:

$$\sigma_{\max} = \{\sigma_h - \sigma_m\} \; MPa \; or \; psi \tag{9.21}$$

The security of the structure can be checked using the relation $n_c = \{\sigma_A/\sigma_a\}$. The so-called safety condition must satisfy $\{n_f = n_c\}$. The corrected endurance limit is found at the stress σ_e. This limit takes account of the design conditions, in connection with the materials and loading:

$$\sigma_e = \sigma_e^{'} \times k_e \times k_f \; ; \; MPa \; or \; psi \tag{9.22}$$

where:

- σ_a or τ_a amplitude of the normal stress or (in shear) in MPa or psi;
- σ_e or τ_e endurance limits subject to a constant force in MPa or psi;
- σ_m or τ_m mean stress cycles in MPa or psi;
- σ_F or τ_F virtual mean stress $\sigma_F = \sigma_e/\psi \; or \; \tau_F = \tau_e/\psi$;
- ψ Haigh diagram coefficient;
- ψ in traction and bending 0.15-0.3 or over;
- ψ in shear 0.10-0.25 or over;
- σ_e basic endurance limit (found by endurance testing of the material);

– k_e is the modified stress intensification factor;

– k_f is the factor of differents effects (root, inclusions etc.).

The basic endurance limit $\sigma_e^{'}$ is ordinarily expressed as $\left\{\sigma_e^{'} = 0.4 \times \sigma_u\right\}$, where σ_u is the ultimate traction stress. The fatigue safety factor for bolted assemblies is: $\left\{n_f \leq \sigma_A / \sigma_a \text{ or indeed } n_f \leq \tau_A / \tau_a\right\}$. The recommended design value for such structure is between n_f = [1.5 to 3]. The fatigue curve can be plotted using the following equations:

$$\left(\frac{\sigma_a}{\sigma_e} + \frac{\sigma_m}{\sigma_F}\right) = 1 \text{ or indeed } \left(\frac{\tau_a}{\tau_e} + \frac{\tau_m}{\tau_F}\right) = 1 \qquad [9.23]$$

On the basis of the endurance limit, we calculate the maximum stress σ_{max}. The calculation procedure may be graphical or grapho-analytical. The Haigh diagram represents the normal stress for different stress ratios $\{\sigma_a / \sigma_m\}$.

9.4. Springs in mechanical design

Under the influence of a load, the spring deforms. It restores the stored energy as it returns to its initial form. The spring can be curved, elongated, compressed or stretched. Once the force is removed, the spring returns to its initial configuration. This is what is known as elasticity. Hooke showed that many materials deform proportionally to the applied force. This applies to linearly elastic materials, known as *Hookean* materials, such as steel bars, springs or springboards. The law is written as follows: F = k.|S|, where *k* is a constant based on the characteristics of the spring (length, wire diameter) and **S** is the displacement. There are numerous types of springs. We can distinguish between compression-, traction- and torsion springs:

– compression springs are the most commonly used. The stress is primarily axial, and directed towards the spring. To facilitate the application of the force, the extremities of the springs are often brought closer together and milled;

– traction springs are subjected to longitudinal stress which increases their length. They are found in windshield wiper blades and even in clothes pegs. The most common method of attaching them is to deform the end spires of the spring to form a hooked loop;

– torsion springs, also rolled up in helix form, have the essential role of restoring a force torque. They have a vast range of applications.

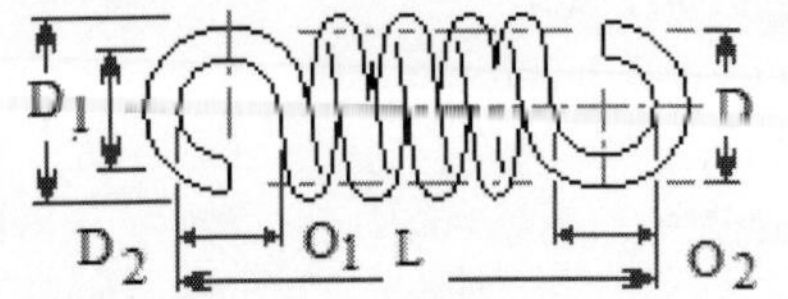		Full Loop Hook Type  $\omega = 0 = 0.8 = 1.1D2$	
		Hooke length	$0_2 = 0.718$ in
		Input length n	$0 \rightarrow L_0$
Length	5.500 in	L_0 of released spring $L_0 = 5.250$ in	
Direction of spires	Right	Total Hooke length = 0 $[0_1+0_2]$ 1.436 in	
∅ of wire	D = 0.244 in	Number of active spires, $n = 10$ ul	
∅ outside	$D_1 = 1.264$ in	Pitch, step	t = p = 0.357 in

Figure 9.3. *States of a traction spring, drawn with Inventor Pro. For a color version of this figure, see www.iste.co.uk/grous/design.zip*

9.4.1. *Materials and geometry of compression springs*

The function of the spring is to store and then restore energy. In order for the spring to *perform* well, the material of which it is made must combine a high elastic limit, good fatigue resistance and good corrosion resistance. Often, such springs are made of steel:

– drawn steels have good fatigue resistance. Their elastic limit is increased when the forming of a spring is followed by a heat treatment (tempering);

– these steels have poor corrosion resistance. They can be given a protective coating. They are able to work at low temperatures, between –80°C and 150°C, and are liable to relaxation at high temperatures;

– tempered steels exhibit good fatigue resistance. They are less vulnerable to relaxation than drawn steels. Depending on the degree of relaxation allowed, the operational range varies between –20°C and 160°C. These steels have poor corrosion resistance.

Springs can also be distinguished on the basis of their use: we have axial springs, blade springs, helical springs, spiral springs, torsion bars. We require materials for a spring with a:

$$\begin{cases} \textit{Minimal volume with a high value index } : \sigma_f^2 / E \\ \textit{Minimal weight with a high value index } : \sigma_f^2 / E\,\rho \end{cases} \qquad [9.24]$$

The main diagrams used highlight the elasticity modulus as a function of the resistance $E = f(\sigma_f)$ and $E/\rho = f(\sigma_f/\rho)$. As the primary function of a spring is to store the potential elastic energy and then release it for the purposes of operation, its energy is expressed by the ratio based on the maximum energy density, written as: $\varphi_{energy} = \sigma^2/2E$. Here, E represents the elasticity modulus and σ its stress. The ratio is therefore a dimensionless coefficient ψ_e. Damage to, or failure of, the spring will occur when the σ goes beyond the resistance (strength) of the material σ_f. We therefore set: $\varphi_{energy} = \left\{\sigma_f^2/2E\right\} \, for \ \sigma \leq \sigma_f$.

Recommendation for specifying the design of springs:

Aims: To store the maximum possible amount of elastic energy per volume.
To store the maximum possible amount of elastic energy per mass.

Constraints: No plastic deformation and absence of cracks (no fatigue).
Choose a sufficiently high failure energy: U > 1 kJ/m^2.

Technical documentation indicates that torsion bars and blade springs offer less in the way of performance than do axial springs. This is due, amongst other things, to the less pronounced loading. We can see that the neutral fiber is never subjected to any loading at all. For example:

$$\begin{cases} Torsion\ bar: \varphi_{energy} = \left\{\sigma_f^2/3E\right\} for\ \sigma \leq \sigma_f \\ Blade\ bar: \ \varphi_{energy} = \left\{\sigma_f^2/4E\right\} for\ \sigma \leq \sigma_f \end{cases} \qquad [9.25]$$

Thus, in mechanical design [HAY 90], the best choice of material is guided by the best material with the largest value of the index:

$$\begin{cases} \varphi^{\uparrow}_{volume} = \left\{ \sigma_f^2/E \right\} \rightarrow volume\ criterion \\ \varphi^{\uparrow}_{weight} = \left\{ \sigma_f^2/\rho E \right\} \rightarrow weight\ criterion \end{cases} \qquad [9.26]$$

The index *rolling ratio* (i) is the ratio between the mean diameter of the spring and the diameter of the wire. A value between 8 and 10 is recommended to facilitate the manufacture of the spring. The DIN standard (German Institute for Standardization) states that the rolling ratio must always be between 4 and 20. For large values of i, the wire is very slightly deformed, leading to significant dispersions in the geometry of the spring. A very low rolling ratio gives rise to powerful inside stresses and increases the stress intensifications during use.

The *minimum functional length* L_n is based on geometric considerations. The dispersions occurring during the manufacture of the spring mean that certain spires touch before the theoretical length with joining spires (Lc) leading to an increase in stiffness and forces. The characteristic curve for the spring is therefore no longer linear. In case of dynamic application (N > 104), S_a is multiplied by 1.5. Thus, we have: $L_n = [L_c+0.15\ (L_0-L_c)]$. In special cases where we do not wish to impose any particular maximum force value, we simply have: $L_n = L_c$:

$$L_n = \{S_a + L_C\}\ where\ S_a = n\{0.1\times d + 0.0015(D_2/d)\} \quad [9.27]$$

The *angle of twist of the spires*: there are numerous possible considerations for determining the angle of twist of the spires. The general method is to base our reasoning on springs with tightly wound spires. The upper limit often used is 7.5°. An approximate value of this limit can be obtained when we consider that the step of the spring must not be greater than the mean diameter divided by 2.5 → $m < D / 2.5$:

– d, D, diameter of the wire in mm or in;

– $D_m = d_m$ is the mean diameter;

– D_1 is the outside diameter of the spring in mm or in;

– D_2 is the inside diameter of the spring in mm or in;

– p (step): mean distance between two successive active spires of a spring;

– G, transverse elasticity modulus or (rigidity) of the spring in MPa;

– τ_0 is the stress in the spring in the free state, in MPa;

– τ is the torsion stress, in MPa;

– W_x is the energy of deformation in *J* or *lb.ft*;

– S = spit of the spring = $S_x = f = H$, , $H = L_1 - L_8$ (mm);

– S_8 is the spit of the spring under full loading in mm or in;

– S_1 is the spit of the prestressed spring in mm or in;

– *n*, = *t* = number of active spires and n_t, total number of spires;

– F, load applied (N);

– F_8 is the workload of the fully stressed spring in MPa;

– F_1 is the workload of the spring under minimum stress in MPa;

– R or k, the stiffness determines the spring's compression resistance. It is measured as 10 N/mm. The tolerance for this parameter is ± 15 %;

– K_w is Wahl's correction factor;

– L_n; L_x maximum authorized length of a spring after torsion. If the spit is greater, there is a risk of plastic deformation (irreversible alteration). There is no risk of deformation of the spring. Ln = [L_C + S_a], where S_a is the sum of the minimum authorized distances between active spires;

– L_c (block length, unstressed): maximum length of the spring after complete blocking. This parameter is on the right of the diagram. The tolerance in relation to this parameter is ± 15 % (an indicative value). F_x is the force of work exerted by the spring;

– L_1 and F_1 (loaded length F=P_1): the force F_1 at length L_1 is $F = Rx\ [L_0 + L_1]$, which gives us the equation of the length $L_1 = Rx\ [L_0 - F_1/R]$;

– L_0 the free length is measured in the non-compressed state of the spring after initial blocking (if necessary). The tolerance in relation to this parameter is ±2 %;

– *i* is the index of the spring i = C = D/d;

– ω is the loop (hooked end part) in mm or in.

Hereinafter, we discuss springs in extension and compression. An extended spring is illustrated below. The springs are identified by a unique reference. With compression springs, the type corresponds to the letter *C*.

Wires and strings	Article code "C"	Article code "D"
Piano wire	ASTM A228 or AMS 5112	DIN 17223 or JIS G4314 SWP-A/B or AMS 5112
Springs made of piano wire	Not recommended for applications where the temperature exceeds 121°C (250°F)	
Extreme values per article code	∅out. between 1.45 and 2.24 brought together (non-milled) ∅out. > 2.24, brought together (milled)	Wire up to 0.8 mm, brought together (non-milled) Wire thicker than 1.0 mm, brought together and milled.
Dimensions – forces		Level 2: DIN 2095
∅ outside, mm	Tolerance mm Code "C"	
1.45-3.02	± 0.08	
3.05-6.10	± 0.13	
6.12-12.70	± 0.20	
12.73-25.40	± 0.38	
25.43-31.12	± 0.51	

31.14-37.08	± 0.76	
37.11-50.08	± 1.02	
Load P	± 10 %	
Stiffness R (k)	± 10 %	

Table 9.4. *Tolerances and article codes according to manufacturer data*

Ordinarily, the spires are on the right. If they are on the left, it is important to take note of this. A non-stressed spring has an index of (0). If it is prestressed, its index is (1). If it is fully loaded, meaning that a maximum work load is applied, it has an index of (8). The limit means that the spring is released at the point where the spires touch, giving an index of (9). The calculation of the spring's inside diameter is $D_2 = (D\text{-}d)$. The hooked part hooked of the spring is calculated by $\omega = (L_0\text{-}L_z) / 2$. L_0 is the free length of the spring and L_Z is the length of the spires.

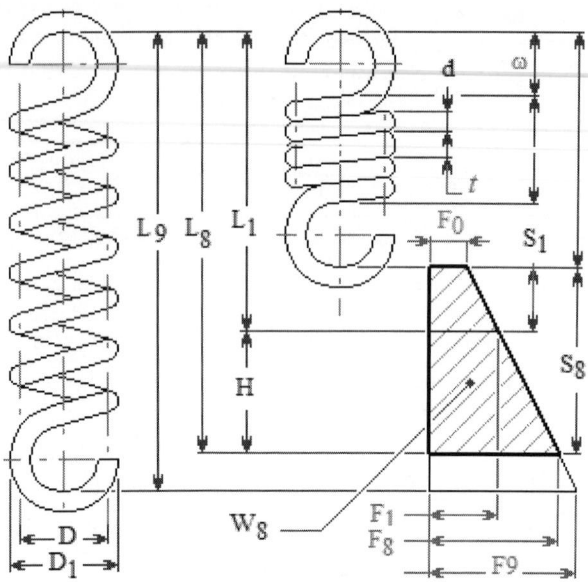

Figure 9.4. *Dimensional characteristics of an extension spring. For a color version of this figure, see www.iste.co.uk/grous/design.zip*

Applying Wahl's theory to the spring, we introduce the so-called Wahl factor:

$$K_W = \left\{ \frac{4C-1}{4C-4} + \frac{0.615}{C} \right\} \quad [9.28]$$

The material usage factor for compression springs is important when making the relevant choices, because it expresses the relation which exists between the torsion stress in the fully loaded state and the admissible torsion stress: the initial tension exerted on the spring is expressed thus:

$$F_0 = \pi d^3 \times \tau_0 / 8D \times K_W \quad N \text{ or } lbf \quad [9.29]$$

In general, the force exerted by the spring is calculated by the following relation:

$$F = \left(\frac{\pi d^3 \times \tau}{8D \times K_W} \right) = \frac{G \times S \times d^4}{8D^3 n} + F_0 \text{ , } N \text{ or } lbf \quad [9.30]$$

The *stiffness* (k or R, which is a constant of the spring) is expressed thus:

$$k = \frac{G \times d^4 \times \tau}{8D^3 n} = \frac{F_8 - F_1}{H} \text{ , } \frac{N}{mm} \text{ or } \frac{lbf}{in} \quad [9.31]$$

$$\tau \left\{ {}^{ISO}_{0} \right\} = \left\{ \begin{array}{l} \left(\left(\frac{300}{i} + 30 \right) \text{ , } \left(\frac{300}{i} \le \tau_0 \frac{300}{i} + 60 \right) MPa \right. \\ \left(\left(\frac{43500}{i} + 4350 \right) \text{ , } \left(\frac{43500}{i} \le \tau_0 \frac{43500}{i} + 8700 \right) psi \right. \end{array} \right\} \quad [9.32]$$

When the spring has no initial tension, the design value of the step is [step = t = 0.35.D, mm]. If the designed spring is not suitable with that step, we must proceed using the relation [0.3 ×D ≤ t ≤ 0.4 ×D]. The design of the spring is based on thus condition [$\tau_8 \le u_s\ \tau_A$]. Thus, the sizing is as follows: [$L_0 \le$ D and $L_0 \le$ 31.5 in and [4 ≤ D/d ≤ 16 and n ≥ 2]. In a free state, the length L_0 is expressed by:

$$L_0 = \left[L_1 (F_8 - F_0) - L_8 (F_1 - F_0) \right] / (F_8 - F_1) \ mm \text{ or } in \quad [9.33]$$

The hook, in the form of a fish hook, of the spring must be between [d ≤ ω ≤ 30d]. For a fully loaded spring, the diameter of the wire is calculated by:

$$d = 2 \times \sqrt[3]{F_8 D K_W / \pi \tau_8} \ mm \text{ or } in \text{ under torsion } \left\{ \tau_8 = 0.85 \times \tau_A \right\} \quad [9.34]$$

It is desirable that these calculations should "fall" on normalized values. Unfortunately, it is not always the case. The role of the designer is to get as close as possible to the normalized values. Thus, the lengths of a preloaded spring L_1 and a fully loaded spring L_8 are expressed as follows. The same is true for the applied forces (F_1 min and F_8 max in N):

$$\begin{cases} L_1 = L_0 + \dfrac{8nD^3(F_1 - F_0)}{Gd^4}\ ;\ L_8 = L_0 + \dfrac{8nD^3(F_8 - F_0)}{Gd^4}\ mm\ or\ in \\ F_1 = F_0 + \dfrac{Gd^4(L_1 - L_0)}{8nD^3};\ F_8 = F_0 + \dfrac{Gd^4(L_8 - L_0)}{8nD^3}\ N\ or\ lbf \end{cases} \quad [9.35]$$

The spring's output parameters are calculated using Hooke's theory on the subject. Thus, the Hooke factor k_0 and the stiffness k are expressed by:

$$\left\{k_0 = \omega/D_2 \text{ dimensionless } and\ k = G \times d^4/8nD^3\ in\ N/mm\right\} \quad [9.36]$$

The length of the spring L_Z with no initial tension, the spit of the preloaded spring S_1, and the total spit S_8 are calculated respectively by:

$$\left.\begin{cases} L_Z = tn + d \text{ and } L_Z = 1.03(n+1)d \\ S_1 = (L_1 - L_0)\ and\ S_8 = (L_8 - L_0) \end{cases}\right\} mm\ or\ in \quad [9.37]$$

The resistance is calculated as follows:

$$\left.\begin{cases} \tau_1 = 8F_1DK_W/\pi d^3 \leftarrow Torsion\ in\ prestressed\ state \\ \tau_8 = 8F_8DK_W/\pi d^3 \leftarrow Torsion\ in\ prestressed\ state \end{cases}\right\} MPa\ or\ psi \quad [9.38]$$

Knowing that k is a Hooke factor, also known as the constant of the spring, F_9 the force of work of the loaded spring in the limit state and F_0 the initial force of tension in N, by analogy, F_9 and the spit S_8 are calculated, respectively, as follows:

$$\left.\begin{cases} F_9 = \pi d^3\tau_A/DK_W,\ N\ or\ lbf \leftarrow \text{Force on the spring in the limit state} \\ S_8 = (F_9 - F_0)/k,\ mm\ or\ in \leftarrow \text{Deflection on the spring in the limit state} \end{cases}\right\} \quad [9.39]$$

We can see that the limit length L_9 of the spring is expressed as $L_9 = (L_0 + S_9)$ mm.

The deformation energy resulting from the stresses is expressed by this relation:

$$W_8 = S_8 \times (F_8 + F_0)/2\times10^3 \ [J]\text{: } \textit{Energy of deformation of the spring} \qquad [9.40]$$

where:

M_1 Torque from the preloading of the spring [Nm, lbf ft]
M_8 Torque due to total preloading of the spring [Nm, lbf ft]
W_8 Energy of deformation of the fully loaded spring [J, ft lbf]
ϕ_1 Deflection angle of the work lever in the preloaded state [°]
ϕ_8 Deflection angle of the work lever in the fully preloaded state [°]
ϕ_h Angle of the working course (stroke) [°]
ρ_0 Working angle in the free state [°]
F Force of work induced by the lever R_1 [N, lbf]
ω, Loop (hooked end part) in mm or in [mm, in])
L_Z Length of the spires of the spring mm or in
l Length of the spires developed in mm or in
F_x Force exerted by the spring in N or lbf
W_x Energy of deformation in J or lbf.ft

Dimensions of a Hooke spring	Diagrams
Half-hook, $\omega = 0.55$ to $0.8\ D_2$	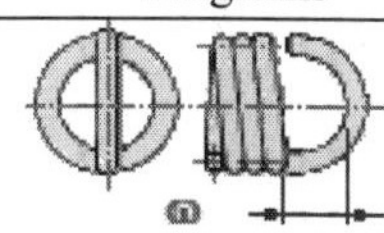
$d \leq 6.3$ mm, $D \geq 3.15$ mm, $i \geq 9$ Complete loop, $\omega = 0.8$ to $1.1\ D_2$	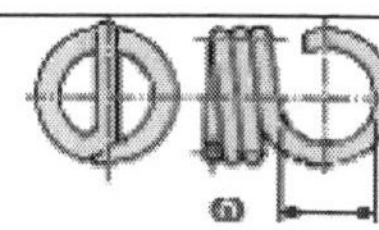
Complete loop on one side, $\omega \approx D_2$	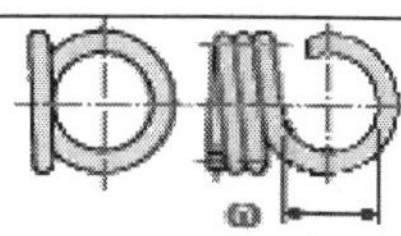
With no force and without reference to the axis of the spring Complete inside loop, $\omega = 1.05$ to $1.2\ D_2$	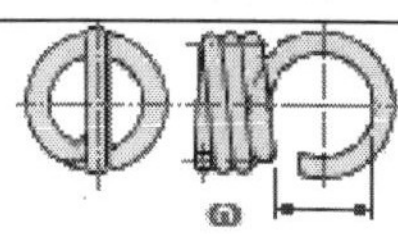
$d \leq 10$ mm, $i \geq 7$ Hook, $\omega = 1.2\ D_2$ to $30\ d$	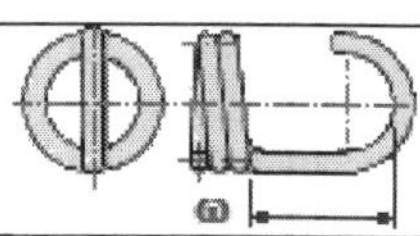
Double torsion Complete loop on one side, $\omega \approx D_2$	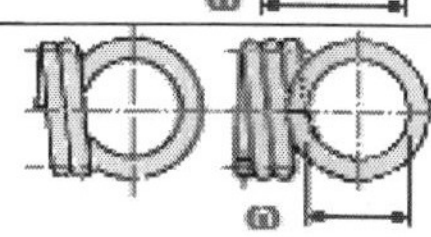

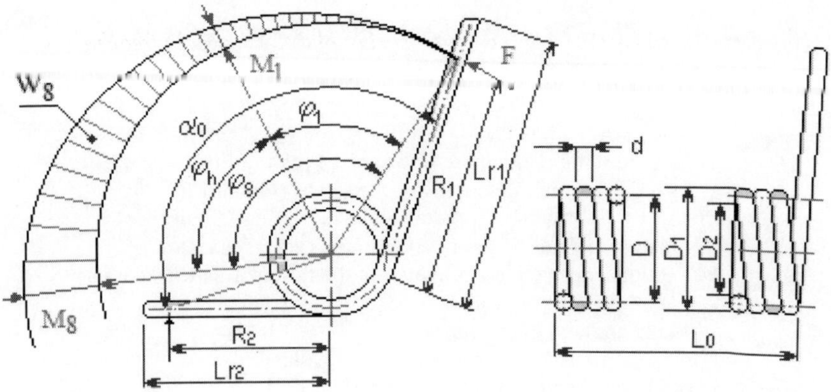

Figure 9.5. *Usual relations used for the design of the looped part (hook, ω). For a color version of this figure, see www.iste.co.uk/grous/design.zip*

For a developed length of wire: $\{l = 3.2 \times D \times n + l_0\}$ mm, we consider the following cases:

Hook affixed at the halfway point (1/2)	$\{l_0 = \pi D + 4 \times O - 2D - 2d\}$
For one length of the integral loops	$\{l_0 = 2(\pi D - 2d)\}$
For an integral loop on only one side	$\{l_0 = 2(\pi D - 2d)\}$
For an integral loop on only one interior side	$\{l_0 = 2(\pi D - d)\}$
For a loop (hook) raised to	$\{l_0 = \pi D + 2 \times O - D + 3d\}$
Double twisted hook	$\{l_0 = 4\pi D\}$

Table 9.5. *Calculation of the l_0 values as a function of the designed of the loops (hooks)*

Calculation of the *mass m* of the spring in kg and its *frequency f* in Hz:

$$\left\{\begin{array}{l} m = \left(\pi \times d^2 \times l \times \rho\right) / 4 \times 10^9 \ kg, \ \text{Mass of spring and} \swarrow \\ f = \left(d / 2\pi n D^2\right)\sqrt{G/2\rho} \times 10^6 \ Hz, \ \text{its specific frequency} \end{array}\right\} \quad [9.41]$$

If the calculation is performed in imperial units – i.e. *lbf*, *lb*, *in*, etc., we set:

$$\begin{cases} m = \left(\pi \times d^2 \times l \times \rho\right) / 4 \times 12^3 \ [lb], \ \textit{mass of spring and } f \textit{ its specific frequency} \downarrow \\ \rightarrow f = \dfrac{d}{2\pi n D^2}\sqrt{\dfrac{Gg}{2\rho 12^{-3}}} = 816.7056 \dfrac{d}{2\pi n D^2}\sqrt{\dfrac{G}{2\rho}} \ [Hz] \end{cases} \quad [9.42]$$

According to the data produced by the software "Inventor", the load resistance condition of the spring is: $[\tau_8 \leq U_{5x}\tau_A]$, for the free length L_0 and L_z the length of the spires. For a height of extension of the recoil spring, we set $\omega = (L_0\text{-}L_z)/2$. The torque of a preloaded spring M_1 and the total preload torque of the spring M_8 in N.m are calculated as follows:

$$\left\{ M_1 = \frac{1}{10^3} F_1 \times R_1 \ and \ M_8 = \frac{1}{10^3} F_8 \times R_1 \right\} [\text{N}.m] \ or \ [\text{lbf}.in] \quad [9.43]$$

where:

– F_1 is the force applied to the spring in N or *lbf* and R_1 the lever arm in mm;

– F_8 is the force applied to the spring in full preloading in N or *lbf*;

– $C = i$ is the index (dimensionless) of the spring C = D/d;

– D, is the mean diameter of the spring and d the diameter of the wire;

– r is the radius of flexion of the lever arm (inside) of the spring;

– d is the diameter of the spring;

– M is the (torque) moment on the spring in N.m;

– k_f is the stress intensification factor of the spring (SIF);

– M is the torque of the spring in N.m or lbf.in;

– R_1 is the lever arm of the working force in mm or in;

– R_2 is the support lever arm in mm or in;

– M_1 is the torque on the preloaded spring in N.m;

– M_8 is the torque on the fully loaded spring in N.m;

– ϕ_8 is the angle of deflection of the lever arm of the fully loaded spring;

– ϕ_1 is the angle of deflection of the lever arm of the preloaded spring;

– ϕ is the angle of deflection of the lever arm.

The angle of the working course ϕh is expressed thus: $\phi_h = (\phi_8 - \phi_1)^{\circ}$. The respective minimum and maximum deflection of the angle of the lever arm are:

$$\left\{ \varphi_1^{min} = \frac{M_1}{k \times \varphi} = \frac{M_1 \times \varphi_h}{M_8 - M_1} \; and \; \varphi_8^{max} = \frac{M_8}{k \times \varphi} = \frac{M_8 \times \varphi_h}{M_8 - M_1} \right\} deg \qquad [9.44]$$

The *stress intensification factor* (SIF):

$$K_f = \left\{ \left(4i^2 - i - 1\right) / 4i\left(i-1\right) \right\} \leftarrow \{ \text{ dimensionless } \}$$

$$i = \{D/d\} \leftarrow \textit{for the bending stress in the active spires} \qquad [9.45]$$

$$i = \{(2r/d) + 1\} \leftarrow \textit{for the bending stress on the lever arm}$$

The stress applied to the spring, in general, is calculated thus:

$$\sigma = 32/\pi d^3 \times \left(M \times k_f\right) \times 10^3 \; MPa \qquad [9.46]$$

The number of active spires (n) of the spring is calculated as:

$$n = \left[\pi \varphi E d^4 - 1220 M \left(R_1 + R_2\right) 10^3 \right] / 3660 \pi D M 10^3 \qquad [9.47]$$

The minimum load is used to calculate the assembly by: $M_1 = M_8 \times \varphi_1 / \varphi_8$. The minimum deflection angle of the lever arm would be:

$$\left\{ \begin{array}{l} \varphi_1^{ISO} = 3660\, M_1 \left(\pi D n + \dfrac{R_1}{3 + (R_2/3)} \right) 10^3 \Big/ \pi D^4 E[^{\circ}] \\ \varphi_1^{Imperial} = M_1 / k\varphi = M_1 \varphi_h / (M_8 - M_1)[^{\circ}] \end{array} \right. \qquad [9.48]$$

The working angle of the course is $\varphi_h = (\phi_8 - \varphi_1)^{\circ}$. The moment M_1 of the minimum load and the preloading force F_1 are respectively expressed as follows:

$$\begin{cases} M_1 = \dfrac{\pi \times \varphi_1 \times E \times d^4}{3660\left(\pi nD + \left(\dfrac{R_1 + R_2}{3}\right)\right)10^3}; F_1 = \dfrac{M_1 10^3}{R_1}; F_8 = \dfrac{M_8 10^3}{R_1} \; [N] \\ M_8 = \dfrac{\pi \times \varphi_8 \times E \times d^4}{3660\left(\pi nD + \dfrac{R_1 + R_2}{3}\right)10^3} \; [Nm] \; ; M_8 = \dfrac{F_8 R_1}{12} \; [lbf.in] \end{cases} \quad [9.49]$$

The output parameters of all springs obey this order of calculations. For example, the ratio (rate) of torsion in the spring would be:

$k_\varphi = \dfrac{M_1}{\varphi_1} = \dfrac{M_8}{\varphi_8} = \dfrac{M_8 - M_1}{\varphi_h} \left\{N.m/_\circ\right\}$. The spacing of the spires would be: $\rightarrow a = (t - d)$, mm

$$\begin{cases} L_0 = 1.05(n+1) \; [mm \text{ or } in] \leftarrow \text{for wrapped spires} \\ L_0 = t \times n + d \; [mm \text{ or } in] \leftarrow \text{for unwrapped spires} \end{cases} \quad [9.50]$$

The bending stress of the active spires, under a minimum load, σ_1 and at the lever arm for a minimum load, σ_{1r} is written:

$$\begin{cases} \sigma_1 = \left(32/\pi d^3\right) M_1 K_f \times 10^3, \text{ The SIF } K_f \text{ calculated with } i = D/_d \\ \sigma_{1r} = \left(32/\pi d^3\right) M_1 K_f \times 10 \,, K_f \text{ calculated with } i = \left(\dfrac{2r}{d+1}\right) \end{cases} \quad [9.51]$$

When the spring is fully loaded, the stress σ_8 is also calculated as follows:

$$\begin{cases} \sigma_8 = \dfrac{32 M_8 K_f}{\pi d^3} 10^3 \; \{MPa \text{ or } psi\}, \text{The SIF } K_f \text{ calculated with } i = D/_d \\ \sigma_{8r} = \dfrac{32 M_8 K_f}{\pi d^3} 10^3 \; \{MPa \text{ or } psi\}, K_f \text{ calculated with } i = \left(\dfrac{2r}{d+1}\right) \end{cases} \quad [9.52]$$

The length of the spires L_{z8} of the spring in the fully loaded state for wrapped spires, D_{18} and D_{28}, are respectively the outside and inside diameters (mm)

$$L_{Z8} = L_Z + \frac{d\varphi_8}{360}\ \{mm\};\ D_{18}^{outside} = \frac{D_1}{1-\frac{\varphi_8}{360n}}\ and\ D_{18}^{inside} = \frac{D_2}{1+\frac{\varphi_8}{360n}}\ \{mm\} \quad [9.53]$$

The test limit of the deflection angle of the lever arm φ_{max} and the energy of deformation W_8 of the spring are, respectively:

$$\left\{ \varphi_{max} = \varphi_8 + \frac{\sigma_A}{\sigma_8}\ (^\circ);\ W_8 = \frac{\pi}{360}\varphi_8 \times M_8\ \{J\} \right. \quad [9.54]$$

The length l and the mass m of the spring are:

$$\begin{cases} l_{rt} \approx \sqrt{R^2 - (D/2)^2} + 2d\ \{mm\},\ Lever\ arm\ under\ simple\ torsion \\ l_{Rr} \approx R - (D/2) + 2d\ \{mm\},\ Hooke\ lever\ arm \\ \{l = 3.2Dn + l_R\}\ and\ m = \pi \times l \times d^2 \times \rho / 4\times 10^9\ kg, mass \end{cases} \quad [9.55]$$

The condition of verification of the loaded spring's resistance [1] is:

$$\sigma_8 \le U_S \times \sigma_A \text{ and } \sigma_{8r} \le U_S \times \sigma_A\ [\text{dimensionless}] \quad [9.56]$$

[1] This is a safety factor (stress ratio) which guides the choice of appropriate materials for the resistance condition. If that calculated ratio is greater than that chosen in the specialized tables, we deduce that we will need less material than chosen in the design, which suggests the following: $\sigma_8 \le U_S \times \sigma_A\ and\ \sigma_{8r} \le U_S \times \sigma_A \leftrightarrow U_S \cong 0.75\ to\ 0.95$. We can also choose a smaller value of U_S for springs in a less harmful environment at a high temperature or for springs subjected to repeated shocks.

9.4.2. *Case study of helical springs in mechanical design*

Helical compression springs [BAU 67, WAL 63] often have rounded cross-sections. They are characterized by the force–deflection *ratio*.

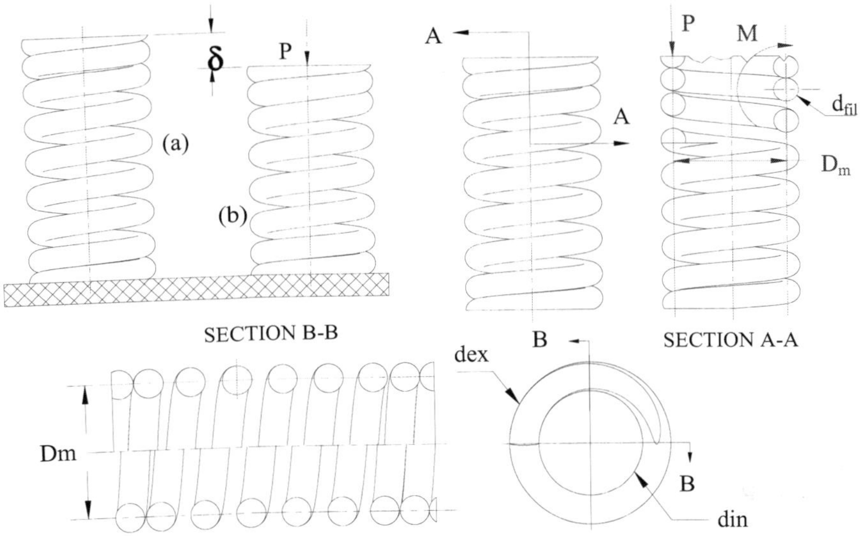

Figure 9.6. *Characteristics of the helical compression spring. Load (P) and deflection (δ). For a color version of this figure, see www.iste.co.uk/grous/design.zip*

The maximum stress on such springs is expressed by the sum of the two types of stresses – namely:

– Mτ is the twisting moment of the spring;

– *J* is the moment of inertia of the cross-section;

– $[\tau]_g$ is the practical slip resistance stress of the material.

$$-\ \tau_{shear} = \left(\frac{M_\tau \times c}{J}\right) \pm \left(\frac{P_{app}}{A}\right) \ and\ \tau_{\max} = \left(\frac{8P_{app}D_m}{\pi d^3}\right) \leq [\tau]_g \qquad [9.57]$$

$$-\ k = \left(\frac{Gd^4}{8nD^3}\right) \ and\ \Delta P = k \times \Delta L = k\frac{Gd^4}{8nD^3} \qquad [9.58]$$

From Hooke's law, ΔL is the elongation of the spring. (k) is the stiffness of the spring. For steel, $G \cong 8\times10^4$ MPa is the rigidity and D the mean diameter. (n) is the number of usable spires. If we introduce the index of the spring which represents the ratio of the diameters $i = C = D/d$, the previous equation is rewritten as follows:

$$\tau_{shear} = \frac{8P_{app}D_m}{\pi d^3} \times \left(1 + \frac{1}{2C}\right) = K_w \frac{8P_{app}D_m}{\pi d^3} = K_w \frac{8P_{app}C}{\pi d^2} \quad [9.59]$$

This, which is known as Wahl's equation [WAH 29] was developed in 1929. K_w is the so-called Wahl stress intensification factor, illustrated below. For a certain number of active spires (n), we propose the equation of the deflection of the spring f, ($f = H = \delta$):

$$\delta = Gd^4 / 8P_{app}D^3 n = Gd / 8C^3 n \quad [9.60]$$

"Some time" passes before the spring exhibits a deflection of some spires. We must consider the wave which progresses in the surface of the fit created between the spires because of buckling. The frequency of the applied loads coincides with the natural frequency (see Chapter 6).

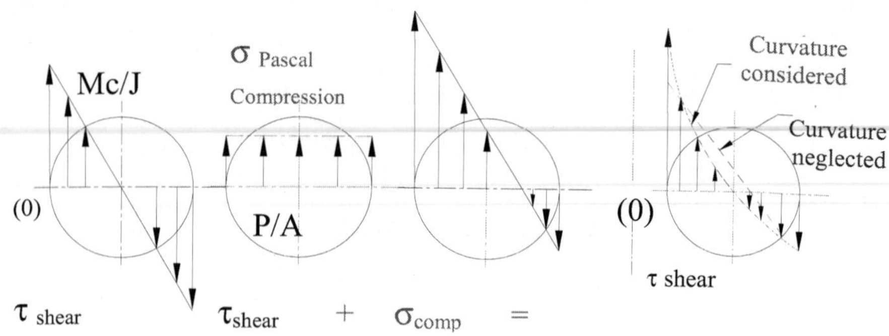

Figure 9.7. *Buckling conditions, taken from the Handbook of Mechanical Spring Design. Bristol. For a color version of this figure, see www.iste.co.uk/grous/design.zip*

9.4.3. *Case study of a spring in a rocker switch*

DESIGN QUESTION.– Design and determine the dimensions for a spring with medium use, made of galvanized steel. It is strongly recommended to consult manufacturer data. We suggest La Cie Bristol (USA) or Vanel (Europe).

SOLUTION WITH DISCUSSION.– Having read the design data, we set out the first choices made by the manufacturers (Bristol and Vanel):

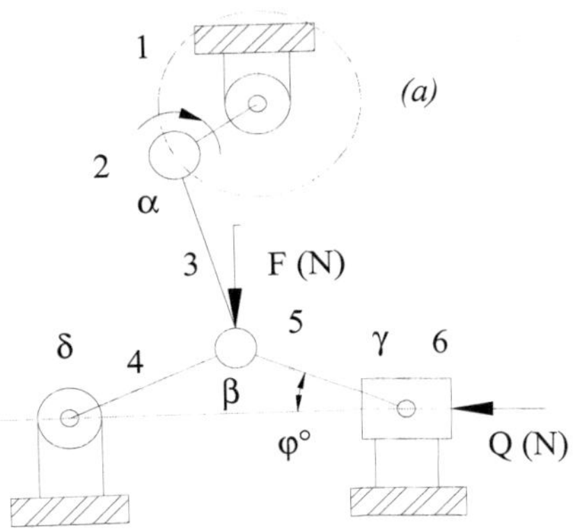

– P_{app} = 2.248715 lb = 1.02 kg is the force exerted on the rocker switch and therefore on the compression spring;

– L_{max} = 0.492 in = 12.5 mm is the maximum working length of the spring;

– P_{max} = 4.409 lb = 2 kg is the force that must not be exceeded on the rocker switch, and therefore on the compression spring;

– L_{comp} = 0.399 in = 10.125 mm is the compressed working length of the spring – i.e. in normal conditions of being switched on.

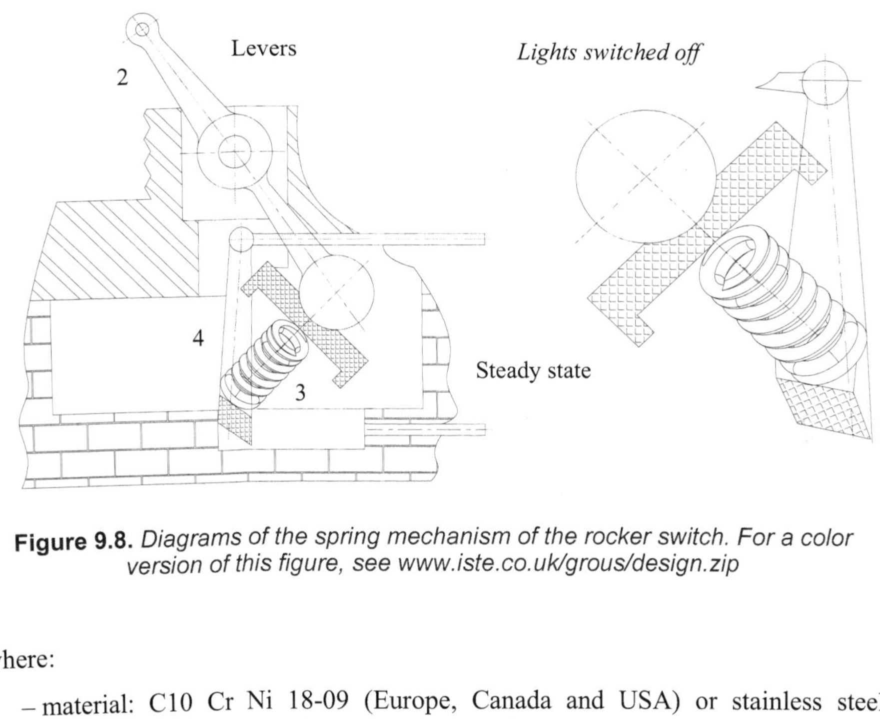

Figure 9.8. *Diagrams of the spring mechanism of the rocker switch. For a color version of this figure, see www.iste.co.uk/grous/design.zip*

where:

– material: C10 Cr Ni 18-09 (Europe, Canada and USA) or stainless steel 17/7PH (Europe). Data from the European manufacturer Vanel. Avoid galvanization of the steel springs. Slightly magnetic in the initial state;

– characteristics of the material chosen: $[\tau]_{adm}$ = 500 MPa = 72520 psi;

– in operation for typical use, the rigidity of the spring would be: G = 75500 MPa;

– Wahl stress intensification factor, K_{wahl} = 1.2 and with index C = 7;

– the operating conditions are typically between N = (1000 and 100,000) cycles under the maximum load. The loads and the deflections are illustrated in Figure 9.8, above.

The switch mechanism illustrated below (Figure 9.9(a)) enables us to calculate the force ratio (Q/F) as follows:

$$\frac{Q}{F} = \frac{\cos(\varphi)}{2\sin(\varphi)} = \frac{\text{cotan}(\varphi)}{2} = \frac{1}{2\tan(\varphi)} \qquad [9.61]$$

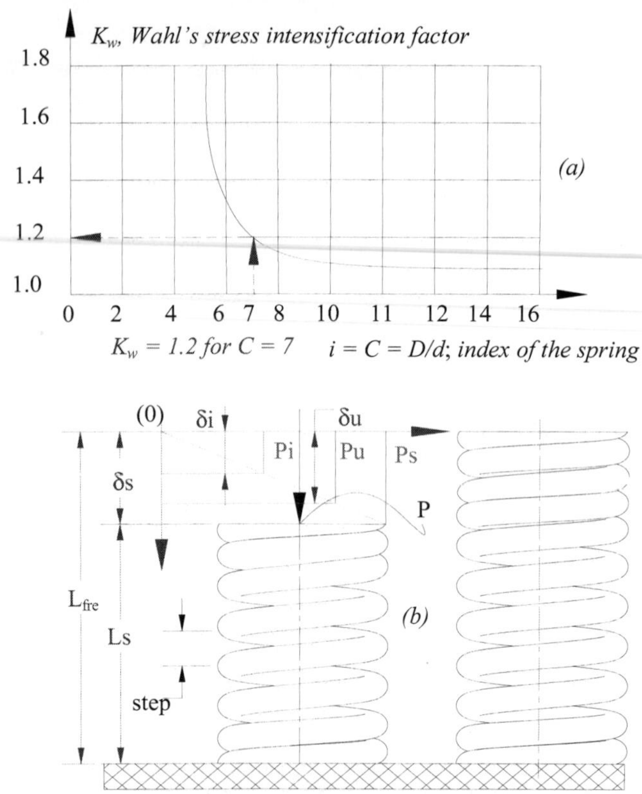

Figure 9.9. *Wahl's stress intensification factor (K_w) (a) and load–elongation diagram for the helical spring under compression (b). For a color version of this figure, see www.iste.co.uk/grous/design.zip*

Based on the mechanical formulae applying to compression springs, we set the expression of the deflection $\delta = L_{max} - L_{comp} = 2.375$ mm = 0.094 in. The load is expressed by $P = P_{max} - P_{app} = 0.98$ kg = 2.16 lb. The load in relation to the deflection is expressed by: (P/δ) is: $(P/\delta) = 0.413$ kg/mm = 23.104 lb/in. The deflection and its verification are as follows:

$$\delta_i = \frac{P_{app}}{P/\delta} = 2.472\ mm = 0.097\ in\ and\ \delta_u = \frac{P_{max}}{P/\delta} = 2.472\ mm = 4.847\ mm$$

The free length of the spring is: $L_{free} = L_{max} + \delta_i = 14.972\ mm = 0.589\ in$

The dimensions of the spring are chosen by calculations based on the tables published by the manufacturers. The choice of dimensions of the spring was made at the start of the project. Here is the confirmation of the chosen dimensions:

– characteristics of the chosen material: $[\tau]_{adm} = 500$ MPa = 72,520 psi;

– in operation for a typical use, the rigidity of the spring would be G = 75,500 MPa;

– Wahl's stress intensification factor, $K_{wahl} = 1{,}2$ and with the index C = 7;

– the average operating conditions are between: N = (1000 and 100,000) cycles at the maximum load. The loads and deflections are illustrated above;

– *block* $\tau = 850$ MPa = 1.233×10^5 psi.

$$\begin{cases} From\ \tau_{shear} = K_w\left(\dfrac{8P_{max}C}{\pi d^2}\right) \le [\tau]_{adm} \rightarrow d \succ \sqrt{K_w \dfrac{8P_{max}C}{\pi d^2}} \rightarrow d = ? \\ d = \sqrt{1.2\dfrac{8\times4.409\times7}{\pi 7.252\times10^4}} = 0.036062\ in = 0.916\ mm \rightarrow d_{chosen} = 0.41\ in = 1.0414\ mm \end{cases}$$

Number of active spires of the spring: n

Based on the material (stainless steel) G = 75500 MPa = 1.095×10^7 psi (*Cie Vanel*), we set:

$$From\ \frac{P}{\delta} = \frac{Gd_{chosen}}{8C^3\times n} \rightarrow n = \frac{G\times d_{chosen}}{8C^3(P/\delta)} = \frac{1.095\times10^7\times(0.0410)}{8\times7^3(23.104)} = 7.081523\ spires$$

From the manufacturers' recommendations, we choose nine spires (number rounded up):

$$n_{total} = n + 1.75 = 7.081523 + 1.75 = 8.831523 \; spires \rightarrow 9 \; spires$$

The calculation of the free length of the spring is then reviewed as follows:

$$\begin{cases} L_{design} = \{L_{solid} - (d_{chosen}/2)\} = 0.348 \; in = 8.852 \; mm \\ \delta_{design} = \{L_{free} - L_{design}\} = 0.241 \; in = 6.12 \; mm \rightarrow \dfrac{\delta_{design}}{\delta_u} = 1.262645 \end{cases}$$

The shock of the spires does not come into question. We can design a smaller diameter and employ an index $i = C = 7$, or $C = 8$.

9.4.4. *Verification of buckling of compression spring*

$$\begin{cases} Cond_{buckling} = \{\delta_u / L_{free}\} = 0.323748 = 0.324 \rightarrow Hence \; no \; buckling \\ Verifications: D = Cd_{chosen} \rightarrow (L_{free}/D) = 2.054 = (L_{free}/Cd_{chosen}) = 2.054 \end{cases}$$

$$\begin{cases} Non\text{-}buckling \; condition \; verified: (D_{mean}/L_{free}) \succ 0.2; (D_{mean}/L_{free}) = 2.054 \succ 0.2 \\ Non\text{-}buckling \; condition \; verified: (d_{wire}/D_{mean}) \succ 0.1; (d_{chosen}/D) = 0.143 \succ 0.1 \end{cases}$$

Recap of the choice of spring for the rocker switch: based on the specification documentation of the chosen manufacturer, we set:

$$\begin{cases} d_{wire} = 0.041 \; in = 1.041 \; mm \cong 1 \; mm \; and \; d_{outside} = d_{wire} + D = 0.328 \; in = 8.331 \cong 8.5 \; mm \\ \text{Proven verification} \rightarrow d\begin{Bmatrix} verified \\ outside \end{Bmatrix} = d_{wire} \times (C+1) = 0.328\,000 \; in = 8{,}331 \cong 8.5 \; mm \end{cases}$$

The total number of spires (without buckling) N_{total} = 9 spires, so the step P would be, for a designed free length L_{free} = 0.589 in =14.4972 mm (in design).

CONCLUSION.– There is neither shock nor percussion between the spires, which do not buckle. The design calculated above is therefore used.

9.5. Simple blade and spiral blade springs

CASE STUDY 1.– Consider a simple blade spring anchored in its place, as illustrated by Figure 9.10. The maximum admissible load on the free part of the spring and its maximum deflection are taken from the mechanical construction documentation.

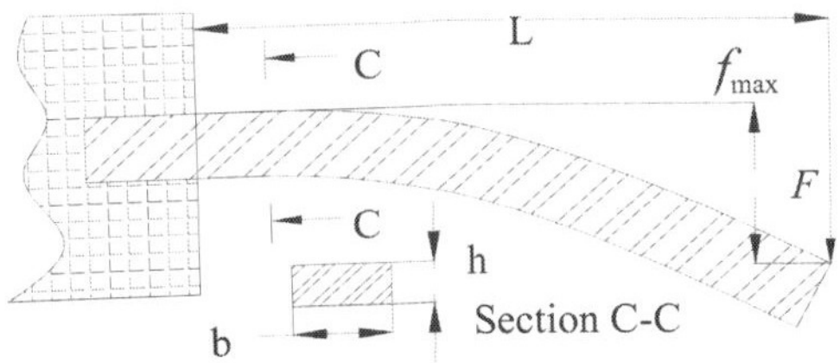

$E = 2\times10^5$ MPa $= 2.901\times10^7$ psi
σ_{re} =156 MPa =22340 psi
P_{app} = 12.25 N = 2.754 lbf
b = 12 mm = 0.472 in
h = 3 mm = 0.117 in
L = 125 mm = 4.921 in

Figure 9.10. *Diagram of a simple blade spring*

Calculate the maximum load in *N* and *lbf* and the corresponding deflection:

$$[P]_{\max}^{adm} = \frac{bh^2\sigma_{re}}{6L} = 22.464\ N\ and\ f_{\max} = \frac{4PL}{bh^3E} = 1.477\ mm = 0.058\ in \quad [9.62]$$

CASE STUDY 2.– Spiral blade spring. For the following data, calculate the applied load and the respective angles of deviation for the rectangular and circular sections. We know that the moment M = P_{app}.R (N.mm):

where:

$$\begin{cases} [P]_{\max}^{adm} = \dfrac{bh^2\sigma_{re}}{6R}; L_{active} = \pi n(R+r) \\ \varphi\left\{\begin{smallmatrix} sec \\ rect \end{smallmatrix}\right\} = \dfrac{12ML}{bh^3E}; \varphi\left\{\begin{smallmatrix} sec \\ circ \end{smallmatrix}\right\} = \dfrac{64ML}{\pi d^4 E} \end{cases}$$

– $E = 2\times10^5$ MPa;
– σ_{re} =25 MPa;
– P_{app} = 15N;
– b = 12 mm;
– h = 1.25 mm;
– n = 8 spires.

SOLUTION.– For a rectangular cross-section we set the following:

$$\begin{cases} [P]_{\max}^{adm} = \dfrac{bh^2\sigma_{re}}{6R} = 6.22\ N\ ;\ L_{active} = \pi n(R+r) = 832.396\ mm \\ M = P_{app} \times R = 308.4\ N.mm \end{cases} \quad [9.63]$$

9.6. Main expressions of design calculations for Belleville washers

The main expressions of design calculations for Belleville washers for springs are valid for both the metric and imperial (ANSI) systems. The parameters used below are presented in the tables given in the appendices. The respective calculations of the no-load height, the (finite) cone, truncated on the free side of the spring and the diameter ratio D/d = δ and h = H – e (mm) are:

SI and imperial	Description of parameter	Formula
χ [ad]	Total number of springs defined at	$\chi = n \times i$
Z [mm, in]	Course or deflection of the spring raised to	$Z = S \times i$
L_0[mm, in]	Length of spring raised with no load	$L_0 = i(h - n \times e)$
L [mm, in]	Length of loaded spring set	$L = (L_0 - Z)$
F [N, *lbf*]	Force exerted by the spring	$F = n \times F_1$

Table 9.6. *Relations of calculations for the spring*

If Weibull's law is applicable, below is how to calculate the parameters (α, β and γ) in that law, respectively of form, scale and position:

$$\alpha = \left[\left(\frac{\delta-1}{\delta}\right)^2 \Big/ \pi\left(\frac{\delta+1}{\delta-1}\right)\right] - \frac{2}{\ln\delta};\ \beta = \frac{6}{\pi\ln\delta}\left(\frac{\delta-1}{\ln\delta} - 1\right);\ \gamma = \frac{3(\delta-1)}{\pi\ln\delta} \qquad [9.64]$$

The force and limiting deflection F_{max} [*N, lbf*]; ν is Poisson's ratio:

$$For\ S_m = h,\ we\ have \rightarrow\ F_{\max} = 4Ee^3 S_m \big/ \alpha D^2\left(1-\nu^2\right) \qquad [9.65]$$

The force (load) applied by the spring to the spit *S* is written as:

$$F = \frac{4\times E\times e^4}{\alpha D^2\left(1-\nu^2\right)}\left(\frac{S}{e}\right)\left[\left(\frac{h}{e}-\frac{S}{e}\right)\times\left(\frac{h}{e}-\frac{S}{2e}\right)+1\right]\ [\mathrm{N}]\ or\ [\mathrm{lbf}] \qquad [9.66]$$

The maximum pressure stress of the spring at the spit S would be:

$$\sigma = \frac{4\times E\times e\times S}{\alpha\times D^2\left(1-\mu^2\right)}\times\left[\beta\times\left(\frac{h}{e}-\frac{S}{2e}\right)+\gamma\right]\ [\mathrm{MPa}]\ or\ [\mathrm{psi}] \qquad [9.67]$$

9.7. Power transmission. Case study: hoist

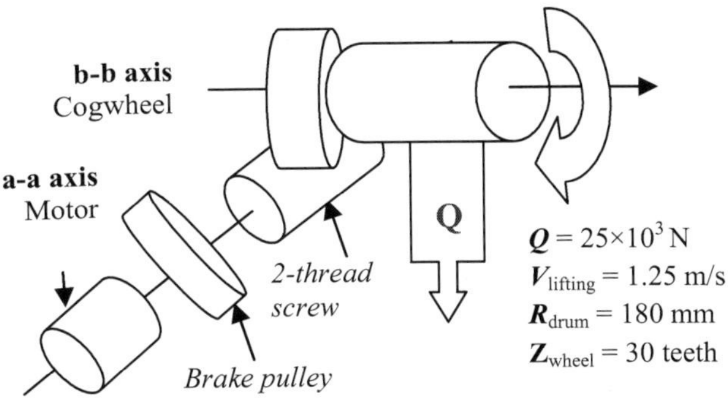

Figure 9.11. *Technological diagram of a hoist*

Consider:

– the yield between the drum and the engine	$\eta_1 = 70\ \%$	$\eta_1 = 70\ \%$;
– the yield between the drum and the brake	$\eta_0 = 70\ \%$	$\eta_0 = 70\ \%$;
– the yield between the engine and the brake	$\eta_2 = 100\ \%$	$\eta_1 = 70\ \%$;
– the reduction ratio between the drum and the engine	$U_1 = 1/15$	$U_1 = 1/15$;
– the reduction ratio between the drum and the brake	$U_0 = 1/15$	$U_0 = 1/15$;
– the reduction ratio between the engine and the brake	$U_2 = 1/1$	$U_2 = 1/1$;
– the load in N	Q = 25,000 N	$Q = 5.62 \times 10^3$ *lbf*;
– the radius of wrap of the drum	$R_t = 180$ mm	$R_t = 7.087$ *in*;
– $V_{lifting}$ is the speed of lifting	$V_{lifting} = 1.5$ m/s	$V_{lifting} = 59.055$ in/s;
– h_1 hoist height	$h_1 = 0.5$ m	$h_1 = 19.685$ in.

Calculation of the load when the hoist is stopped: the minimum resistant torque opposed by the brake is expressed as follows:

$$M_f = \eta_1 U_1 \times QR = (70\%/15)\left(25.10^3 \times 180\right) = 210\ Nm = 1.859.10^3\ lbfin \qquad [9.68]$$

Calculation of the rising load at constant velocity: the driving torque delivered by the engine during rise is written:

$$M_{engine}^{Vconst} = (QRU)/\eta = 428.571[\text{N.m}] = 3.793 \times 10^3 [\text{lbf.in}] \quad [9.69]$$

On startup, the driving torque delivered by the engine is written:

$$M_{engine}^{startup} = \frac{1}{\eta}(Q \times R \times U) + \sum_{i=1}^{n} M_{inertia} \; [\text{N.m}] \; or \; [\text{lbf.in}] \quad [9.70]$$

$\sum_{i=1}^{n} M_{inertia}$ is the sum of the resistant torque values due to the forces of inertia.

As the uniformly accelerated motion is expressed thus by the time τ_1, we can set:

$$For \; h_1 = 0.5m \; h_1 = V_{hoisting} \, \tau_1/2 \rightarrow \tau_1 = 2h_1/V_{hoisting} = 0.667s \quad [9.71]$$

Calculation of the forces F_i of inertia and $M_{inertia}$ reduced at the axis of the engine (a-a):

$$\begin{cases} F_{inertia} = m \times V_{\text{hoisting}}^2 /2h_1 = 2.551 \times 10^3 \; N = 573.59 \; lbf \\ M_{inertia} = F_{inertia} \times R_{drum} \times U_0 = 3.062 \; N.mm = 270.988 \; lbf.in \end{cases} \quad [9.72]$$

9.7.1. *Power transmission and simple drum brake*

CASE STUDY.– Below are the initial data for a case study of a drum brake:

Consider the following:

– $\mu.N$ is the force of friction in N. Leather/metal at the temperature of 150°C;

– $\mu.$ = 0.35 (leather on metal);

– N is the normal force 10 N = 2.248 lbf;

– P_a is the maximum pressure exerted =120 MPa, with $P_a \leq P_{admissible}$;

– b is the width of the jaw 10 mm = 0.394 in;

– r is the inside radius of the drum, 10 mm = 0.394 in;

– ρ_a is the angle at 90 degrees for $(\rho_2\text{-}\rho_1) > 90$ deg ρ_a = 1 deg, 5 deg, …, 90 deg;

– ρ_1 is the angle at 90 degrees for $(\rho_2\text{-}\rho_1) > 90$ deg; in our case 15 deg;

– ρ_2 is the angle at 90 degrees for $(\rho_2\text{-}\rho_1) > 90$ d°→115 deg; $(\rho_2\text{-}\rho_1) = 100° > 90°$;

– ρ is the angle corresponding to the pressure = 45 deg.

Braking moment and normal (radial) contact pressure

$$M_{brake}(\rho_a) = \mu \frac{P_a b r^2}{\sin(\rho_a)} \cos(\rho_1)\cos(\rho_2); P_{Norm}(\rho_a) = P_a \frac{\sin(\rho)}{\sin(\rho_a)} \quad [9.73]$$

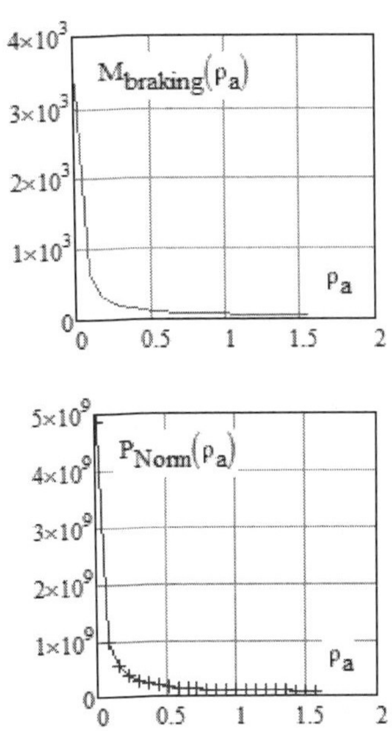

$M_{braking}(\rho_a) =$ ·N·mm

$3.342 \cdot 10^5$
$6.691 \cdot 10^4$
$3.728 \cdot 10^4$
$2.593 \cdot 10^4$
$1.995 \cdot 10^4$
$1.627 \cdot 10^4$
$1.380 \cdot 10^4$
$1.203 \cdot 10^4$
$1.071 \cdot 10^4$
$9.690 \cdot 10^3$
$8.889 \cdot 10^3$
$8.248 \cdot 10^3$
$7.727 \cdot 10^3$
$7.302 \cdot 10^3$
$6.954 \cdot 10^3$
...

$P_{Norm}(\rho_a) =$ ·MPa

$4.862 \cdot 10^3$
973.577
542.418
377.206
290.222
236.776
200.779
175.023
155.796
140.995
129.337
120.000
112.431
106.247
...

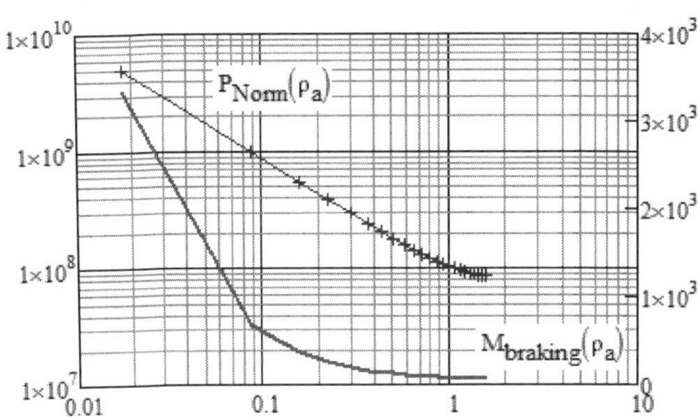

Figure 9.12. *Numerical and graphical results for the moments and normal pressure. For a color version of this figure, see www.iste.co.uk/grous/design.zip*

9.8. Case study on couplings

Consider the following:

– ultimate traction stress $\sigma_t = 520$ MPa for a coated SAE 1045 steel;

– failure threshold due to traction $\sigma_{ult} = 4.641\times10^4$ psi;

– diameter of shaft $d_{inc} = 1.7575$ in = 40 mm;

– number of bolts $N_b = 4$ bolts;

– diameter of bolts $d_b = 0.56$ in = 14.224 mm;

– inside contact diameter $d_{inc} = 7.25$ in = 184,15 mm;

– outside contact diameter $d_{outc} = 8.25$ in = 209.55 mm;

– friction coefficient, $\mu = 0.15$;

– load on each bolt $F_b = 7000\ lb = 3175\ kg$;

– speed of rotation of the coupling $\omega_{acc} = 360\ rpm$, Coefficient = 63024;

– axial force of the coupling, due to the loading of the bolts $F_{ax} = 30000\ lb$ = 13610 kg.

SOLUTION WITH DISCUSSION.–

$$\begin{cases} R_{friction} = \dfrac{2}{3}\dfrac{R^3_{outside} - R^3_{in}}{R^2_{outside} - R^2_{in}} = 3.88\ in = 98.562\ mm\ (radius\ of\ friction) \\ Moment \rightarrow M_\tau = R_{friction}\mu F_{ax} = 1.746\times10^4\ lb.in = 201.17\ kg.m \end{cases} \quad [9.74]$$

We accept the hypothesis that the pressure is uniformly distributed and calculate the power P_t:

$$\begin{cases} P_\tau = M_\tau \dfrac{R\,\omega_{acc}}{Coef} = 17460\dfrac{360}{63024} = 99.733\ hp = 74370\ W \rightarrow \sigma_s = \sigma_{ultimate} 18\% \\ \sigma_s = 93.6\ MPa,\ finally\ \sigma_{syp} = \sigma_{threshold} \times 30\% = 1.392\times10^4\ psi = 93\ MPa \end{cases} \quad [9.75]$$

Torque and power on the shaft P_{shaft}:

$$\begin{cases} M_{shaft} = 0.75\sigma_{syp} \times \pi\, d^3_{shaft}/16 = 8008\ lbfin = 9048\ Nmm \\ P_{shaft} = M_{shaft}\left(\omega_{acc}/Coeff\right) = 8008(360/63024) = 45.743\ hp = 34110\ W \end{cases} \quad [9.76]$$

9.8.1. *Case study: analysis in design of brake elements*

STATEMENT OF THE PROBLEM.– Consider a drum brake, made of cast iron, weighing 94.42 lbf = 420 N, designed to absorb 18 hp. It must not experience a temperature increase greater than 280°F = 137.778°C, during its normal operating cycle. We wish to come up with a design such that the radiation relative to the surface can be absorbed within the working limits, i.e. 18 hp. How much time would it take for the drum brake to return to its original temperature if it is installed in a space where the normal temperature is 68°F = 20°C?

where:

– power absorbed by the brake, E = 18 hp. Brake load factor, f_{load} = ½;

– temperature increase ($\tau_{increase}$) in °F = 280°F and °C = 137.778°C;

– specific heat of the cast iron drum, Cs = 0.13 (in^2/lb)×min = 01.849^{-4} (m^2/kg).min;

– specific weight of the cast iron drum, P = 94.42 *lbf* = 420 *N*;

– constant Kc = 0.4.

Radiation factor ξ	C	F
0.00060	55.6	100
0.00075	111.1	200
0.00083	166.7	300
0.00090	222.2	400

Table 9.7. *Increase in brake temperature*

SOLUTION.– From the table published by the drum brake manufacturer, we can read the referenced value of the radiation factor (ξ = 0.00083). The factor of heat increase in the brakes is calculated by the following relation: constant = $\xi.\tau_{increase}$ = 0.232. Knowing, from our initial hypothesis, that the temperature increase ($\tau_{increase}$) is of the order of 280°F = 137.778°C, let us return to the table above, read the value of the thermal radiation and calculate the area (A) affected by the heat as follows:

$$A_{irr} = \left(42.4\times in^2/hp\right)E\times f_{load}/\text{Constant} = 1.059\ m^2 = 1642\ in^2 \quad [9.77]$$

The cooling time for the brake will be given by the following:

$$T_{cooling} = C_s P \ln\left(\tau_{increase}\right)/K_c A = 61.962\ m/s = 2439\ in/s \quad [9.78]$$

COMMENT.– We use the results calculated above to better assess the friction, with a view to reducing energy losses, and controlling the normal operation of the brake, as is often the case in vehicles, hoist systems and handling systems.

9.9. Case study on power transmission: external spring clutch

Consider a clutch mechanism with an external steel spring, whose characteristics are as follows:

– the diameter of the spring, d_r is = 1.25 mm = 0.049 (≈ 0.5 in);

– the diameter of the shaft, d_a = 28 mm = 1.102 in;

– the elasticity modulus of the spring, E = 2×10^5 MPa = 29.01×10^6 psi;

– the number of spires per shaft, N_s = 7;

– the diametral interference between the helix of the spring and the shaft, i = [1 in to 0.985 in].

Calculation of the *moment (torque)* transmitted by the clutch:

$$M_t(i) = \left[\pi E d_r^4 i\right] / 32 d_a^2 \left(Exp^{2\pi\mu N_s} - 1\right) = 302.799\ N.m = 2680\ lbf.in \quad [9.79]$$

$M_t(i)$ is the moment transmitted by the clutch. It indicates the work of the spring. The influence of friction is considerable, as we can see from the following formula, where the exponential term (Euler's law) contains the friction coefficient (μ). In design, it is quite commonplace to overlook the considerable influence of springs in clutch systems, even when they are small springs. In the majority of cases, both modes of operation involve the springs.

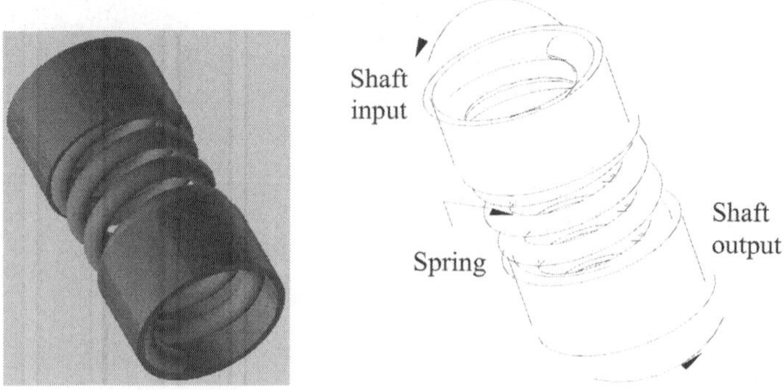

Figure 9.13. *Inside spring wrapping the clutch. For a color version of this figure, see www.iste.co.uk/grous/design.zip*

Slippage moment of the clutch, by virtue of Euler's law

$$M_G(i) = \frac{\pi E d_r^4 i}{32 d_a^2}\left(\frac{1}{Exp^{2\pi\mu N_s}} - 1\right) = -1.545\ N.m = -13.677\ lbf.in \qquad [9.80]$$

The negative sign of the slippage moment indicates slip, which must be taken into account. The diameter of the helix is slightly larger than the inside diameter, to allow for insertion of the steel coils which remain in constant contact. The design inside the system is important: when the engine is spinning in the same direction as the helix, under torsion, the rotational motion tends to unwind the inside spring, and the coils stretch it out. This produces more energy. During gear changes, in relative rotation, the motion tends to wind up the spring, whose spires become more tightly pressed together. The inside spring is then stopped or slowed, allowing the system to rotate freely. The helix is unwound. The residual friction is negligible, because it is largely reduced. The rate of wear of the lining is also reduced, which makes the system fairly reliable. This design is highly advantageous, preventing wear and tear on the system by the centrifugal forces. Unlike the system with external springs, this design is capable of operating at very high rotation frequencies without unwinding the spring. By varying the dimensions of the spring and the surface state of the material, we gain a good understanding of the drag force generated by the work. This drag offers us better control of clutch failures, because the end of the spring is arched on one side and pressed on the other to prevent energy loss.

9.9.1. *Case studies: power transmission. Bolted assembly*

Design a steel coupling with flange to transmit a twisting moment of 15×10^5 N.mm, using a steel shaft 62 mm in diameter. The load is uniform, without shocks. Let us determine how many bolts are needed in the coupling if the admissible shear stress on the bolts is 25 MPa. Quickly calculate the thickness of the flange on the coupling, the length, the center of the coupling at the hub if the admissible stress of the material for the center is 150 MPa. The permissible shear stress in the wrench is 90 MPa. There is no thrust imposed on the coupling:

– the diameter of the shaft, d_a is = 62 mm = 2.441 in;

– the moment (torque) M = T = 15×10^5 N.mm = 1.328×10^4 lbf.in;

– the stress on the wrench σ_s = 90 MPa = 1.305×10^4 psi;

– the shear stress on the bolts, τ_b = 25 MPa = 3.626×10^3 psi;

– the shear stress on the wall, τ_p = 150 MPa = 2.176×10^4 psi;

– the number of bolts, N_s = ?

Choice of diameter of the bolts used to hold the coupling: we suppose a circle diameter of the bolt for the coupling. As an initial choice, suppose that the diameter of the bolt circle is three times the diameter of the wall: $d_b = 3d_a = 7.323$ in = 186 mm. This is a reasonable assumption, given the couplings available on the market. The radius r_b will therefore be: $r_b = d_b/2 = 93$ mm = 3.661 in.

Determine the shear force acting on the bolts:

$$F_{shear} = T/r_b = M/(d_b/2) = 1.613 \times 10^4 \; N = 3.626 \times 10^3 \; lbf \quad [9.81]$$

Determine the number of bolts needed to fit the coupling:

$$N_b = 8F_{ci} / \pi d_b^2 \times \tau_{shear} = 10.186 \cong 10 \; bolts \quad [9.82]$$

The diameter of the bolts for a rigid coupling varies between (1/4 and 2) in. For our purposes, let us choose: d_b = ½ in, and calculate the number of bolts required. Rigid couplings are often held together by two to eight bolts, depending on the moment they are required to transmit. A coupling held with fourteen bolts represents a design that needs to be re-examined. To reduce the number of bolts, we can work with a larger diameter, such as: d = 3/4 in (0.75 in), and recalculate the number of bolts needed. The number of bolts chosen is N = 8 bolts, each 0.75 in (19.05 mm) in diameter. It is rare for an odd number of bolts to be used in flange couplings. We calculate the shear stress on each bolt in the knowledge that the admissible stress is 25 MPa. The bolts are therefore capable of withstanding the stress:

$$\tau_{shear} = 8F_{shear} / \pi d_b^2 \times N_b = 14.147 \; MPa = 2052 \; psi \quad [9.83]$$

The thickness of the flange is: $\varepsilon = 2F_{shear} / \pi d_a \tau_{wall} = 1.108 \; mm = 0.0436 \; in$. The required length of the hub will be calculated as a function of the key, as follows: $l_{hub} = 2F_{shaft} / \varepsilon_{key} \tau_{wall}$, where the inside radius of the hub is the same as that of the diameter of the shaft – i.e. r_{hub} = da/2 = 31 mm = 1.22 in. The force acting on the shaft is therefore: Fa = M/r_{hub} = 48390 N = 10880 lbf.

Considering a square key with a side length of ¾ in = 19.01 mm, we set: ε_{key}. = 0.75 in = 19.01 mm and calculate the length of the hub:

$$l_{hub} = 2F_{shaft} / \varepsilon_{key} \times \tau_{wall} = 33.867 \; mm = 1.333 \; in \quad [9.84]$$

The stress of the material initially chosen was 150 MPa (21760 psi). The calculated stress on the wall of the protruding hub is barely 14 MPa (14.147 MPa = 2052 psi). This means that the system's operation is very stable. For an optimized and economical design, it is helpful to reconsider the choice of the initial material to choose another affordable and safe material. It is also important to calculate the necessary length of the hub (l) and its thickness (ε). In fact, it is the duality between geometry and stress which is at the very heart of design. We must examine the geometry to offer safe resistance to the coupling, and *vice versa*. To do so, we recalculate the force acting on the coupling at the hub:

$$F_{hub} = M/r_{hub} = 4.839\times10^4\ N = 1.088\times10^4\ lbf \quad [9.85]$$

Next, we calculate the necessary thickness of the hub ε:

$$d_{hub} = 2r_{hub} \rightarrow \varepsilon_{nec} = \frac{F_{hub}}{\pi d_{hub} \times \tau_{shear}^{shaft}} = 10.361\ mm = 0.408\ in \quad [9.86]$$

The dimensions of couplings are normalized (standardized) to facilitate interchangeability. Therefore, they are fairly reliable.

9.9.2. Computer-assisted design of a hub (bolted assembly)

In this section, we present a concrete example of a coupling hub, assembled with four bolts. We shall calculate the parameters set out below. Computer-assisted design is indubitably reliable, fast and elegant. However, for purely pedagogical reasons, in this author's publications, he has tended to set out his own calculations, step by step.

Hub of a coupling designed with Inventor Pro

Consider a coupling hub (see results below) and the results of the calculations.

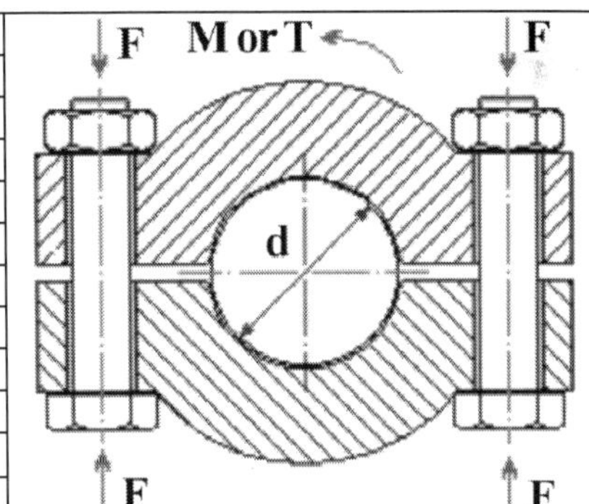

Design of the length of the hub		
Torque (Moment) T	40,000 lbf.ft	
Axial force, F_a	300,000 lbf	
Dimensions		
Shaft diameter	d = 1 in	
Hub length	L = 1.500 in	
Joint Properties		
Loading	Static load	
Material	Steel gr. 37; 42	
Properties of bolts	Steel SAE 1015	
Admissible pressure	$0.25\times\sigma_y = 11375$ psi	
Number of joints	Bolts N = 4	
Admissible pressure	$P_A = 12800$psi	$I_{Min} = 0.469$ in; $M_{Min} = 0.438$ in
Press factor	$\nu = 0.080$	$P_{AC} = 4001.1885$ psi
Safety factor	$K_s = 1.500$	F = 1500.707 lbf; $d_{mini} = 0.55$ in

Figure 9.14. *Results of calculations of computer-assisted design. For a color version of this figure, see www.iste.co.uk/grous/design.zip*

Below, we present a series of case studies. For pedagogical purposes, the calculations are set out manually.

CASE STUDY.– What is the necessary length of a helical spring having a mean diameter $d_m = 20$ mm wrapped in n = 7 spires? That spring, made of chrome-vanadium, has been heat tempered and coated with a thickness ε = 0.056 in = 0.056 mm in diameter. The constant K of the spring is K = = 1.55×10^{-6} kg/mm^3:

– mean diameter du spring, $d_m = 25$ mm =0.984 in;

– number of spires, n = 7 spires;

– diameter of the spring (solid tube), $d_r = 1.422$ mm = 0.056 in;

– constant of the material (density) for the spring, $\rho = 1.56\times10^{-6}$ kg/mm^3 =0.056 lb/in^3.

SOLUTION.– Calculation of the length of the spring: $l_{r\rightarrow} l_r = \pi d_m n = 549.8\ mm = 21.645\ in.$

The load (weight) on the spring, P will be:

$$P_r = \rho \times l_r \times d_r^2 = 1.724\times10^{-3}\ kg = 0.0038\ lb.$$

9.10. Couplings and machine elements subjected to stress at high speeds

STATEMENT OF THE PROBLEM.– To determine the power transmitted in machine elements such as couplings, let us say: P = 50 hp at an RPM = 200 rpm, accepting that there is an angular disalignment varying from the minimum to 45°, i.e. α = [0° to 45°]. What would be the effect of that angular disalignment on the localization of the shaft, its speed, its acceleration, if we noted a disalignment varying between [30° and 45°]? The shaft rotates at 200 rpm.

From the table below, drawn from the work of N.B. Rothfuss [ROT 73], we choose the appropriate characteristics for the type of coupling.

Characteristics	Characteristics of couplings *			
Main characteristics	Profiled diaphragm	Axial spring	Laminated disc	Universal joint
Speed range, rpm	0-60,000	0-8,000	0-20,000	0-8,000
Power, kW/100	1-500	1-9,000	1-100	1-100
Angular disalignment [°]	0-8.0	0-2.0	0-1.5	0-45
Parallel disalignment, mm	0-2.5	0-2.5	0-2.5	None
Axial displacement, cm	0-0.5	0-2.5	0-0.5	None

Ambient temperature, °C	900	Variable	900	variable
* *Product Engineering*	Rods	Pinion	Chain	Elastomer
Angular velocity rpm	0-8,000	0-25,000	0-6,300	0-6,000
Power kW/100 rpm	1-100	1-2,000	1-200	0-400
Angular disalignment, [°]	0-40	0-3	0-2	0-4
Parallel disalignment, mm	None	0-2.5	0-2.5	0-2.5
Axial displacement, cm	None	0-5.1	0-2.5	0-0.8
Ambient temperature, °C	variable	Variable	variable	variable
Ambient pressure, kPa	variable	Variable	variable	variable

Table 9.8. *Working characteristics of couplings*

9.10.1. *Determination of the error in position of the shaft*

From the specialized documentation (*Handbook, Shaft Alignment*) [PIO 06], we have *pulled together* our own tables, showing the variations caused by disalignment between the coupling and the shaft. For example, a fault of 30° in the angular alignment leads to an error in position of the shaft [from – 4° 6' 42" to + 4° 6' 42"]. This means that the position of the shaft undergoes that error, every two revolutions. Thus, at a 45° alignment error, the error in position of the shaft in relation to the coupling, gives us a fault of 9° 52' 26" (Table 9.10), etc.

Table 9.9.	Characteristics of couplings			
Main characteristics	Profiled diaphragm	Axial spring	Laminated disc	Universal joint
No lubrication	✓		✓	
No clearance	✓	✓	✓	*
Constant velocity ratio	✓	✓		**
Confinement			*	
Angular	✓		✓	✓
Axial and angular	✓		✓	
Axial and parallel	✓	✓	✓	
High temperature	✓		✓	
High rate of torsion of the spring	✓			
Low bending moment	✓	✓	✓	✓
No relative motion				✓
Main characteristics	Rods	Clutch	Chain	Elastomer
No lubrication	*			✓
No clearance	✓	**	**	✓
Constant velocity ratio	*	✓		
Confinement	✓	✓		

Axial and angular		✓	✓	✓
Axial and parallel			✓	✓
Axial, angular, parallel				
High rate of torsion of the spring	✓	✓	✓	✓
Low bending moment				✓
* No clearance. Confinement can be achieved by special design ** Constant velocity ratio with approximate angles				

Table 9.9. *Working characteristics of the couplings*

Table 9.10, below, shows that at 30°, the alignment error of the shaft in relation to the coupling is ± 15.47 %. The output speed is 200 rpm. The fault corresponds to (1.00° 0.1547) = 230.94 and 169.06 rpm. This variation in velocity occurs every two rotations. For example, at a 45° alignment default, the rotation velocity would range from 117.16 to 282.84 rpm every two rotations of the shaft.

Table 9.10	Variation of output velocity from universal joints [*]		
Alignment angle fault	Maximum position error	Maximum % velocity error	Ratio A/ω^2
05°	0° 06' 34"	0.382	0.011747
10°	0° 26' 18"	1.543	0.030626
15°	0° 59' 36"	3.526	0.069409
20°	1° 46' 54"	6.418	0.124966
25°	2° 48' 42"	10.338	0.198965
30°	**3° 06' 42"**	**15.470**	**0.294571**
35°	4° 42' 20"	22.077	0.417232
40°	5° 36' 43"	30.541	0.576215
45°	6° 52' 26"	41.421	0.787200
* Caused by the alignment of the shaft.			

Table 9.10. *Machine elements in mechanical design. Universal joints*

9.10.2. *Determination of the output velocity of the shaft*

We can easily read the maximum acceleration ratio. By conventional calculation, this can be written as: $\omega = 2\pi R$ (rad/s). For an angular alignment fault of 30°, we read: $A/\omega^2 = 0.294571$. Thus, for a constant angular velocity of 200 rpm, an acceleration ranging [from -129.2 to + 129.2] rad/sec² is produced, at a frequency of

2×200 rpm = 400 cycles/minute. For an angular alignment fault of 45°, we read [-346 to +346] rad/sec². The range of the acceleration allows for a shaft alignment fault 2.67 times smaller than if we had a 30° angular alignment fault. The values contained in the tables are absolutely not limitative. It is even possible to use them for couplings with shock loads, by assigning corrective coefficients to them.

9.11. Design of spring rings

CASE STUDY 1.– What is the maximum dimension of the *ring of the spring* if the admissible stress is $[\sigma]_{adm} = 1.5\times10^3$ MPa = 2.176×10^5 psi? The elasticity modulus of the material, E = 1.5×10^5 MPa = 2.176×10^6 psi. The friction coefficient $\mu = 0.12$ and the outside diameter $d_{outside}$ = 225 mm = 8.858 (≈ 9 in); the inside diameter d_{in} = 180 mm = 7.087 in. The angle of the conicity is $\theta = 14°$, and the axial load applied to the spring is F_a = 50,000 kg =110,200 lb.

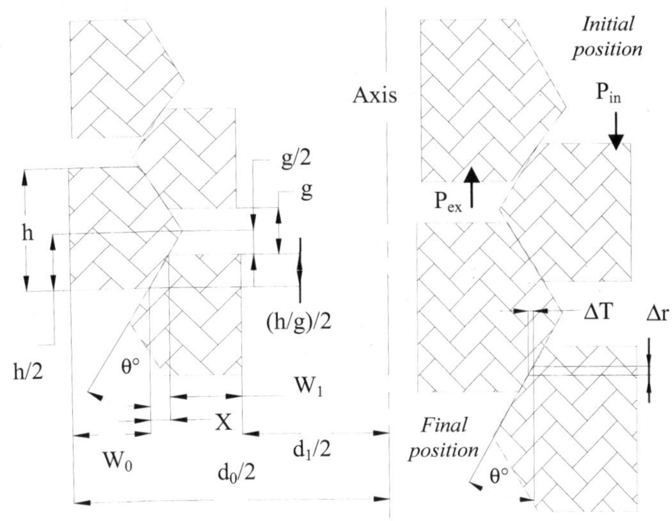

Figure 9.15. *Ring/spring positions and their dimensions. For a color version of this figure, see www.iste.co.uk/grous/design.zip*

SOLUTION.– The inside dimensions of the ring, for an ordinary spring, correspond to 15 % of the outside diameter, meaning:

$$h_{in} = 15\%\times d_{outside} = 33.75\ mm, we\ set\ \Delta T = d_0\left(S_{in} + S_{outside}\right)/2E \qquad [9.87]$$

The calculation of Kc confirms our reading of the above table, meaning that:

$$K_c = \lfloor \sin(\theta) + \mu\cos(\theta) \rfloor / \lfloor \cos(\theta) - \mu\sin(\theta) \rfloor = 0.38072 \leftarrow QED \qquad [9.88]$$

Calculation of the inside area A_{in} of the ring and its width l_{in}:

$$\begin{cases} A_{in} = F_a g / \pi K_c [\sigma]_{adm} = 273.303\ mm^2 = 0.424\ in^2 \\ l_{in} = \left[A_{in} - \left(h_{in}^2 \tan(\theta)/4 \right) \right] / h_{in} = 5.994\ mm = 0.236\ in \end{cases} \qquad [9.89]$$

The axial clearance between the rings is usually 25 % of the product of the height. To find the area A*in* of the ring, we begin by finding the constant $K_c = 0.38$ on the figure below, where, for an angle $\theta = 14°$, we read a friction value of $\mu = 0.12$. To calculate the dimensions of the outer ring, we use a rule of three which takes account of the stress in accordance with the requirements of the TS:

$$\begin{cases} h_{outside} \equiv h_{in} = 33.75\ mm = 1.329\ in, l_{outside} = \left(A_{outside} - h_{outside}^2 \tan(\theta)/4 \right) / h_{outside} = 9.6\ mm = 0.378\ in \\ A_{outside} = 395\ mm^2 = 0.612\ in^2\ and\ \sigma_{outside} = F_a g / \pi A_{outside} K_c = 1038\ MPa = 150529\ psi \end{cases}$$

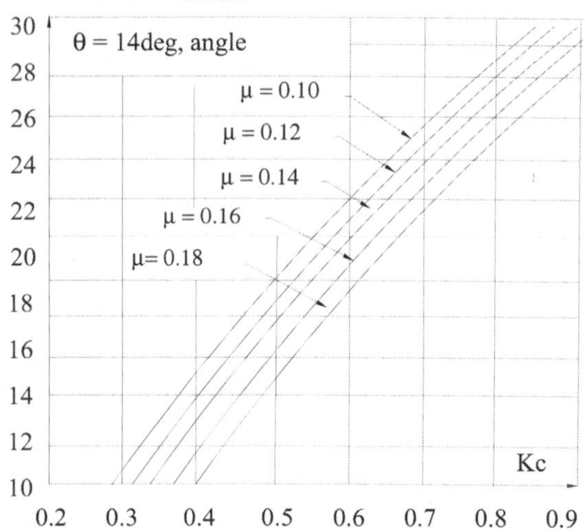

Figure 9.16. *Constant K_c of the ring/compression spring. For a color version of this figure, see www.iste.co.uk/grous/design.zip*

The outside diameter of the ring including the clearance would be:

$$\begin{cases} For\ a\ clearance = h_{in} \times 25\% = 8.438\ mm = 0.332\ in \rightarrow we\ have: \swarrow \\ D_{outside} = d_{in} + 2l_{outside} + 2l_{outside} + (h_{in} - gap)\tan(\theta) = 217.499\ mm = 8.563\ in \end{cases} \quad [9.90]$$

This result is close to the maximum dimensions allowed for the $\varnothing_{outside}$ of around 228.6 mm = 9 in. This is acceptable. To calculate the number of rings required for this design, we must first calculate the mean diameter of the spring, as follows:

$$d_{mean} = 0.5 \times \left[(D_{outside} - 2l_{outside}) + (d_{in} + 2l_{in}) \right] = 195.144\ mm = 7.683\ in \quad [9.91]$$

The axial deflection per ring would be:

$$f = d_{mean} \left[(\sigma_a + \sigma_{outside})/2E \right] \text{cotan}(\theta) = 6.621\ mm = 0.261\ in \quad [9.92]$$

As the deflection is no more than 203.2 mm = 8 *in*, the number of rings required would be:

$$N_{rings} = \delta_f / f = 30.206 \equiv 31\ rings \quad [9.93]$$

COMMENTS.– The rings for springs are designed to damp vibrations and absorb shocks. Their sizing is based on the compressed height of around four times the deflection of the spring. The height of the ring corresponds to 15 % or 25 % of the outside diameter (of the ring). The friction coefficients range between 0.10 and 0.18.

9.12. Principle of calculations for a Belleville washer: case study

Design the sizing of a Belleville washer for springs for AISI 1074 steel – high carbon content (ASTM A 684, AMS 5120H).

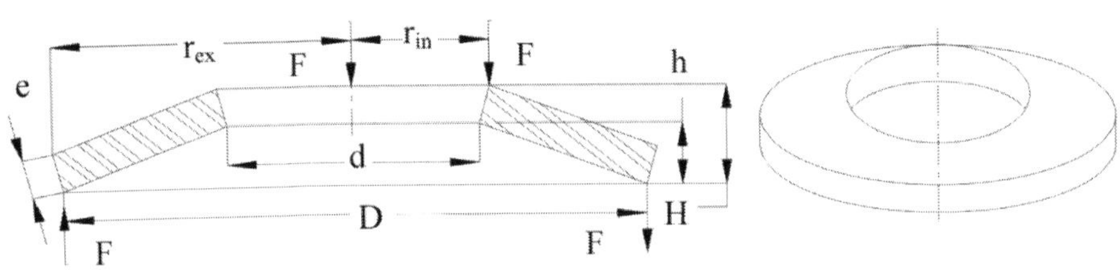

Figure 9.17. *Belleville washer for compression springs. For a color version of this figure, see www.iste.co.uk/grous/design.zip*

Constants in design of Belleville washers for springs		Constants in design of stresses on Belleville washers for springs		Data for the exercise
		$R_{outside}/r_{in} - 1.75$		$F = 4500$ N $= 1012$ *lbf*
h/e	**$R = r_{outside}/r_{in}$**	**K**	**$r_{outside}.\sigma_c/L^{1/2}$**	Compression stress, σ_c = 150 MPa = 21760 psi
1.00	1.25	-3.2455	-1346	Elasticity modulus, E = 2×10^5 MPa = 2.9×10^7 psi
1.25	1.50	-4.3734	-1622	Poisson's ratio $\nu = 0.3$
1.50	1.75	-5.6279	-1905	
1.75	2.00	-7.0090	-2197	
2.00	2.50			

Table 9.11. *Constants in design of Belleville washers for springs*

From the above table, the ratio of the radii R= 2. The outside radius $r_{outside}$ would be:

$$\frac{r_{outside}\sigma_{comp}}{F} = -2197\frac{\sqrt{lbf}}{in}\ simplify \rightarrow 100000.Pa.r_{outside}/3N=\frac{2197\sqrt{lbf}}{in} \qquad [9.94]$$

Let us isolate $r_{outside}$ as follows:

$$r_{outside} = -2197\left(\sqrt{lbf}/in\right)\times\sqrt{F}/-\sigma_{com} = 81.584\ mm = 3.21\ in.$$

The sign (-) indicates that the spring is compressed. The inside radius of the washer (r_{in}) is calculated as follows: $+r_{outside}/r_{in} = R_r \rightarrow r_{in} = r_{outside}/R_r = 40.792 = 1.606\ in.$ From the above table, we can read the value of K = -7.0090 (stress constant), for (h/e) = 1.75 and R = 2. Then, we set the expression of the thickness of the washer (e) as follows:

$$e = \sqrt{-\sigma_{comp}\left(r_{outside}\right)^2/K\times E} = 0.844\ mm = 0.03323\ in \qquad [9.95]$$

From the ratio (*h/e*) = 1.75, we deduce the height: $h = 1.75\times e = 1.477\ mm = 0.058\ in.$

9.13. Determination of the pressing moment for a bolted assembly

CASE STUDY.– Suppose we wish to calculate the moment (torque) required to press an 11-NC bolt of diameter D_b = 15.875 mm (i.e. 0.625 in). The bolt is pre-stressed with 80,000 N (17980 lbf). The friction coefficient between the threads is $\mu_f = 0.10$, and that between the nut and the washer $\mu_{e\text{-}r}$ = 0.12. From the standard tables [OBE 16] for threads (USA), we can read:

– the minor diameter, $D_M = 13.043$ mm = 0.51135 in;

– the *angular pitch diameter* $D_{pitch} = 14.376$ mm = 0.5660 in;

– the mean diameter, $D_{ou} = 1.25 \times D_b = 19.844$ mm = 0.781 in.

Calculation of the helix angle (β) of the 11-NC threads: knowing the value of the friction coefficient, we set: $\tan(\varphi) = \mu_f \rightarrow \varphi = arc\tan(\mu_f) = 5.711°$. The helix angle would be:

$$\sin(\beta) = 1(in)/11\pi D_{pitch} \rightarrow \beta = arc\sin\left(1(in)/11\pi D_{pitch}\right) = 2.931°$$

To find the pressing moment (torque), it is necessary first to find the angle factor of the threads, as follows:

$$\begin{cases} \alpha = 30° \rightarrow f_{ang} = \tan(\beta + \varphi)/\cos(\alpha) = 0.2 \rightarrow \textit{Hence the moment } M_{app} \\ M_{app} = 0.5 \times F_{ch}\left(D_{outside} \times \mu_{e.r} + f_{ang} D_{pitch}\right) = 196.161\ N.m = 1736\ lbf.in \end{cases} \quad [9.96]$$

The combined stress between the nut and the bolted joint (washer) would therefore be:

$$\tau_{comb} = M_{app} \frac{0.89 + 1.66\sqrt{5.2 F_{ch} \dfrac{D_{pitch}}{D_M}}}{D_M^2\left(\mu_{e.r} D_{outside} + f_{ang} D_{pitch}\right)} = 762.863\ MPa = 110644\ psi \quad [9.97]$$

From the specialized literature, we can see that 68 % of the ultimate stress affects the combined stress. Therefore, we can set:

$$1/68\% = 1.4706 \rightarrow \tau_{ult} = 1.47 \times \tau_{comb} = 1122\ MPa = 162713\ psi$$

COMMENTS.– The approach employed here can be used to calculate the pressing moment of the bolts with joints and other articulations, such as washers. We have chosen to use the formulae found in the *Product Engineering* manual.

9.14. Power transmission by epicyclic gear system

HYPOTHESIS 1.–

– RPM of the driving wheel: $RPM_{RM} = 50$ rpm;

– number of teeth of the steady wheel $Z_{sa} = 80$ teeth;

– number of teeth of the driving wheel, $Z_{da} = 18$ teeth.

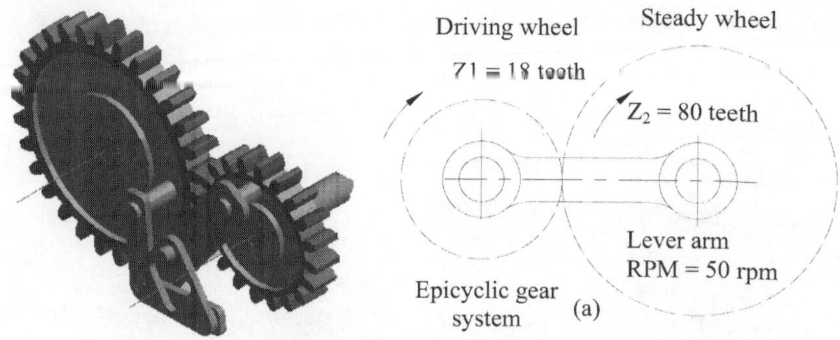

Figure 9.18. *Straight-toothed epicyclic cogwheels. For a color version of this figure, see www.iste.co.uk/grous/design.zip*

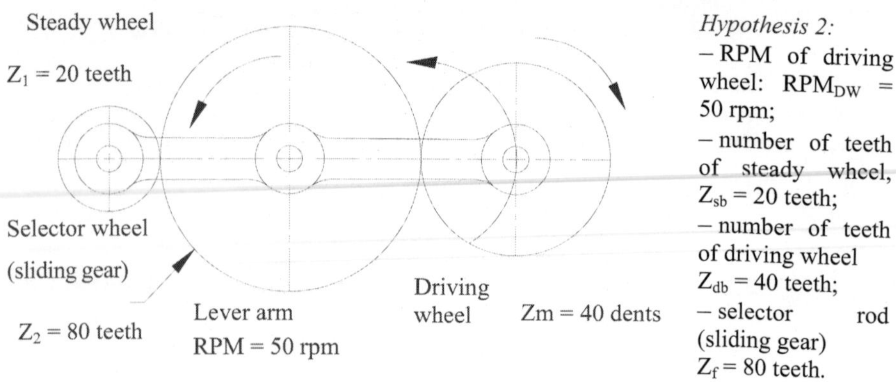

Figure 9.19. *Intermediary straight-toothed wheel. For a color version of this figure, see www.iste.co.uk/grous/design.zip*

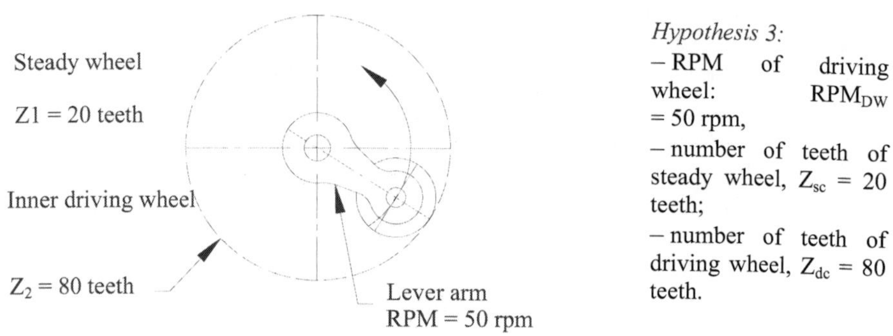

Figure 9.20. *Internal straight-toothed wheel. For a color version of this figure, see www.iste.co.uk/grous/design.zip*

SCENARIO 1.– With the configuration shown in Figure 9.18, the driving wheel rotates at an $RPM_{da}(1+Z_{sa}/Z_{da})$ = 272.222 rpm. The side gear transmission ratio of the steady mechanism is Z_{sa}/Z_{da} = 4.444. Turning axially with a ratio of 4.444, the mechanism carries out an additional, steady rotation. Its total axial and side-gear rotation will be (4+1)/rpm of the arm. We have $i_R = Z_{da}/Z_d = 0.184 = 1:5$.

SCENARIO 2.– With the configuration in Figure 9.19, the sliding wheel spins at the following RPM: $RPM_{db}x(1\text{-}Z_{sb}/Z_{db}) = 25$ rpm.

SCENARIO 3.– With the configuration shown in Figure 9.20, the steady wheel rotates at an RPM as follows: $RPM_{dc}x(1\text{-}Z_{sc}/Z_{dc}) = 37.5$ rpm. The inner wheel drives the gear system, which turns the lever arm at the ratio of: $i_d = Z_{sc}xN_d/(N_d+N_s)$.

9.15. Conclusion

The case studies presented here pertain to bolted assemblies, springs, washers and gear systems. We have seen that the calculations are not problematic, because we have effective formulae developed previously in the specialized literature [HAY 90, ROT 73, RIC 70, ROA 75]. The question remains of choosing the right materials to serve the functions prescribed by the calculations.

When designing parts to make a machine, or even a simple mechanism, it is always advisable to optimize their performances. This is done by adapting the properties of the material and its geometry (often its cross-section). In this book, we have presented case-studies where the macroscopic geometry is the primary focus. Microscopic geometry, although it is highly important, is dealt with by the choice of materials (isotropy, homogeneities, etc.).

10

Design of Plastic Products

10.1. Calculations for the design of plastic parts

As synthetic products, plastic materials are made of long chains of large molecules. Each molecule is made up of numerous organic chemical units, and thus the material is referred to as a *polymer (multiple units),* or a *macromolecule.* The material is solid and rigid. At ambient temperature, it is capable of withstanding considerable structural loading. Plastic materials are classified as follows:

Thermoplastics		**Heat-hardening materials**
Crystalline	*Amorphous*	Epoxy
PP, PE, PET, PBT, PA (Polyamide), POM (Actal)	PC, PS, PPS, PPO, PVC, ABS, SAN, Polysulfone, Polyacrylate, Polyetheremide, Acrylic PMMA	Melamine Phenolic Unsaturated polyesters Butyl elastomers

When designing plastic products, the methodology used can sometimes be rather complex, because these materials behave differently to metals due to the internal morphologies. In this chapter, we examine the principles behind the calculations used when designing with thermoplastics and heat-hardening materials. The design methodology of these products focuses on the choice of material, which determines the quality of the part, the process used to make it, its recycling and therefore its environmental impact. The computational method is founded on the definition of the design based on a set of functional specifications (FS) hinging on the choice of material:

– estimated drawing and rendering, forming a virtual prototype;

– choice of material to be used for the design, which takes account of its rheological properties;

– esthetic choice made, its adjustment and ultimately its approval;

– machining by computer-aided manufacturing (CAM) and tests;

– conditioning and delivery.

The Technical Specifications (TS) of materials, in plastic design, must be most rigorous indeed. A mistake in choosing the right material would lead to manufacturing failure. There are numerous criteria which need to be optimized, including:

– flexibility, hardness, fatigue, stiffness and Mohs scratchability, and wear;

– physical properties such as its density, moldability and permeability;

– chemical properties determine the material's stability in its environment;

– thermal properties such as fire resistance and swelling;

– electrical properties like insulation;

– feel, hardness, appearance, hygiene, heat, cold and radiation;

– the way in which the elements are assembled, such as gluing, screwing, clamping and welding;

– recycling (reuse), ageing, degradation, etc.;

– likelihood of deforming when in use;

– conformity to the applicable standards in terms of health, medical standards, toxicity, etc.;

– silkscreen printing, painting and other types of marking or metallicization;

– storage, hygrometric properties, radiation, etc.

10.1.1. *Mechanical parameters used during traction tests*

In accordance with ISO 527 (DIN527), traction tests are conducted to determine the behavior of plastic materials when subjected to short-term uni-axial stresses.

The parameters are as follows:

– σ_b is the maximum tension in MPa;

– σ_r is the failure resistance in MPa;

– σ_T is the traction stress in MPa;

– ε_b is the swelling (relative elongation) in mm due to the maximum traction force;

– ε_r is the swelling (relative elongation) in mm due to failure;

– ε_T is the traction force;

- the relative traction stress (σ) corresponds to the smallest initial cross-section measured for the part at any time during the test;

- the traction resistance stress (σ_b) corresponds to the most powerful load;

- the failure resistance stress (σ_r) corresponds to the strongest traction load at the time of failure (tearing) of the material;

- the elastic stress to failure (σ_T) corresponds to the strongest traction load. It is equal to the slope of the force–elongation plot when it first touches zero;

- E is the elasticity modulus of the plastic material. Hooke's law is applicable in the linear part of the curve, so $E = \sigma/\varepsilon$ in MPa;

- ε is the elongation or swelling in mm. L_0 measured relatively to the test piece gives us the value of ΔL at any time during the test. The failure swelling when subjected to the greatest force is characterized by ε_r under the traction force.

In order to choose more appropriate plastic materials in mechanical design, it is wise to take account of the behavior under traction, swelling, temperature, environment and also the duration of the test.

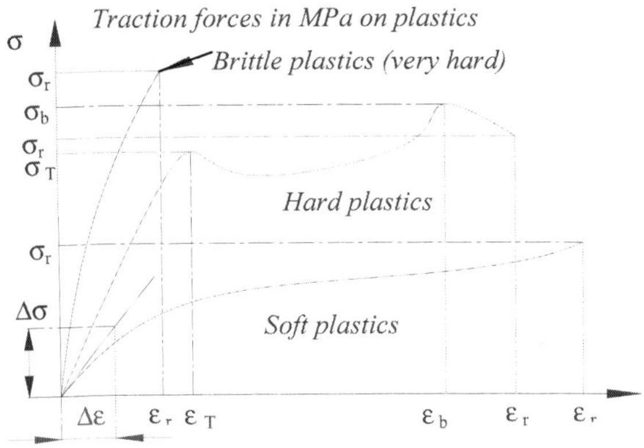

Figure 10.1. *Force–elongation test graphs. Source: ENSINGER UK-USA. For a color version of this figure, see www.iste.co.uk/grous/design.zip*

10.2. Jointing of a ball bearing in a metal casing

The forced jointing of ball bearings is an important part in mechanical assemblies. Aside from calculating the classic fits, we focus here on calculating the resistance as the duality between load and geometry. Designers of plastic products are then faced with environmental issues, questions of form and dimensions. This duality is at the heart of the calculations and the tolerance values are tricky, depending on the dimensions. By applying relation [10.1], we find an assembly pressure as a function of the defined over-dimensioning (= 1.67). Below is a solved example:

$$P_2(S) = E^t_{relaxation} \times \kappa_{form} \left(S/d_1\right) \qquad [10.1]$$

where:

– $E^t_{relaxation}$ is the material's relaxation modulus over time, POM = 3800 MPa;

– S = 3 mm, 3.5 mm, …, 6 mm. S is the over-dimensioning as a function of the material and the ∅ d_1;

– d_1 = 25 mm, d_0 = 21 mm, $\varphi = d_1/d_0$ is the ratio between the diameters ≥ 1.2 for plastic ball bearings assembled by force in a metal bore $\delta = (d_1/d_0) = (25/21) = 1.19 > 1.29$ (source of documentation: Hoechst);

– K_{form} is the form coefficient = 1.670 (POM according to Hoechst);

– the assembly pressure $P_2(S)$ as a function of the over-dimensioning S.

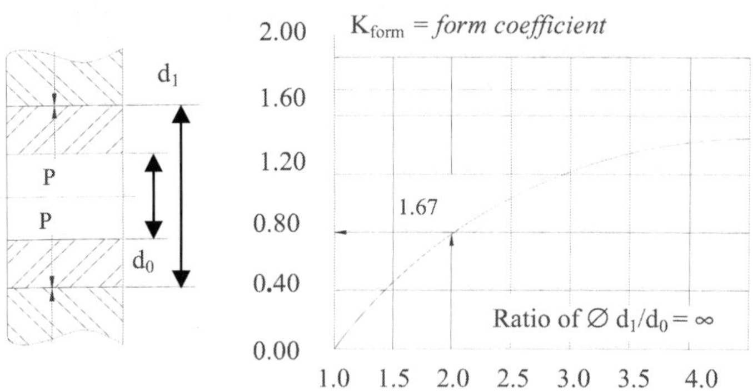

$P_2(S) =$ MPa

	0
0	781.56
1	911.82
2	$1.042 \cdot 10^3$
3	$1.172 \cdot 10^3$
4	$1.303 \cdot 10^3$
5	$1.433 \cdot 10^3$
6	$1.563 \cdot 10^3$

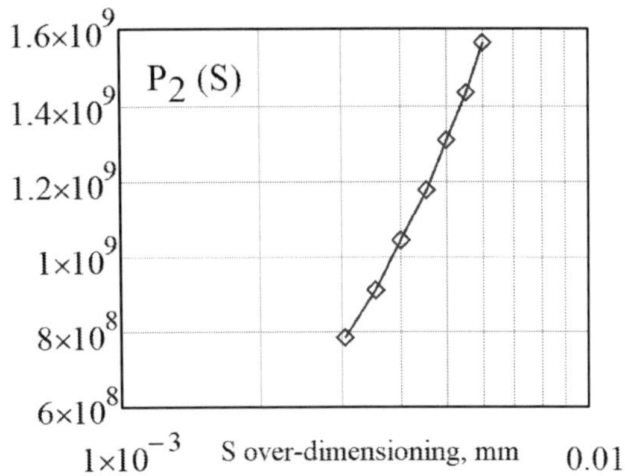

Figure 10.2. *Results of the assembly pressure as a function of the over-dimensioning. For a color version of this figure, see www.iste.co.uk/grous/design.zip*

10.3. Cylindrical clip of PP (e.g. blinds): force exerted

According to the documentation from the manufacturer Hoechst, if the strain of the components is imperfectly known, it is difficult to devise the calculations of a cylindrical clip. We present the calculations of the forces $F_{1\text{-}2}$ (*N* and *lbf*) and pressure (P, in MPa and psi):

– E is the elasticity modulus of the material under stress (pp) = 3600 MPa = 5.221×10^5 psi;

– ε_{adm} = (0.01 to 0.04) is the maximum admissible strain to the housing (pp);

– $\mu = 0.4$ is the pp to pp friction coefficient;

– $K = 3.0$ is the form coefficient (below table inspired by Hoechst for the pp values);

– l_1 = 1.25 mm is the click width;

– l_2 = is the total click width (?);

– d_2 = 12 mm is the external ∅ of the axis;

– d_1 is the minimum ∅ of the click fitting (?);

– H is the click height (?).

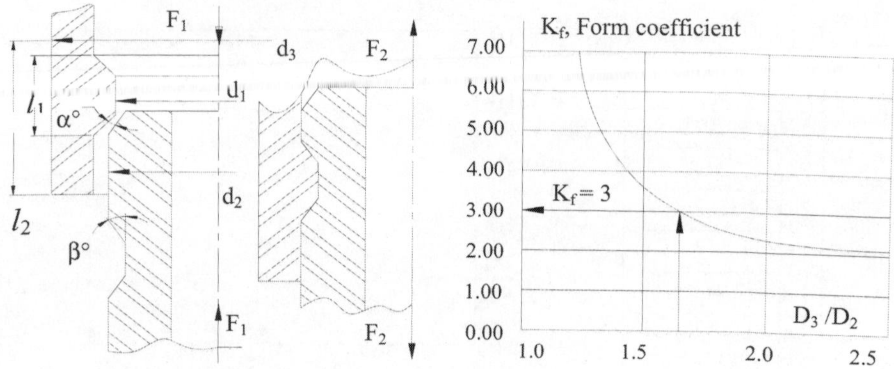

Figure 10.3. *Assembly forces for a cylindrical clip. Form coefficient K_f. For a color version of this figure, see www.iste.co.uk/grous/design.zip*

Indicative characteristics of (α) and (ε) according to data from Hoechst						
Angle, (α)	8°	11.5°	14°	16.3°	20°	23°
Strain (ε)	1/100	2/100	3/100	4/100	6/100	8/100

Table 10.1. *Indicative characteristics of (α) and (ε) (source: Hoechst)*

SOLUTION.– From Table 10.1 for (α) = 14° → we read ε = 3/100 → ε = ?

a) *Calculation of the real aperture diameter (d_1) (mm, in)*:

$$\varepsilon = \left[1-(d_1/d_2)\right] = 3\% \rightarrow d_1 = d_2(3\%+1) = 12(3\%+1) = 12.36\ mm = 0.472\ in \quad [10.2]$$

b) *Calculation of the click height, H*:

$$H_{click} = \frac{d_1 - d_2}{2} = \frac{12.36-12}{2} = 0.18\ mm = 7.087\times10^{-3}\ in \quad [10.3]$$

c) *Calculation of the click width, $l_2(\alpha)$:*

$$l_{2-click}(\alpha) = \frac{d_2 \sin(\alpha)}{2} = \frac{12\sin(14°)}{2} = 1.452\ mm = 0.057\ in \qquad [10.4]$$

d) *Strain (ε) initially chosen as a function of (α):*

$$\varepsilon_{deformation}(\alpha) = \left|1 - \frac{d_1}{d_2}\right| = 0.03\ and\ \varepsilon_{deformation}(H, d_2) = \left|\frac{2H}{d_2}\right| = 0.03\ \text{QED} \qquad [10.5]$$

e) *Second proof of confirmation of choice of strain (ε):*

$$\varepsilon_{deformation}(\alpha) = 1 - \cos(\alpha) = 1 - \cos(14°) = 0.03 \rightarrow QED \qquad [10.6]$$

$$2l_2(\alpha) = d_2 \times \sin(\alpha) \rightarrow l_2(\alpha) = d_2 \sin(\alpha)/2 = 1.452\ mm = 0.0057\ in$$

f) *Calculation of the pressure exerted on the assembly p:*

$$P_{1/2} = \frac{2H \times E}{d_1 \times K} = \frac{2 \times 0.18 \times 3600}{12.36 \times 3.0} = 34.951\ \text{MPa} = 5.069 \times 10^3\ psi \qquad [10.7]$$

g) *Calculation of the clipping assembly force, $F_{1/2}$:*

$$F_{1/2} = (\pi \times d_2 \times p \times 2l_1) \times \left(\frac{\mu + \tan(\alpha_{1/2})}{1 - \mu \times \tan(\alpha_{1/2})}\right) N\ or\ lbf \rightarrow ... \qquad [10.8]$$

$$F_{1/2} = (\pi \times 12 \times 34.951 \times 2 \times 1.25) \times \left[0.4 + \tan(\alpha_{1/2})\right] / \left[1 - 0.4 \times \tan(\alpha_{1/2})\right] \downarrow$$

Let us vary $\alpha_{1/2}$ from [8° to 23°] and calculate the retention force for the series of assembly angles (retention values) the result of which are shown below. We can see how the retention force increases in proportion to the assembly angle. It depends, in fact, on the angle, pressure and clipping width.

$\varepsilon_1(\alpha_{12}) =$

	0
0	0.03
1	0.03
2	0.03
3	0.03
4	0.03
5	0.03
6	0.03
7	0.03
8	0.03
9	0.03
10	0.03
11	0.03
12	0.03
13	0.03
14	0.03
15	0.03

$l_2(\alpha_{12}) =$

	0
0	0.835
1	0.939
2	1.042
3	1.145
4	1.247
5	1.35
6	1.452
7	1.553
8	1.654
9	1.754
10	1.854
11	1.953
12	2.052
13	2.15
14	2.248
15	2.344

$F_{12}(\alpha_{12}) =$ N

	0
0	$1.887\cdot10^3$
1	$1.964\cdot10^3$
2	$2.043\cdot10^3$
3	$2.123\cdot10^3$
4	$2.205\cdot10^3$
5	$2.29\cdot10^3$
6	$2.376\cdot10^3$
7	$2.464\cdot10^3$
8	$2.555\cdot10^3$
9	$2.649\cdot10^3$
10	$2.745\cdot10^3$
11	$2.844\cdot10^3$
12	$2.945\cdot10^3$
13	$3.051\cdot10^3$
14	$3.159\cdot10^3$
15	$3.271\cdot10^3$

$$F_{1/2} = (\pi d_2 p 2 l_1)\left(\frac{\mu + \tan(\alpha_{1/2})}{1 \ \mu \tan(\alpha_{1/2})}\right)$$

Consider: $l_2(\alpha_{12}) \equiv 1.452\text{mm}$

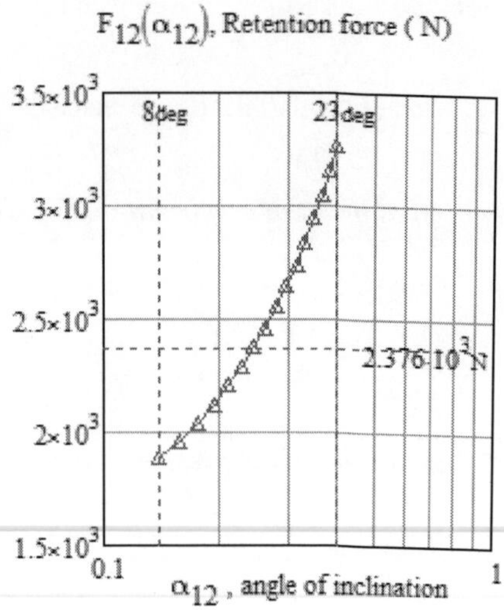

Figure 10.4. *Retention force $f_{1/2}$ as a function of the angle of inclination. For a color version of this figure, see www.iste.co.uk/grous/design.zip*

10.3.1. *Spherical clip of a PP: force exerted*

On a spherical clip in the form of a ball joint made of PP (polypropylene) we impose a force of F = 120 N = 26.977 lbf; E = 4400 MPa = $6.382.10^5$ psi; d_2 = 10 mm = 0.394 in; d_3 = 14 mm = 0.551 in and α = 8 degrees and μ = 0.4 is the friction coefficient (also see Chapter 11, case study in dental hygiene).

a) *Calculation of the retention diameter d_1 in mm and in*:

$$D_1 = d_2 \times (1 - \varepsilon) = 9.9 \ mm = 0.39 \ in \tag{10.9}$$

b) *Retention pressure in MPa and psi*:

$$P = (E/k) \times (d_2 - D_1)/D_1 = 10.582\ MPa = 1.535 \times 10^3\ psi \qquad [10.10]$$

c) *Retention force F (N, lbf) where (l_{click}) is the click width in (mm, in)*:

$$l_{click} = d_2 \times \sin(\alpha)/2 = 0.696\ mm = 0.027\ in \qquad [10.11]$$

$$F_{click} = \pi d_2 \times 2 l_{click} \times P \left(\frac{\mu + tan(\alpha)}{1 - \mu \times tan(\alpha)} \right) = 264.99\ N = 59.572\ lbf \qquad [10.12]$$

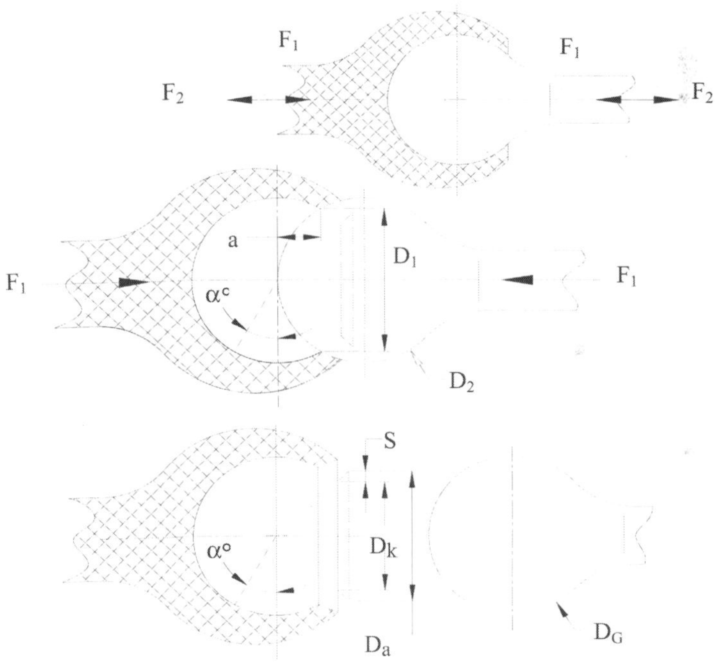

Figure 10.5. *Assembly forces for a spherical clip. For a color version of this figure, see www.iste.co.uk/grous/design.zip*

d) *By varying d_2 between 2 and 10 mm, we plot the graph of $F = f(d_2)$*:

F(d2) =

	0
0	-159.197
1	-238.795
2	-318.394
3	-397.992
4	-477.59
5	-557.189
6	-636.787
7	-716.386
8	-795.984

·N

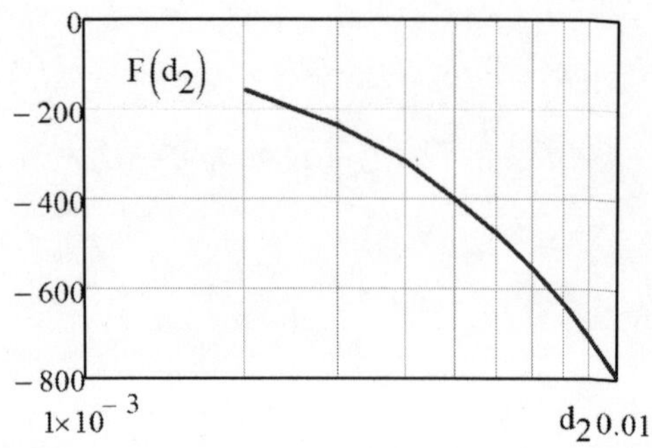

Figure 10.6. *Illustration of the clipping force as a function of the diameter d_2. For a color version of this figure, see www.iste.co.uk/grous/design.zip*

10.4. Types of clip fitting: counter-cylindrical cantilever

Among the simple and less costly methods for holding two parts together, it is quite common to use *clips*. These clips can be repeatedly placed and removed without damaging the whole assembly. Such assembly methods are environmentally friendly. It is simply important to be conscious of the limitations of these attachment systems. We must check the clearance fit, due to the accumulation of the tolerances of the two elements of the clip and an insufficient traction force. The majority of applications use the cantilever configuration (e.g. blinds). The cylindrical config-uration is used on non-reinforced plastic materials. The diagrams below illustrate some types of clip.

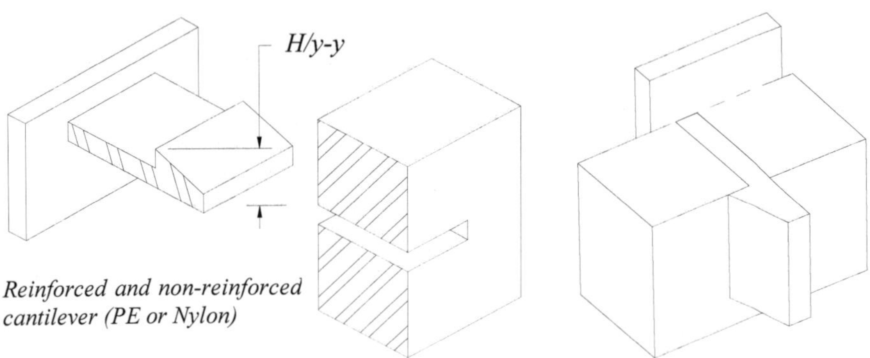

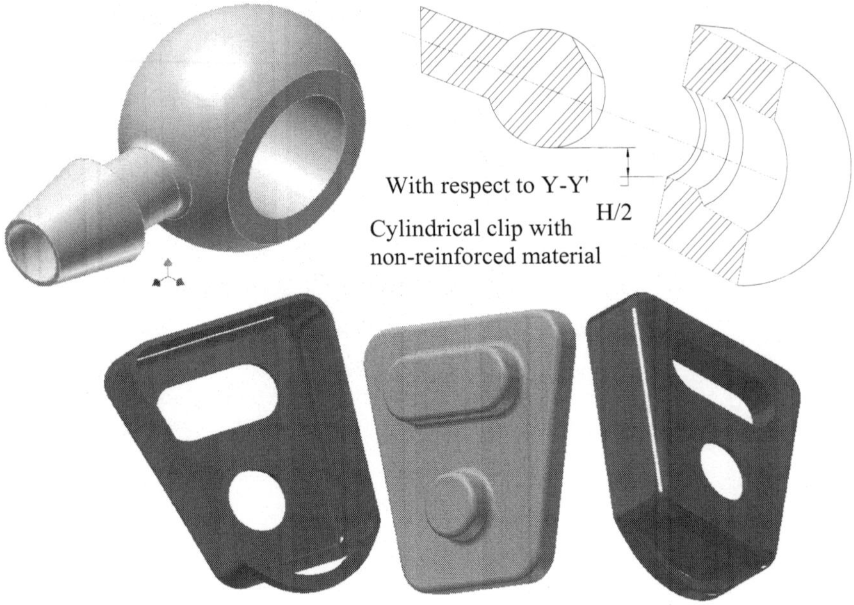

Figure 10.7. *Diagrams of clips (also see Chapter 11, Project for an HD spherical clip). For a color version of this figure, see www.iste.co.uk/grous/design.zip*

In the design of a cantilever attachment, designers need to carry out numerous procedures, such as adjusting the length, thickness, size of the deformations, etc., in order to produce a clip where the stresses are lower than those tolerated by the material.

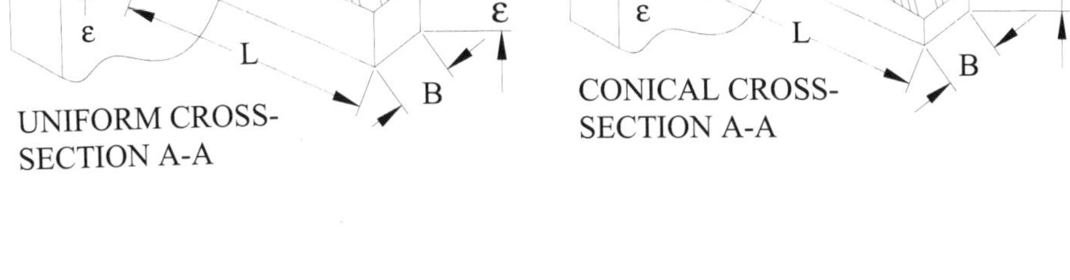

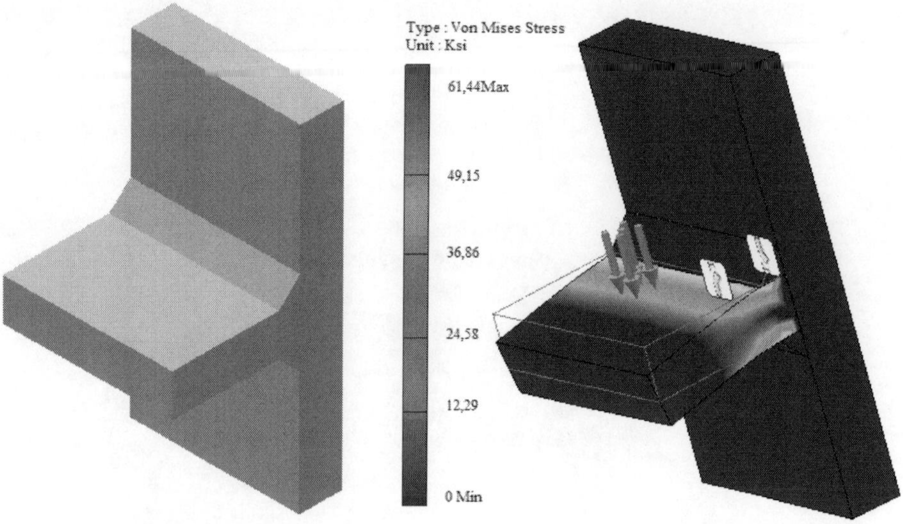

Figure 10.8. *Cantilever compared to the cylindrical clip. For a color version of this figure, see www.iste.co.uk/grous/design.zip*

10.4.1. *Conical cantilever*

The configuration with a uniform cross-section is sufficient in most applications for cantilever clips (see Figure 10.9). It is useful to include a conical element when greater deformation is needed:

$$\left\{\text{Uniform cross-section} \rightarrow H = \frac{3}{2}\frac{\varepsilon}{L^2}\right\}, \left\{\text{Conical cross-section} \rightarrow H = \frac{92}{100}\frac{\varepsilon}{L^2}\right\} \quad [10.13]$$

10.4.2. *Short cantilever*

Conventional cantilever clips are designed for lesser strains (deformations) than those observed in the area of short cantilevers. In the conventional formulae, the wall from which the fixation strip is projected is supposed to be rigid. The same is also true for long cantilevers, but the principle does not actually hold true with short ones. The intersecting wall actually deforms under the load of short cantilevers. The formulae designed to calculate the maximum stress, deformation and force required to assemble the parts are also indicated. There is a dedicated manual for clips.

10.4.2.1. *Configuration of strips and factor (Q). Maximum deformation (Figure 10.9c)*

$$\varepsilon_{max}:\ \varepsilon_{max} = (3/2)\times\left(\varepsilon H/L^2 Q\right) \quad [10.14]$$

where:

- ε_{max}, maximum deformation at the base, mm;
- E, thickness of the strip;
- H, deformation in mm;
- L, Length of the strip, mm;
- λ, coefficient of strain amplification (manufacturers' tables).

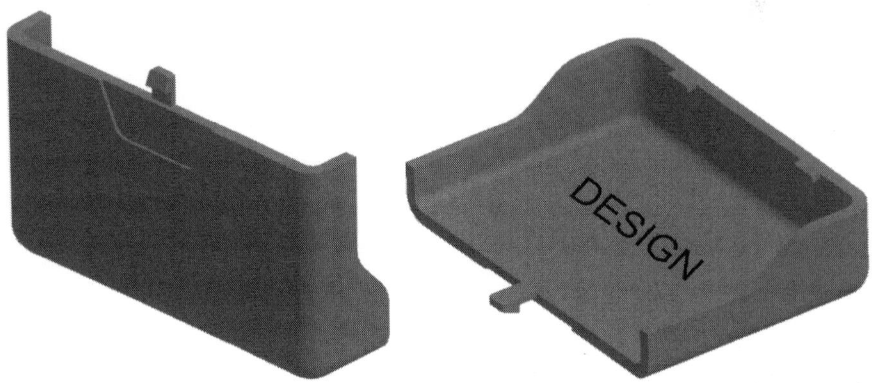

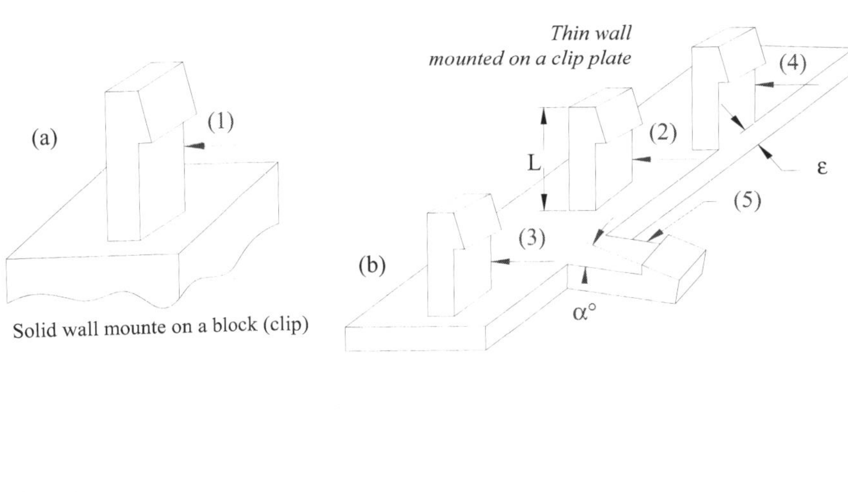

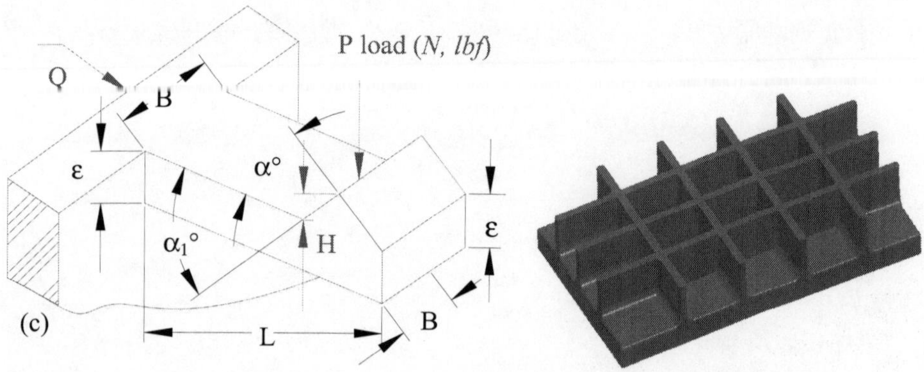

Figure 10.9. *(a) Solid wall. (b) Thin wall. (c) Influence of factor Q. Ribs. For a color version of this figure, see www.iste.co.uk/grous/design.zip*

Consider a PET plastic. Determine the maximum deformation of the strip and the coupling force on it (Figure 10.9c).

Maximum deformation of the strip (in mm or in)

$$\begin{cases} \varepsilon_0 = \dfrac{3}{2}\dfrac{\varepsilon H_{max}}{\lambda L^2} & [10.15] \\ H_{max} = \dfrac{\varepsilon_0 \lambda L^2}{1.5\varepsilon}; \dfrac{L}{\varepsilon} = \dfrac{1}{2} \\ \lambda_{tab} = 2 \rightarrow H_{max} = \\ = \dfrac{0.015 \times 15^5 \times 2.0}{1.5 \times 3} = 1.5\ mm \end{cases}$$

- $Q = F$ force of thrust;
- $Q' = F_1$ force of traction;
- P = perpendicular force;
- μ = friction coefficient = 0.2;
- α = pitch angle = 30°;
- α_1 = return angle;
- B = width of strip = 6 mm;
- ε = thickness of strip = 3 mm;
- E = bending modulus = 9000 MPa;
- σ_{max} = maximum stress;
- L = length of strip = 15 mm;
- λ = amplification of deformation;
- ε_0 = relative elongation = 1.5 %.

SOLUTION.– In a real design, we need to choose a lower deformation value (Y) in order to maintain a greater safety margin. The coupling force (F) is:

$$\left\{ F = P\frac{\mu + \tan(\alpha)}{1 - \mu \tan(\alpha)} = 35.6N \text{ where } P = \frac{\varepsilon_0 B \varepsilon^2 E}{6L\lambda} = 40.5N \right\} \qquad [10.15]$$

Thus, we need a force of F = 35.6 N to drive the pressure in the assembled position.

10.5. Configuration of strips: two-dimensional spline interpolation

The strip illustrated below was actually designed and built in the modular tool design lab by the author. Then, in the metrology lab, the author used spline modeling to carry out an interpolation of 2D functions to obtain "copying", over the chosen section of the original surface, before attempting to juxtapose it with the strip manufactured by machining (of 6061 steel).

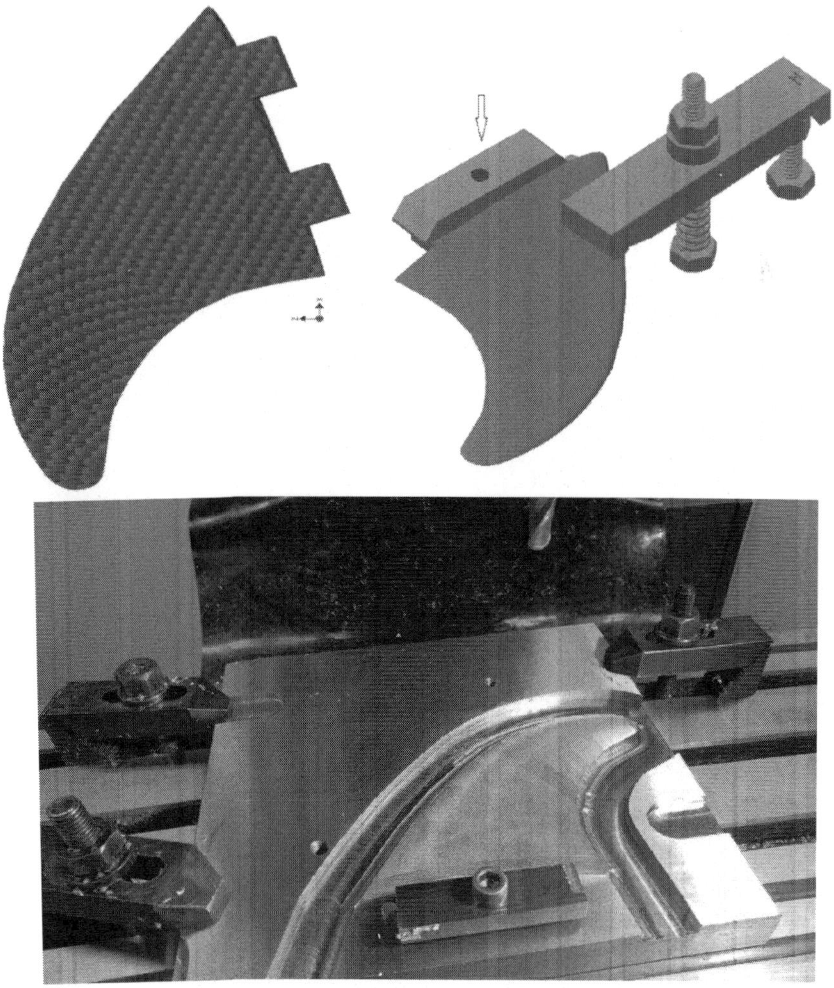

Figure 10.10. *Strip drawn with Inventor Pro by "loft and sweep". For a color version of this figure, see www.iste.co.uk/grous/design.zip*

This photograph shows our actual machining setup. On a CMM (3D Coordinate Measuring Machine), we scanned numerous points on the surface of the original piece (made of carbon fiber). To facilitate the case study, below we present a matrix Mz specifying the scanned surface:

Mz = mesh matrix (real scan)				X	Y
3.99490	6.52920	7.38410	7.67590	0	0
4.07870	6.97300	7.18870	7.74430	1	1
4.93630	7.30500	7.36520	7.63890	2	2
5.81910	7.48230	7.51090	7.58730	3	3

Table 10.2. *Points scanned on CMM using the KT1000 software published by Mitutoyo*

The number of rows in the matrix must be equal to the number of columns: 4 by 4. We used our mathematical program as follows: we specify X and Y and (n) vectors which determine the mesh of the matrix drawn from our measures in metrology [GRO 11, GRO 13].

$$\begin{cases} \{rows(Mz)=4 \quad cols(Mz)=4 \quad n = rows(Mx)\} \\ Mxy = augment\left[sort(X),\ sort(Y)\right] \rightarrow rows(Mxy) = 4 \\ Computed\ spline\ coefficients \rightarrow coef = cspline(Mxy,\ Mz) \\ Fitting\ function\ for\ surface \rightarrow fit(x,y) = \text{interp}\left[coef, Mxy, Mz, \begin{pmatrix} x \\ y \end{pmatrix}\right] \end{cases}$$

$$Values\ of\ interpolated\ samples: \rightarrow \begin{Bmatrix} fit(2.5,3.9)=9.434 & fit(0.1,1.7)=7.218 \\ xlow = Mxy_{0,0} & xhigh = Mxy_{n-1,0} \\ ylow = Mxy_{0,1} & yhigh = Mxy_{n-1,1} \end{Bmatrix}$$

Mesh density of the interpolation: $\{xn = 4nn\}$

$$\begin{cases} i = 0, \ldots \mathrm{x}n-1 \quad xind_i = xlow + i \times \dfrac{xhigh - xlow}{\mathrm{x}n-1} \\ j = 0, \ldots \mathrm{y}n-1 \quad yind_j = ylow + j\dfrac{yhigh - ylow}{\mathrm{y}n-1} \quad and\ FIT_{ij} = fit(xind_i,\ yind_j) \end{cases}$$

10.5.1. *Graphs of the model of the original surface*

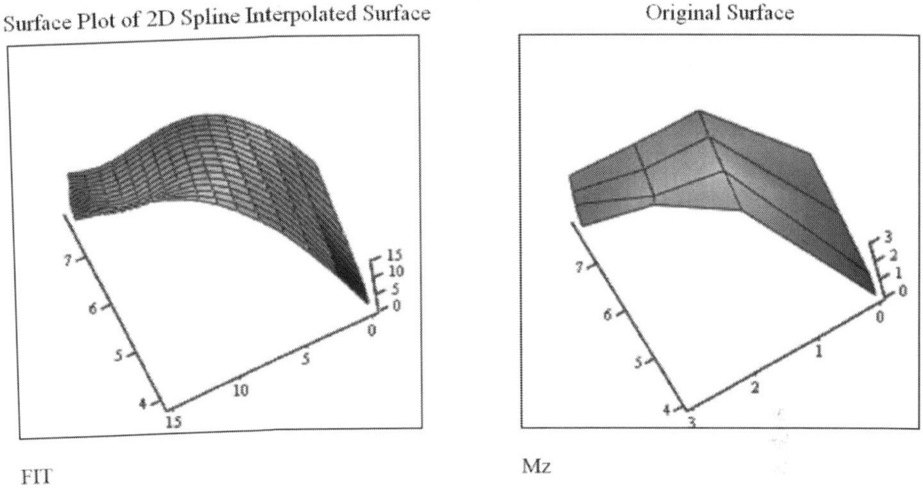

Figure 10.11. *Results of the mathematical interpolation to find the original surface. For a color version of this figure, see www.iste.co.uk/grous/design.zip*

10.6. Press assembly

It is possible to assemble two parts by pressure. This method, which is tried, tested and trusted in assembling metals, is trickier with thermoplastics. The creep of the plastic materials (or relaxation of the stresses) leads the *designer* to take account of a significant reduction in the initial pressing forces of the pressure system. A good design will minimize the stresses imposed on the plastic. It takes account of the accumulation of the tolerances and estimates the residual pressing force resulting from the relaxation of the stresses.

Press fitting for two identical materials (radial deformation):

– the radial deformation (δ) is doubled to adapt to the interference of a ∅;

– (a), (b), and (c) are radii, rather than diameters, ∅;

– the formula is valid only if the shaft and the housing are of the same material;

– a full analysis of the creep is needed.

$$\delta = \frac{2\varepsilon \times b^3\left(c^2 - a^2\right)}{\left(c^2 + b^2\right)\left(b^2 - a^2\right)} \ if \ a = 0 \rightarrow \delta = \frac{2\varepsilon b^3 c^2}{\left(c^2 + b^2\right)} \qquad [10.16]$$

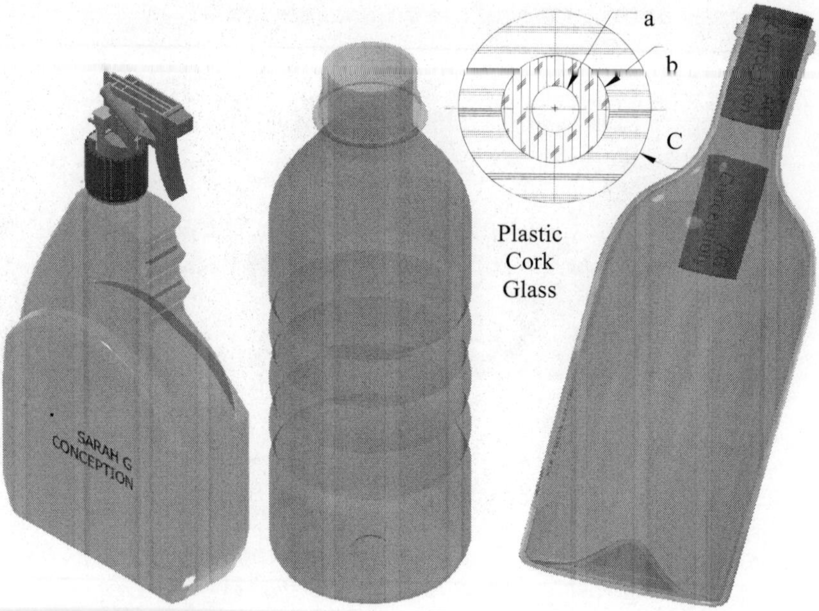

Figure 10.12. *System for press fitting of an axle into a housing. For a color version of this figure, see www.iste.co.uk/grous/design.zip*

Press fitting for two different materials (radial deformation):

– the radial deformation δ must be doubled to fully adapt to the interference ∅;

– (b) and (c) are radii, rather than diameters, ∅;

– the formula is value only if the shaft and housing are made of the same material;

– a full creep analysis is needed.

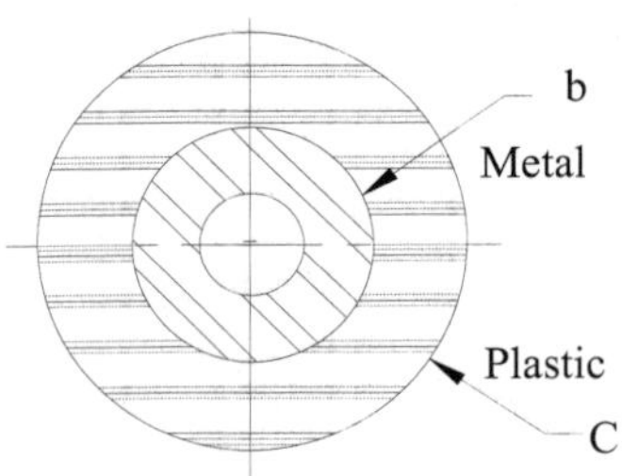

$$\delta = b\varepsilon \frac{\left(c^2+b^2\right)}{\left(c^2+b^2\right)}\left(\frac{c^2+b^2}{c^2+b^2}+\nu_{plast}\right)$$

ε is deformation
δ is radial deformation
ν is Poisson's ratio

Figure 10.13. *System for press fitting an axle into a housing. For a color version of this figure, see www.iste.co.uk/grous/design.zip*

NUMERICAL APPLICATION.– Consider a metal insert (O.D. = 12.5 mm), which needs to be affixed to an embossment made of PA 6/6 (O.D. = 20 mm). Determine the maximum interference in the setup using an acceptable stress of 2 % for the PA 6.

$$\delta = b\varepsilon \frac{c^2 + b^2}{c^2 + b^2}\left(\frac{c^2 + b^2}{c^2 + b^2} + \nu_{plast}\right) = 6.25 \times 0.02 \frac{10^2 + 6.25^2}{10^2 + 6.25^2}\left(\frac{10^2 + 6.25^2}{10^2 + 6.25^2} + 0.35\right) = 0.144\ mm$$

COMMENT.– The PA 6 embossment must be identified by a maximum interference of 0.29 mm (2d) or 12.21 mm minimum diameter.

10.7. Reduction of stress relaxation: bolts and self-tapping screws

Although self-tapping screws are the most commonly used, it is also common to use metal fixings to assemble thermoplastic components. In doing so, we must be careful not to apply too much compressive stress to the plastic material:

– the assembly must remain within controlled torque limits;

– high temperatures will exacerbate the problem of relaxation. A powerful moment increases the compressive stresses. This results in a rapid initial *relaxation* of the stresses, followed by stabilization. The relaxation is very significant in the case of high stress levels;

– a large-diameter screw head, plus a large-diameter metal washer under the screw head and/or the nut increases the load-bearing surface area and decreases the stresses;

– flathead screws and rivets are to be avoided with plastics. These *conical fixings* act as vertices, and can break or tear the part;

– when the stresses are relaxed, the force of *pressing* and *retention* of the moment decrease. The fixation is unpressed. An elastic or spring washer enables the system to maintain acceptable levels of pressure and torque. The majority of the pressing is metal on metal, although a lesser force is exerted on the plastic.

We have designed a fishing reel and noticed that the reduction of relaxation stresses has a not-insignificant role to play, depending on the materials used.

Figure 10.14. *Fishing reel designed to reduce the relaxation of the stresses. For a color version of this figure, see www.iste.co.uk/grous/design.zip*

10.8. Case study: piping link

According to ISO 1043, POM (Poly-Oxy-Methylene, or poly-formaldehyde) is a polymer in the poly-acetyl family. It is used in many industries (automobile, aeronautics, sports, etc.). Its structure and high crystallinity give it excellent physical characteristics:

– good dimensional stability;

– high traction resistance, shock resistance and fatigue resistance;

– good operational temperature range;

– good resistance to chemical agents and to creep;

– good electrical insulation characteristics;

– good abrasion resistance with a low friction coefficient;

– POM cannot be used as a permanent medical implant;

– POM is unstable when exposed to oxidants and to organic and inorganic acids.

Average physical properties of POM		Values
Failure resistance under traction		70 MPa
Failure resistance under bending		110 MPa
Continuous heat resistance (° C)		[–40 to 115]°C
Traction resistance DIN 53455		68 to 700 MPa
Cold resistance temperatures		< [40–50]°C
Elongation to failure according to DIN 53455		[12–40] %
Limiting bending tension according to DIN 53452		97 MPa
Failure resistance under compression		110 MPa
Elasticity modulus under traction, E		3100 MPa
Elasticity modulus under bending, E DIN 53457		2950 MPa
Poisson's ratio, ν		0.35
Linear thermal expansion coefficient, DIN 52328		9 [1/K]
Dry friction coefficient μ		[0.25 to 0.32]
Index of wear through slip		4.6 μm/km
Creep modulus		1400 MPa
Molding shrinkage, L/T (%)		[1.4 to 1.3]
Specific weight according to DIN 53479		1.42 g/cm^3
Resilience according to DIN 53453		6.5 J/m^2
Shore hardness		Scale D
Rockwell hardness with a pressure from the H358 bearing in accordance with DIN 53456		Scale R 120 130 MPa
$\varepsilon = S/d_1$ for material (POM) for casings		Admissible deformation ratio
$d_1 \leq 5$ mm →	0.050	
5 mm < $d_1 \leq$ 30 mm →	0.030	
$d_1 > 30$ mm	0.005	
Clip in with hook	0.080	
Cylindrical assemblies	0.040	

Table 10.3. *Average physical properties of POM. DIN standards*

STATEMENT OF THE PROBLEM.– Consider a connecting pipe made of POM, used in simple piping, of diameter d = 75 mm, at a pressure p = 1 MPa over the course of ten years, at the average temperature of 20°C. Tables and other breakdowns of experimental results show that the resistance over ten years at 20°C is 16 MPa with a safety coefficient of S = 2.5. Knowing the maximum stress σ_{max} of 7 MPa applied to the pipe, calculate the thickness (ε) needed for it to withstand the ravages of time and the environmental conditions:

– d = 70 mm = 2.756 in is the internal diameter;

– S = 2.5 is the safety coefficient;

– p = 1 MPa = 145.038 psi at a pressure p, over ten years at 20°C;

– σ_{max} = 7 MPa = 1.015×10^3 psi is the maximum stress applied.

$\varepsilon_t(d_v) =$

2.857
3.429
4.000
4.571
5.143
5.714

·mm

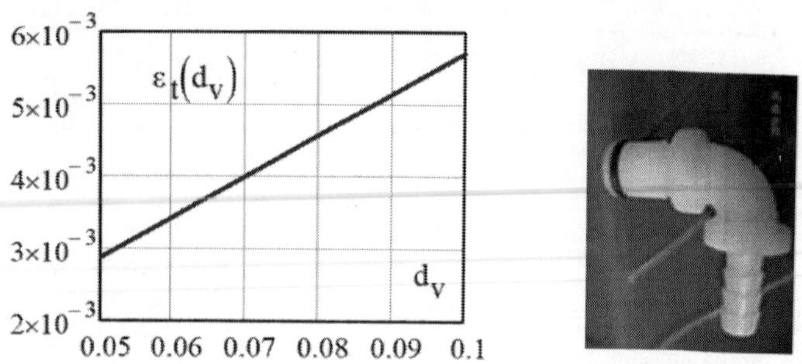

Figure 10.15. *Thickness of connecting pipe (mm) of POM (garden sprinkler system). For a color version of this figure, see www.iste.co.uk/grous/design.zip*

The thickness (ε) is written: $\varepsilon_{tube} = (pd)/(S\sigma_{adm}) = (1\times75)/(2.5\times7) = 4\ mm = 0.157\ in$

or indeed: $\varepsilon_v(d_{\text{variable}}) = pd_{\text{variable}}/S[\sigma]_{adm}$; $for\ d_{\text{variable}} = 50\ mm\ to\ 100\ mm$.

10.9. Assembly by forced jointing

STATEMENT OF THE PROBLEM.– In the design of forced jointing structures, it is important to take account of the axial force. Thus, the maximum axial force is expressed as follows:

$$Maximum\ axial\ force\ F_{\max}^{axial} = \pi\times\mu\times D_1\times L\times p\ [N] \quad [10.17]$$

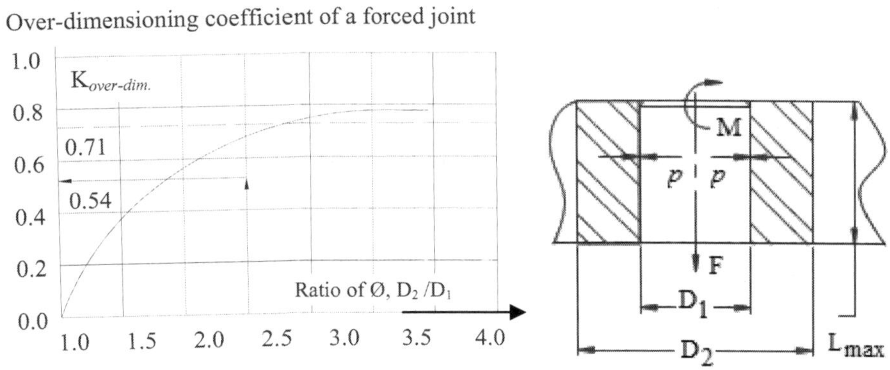

Figure 10.16. *Over-dimensioning coefficient on a forced joint. For a color version of this figure, see www.iste.co.uk/grous/design.zip*

where:

– μ is the friction factor (POM) = 0.3;

– L is the width in mm or in = 15 mm = 0.519 in;

– D_1 is the diameter of the inserted element in mm or in, 12 mm = 0.472 in;

– p is the assembly pressure for two products in MPa or psi, (to be calculated);

– F_{max} is the maximum force applied in N or lbf, (to be calculated);

– M is the torque (of forces) applied in N.m or lbf.in (to be calculated);

– S is the over-dimensioning in mm or in (see Figure 10.16 ≡ D_2-D_1) (to be calculated or deduced);

– E_{relax} is the relaxation modulus of the POM = 800 MPa = 1.16×10^5 psi over ten years.

Maximum torque: $M_{\max} = \pi \times \mu \times \left(D_1^2/2\right) \times L \times p \; [N.m]$ [10.18]

Pressure as a function of the over-dimensioning:

$$p = E_{relax}^{ther} K_{\text{over-dimensioning}} \left(S/D_1\right) \quad [10.19]$$

In joints formed of plastic parts, it is suggested to take the ratio of the diameters as follows: D_2/D_1>1.6. Over the course of around ten years, a torque of 5N is transmitted to a forced joint (POM) from the axle of an engine. D_1 = 12 mm and L_{max} = 18 mm. We wish to devise the dimensions of the joint in such a way that the design is viable.

SOLUTION.– Calculation of the diameter ratio gives us: $(D_2/D_1) = 1.667$, so:

$[D_2/D_1] > 1.6$; as $D_1 = 12$ mm $= 0.472$ in, we isolate D_2 as follows;

– $[D_2/D_1] > 1.6 \rightarrow D_2 > 1.6 \times 12) \rightarrow D_2 > 19.2$ mm $= 0.756$ in;

– from $[D_2/D_1] = 1.667 \rightarrow$ we read $K_s = 0.4$ (see Figure 10.17).

The deformation ε of the POM is 3 %, while the over-dimensioning is estimated at S = 0.36 mm = 0.014 in. $\varepsilon = 3\ \% \rightarrow S = D_1.\varepsilon = 0.36$ mm $= 0.014$ in. The thermal relaxation modulus E_r of POM over ten years of operation is 800 MPa $= 1.16 \times 10^5$ psi. The jointing pressure P_{joint} is:

$$p = E_r \times K_S \times (S/D_1) = 9.6 MPa = 1.392.10^3\ psi\ .$$

The friction coefficient μ = 0.3. The static friction coefficient, of plastic on metal = (0.25 to 0.4) and plastic on plastic is (0.3 to 0.4). Over ten years of operation, according to the hypothesis, the force torque must be greater than or equal to 5 N.m. Let us verify this condition for $L_{max} = 15$ mm $= 0.591$ in, and calculate the force torque.

Calculation of the moment: $M = \pi \times (D_1^2/2) \times L_{max} \times p \times \mu = 9.772\ N.m = 86.486\ lbf.in$

This torque M = 9.772 N.m is almost twice its hypothetical value (5 N.m). The structure will therefore be able to survive for longer. Hence, the joint is safe to use. Finally, the maximum force, F_{max}, exerted on the structure would be:

Calculation of the maximum force: $F_{max} = \pi L_{max} D_1 p \mu = 1.629\ N = 366.124\ lbf.$

As the length increases (L_{max}), the maximum force does too. Let us vary (L_v) from 10 mm to 18 mm and calculate: $F_{max}(L_V) = \pi \times L_V \times D_1 \times p \times \mu\ [N]\ and\ [lbf]$.

$F_{variable}(L_V) =$ N

	0
0	$1.086 \cdot 10^3$
1	$1.194 \cdot 10^3$
2	$1.303 \cdot 10^3$
3	$1.411 \cdot 10^3$
4	$1.52 \cdot 10^3$
5	...

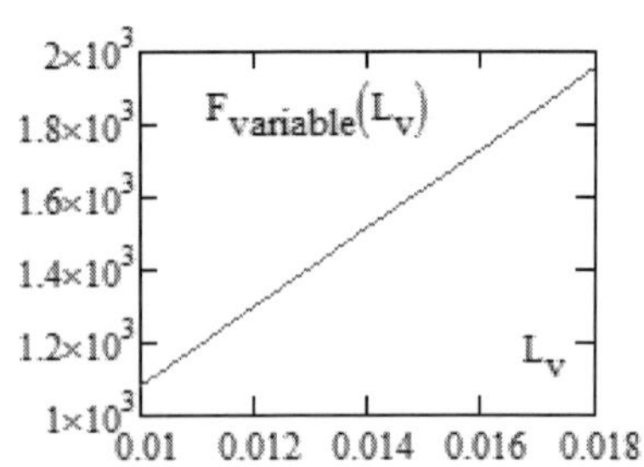

Figure 10.17. *Forces as a function of the L_V and the forced jointing pressure. For a color version of this figure, see www.iste.co.uk/grous/design.zip*

10.10. Stress and thermal swelling in assembled materials

Ordinarily, thermal stresses are not cause for concern, except in the case of dismountable materials forming a structure to withstand variations in temperature, such as a thermoplastic affixed to metal.

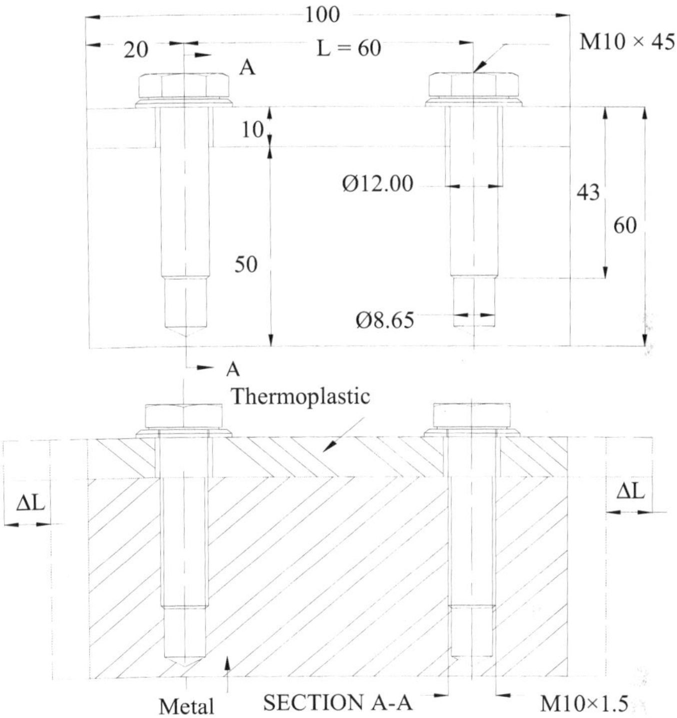

where:

– σ_{cc} is the critical compression stress;

– A_1 is the area of the transverse cross-section of the material of surface state nuance 1;

– J is the moment of inertia of material 1;

– E is the elasticity modulus;

– λ_1 = coefficient of thermal expansion of material 1;

– λ_2 = coefficient of thermal expansion of material 2;

– ΔL = change of length;

– L is the length between the two tips;

– ΔT = is the temperature variation.

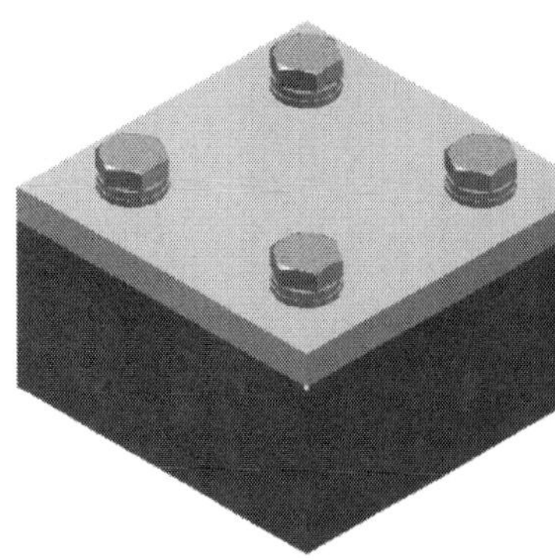

Figure 10.18. *Thermal stress and expansion. For a color version of this figure, see www.iste.co.uk/grous/design.zip*

The material whose expansion rate is higher tends to buckle, because of compression. The resistance tolerance values of the material with the higher expansion rate must be less than the compression loads resulting from the swelling. According to Euler's law, the critical buckling load is written:

$$P_{Euler} = \left(4\pi^2 EJ\right)/L^2 \text{ we set } \sigma_{cc} = P_{Euler}/A = \left[\left(4\pi^2 EJ\right)/L^2\right]/A \qquad [10.20]$$

The difference in thermal expansion between two different materials (1 and 2) is: $\Delta L = (\lambda_1 - \lambda_2)\Delta\tau L$. The thermal stress is expressed as follows:

$$\sigma_{therm} = (\lambda_1 - \lambda_2)\Delta\tau \times E \qquad [10.21]$$

COMMENT.– Buckling occurs if $\sigma_c > \sigma_{cc}$. To prevent it, the designer must add bolts or increase the modulus of the cross-section (*J/c*) of material (1). Sometimes, if a clearance is left between the bolts and the perforations, this can allow for the possibility of some motion. The effects of relaxation of the stresses on plastic parts are, typically, represented in the documentation published by manufacturers. It is wise to take certain precautions when it comes to *shocks*. This situation of stress is applied to the piece for a very short period of time. One of the most important factors is to reduce *stress intensifications*. Even embossments cause stress intensifications and lead to failures due to shocks. If (*h*) is a height of the drop and (y) is the static deformation, we can prevent warping by calculating the static deformation, with a view to deducing an amplification factor:

$$K_1 = 1 + \sqrt{1 + 2h/y_{static}} \qquad [10.22]$$

The next example is the simulation of a shock of P = 6.678 N (m = 0.681 kg) falling from a height of h = 1.25 m at the center of a circular disc of 80 mm, entirely maintained, of thickness e = 5.56 mm. We need to determine the static deformation and the stress using the formulae applied to flat plates, shown below:

$$\begin{cases} y_{static} = \dfrac{-3P\left(m^2-1\right)r^2}{16\pi E m^2 e^3} \text{ and } \sigma_{static} = \dfrac{3P}{4\pi e^2} \\ y_{dynamic} = K_{dyn} y_{static} \text{ and } \sigma_{dynamic} = K_{dyn}\sigma_{static} \end{cases} \qquad [10.23]$$

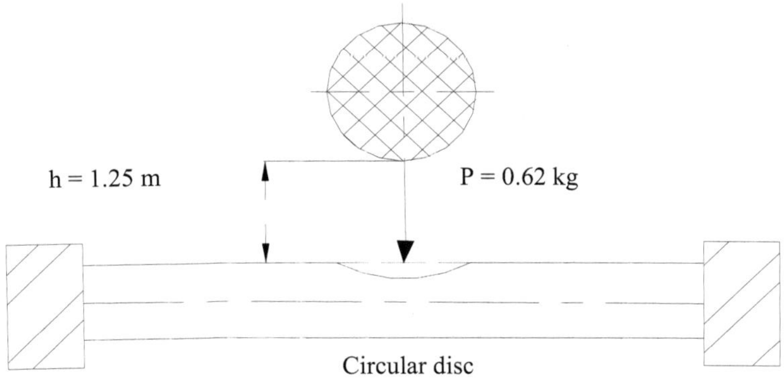

Figure 10.19. *Effect of shock. For a color version of this figure, see www.iste.co.uk/grous/design.zip*

This method is approximate. All the calculations must be verified by tests.

10.10.1. *Stress intensifications [ROA 75]*

Irregularities in a structure are likely to engender rather localized stresses, known as stress intensifications [GRO 94, GRO 09, GRO 13]. These zones include perforations, acute angles, nicks, sudden changes in thickness of the walls and numerous geometric irregularities:

– *before probable* failure Figure 10.20, part (a), the *acute angles* concentrate and intensify the stresses;

– *after* stresses Figure 10.20, part (b) (F or P), the stresses cause the *probable* failure.

When the curve of an angle is fairly small in comparison to the thickness of the wall, this gives a large (SIF) K_f. In many cases, it is tricky to precisely calculate the real stresses, unless we introduce robust moduli by the finite element method. It is now easy to predict the real stress by multiplying the stress thus calculated by the factor K [ROA 75].

{Zones of stress intensification ≡ Sources of stress intensifications}

$$\begin{cases} Example: M = F \times d = 50 \times 50 = 2.5\ Nm;\ \varepsilon = b/2 = 3.125\ mm; \\ J = hb^3/12 = \left(5 \times 6.25^3/12\right) = 101.725\ \text{mm}^4;\ \text{R}/b = 1.25/6.25 = 0.2 \rightarrow K_f = 2.5 \end{cases}$$

$$\begin{Bmatrix} \text{Incorrect method} \\ \sigma = \dfrac{M \times e}{J} = \dfrac{2500 \times 3.125}{101.725} \\ \sigma = 76.8\ MPa \rightarrow to\ be\ rejected \end{Bmatrix} \begin{Bmatrix} \text{Correct method } K_f = 2.5 \\ \sigma = K_f \dfrac{M \times e}{J} = 2.5 \times 76.8 \\ \sigma = 192\ MPa \rightarrow to\ be\ used \end{Bmatrix} \quad [10.24]$$

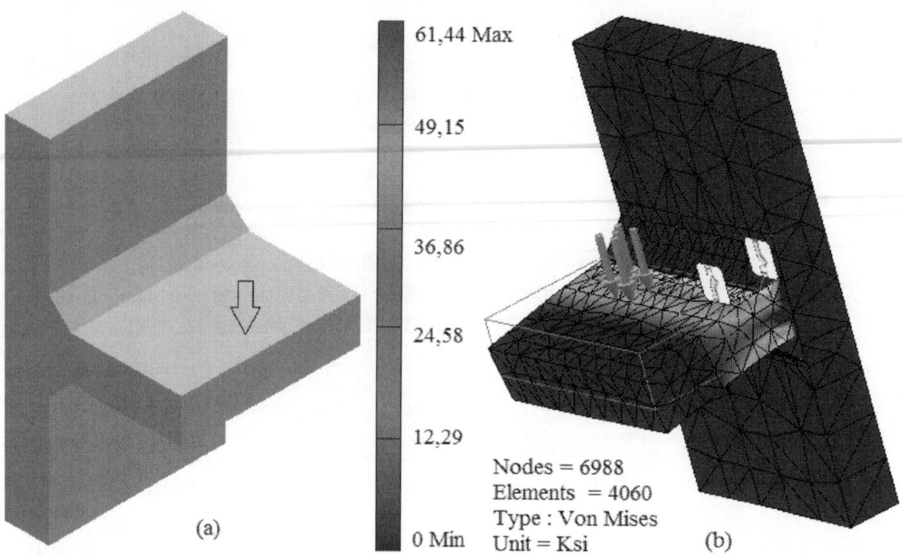

Figure 10.20. *Stress intensifications in the particular zones [K_f, SIF]. For a color version of this figure, see www.iste.co.uk/grous/design.zip*

The designs and the indicated load data must be able to withstand a stress of 192 MPa = 27850 psi. Modifying the radius to 6.25 mm alters the predictable stress intensifications to 96 MPa = 13920 psi.

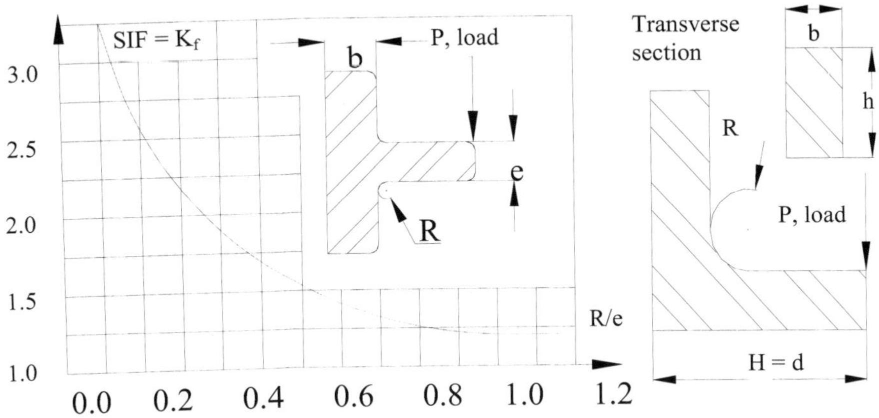

Figure 10.21. *Stress intensifications in the particular zones [K_f, SIF]. For a color version of this figure, see www.iste.co.uk/grous/design.zip*

10.11. Capacity and reliability of roller bearings (plastic and otherwise)

Consider a case in which:

– equivalent radial load L_{equi} = 3000 lbf = 13340 N;

– lifetime in millions of turns of the bearing = R_L = 400;

– reliability, R_e = 0.92.

If the equivalent radial load on a bearing is 3000 *lbf* = 13340 N and that bearing is predestined to operate for 400 million turns with a reliability R = 0.92, what would be the required load ratio? The ratio of the mean lifetime is 5.0. Explain how, in design, we draw the connection between the load ratio, if it happens that the two-parameter Weibull distribution [GRO 94, GRO 13] is the most appropriate law to apply to this case?

SOLUTION WITH DISCUSSION.– The ratio of the base load, using the two-parameter Weibull distribution, if the ratio is 5.0, is calculated as follows:

$$\frac{L_{eq}}{L_B} = \frac{1{,}898}{R_L^{0.333}} \times \left(\ln\left(\frac{1}{R_e} \right) \right)^{0.285} \quad solve \rightarrow 104852{,}40927084132092\ N \qquad [10.25]$$

L_B is the base load required ($C = L_B$) which delivers the desired reliability [MIS 65]:

$$L_B = L_{eq}\frac{R_L^{0.333}}{1.898}\times\frac{1}{\ln\left(1/\mathrm{R_e}\right)^{0.285}} = 1.05\times10^5\,N = 23594\ lbf \qquad [10.26]$$

The bearing must have at least 23,594 *lbf* = 104.951 *N*, for a reliability of 0.92. Typically, we choose the bearing made by the closest manufacturer. The base load ratio of the bearing, by virtue of the two-parameter Weibull distribution, is calculated thus:

$$\frac{L_{eq}}{L_B} = \frac{1,780}{R_L^{0.3}}\times\left(\ln\left(\frac{1}{\mathrm{R_e}}\right)\right)^{0.257}\ solve \rightarrow 0.156 \qquad [10.27]$$

COMMENT.– Here, the Weibull distribution is used when the lifetime of the bearings is within the reliability range of 0.9 and over. The ratio of the mean lifetimes is 5 for the set of bearings.

10.12. Safe stress of the appropriate material for a plastic clutch system

CASE STUDY.– For these data, let us determine the safe stress, the linear velocity and the appropriate material to convey 100 W (0.134 hp) of power at an angular frequency of 360 rpm, for ten hours of loaded work:

- transmission power, Hp = 100 W = 0.134 hp;
- angular velocity, ω_{wheel} = 360 rpm;
- number of teeth, Z = 75 teeth;
- diametral pitch, p = 625 mm = 24,606 in;
- pressure angle, α = 20 deg;
- pitch diameter, D_p = 65 mm = 2.559 in;
- tooth width, B = 28 mm = 1.102 in.

SOLUTION.– Calculation of the velocity at the point of application of the contact of the power transmission, i.e. on contact with the pitch diameter:

$$V = \omega_{wheel}\times D_p\times\pi = 73.513\left[m/\min\right] = 241\left[ft/\min\right] \qquad [10.28]$$

The calculation of the safe stress σ_s is subject to the load factor (C_s) and the factor taken from the following experimental table (see manufacturer catalogs).

Type of loads	Intermittent			Occasional
	8-10 h/d	24 h/d	3 h/d	0.5 h/d
Regular (stable)	1.00	1.25	0.80	0.50
Light shock	**1.25**	1.50	1.00	0.80
Medium shock	1.50	1.75	1.25	1.00
Powerful shock	1.75	2.00	1.50	1.25

Table 10.4. *Service factor* C_S *for the equations to calculate the power*

Number of teeth	14½° Involute	20° Full tooth	20° Stub-tooth (USA)	Full tooth α = 20°	
Z	cycloid	involute	involute	pinion	wheel
50	0.352	0.408	0.474	0.437	0.613
75	0.364	0.434	0.496	0.452	0.581
100	0.371	**0.446**	0.506	0.462	0.581
150	0.377	0.459	0.518	0.468	0.565
300	0.383	0.471	0.534	0.478	0.534
Rack	0.390	0.484	0.550		

Table 10.5. *Form factor* **Y** *for the equations to calculate the power*

Plastics	Without reinforcement	Glass reinforcement
	psi	psi
ABS	3,000	6,000
Nylon	**6,000**	12,000
Polycarbonate	**6,000**	9,000
Polyester	3,500	8,000

Table 10.6. *Safe stresses* σ_S. *Source: Product Engineering*

From Table 10.4, we choose $C_s = 1.25$ and $Y = 0.446$, and then calculate the safe stress (with light shocks) as follows:

$$\sigma_s = \frac{1}{10}\left(\frac{\sec}{ft.in^2}\right)\frac{600\left(\frac{ft}{\min}+V\right)pC_sHp}{B\times Y\times V} = 11.096\ MPa = 1609\ psi \qquad [10.29]$$

We then choose nylon 6/6 or polycarbonate, without reinforcement, to design the plastic gear system. The safe stress is 6000 psi, although our calculations give us a value of 1609 psi: this is amply sufficient. Lubricating the parts involves adding a substance between the two surfaces of pieces in contact to influence the tribology, i.e. the nature of the friction. The parameters which characterize the phenomenon of lubrications are:

– τ_{shear} is the shear stress of the fluid in psi or MPa;

– A is the area of the element in contact in in^2 or mm^2;

– P is the pressure exerted in psi or MPa;

– K is the viscosity of the fluid in centipoises. It is the proportionality constant;

– V is the linear velocity in m/s or in/s;

– ε is the thickness of the (liquid) lubricating film in in or mm;

– ν is the absolute viscosity (in Reynolds' sense) in centipoises;

– ρ is the density of the lubricant used, in g/cm^3;

– η is the kinematic viscosity in centistokes.

$$\tau_{shear} = \frac{P}{A}; \ \frac{P}{A} \propto \frac{V}{\varepsilon} \rightarrow \frac{P}{A} = K \times \frac{V}{\varepsilon} \text{ where } K = \frac{P/A}{V/\varepsilon} \quad [10.30]$$

In the imperial system, the absolute viscosity is expressed by the Reynolds number:

$$\left\{ \nu_{\text{Reynolds}} = K/6.9 \times 10^6 \ centipoises \ or \ \eta_{kinematic} = K/\mu \ centistokes \right\} \quad [10.31]$$

10.13. Case study: plastic ball bearings

Consider a case in which:

– R_1 and R_2 are respectively the loads on the bearings in *N* or *lbf;*

– l_1 and l_2 are respectively the lever arms (distances from the applications of loads to the bearings in mm or in);

– P is the equivalent radial load in N;

– α is the nominal contact angle (between the line of action of the balls) in °;

– Z is the number of balls;

– i is the number of sets of balls in the bearing;

– C is the base load in N or lbf;

– D is the diameter of the ball bearing;

– D_p is the pitch diameter in mm or in;

– f_c is a constant given by the manufacturers as a function of $Dcos(\alpha)/d_p$.

CASE 1.– We wish to calculate R_1 and R_2, in both cases if $L_1 = 0.20$ m and $L_2 = 0.30$ m.

CASE 2.– We wish to calculate R_1 and R_2, in both cases if $L_1 = 0.50$ m and $L_2 = 0.30$ m.

$$\begin{cases} R_2 = l_1 P/(l_1 + l_2) \\ R_1 = l_2 P/(l_1 + l_2) \end{cases} \quad [10.32] \qquad \begin{cases} R_2 = l_1 P/(l_1 - l_2) \\ R_1 = l_2 P/(l_1 - l_2) \end{cases} \quad [10.33]$$

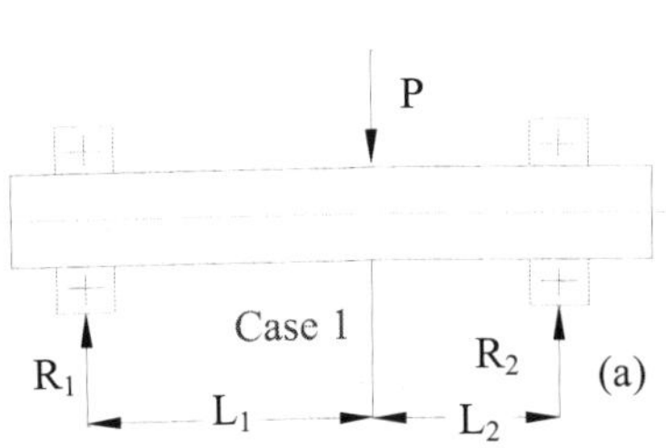

a) Radial load between the bearings

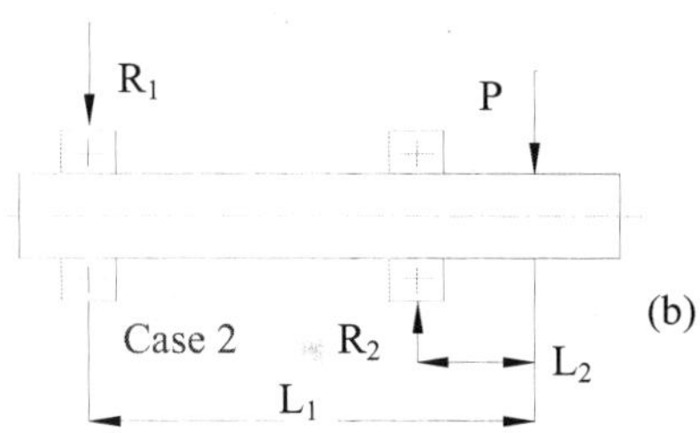

b) Radial load in cantilever

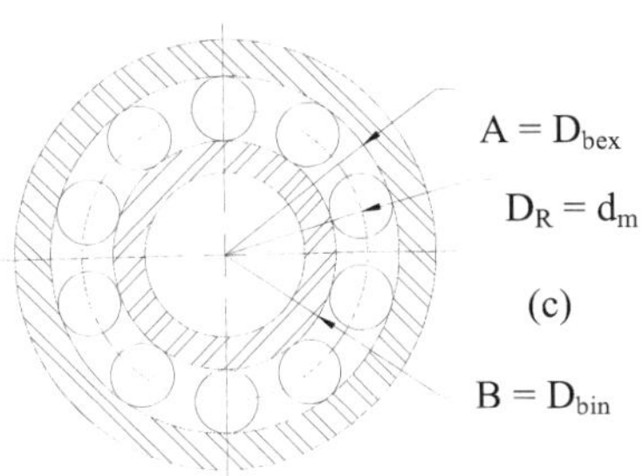

$$\begin{cases} d_p = (A+B)/2 \\ L_{10} = (C/P)^n \end{cases}$$

$$\begin{cases} Dcos(\alpha)/d_p \leftarrow \textit{tableau suivant} \\ C = f_c\left[icos(\alpha)\right]^{7/10} Z^{2/3} D^{1.8} \end{cases}$$

c) Plastic bearings

Figure 10.22. *Plastic bearings. For a color version of this figure, see www.iste.co.uk/grous/design.zip*

The following table is essentially based on the (Hertzian) contact calculations.

$\frac{D\cos(\alpha)}{d_p}$	Single set, radial contact, single and double sets, angular contact		Double set of balls (contact groove)		Auto-aligned	
	mm *	in *	mm	In	mm	in
0.05	46.7	3550	42.2	3360	17.3	1310
0.06	49.1	3740	46.5	3530	28.6	1420
0.07	51.1	3880	48.4	3680	29.9	1510
0.08	52.8	4020	50.0	3810	21.1	1600
0.09	54.3	4130	51.4	3900	22.3	1690
0.10	55.5	4220	52.6	4000	23.4	1770
0.12	57.5	4370	54.5	4140	25.6	1940
0.14	58.8	4470	55.7	4230	27.7	2100
0.16	59.6	4530	56.5	4290	29.7	2260
0.18	59.9	4550	56.8	4310	31.7	1940
0.20	59.9	4550	56.8	4310	33.5	2100
0.22	59.6	4530	56.5	4290	35.2	2680
2.24	59.0	4480	55.9	4250	36.8	2790
0.26	58.2	4420	55.1	4190	38.2	2910
0.28	57.1	4340	54.1	4110	39.4	3000
0.30	56.0	4250	53.0	4030	40.3	3060
0.32	54.6	4160	51.8	3950	40.9	3110

0.34	53.2	4050	50.4	3840	41.2	3130
0.36	51.7	3930	48.9	3730	41.3	3140
0.36	50.0	3800	47.4	3610	41.0	3110
0.40	48.4	3670	45.8	3480	40.4	3070

Table 10.7. *Ball bearings. Source: Quality Bearing & Components*

10.13.1. *Calculation of the lifetime of roller bearings*

Consider the following conventional relations, which express the lifetime of a bearing in millions of rotations and hours of operation:

– *n* is an exponent of the lifetime as a function of the type of bearing (n = 3 for ball bearings and n = 10/3 for roller bearings);

– C_{rate} is the rate of dynamic loading of the bearing in (N or lbf) = 1560 N;

– $\Re_f$ is a reliability factor valued at 90 % (dimensionless), $\Re_f \equiv 1$;

– L is the stroke length of the bearing in (mm or in) = 420 mm;

– ω is the rotation frequency of the bearing in (rad/s) = 30 rad/s;

– HRC is the hardness of the material of the stroke path (rotation) = 58;

– T is the operating temperature in °C or °F;

– 8.33 is a factor derived from the calculations suggested by the manufacturer (source QBC).

The application of these relations gives us the following results:

$$\begin{cases} \ell^{life}_{Millions\ of\ rotations} = \Re_f \left(C_{rate}/C_{dyn}\right)^n \times 10^5 = 63.554\times10^6, \textit{ in millions of rotations} \\ \ell^{vie}_{Hours\ of\ operation} = \ell^{life}_{Millions\ of\ rotations}\left(8.33/L\omega\right) = 42.016\times10^3, \textit{ in hours of operation} \end{cases} \quad [10.34]$$

10.14. Limits of performances of polymer design

Frequently, *and incorrectly*, materials with a low toughness such as K_{IC} < 15 MPa $m^{1/2}$ have been cited as examples of the limitations of polymer design.

This is linked, to a degree, with the metals that have hitherto been used in the industry, where the toughness values are almost K_{IC} = [15 to 105] MPa $m^{½}$. Cast irons exhibit a $K_{IC} \cong 10$ MPa $m^{½}$, while ceramics never achieve more than a $K_{IC} < 7$ MPa $m^{½}$, not to mention polymers, whose K_{IC} values are around (1 to 3) MPa $m^{½}$. From the laws of fracture mechanics, we can set:

$$\sigma_r = \Psi \times K_c / \sqrt{\pi a_c} \qquad [10.34]$$

where:

– Ψ is a factor (constant), intrinsic to the material;

– a is the size of the largest crack in µm, mm or in;

– K_c is the toughness relative to the type of loading.

In design, we are limited by the load or sometimes by the energy and/or deformation. Let us present these three cases, as follows:

$$\left\{\begin{matrix} Load \\ K_c \\ i_{perf} \equiv K_c \end{matrix}\right\} \left\{\begin{matrix} Deformation \\ \varepsilon = \dfrac{\sigma}{E}; \varepsilon_r = \dfrac{\Psi K_{Ic}}{\sqrt{\pi a_c} E} \\ i_{perf} \equiv K_{Ic}^2 / E \end{matrix}\right\} \left\{\begin{matrix} Energy \\ E_{em} = \dfrac{\sigma\varepsilon}{2} = \dfrac{\sigma^2}{2E}; E_{\max} = \dfrac{\Psi^2}{2\pi a_c} \dfrac{K_{IC}^2}{E} \\ i_{perf} \equiv K_{Ic}^2 / E \cong G_r \left[kj/m^2 \right] \end{matrix}\right\} \qquad [10.35]$$

10.15. Case study: fan with plastic blades

Homes, vehicles, airplanes and running machines are all equipped with fans to cool down the radiators when their movements and the environment are not sufficient for that task. Fan blades are often made of plastic or lightweight materials which, sometimes, have to withstand powerful centrifugal forces. Bending stresses are due to the sudden accelerations of the motors. There is even the risk of disintegration of the fan as a result of sudden runaway. Safety is therefore a primary concern for designers, and it is important to choose the materials very astutely. The author has carried out numerous design projects; some of the results are shown below.

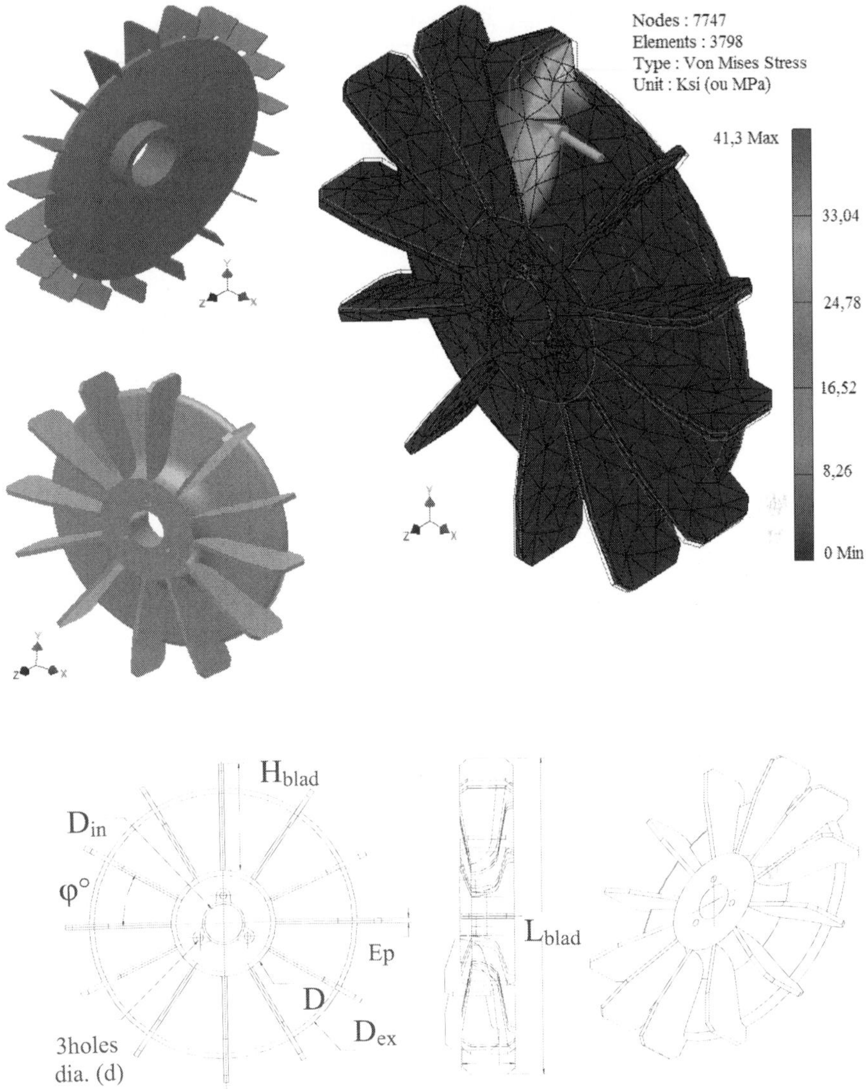

Figure 10.23. *Simulation of the bending of plastic fan blades. For a color version of this figure, see www.iste.co.uk/grous/design.zip*

The material chosen must be safe, inexpensive and environmentally friendly. The objective, then, is to control the angular velocity with a predetermined blade radius. If A is the surface area of the blade of length L = φ.R, (φ) is the radius of the fan where the blades are spread out, because the rest of the surface is occupied by the hub (center with three radiating blades, in our case). Thus, the volume of the fan is expressed by: V = $[\varphi RA]$. The centrifugal force is as follows:

$$F_{centrifugal} = \rho VR\omega^2 = \rho R\omega^2 \left[\varphi RA\right],\ N\ or\ lbf \qquad [10.36]$$

The stress (our models use the Von Mises criterion in our fan project case studies) will therefore be:

$$\sigma = F/A = \rho VR\omega^2 = \varphi\rho R^2\omega^2,\ MPa\ or\ psi \qquad [10.37]$$

In order to respect the required safety, this stress must not exceed σ_F, expressed in relation to a safety factor (s ≅ 3). In our designs, this coefficient does not have a significant role to play, because the expression of the angular velocity is as follows:

$$\omega \prec \left(1/\sqrt{\varphi R}\right) \times \sqrt{\sigma_F/\rho},\ \left[rad/s\right] \qquad [10.38]$$

Reading this equation, we observe that the importance is attached to the material (ρ and σ_F) used for the fan (blades) and to its geometry (φ and R). By enlightened design, by setting (φ and R), the rate of rotation which ensures and optimizes safety, we now simply need to choose the appropriate material – i.e. the material which has the best values of the ratio between (ρ and σ_F). Fans are designed with heavy or lightweight metal, polymer or by plastic injection. It is not the fans which are costly, but the sudden variations in speed:

$$M \prec \left\{\sigma_F/\rho\right\},\ MPa \times m^3/mg \qquad [10.39]$$

10.16. Conclusion

Having come to the end of this chapter, what must be noted is the peculiarity of the design of products based on the materials, and specifically plastics. The calculation principles include specific coefficients which take account of the rheology of the materials, and these must be observed. In mechanical design, the scant interest paid to materials such as ceramics is attributable to their (lack of) toughness.

11

Mechanical Design Projects

11.1. Proposed projects in mechanical design

In design, modeling is an absolutely unavoidable key step. In the preliminary stages of the project, it is imperative to make sketches, along with a rough *graphic-analytical* model. If the design sketched at the preliminary stage is found to be viable, we would continue to the other factors of progressive design. The principles behind the calculations involve combinations of properties to achieve the best possible performances of the project: the surface states of the materials, the sizing in connection with the intervals of values for the forces and other loads.

Multi-criterion design: throughout the chapters of this book, we have constantly reiterated the importance of the material–geometry couple. The choices made in designs are a compromise, struck in order to satisfy numerous constraints: weight, restrictions on stiffness and strength, fatigue toughness, esthetic beauty, cost, price, etc. These are multi-criterion – or multi-objective – designs. Examples might include crank/rod systems and vehicle motors. The lectures on mechanical design have looked at numerous cases of machines and other mechanisms. A few similar examples are examined in this chapter. The idea of performance indicators seems helpful, but closer attention must be paid to the given mathematical expressions [ROA 75]. It is not always possible to model everything, because qualifying factors (shape, colors and esthetic aspect) also play a crucial role in the delivery of the product.

11.2. Case studies of hoisting and handling devices

In this chapter, we present three mechanical design projects. Given the enormous range of subjects touched upon, we have selected the salient points dealt with in supervised projects and workshops. The project to which we first turn our attention to is a simulated case concerning the design of hoisting apparatus, used in civil engineering, workshop mechanics and construction, for example. The parameters have a major influence on cranes, winches and other hoists with a lifting hook.

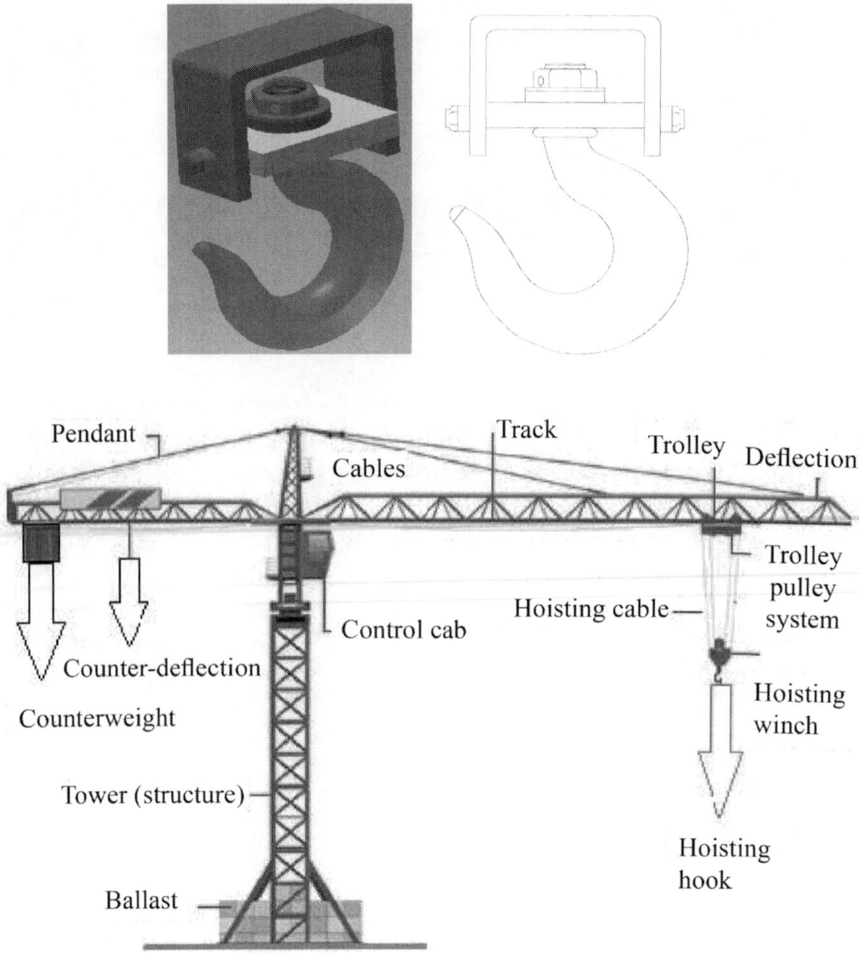

Figure 11.1. *3D sketch of the principles of calculations for a lifting hook made of SAE 1010 or 1018 steel. For a color version of this figure, see www.iste.co.uk/grous/design.zip*

11.3. Projects design proposal for a lifting winch

The problem at hand is to cater for the usage needs of a winch. The machine elements to which the design will pertain are the winch mechanism itself and the lifting hook. To succinctly examine the project, it is worthwhile using the same approach as presented in Chapter 1.

Pre-project proposal		
Subject	Design of a crane-mounted winch	
Client	*Département de Techniques de Génie Mécanique* (TGM – Mechanical Engineering Techniques Dept.)	
Initial data	Just the idea developed when examining the problem	
End goal	Support that is enlightening for TGM	
Creation date	....	Obs.
Delivery date	....	Obs.
Project manager	....	Mechanical design
Requirements specification		
Requirements	Raise the mold from the injection machine	
Technical specifications		
Validation of requirement		
Client	Département de TGM	
Date of expression	....	
Validation of requirement		
Requirement(s)		
Technical specifications		
Validation of requirement		
Client	Département de TGM	
Date of expression	....	
Date of expression	....	Dossier #: 00 11 22

Table 11.1. *Dossiers for analysis of the project*

11.3.1. *Case study: parameters in sketching a lifting hook*

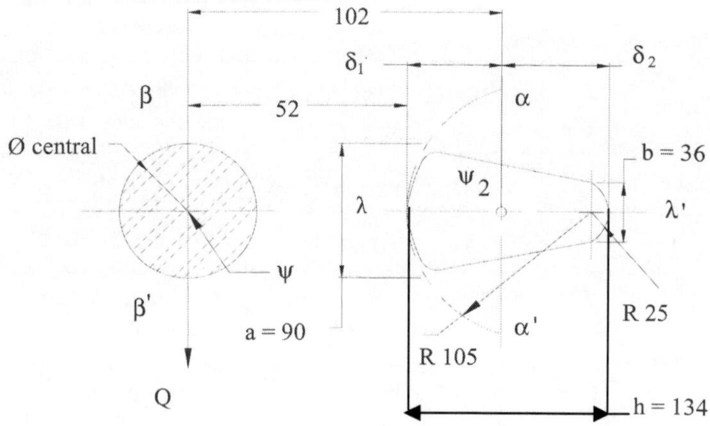

Figure 11.2. *Diagram of calculation of the moment of inertia. For a color version of this figure, see www.iste.co.uk/grous/design.zip*

Where:

– $J._{\alpha\,\beta}$ is the modulus (moment) of inertia of the hook;

– δ_1 is the distance between the center of inertia ψ_2 and the inside right extremity of the circle of the hook's concave part (inside neutral fiber);

– δ_2 is the distance between the center of inertia ψ_3 and the outside right-hand extremity;

– ψ_3 is the area of the trapezoidal form;

– $Ø_1$ and $Ø_2$ are the diameters of the sections ψ_1 and ψ_2;

– Δ is the distance between the center of inertia ψ_2 and the center of the concave section of the hook;

– Q is the suspended load, in N;

– $M_{bending}$ is the bending moment, in N.mm;

– $\sigma_{extension}$ is the traction stress applied in MPa;

– $M_{compression}$ is the compression moment, in N.mm;

– $\sigma_{compression}$ is the compressive stress, in MPa;

– $M_{bending}$ is the calculated bending moment, in N.mm.

Calculation and verification of the sections ψ_1 and ψ_2 of the lifting hook

The load Q, in N, applies stress to the lifting hook at point ψ_2 along the axis {α-α'} located on the threaded stem {y-y'}. We observe that the section ψ_1 is subject to shear. The section ψ_2, for its part, is subjected to traction. Thus, we can calculate the cross-section with the diameter $\varnothing_2$ found from the cross-section ψ_2 of the threaded stem.

11.3.2. *Principles of calculations of the resistance of a lifting hook*

Initial data: $\left\{\begin{array}{l} J_{\alpha\beta} = 400\ mm^4; \delta_1 = 20\ mm\ ; \delta_2 = 24\ mm\ \ \psi_3 = 600\ mm^2 \\ Q = 4\times10^4\ N;\ \Delta = 45\ mm\ ;\ d_1 = 30\ mm\ ;\ d2 = 40\ mm \end{array}\right\}$

Calculate the center of gravity G and the distances v and v' from G to the end fibers.

$$J_{\alpha\beta} = \frac{1}{36}\left(\frac{a^2+4\times a.b+b^2}{a+b}\right)\times h^3 \qquad [11.1]$$

For λ= 9 cm = 3.543 in, b = 3.6 cm = 1.417 in and h = {δ_1+δ_2} = 13.4 cm = 5.276 in, the moment of inertia would be: $J_{\alpha\beta} = 1.186\times10^3\ cm^4 = 28.491\ in^4$.

QUESTION 1.– Calculate:

$\sigma_{extension}$ *in MPa; d_1 in mm and in;* $M_{bending}$ *in $N\times mm$ and $lb\times in$.*

SOLUTION.– The fiber (λ) is in extension δ_1 = 0.6 d_1 → d1 = δ_1 /0.6 = 33.33 mm = 1.212 in. The bending moment is the product of the force over the distance of the lever arm. With the bending moment, we can calculate the traction stress by applying the following formula. The ratio (Q/ψ_3) must be added, because the weight exerts a further force of traction beyond the real load (to be calculated) [ROA 75]:

$$\left\{\begin{array}{l} M_{bending} = \Delta\times Q = 1.8\times10^3\ N.m = 1.593\times10^4\ lbf.in \\ M_{extension} = \dfrac{M_{bending}\delta_1}{J_{\alpha\beta}} + \dfrac{Q}{\psi_3} = 90070\ MPa = 1.306\times10^7\ psi \end{array}\right. \qquad [11.2]$$

SOLUTION WITH DISCUSSION.– The fiber (λ') is under compression: the compression moment is the same as the bending moment, because it is the same load that the hook has to support.

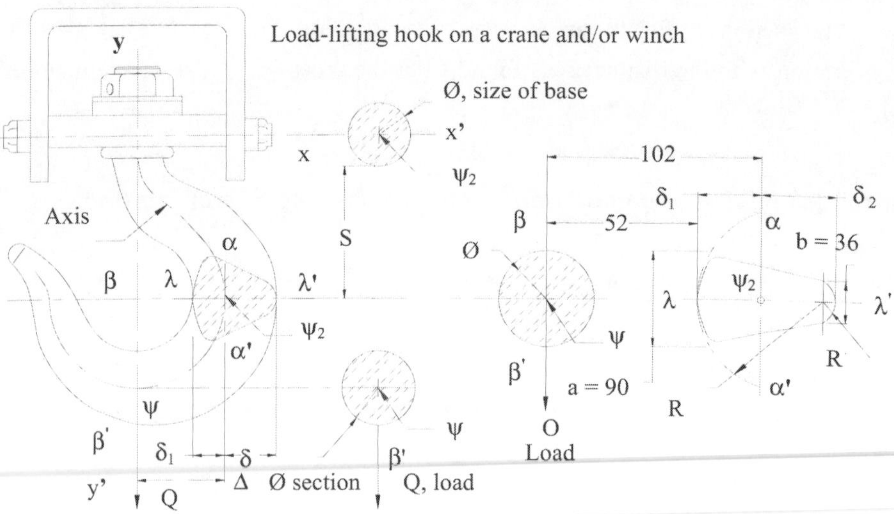

Figure 11.3. *Illustration of the principle of calculations for a lifting hook made of SAE 1010 or 1018 steel. For a color version of this figure, see www.iste.co.uk/grous/design.zip*

$$\begin{cases} M_{Compression} = \Delta \times Q = 1.8\times10^3\ N.m = 1.593\times10^4\ lbf.in \\ \sigma_{compression} = M_{fl}\delta_1 / J_{\alpha\beta} - Q/\psi_3 = 89930\ \ MPa = 1.304\times10^7\ psi \end{cases} \quad [11.3]$$

STRENGTH CONDITION.– For this formula, the Q/ψ_3 must be subtracted, because the weight is counter to the force we wish to calculate.

$$\frac{M_{fl}(\delta_2-\delta_1)}{J_{\alpha\beta}} = 2\frac{Q}{\psi_3}\ \ Hence\ M_{fl} = 2\frac{Q\times J_{\alpha\beta}}{\psi_3(\delta_2-\delta_1)} = 1.333\ Nm = 118.01\ lbfin \quad [11.4]$$

Plotting the bending moment of the hook as a function of the lifting load:

Consider the section {λ-λ'}, taken from a trapezoidal form (Figure 11.3) whose dimensions are a = 90 mm, b = 36 mm, h = 120 mm. We can ignore the rounded end sections, because this is not a finite-element calculation. The cross-section S of {λ-λ'} is 7500 mm^2. Under the influence of the work exerted by the hook, we consider traction under normal force to be 12×10^4 N; the bending stress is $\sigma_1 = 16$ MPa and the bending moment: $Q_x OG$.

(*d*) is the dimension of the base, calculated below. It is helpful, when determining the dimension (d), to consider that the core of the screw opposes a safety traction resistance of around 56 MPa. Plot the dimensions of the hook which satisfy the resistance criteria in different sections. To do this, let us perform the calculation with the following initial data: by varying Q from 0 N to 104^3 N, we produce the plot below.

NOTE.– The material (SAE 1010 or SAE 1018) must be handled rationally. The relation shows that the section (δ_2) must be greater than (δ_1). It is for this reason that we see dissymmetry in the cross-section on the two sides of the axis.

$M_{flcalcul}(Qvaria) =$

	0
0	0.000
1	33.333
2	66.667
3	100.000
4	133.333
5	166.667
6	200.000
7	233.333
8	266.667
9	300.000
10	333.333
11	...

·N·m

etc. ...

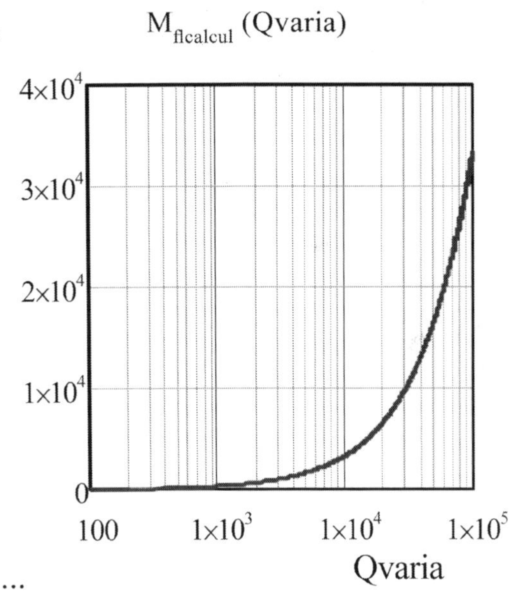

Figure 11.4. *Plot for the lifting hook of SAE 1010 or 1018 (forged) steel. For a color version of this figure, see www.iste.co.uk/grous/design.zip*

11.3.3. *Calculation and design (choice) of the round-wire coil spring*

With the following data:

– r_{mean}, mean radius 5 mm or 0.197 in;

– d_{spring}, diameter of spring 3 mm or 0.118 in;

– n_u, number of useful spires = 7, step = 3.25 mm = 0.128 in;

– σ_{adm}, practical applied resistance stress 340 MPa or 4.931×10^4 psi;

– G_{adm}, practical applied stiffness stress (total deflection considered for the spring) 8×10^4 MPa or 1.16×10^7 psi.

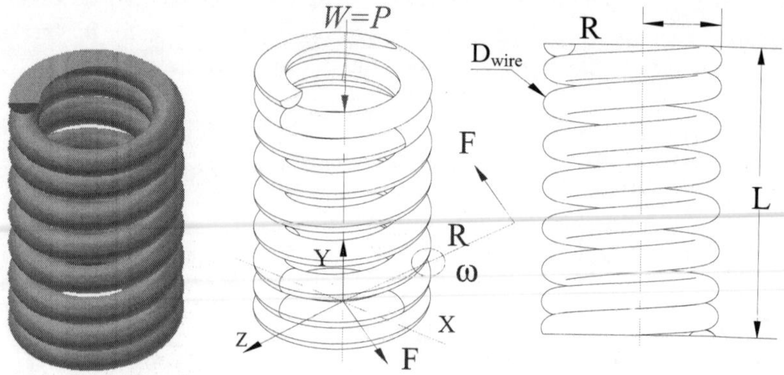

Figure 11.5. *Calculating the resistance of the spring.*
For a color version of this figure, see www.iste.co.uk/grous/design.zip

Calculate the maximum compression load that can be safely applied, W_{safe}. By virtue of the law of SOM, the torsion modulus is written as follows: $J_0/v = 0.2d^3$; which enables us to verify the resistance condition: $J_0/v \cong 0.2\times d^3 \le M_{torsion}/[\sigma]_{admissible}$.

The total load W is calculated as follows:

$$W_{total} \le 0.2\times d^3_{spring}\times[\sigma]/R_{mean} = 13.6\ N = 3.057\ lbf \qquad [11.5]$$

The deflection of the spring, per useful spire, is expressed thus:

$$f_{deflection} = 64\times W_{safe}R^3_{mean}/[G]d_{spring} = 1.36\ mm = 0.0535\ in \qquad [11.6]$$

Calculate the total deflection (n_{su} = 7) experienced by the spring, f_{tot}. Note that the step is always measured when the spring is in the free state:

$$f_{total} = f_{deflection} \times n_{su} = 1.36 \times 7 = 9.52\ mm = 0.3748\ in \qquad [11.7]$$

$f(R_v) =$	$W(R_v) =$
1.36	13.6
1.81	12.364
2.35	11.333
2.988	10.462
3.732	9.714
4.59	9.067
5.571	8.5
6.682	8
7.932	7.556
9.328	7.158
10.88	6.8
mm	N

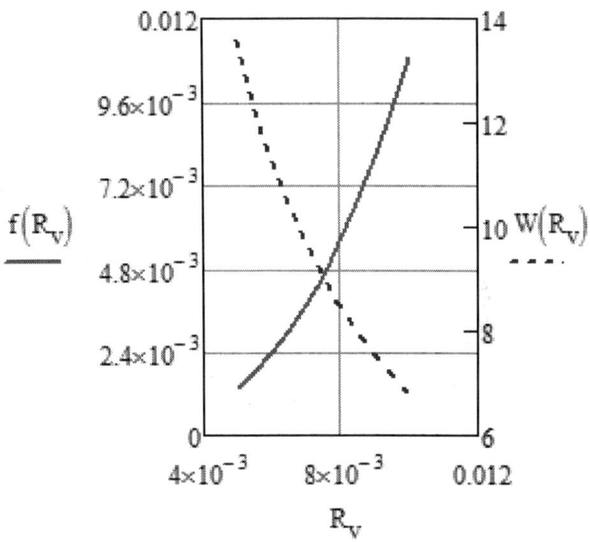

Figure 11.6. *Illustration of the force W (N, lbf) and deflection f (mm, in). For a color version of this figure, see www.iste.co.uk/grous/design.zip*

Two consecutive spires are spaced a distance (3.25 – 1) = 2.25 mm apart. Thus, contact between the spires is prevented by the fact that two spires sit 2.25 mm (0.0535 in) apart. The graph below is plotted with a deliberate variation of R_{mean} between 5 mm and 10 mm, to clearly show the evolution of the deflection (mm) and the load (N).

COMMENTS.– The spring has tightly wound spires. The load *W* (N) follows the direction of the geometric axis. It tends to squash the spring. At the point (G), the reduction of the straight section of a single spire gives rise to a force W_1 equivalent to W and a torque (τ and τ_1). The latter tends to make the spring slip within that area, by rotation, around G. That moment is $M_\tau = R_x W$. Ordinarily, the shear force has very little effect in comparison to the twisting moment M_τ. The calculation condition is satisfied, by the M_τ, whose resistant modulus is written as: $(J_0/v) = 0.2d^3$, and must satisfy the resistance condition: $0.2d^3 \leq M_\tau\ /[\sigma]_{adm}$.

11.4. Calculation and design of a bolted assembly

Variables and statement of the problem of stress intensification at the root:

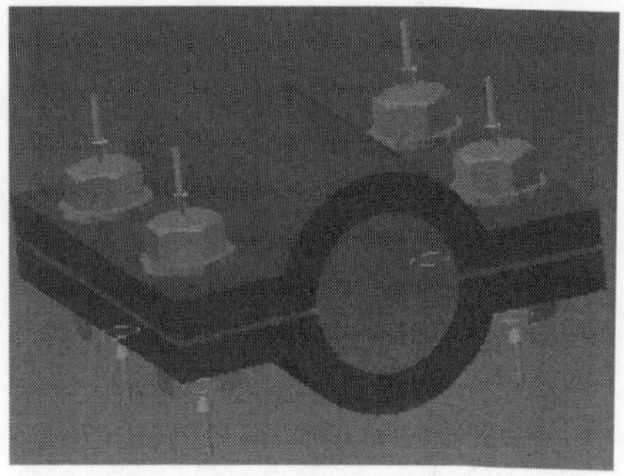

T_1 is the initial traction applied directly to the threaded stem – bolt

T_2 is the outside traction caused by the load T_1; applied to the supporting metal (gusset plates)

$m = \Sigma\, m._i$ is the sum of the deflections – displacements in in.lb or kg.mm – of the assembled components, including the washer or any joint between the components

f is the deflection corresponding to the displacement of the threaded stem (bolt)

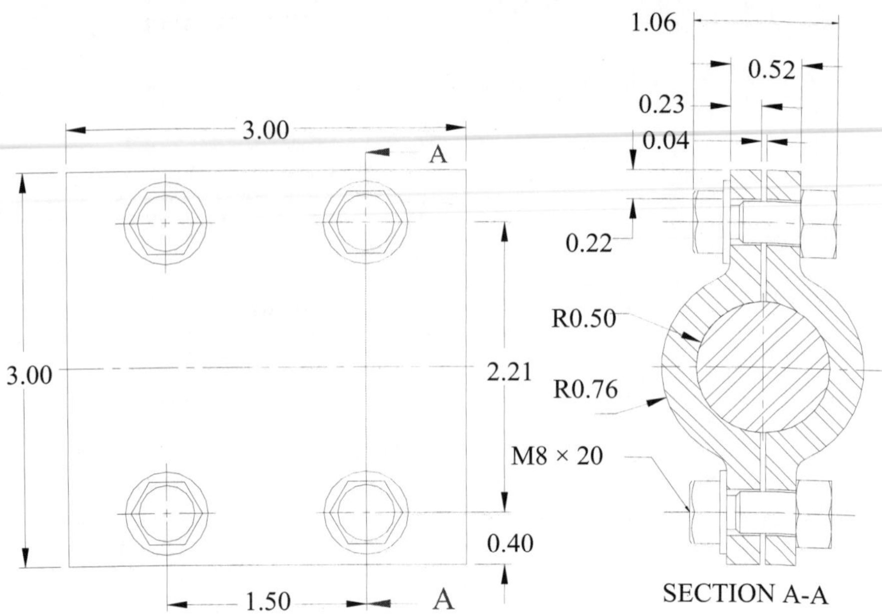

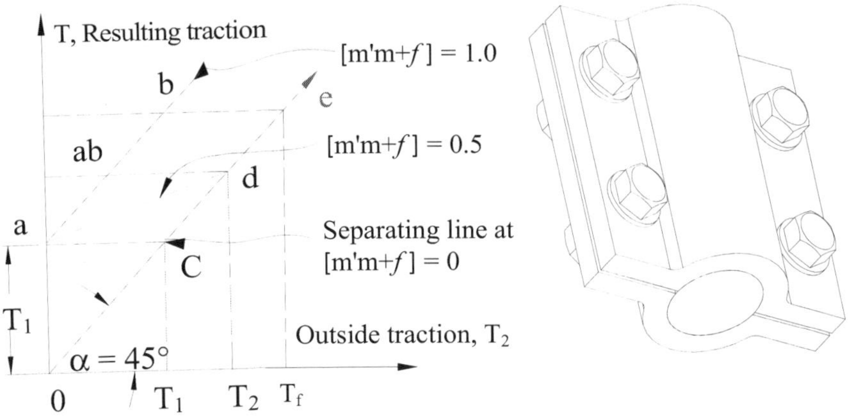

Figure 11.7. *Illustration of the assemblies connected by threading*

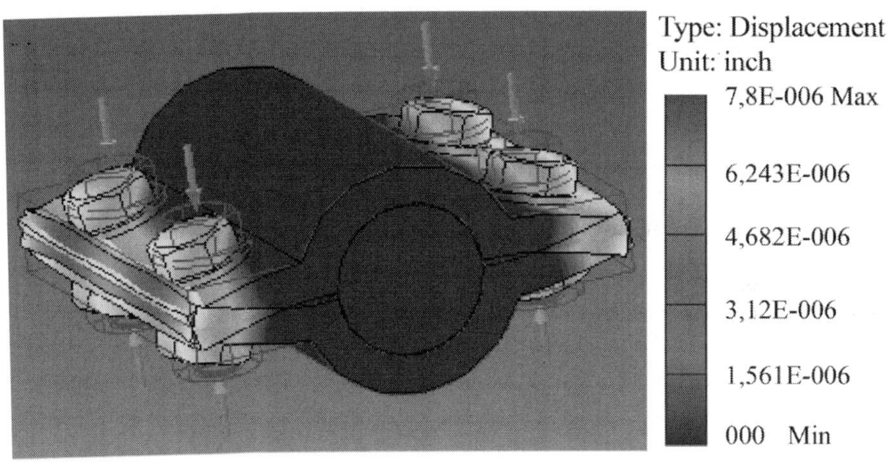

Figure 11.8. *Simulation of the displacement of bolted assemblies. For a color version of this figure, see www.iste.co.uk/grous/design.zip*

The relation used to calculate the total traction is written as:

$$T = T_1 + T_2 \left(\frac{m_1 + m_2 + \ldots + m_i}{f + (m_1 + m_2 + \ldots + m_i)} \right) \; [lb \ or \ kg] \qquad [11.8]$$

The initial traction on the threaded stem (bolt or stud) would generate (a force of work) a moment. This fact is confirmed by experimental observations, so we can set:

$$T_1 = +\Delta\kappa \times d \; [lb.in \text{ or } kg.mm] \qquad [11.9]$$

Δ κ is the range of the traction forces. For instance, from [9000 lb = 4.082×10^3 kg to 18,000 *lb* = 8165 kg], d is the nominal thread diameter. Experimentally, we find that if the moment is employed to tighten the bolt, then the above expression of T_1 is rewritten as:

$$T_1(M) = M/0.2 \times d, \; [lbf \text{ or } N] \qquad [11.10]$$

M is the moment applied to tighten the bolt. Equation [11.9] is considerably condensed, given that we have neglected α, the helix angle of the thread. We suppose that the friction is roughly 0.15 for the thread of the stem and for the bolt. We suppose that the radius of the spires of the bolt is around 2/3 the pitch diameter of the bolt.

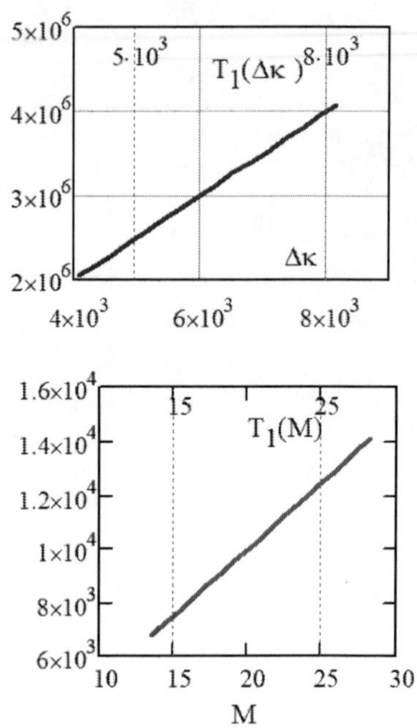

$T_1(\Delta\kappa)$ =	$T_1(M)$ =	
$2.04\cdot10^6$	$6.774\cdot10^3$	N
$2.176\cdot10^6$	$7.338\cdot10^3$	
$2.312\cdot10^6$	$7.903\cdot10^3$	
$2.448\cdot10^6$	$8.467\cdot10^3$	
$2.584\cdot10^6$	$9.032\cdot10^3$	
$2.719\cdot10^6$	$9.596\cdot10^3$	
$2.855\cdot10^6$	$1.016\cdot10^4$	
$2.991\cdot10^6$	$1.073\cdot10^4$	
$3.127\cdot10^6$	$1.129\cdot10^4$	
$3.263\cdot10^6$	$1.185\cdot10^4$	
$3.399\cdot10^6$	$1.242\cdot10^4$	
$3.535\cdot10^6$	$1.298\cdot10^4$	
$3.671\cdot10^6$	$1.355\cdot10^4$	
$3.807\cdot10^6$	$1.411\cdot10^4$	
$3.943\cdot10^6$		
$4.079\cdot10^6$		

Figure 11.9. *Results of calculations of $T_1(\Delta K)$ and $T_1(M)$. For a color version of this figure, see www.iste.co.uk/grous/design.zip*

The lubrication is also accepted to be normal (appropriate grease), for: M = 120 *lbf.in* =*13.558 Nm*, [130 to 250] *lbf.in* = [14.688 to 28.246] *Nm* and d = 0.394 *in* = *8.865 mm*, we calculate T_1: the moment applied to turn the screw or the bolt is then expressed thus:

– r_m mean radius of the thread, in or mm = 0.4617;

– r_s radius of the spires in frictional contact, in or mm;

– μ is the friction coefficient of the threads;

– μ_s is the friction coefficient of the spires;

– α is the helix angle in degrees;

– θ_n is the helix half-angle in degrees;

– T_{01} is the initial pressing load of the thread in lb or kg.

The moment may be (+) or (-), meaning that it is (+) if the work drives the screw forward and (-) if the work delays (slows) the rotation. This also explains the equilibrium. For example, which the axial load, alone, causes rotation, we say that there is *overhauling* of the motion. The moment required to advance the screw – as is the case with bolting – the bolt, stud or any stem threaded in the direction of the load (lowering or slowing). As this moment may be negative, we set the following:

$$\begin{cases} M_{thread}^{positive} = T_{01} \times r_m \left[(+)\tan(\alpha) + \mu/\cos(\theta_n)\right] \Big/ \left(1 - \mu \dfrac{\tan(\alpha)}{\cos(\theta_n)}\right) + \mu_s \times r_s \\ M_{thread}^{negative} = T_{01} \times r_m \left[(-)\tan(\alpha) + \mu/\cos(\theta_n)\right] \Big/ \left(1 + \mu \dfrac{\tan(\alpha)}{\cos(\theta_n)}\right) + \mu_s \times r_s \end{cases} \qquad [11.11]$$

11.5. Yield of power transmission of a screw mechanism

There is sometimes a tendency to overlook the yield in academic calculations. The yield, in fact, is the ratio of work on output of the motion to the work on input:

$$\eta = \frac{100 T_{01}}{2\pi M_{thread}} = \frac{100\tan(\alpha) + \mu/\cos(\theta_n)}{\left(\tan(\alpha) + \dfrac{\mu}{\cos(\theta_n)}\right) \Big/ \left(1 - \mu \dfrac{\tan(\alpha)}{\cos(\theta_n)}\right) + \dfrac{\mu_s r_s}{r_m}} \% \qquad [11.12]$$

WORKING CONDITIONS	Steel screw and bronze or cast-iron bolt μ is the mean friction coefficient		Contact bush ring Materials	μ_s, mean friction coefficient	
	Startup	Motion		Startup	Motion
Good transmission condition	0.14	0.10	Mild steel on cast iron	0.17	0.12
Average conditions	0.18	0.13	Tempered steel on cast iron	0.15	0.09
Poor conditions: ↓ Newly machined part before final grinding			Mild steel on bronze	0.10	0.08
0.21	0.15		Tempered steel on bronze	0.08	0.06

Table 11.2. *Experimental data for friction coefficients of the threads*

This expression can also be written as follows:

$$\eta_{\lambda,\mu} = \tan(\lambda)\left[\left(\cos(\alpha) - \mu\tan(\lambda)\right)/\left(\cos(\alpha)\tan(\lambda) + \mu\right)\right]\% \qquad [11.13]$$

For the pressure angle α = 14.5° (20°, ISO). Λ is the directional angle and μ is the friction coefficient. We find the influence of the yield by varying the directional angle λ between [0° and 90°]. We have plotted four curves with different friction coefficients to observe the evolution of the yield as a function of the materials in contact.

$\eta(\lambda,\mu_1) =$ ·%

0.000
7.784
14.430
20.170
25.175
29.577
33.478
36.957
40.078
...

continue

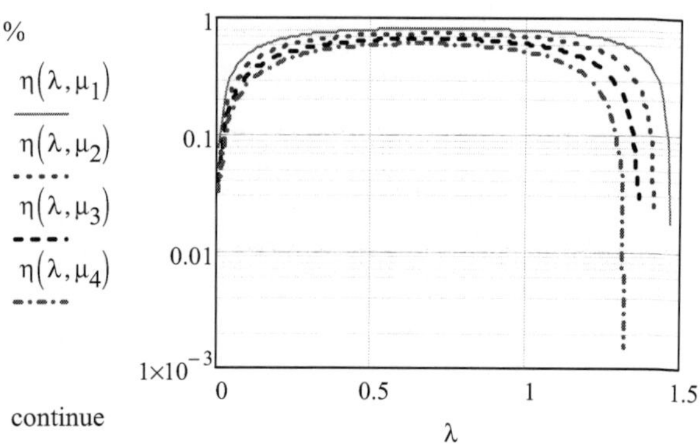

Figure 11.10. *Yield of a screw power transmission (ACME). For a color version of this figure, see www.iste.co.uk/grous/design.zip*

11.5.1. *Calculations of stresses on the threads of a screw mechanism*

Our calculations can be assimilated to a short beam. The beam is subjected to loading along the axis of the screw. The loads here are concentrated around the middle radius, i.e. over half the depth of the thread (h/2). The width corresponds to the length of the thread.

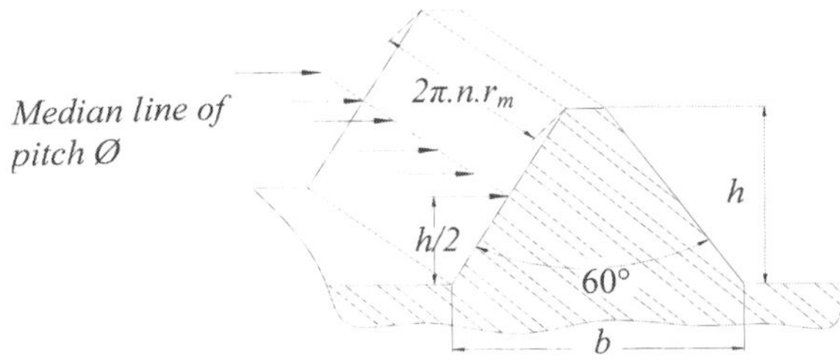

Figure 11.11. *Diagram of a thread subjected to loading. For a color version of this figure, see www.iste.co.uk/grous/design.zip*

The bending and shear stresses due to the transverse force are:

$$\sigma_{bending} = 3T_{01}h/2\pi nr_m b^2 \text{ and } \sigma_{shear} = T_{01}/2\pi nr_m b \qquad [11.14]$$

where:

– r_m is the mean thread radius, in or mm = 0.4617;

– M_{thread} is the moment applied to the screw in N.mm or lbf.in;

– h is the height of the threads in mm or in. *n* is the number of threads engaged;

– b is the width of the threads in mm or in (section at the root);

– T_{01} is the initial load of pressure on the thread in lb or kg.

Between the surface of the screw and the nut (Figures 11.7 and 11.8), there is a critical factor which must be taken into account in designing screw power transmissions. The estimation of this background factor of the thread is not taken into account by the total height. It must be taken into account when conducting the design work. The other reason is that the load is not distributed along the full length of the thread:

$$F_{critical} = T_{01}/2\pi nr_m h \qquad [11.15]$$

11.5.2. *Calculations of stresses at the root of the thread in a screw mechanism*

If r_i is the radius at the root of the thread, and $M_{twisting}$ is the twisting moment applied to the loaded area of the screw. The twisting and compression stresses are:

$$\left\{\tau_{twisting} = 2M_{twisting} \big/ \pi r_i^3 \text{ and } \tau_{compression} = T_{01} \big/ \pi r_i^2\, i \right\} \qquad [11.16]$$

11.5.3. *Case study: numerical applications*

The kinematic diagram of the winch studied here is a real diagram.

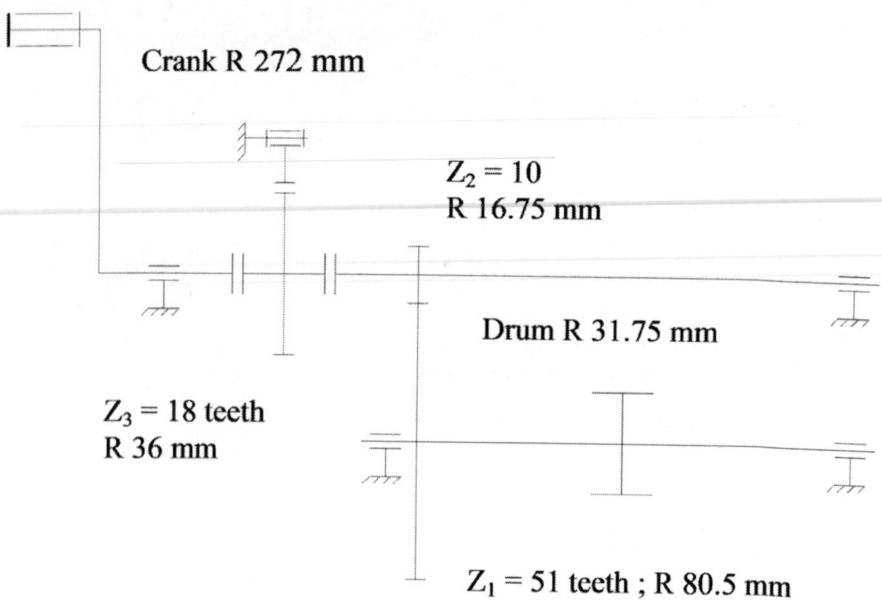

Figure 11.12. *Kinematic diagram of the winch. For a color version of this figure, see www.iste.co.uk/grous/design.zip*

We can draw inspiration from this diagram to design another winch if needed, in pedagogical case studies. The mold of the injection machine is lifted by a simple winch, illustrated by the below three photographs.

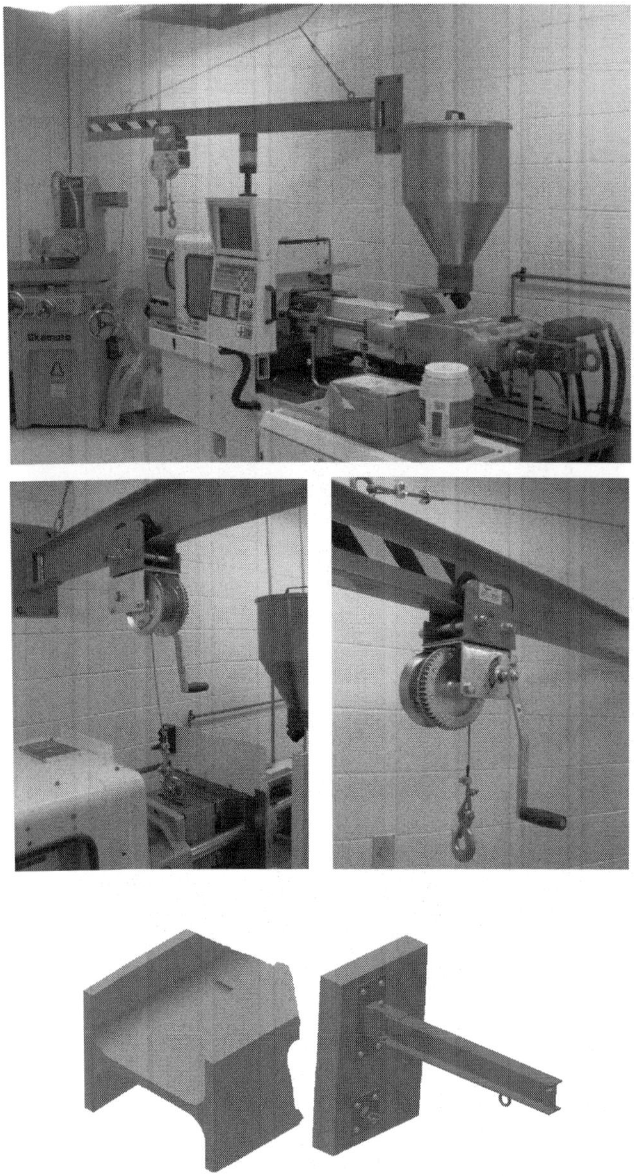

Figure 11.13. *Photographs*[1] *of the winch project, and diagrams of I-beams and T-beams. For a color version of this figure, see www.iste.co.uk/grous/design.zip*

1 This is a real case encountered in our manufacturing processes workshop.

QUESTIONS.– Completely dismantle a winch. Draw all the parts, including the standard ones (springs, screws, bolts, etc.). Devise a set of technical specifications, calculate the strengths, geometries and assemble the winch once the design is completed. What is the maximum acceptable load the beam can withstand for the winch to be able to lift and deposit the mold (see real photographs from construction process) and the safety coefficient (s)?

The mold is lifted by hook, as shown in the photographs (Figure 11.13)
– I-beam (or double-T-beam)
– L-beam
Effective length, Le = 4 m
Area of the cross-section, A = 149.1 cm^2
$J_x = 8563\ cm^4$
$W_x = [J_x / V_x] = 571\ cm^3$
$J_y = 25166\ cm^4$
$W_y = [J_y / V_y] = 1680\ cm^3$
$\sigma_e = 160$ MPa
$E = 180 \times 10^9$ MPa
$F_{excentered} = F_{app} = 2.2 \times 10^5$ N
Excentricity ε = 250 mm

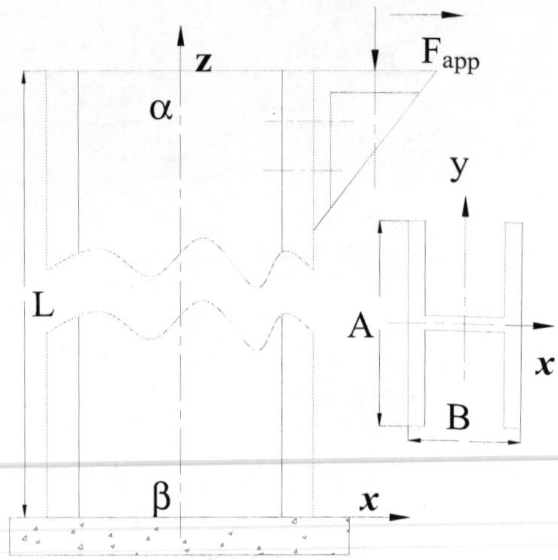

Figure 11.14. *U-beam. For a color version of this figure, see www.iste.co.uk/grous/design.zip*

SOLUTION WITH DISCUSSION.–

The effective length L_e is $L_{equivalent} = 2\ L = 8$ m = 26.247 ft.

$$\textit{The radius } r_y \textit{ is}: \ r_y = \sqrt{J_y / A} = 0129918\ m = 5.115\ in$$

$$\left(L_e/2r_y\right) = 30.789;\ \textit{Degree of excentricity} = \left(\varepsilon V_y / r_y^2\right) = \left(\varepsilon A / W_y\right) = 2.21875$$

$$\textit{From } \left(\varepsilon V_y / r_y^2\right) = \left(\varepsilon A / W_y\right) \rightarrow V_y = \left(\varepsilon A / W_y\right) \times \left(r_y^2 / \varepsilon\right) = 0.15\ m = 5.898\ in$$

$$K = \sqrt{F_e / EJ_y} = 6.969 \times 10^{-5}\left(m^{-1}\right) = 1.77 \times 10^{-6}\left(in^{-1}\right)$$

The applied stress will therefore be: $\sigma_{app} = F_e / A = 14.755198\ MPa = 2140\ psi$:

$$\sigma_{\max} = \frac{F}{A} + \frac{M_{bending}^{\max} \times V}{J_z} = \frac{F}{A} + \frac{F_e V \varepsilon}{J_z} \sec\left(\frac{KL_e}{2}\right) = ?\ MPa\ and\ psi \qquad [11.17]$$

The maximum acceptable load must be calculated by the so-called secant formula. Thus, it is important not to use σ.e/σ max, because of the *lack of linearity between the effective load (F_e) and the stress σe σ_e*. We therefore apply the following equation:

$$\sigma_{\max} = \sigma_{app} \left| 1 + \frac{V\varepsilon}{r_y^2} \right| \sec\left(\frac{L_e}{2r_y}\sqrt{\frac{F}{EA}}\right) = 47.493294\ MPa = 6888\ psi \qquad [11.18]$$

$$\text{Recap of trigonometry: } \sec\left(\frac{L_e}{2r_y} \times \sqrt{\frac{F}{EA}}\right) = 1 \Big/ \cos\left(\frac{L_e}{2r_y} \times \sqrt{\frac{F}{EA}}\right)$$

Condition of strength is satisfied if $\sigma_{\max} \leq [\sigma]_e \rightarrow 47.493294\ MPa \prec 160\ MPa \rightarrow OK$

We then deduce that we need to calculate the maximum force which respects the stress σ_e:

$$\sigma_{\max} = \frac{F_{\max}}{A} \times \left| 1 + \left(\frac{V\varepsilon}{W_y}\right) \times 1 \Big/ \cos\left(\frac{L_e}{2r_y} \times \sqrt{\frac{F_{\max}}{EA}}\right) \right| \rightarrow F_{\max} = ?$$

From this, we deduce $F_{\max} = 741.158 \times 10^3\ N = 1.666 \times 10^6\ lbf$. In fifteen years' work, we have never exceeded that load to lift the mold in the workshop. The verification for $F_{\max}$ = 741158 N must prove the veracity of the stress initially applied:

$$\sigma_e = \frac{F_{\max}}{A} \times \left| 1 + \left(\frac{A\varepsilon}{W_y}\right) \times 1 \Big/ \cos\left(\frac{L_e}{2r_y} \times \sqrt{\frac{F_{\max}}{EA}}\right) \right| = 160.000\ MPa \leftarrow QED$$

The safety coefficient is:

$$S = F_{\max} / F_e = \left(741.158 \times 10^3 / 2.2 \times 10^5\right) = 3.368900$$

For an I-beam (also known as a double-T-beam) of the type HE 250B, the structure is used, at most, ten times per study session. The load simulated here is actually exaggerated. We have never exceeded a load of $F_{\max}$ = 500 N. Thus, the safety coefficient should never exceed 2 instead of the calculated value of S = 3.3689, which is around 3.5.

NOTE.– This case study, like many more presented herein, is a real-world scenario modeled in the author's workshops. The above calculations still hold true, and the mold-winch is still in very good working order. We have presented only a tiny part of the project. However, readers can draw inspiration from this to inform their own designs. The other results will be made available online on our website, if readers wish to consult them. The same is true for all the design projects discussed in this book, including the scooter presented below. In fact, this book is a condensed version of all our case studies in applied mechanical design.

11.6. Project 2: case studies: scooter

DESCRIPTION OF THE SITUATION.– The summer is a time to take advantage of the weather and leave the car at home. Samy is very aware of that. He chooses a different means of transport to get around between his home and his friends' houses. Samy is not heavily built. His scooter must respect certain conditions, such as safety, stability, lifetime, esthetic beauty and easy storage. Here is a brief rundown of the technical specifications.

General functions	Main functions	Assessment criteria
Safety	Be able to drive safely Conform to the safety standards Safety-durability Be suited to its user	Wheel diameter Non-skid platform Materials: handles Height and width of steering column
Educational value	Maintain balance while moving Steer easily	Stability
Lifetime	Withstand intensive use Withstand the loads	Choice of materials Principle of design
Esthetic	Be lightweight and elegant	
Easy to store	Compact and foldable	Design
Purchase conditions	Acceptable price	Materials – Design

Table 11.3. *Technical specifications of Samy G's scooter*

In order to satisfy the request, we proposed an electric scooter as a means of transport. The scooter conforms to the selection criteria on all points.

PURCHASE CONDITIONS.– In view of the financial limitations, the electric scooter is an excellent choice. The new price is around $200. If we consider the price of a new bicycle, which is around $400, we can see that there is a good quality-to-cost ratio. The scooter does not require a great investment in terms of energy, because it has an electric motor.

Safety: for getting around, Samy is looking for something solid, which performs well on the road and conforms to the safety standards. The parts are welded together. Both tires are made of rubber to ensure good grip on the asphalt. Their width allows good adherence on turns. The handles play an important role in safety. For that reason, they are of adjustable height to ensure a safe stance. It is a safe, ergonomic means of transport.

Educational value: whilst getting around, Samy wants to have fun. The aim is to take advantage of the good weather. The scooter will represent an easy-to-handle means of transport. Thus, it is crucial that he be able to operate the vehicle whilst maintaining reasonably easy balance. In order to ensure this stability, the handles are adjustable, the standing platform is fairly wide so the rider can fit both feet on, and thus be able to maintain good bodily balance.

Lifetime: Samy wants to buy a *durable* product which he can reuse. The welds and bearings can support loads up to a weight of 140 lbs. The axial forces exerted on the bearings and the shear forces and cracking inflicted on the welds can support Samy's slight weight, because the limit is 140 lbs.

Esthetic: to facilitate his movements and limit the effort required, Samy's means of transport must be lightweight. The use of hollow steel greatly lessens the weight of the scooter. In terms of esthetics, the scooter is handsome, with visible and attractive color. Thus, Samy can be proud of the way his scooter looks.

Easy to store: as it is compact, the scooter is an easy-to-store means of transport.

Initial data: the scooter will achieve a speed of 18 km/h whilst supporting a weight of up to 160 lb. The bearings for the wheels will need to have a lifetime of over 200 hours of use.

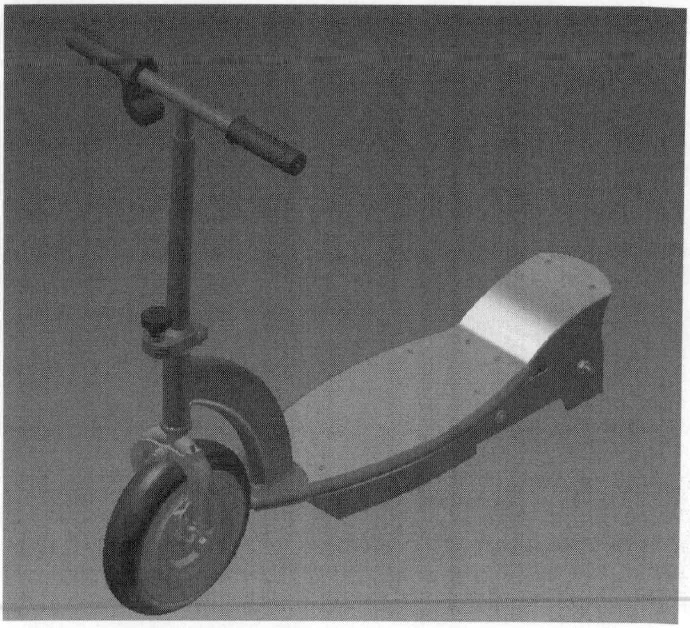

Figure 11.15. *Drawing of a scooter (Inventor Pro).*
For a color version of this figure, see www.iste.co.uk/grous/design.zip

11.6.1. *Presentation of the main parts*

Introduction: The project is to make improvements to an electric scooter in order to satisfy the technical specification. We wish to design the strongest, best-performing scooter possible.

Presentation of the mechanism: The scooter involves two mechanisms to be analyzed. The electrical system is reasonably simple. The motion is transmitted by a belt linking a cogwheel (at the rear) to the cogwheel of a motor which runs the mechanism. It also has a braking mechanism. The handbrakes are activated from the handlebars, comprising a wire which pulls the braking system onto the front wheel. This wire reduces the distance between the two pads that press against the wheel to immobilize it.

Materials: To make the scooter as light as possible, we change the chassis to aluminum. However, the fork must be made of steel, given the forces of pressure exerted by steering, the resultant of the ground and Samy's weight.

Increasing the power of the motor will mean that Samy can quickly get around, from home, to school and to his friends' houses. Increasing the admissible load will prevent premature wear on the *hot rod*. The braking distance we determined is that which we judge to be safest.

Design: SAMY's scooter	**Parts**	**Materials**
	Wheels	Rubber
	Bearings	1040 steel
	Handlebar	Aluminum
	Battery casing	Plastic
	Fork	1040 steel
	Shaft	1040 steel
	Cogwheel	Rubber
	Drive belt	Rubber
	Screw	1040 steel
	Washer	1040 steel
	Nut	1040 steel
	Handle	Rubber
	Brake spring	Aluminum
	Brake pad	Rubber

Figure 11.16. *Design of the assembled scooter. For a color version of this figure, see www.iste.co.uk/grous/design.zip*

NOTE.– We conducted the scooter project from start to finish (design, sketches and calculations), but in view of the volume of results which would need to be presented and explained, we shall limit this discussion to the bare bones. Readers can draw inspiration from the guidelines presented here, or contact the author to download the results.

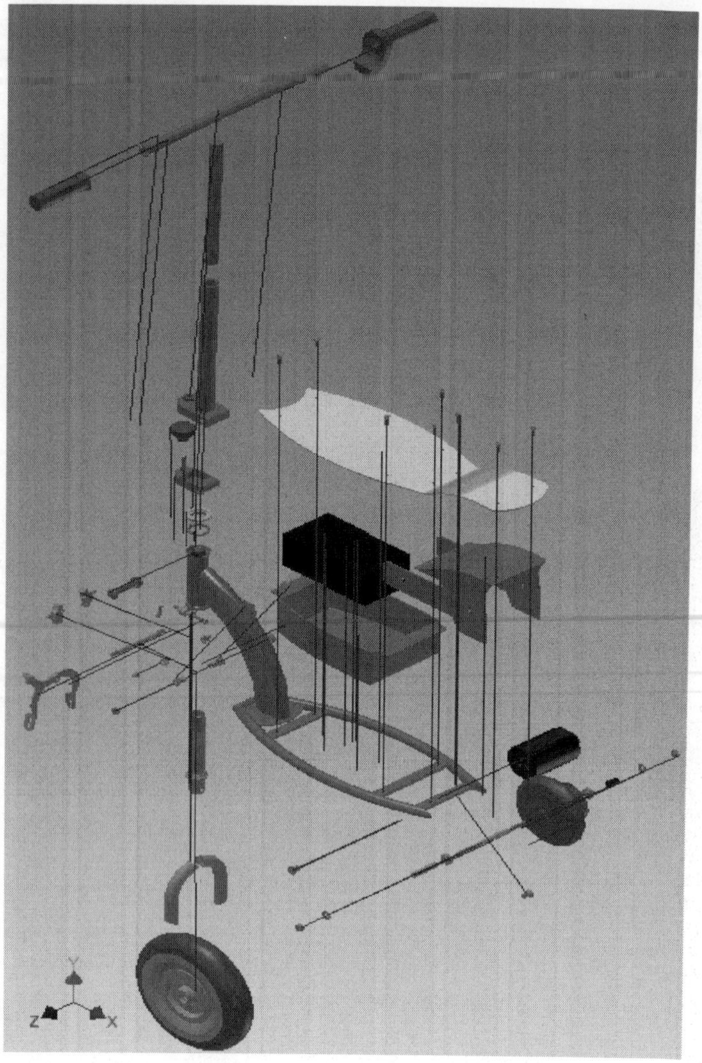

Figure 11.17. *Exploded diagram of the scooter designed for G. SAMY. For a color version of this figure, see www.iste.co.uk/grous/design.zip*

11.7. Project 3: dental hygiene dummy

This project is a real one, which caters to a pedagogical need expressed by the dental hygiene department. The goal was to design and create a system which the professors and students can use to imitate the *skull* and the *jawbone* (Figures 11.19 and 11.20).

Technical specifications

The below files are an overview of the steps required to bring the project to completion – i.e. from sketching/design and definition drawings to final production. Calculations of geometric specifications support the elements of the system's assembly.

Pre-project proposal dossier		
Subject: design and production of a *dummy* (Figure 11.20) used for demonstrative teaching to benefit both teachers and students in dental hygiene.		
Client	Dental hygiene department	
Initial data	A simple idea developed in response to the problem	
End goal	Teaching prop for dental hygiene (DH)	
Creation date	....	Obs. ....
Delivery date	....	Obs. ....
Project manager	.... and	Teacher in DH Lab.
Dossier # 05		

Table 11.4. *Pre-project proposal*

In this mechanical system, the goal is to be able to, precisely and with stability, manipulate the jaws as part of the skull (Figure 11.18). The objective for the dental hygienists is to mount jaws on stem number 3 so that this sort of 360° manipulation can easily be done. The system is a mechanism for educational demonstrations in their workshops and other DH labs. The problem at hand was dealt with by the department of mechanical engineering techniques.

Client requirement analysis		
Client requirement (s)	Dossier #	
Definition of technical specifications		
Client	DH department	
Date of expression	....	**Dossier 10**
Analysis and validation of requirement		
Requirement(s)	Dossier # (see technical specifications)	
Validation of requirement	Requirement validated after consultations with the client (DH)	
Client	Dental hygiene department	
Date of expression	....	**Dossier 15**

Table 11.5. *Project analysis*

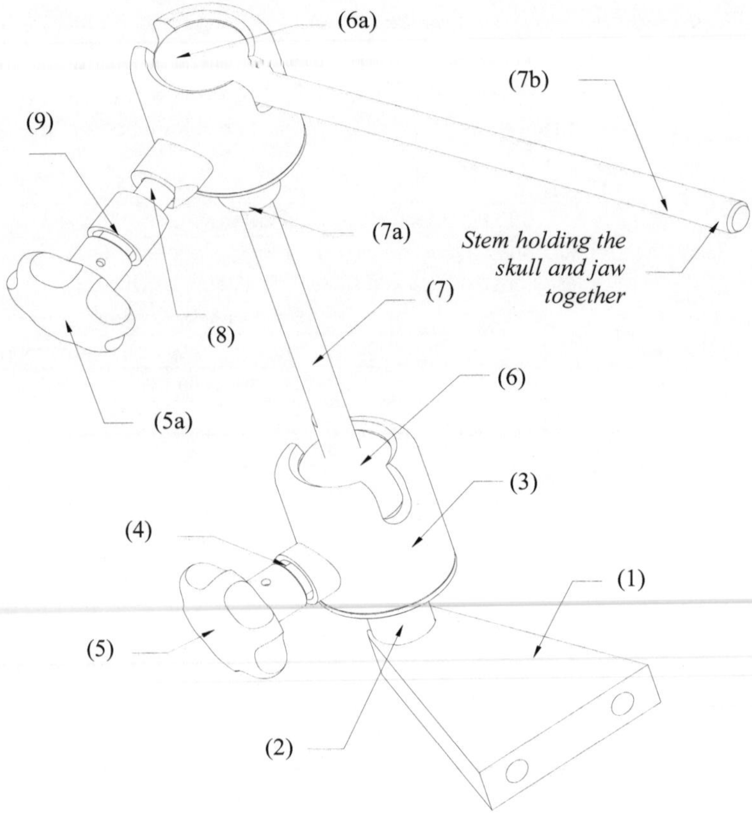

Figure 11.18. *Skull and jaws drawn in 3D with Inventor Pro*

Teachers and students in the DH department often use educational demonstrations in their lab activities. To use the support presented above, they manipulate the patients' jaws to demonstrate activities in the area of dental hygiene. The system is conditioned by:

– the flexibility of the setup and the ability to remove the jaws from the stem (3);

– the system's light weight (total mass);

– its versatility and the ease with which it can be transported (this system will be lent to students to take home). The system must be able to adapt to represent patients of different sizes.

Similar mechanical systems are already in place in the circles of industrial dentistry. For purely pedagogical and economic reasons, we decided to design a simple mechanical system, suited to the requirements of the DH department. These requirements could evolve over time based on new usage conditions expressed by the client. For example, we can imagine more versatile manipulation with pneumatic controls. It is helpful to follow this organigram.

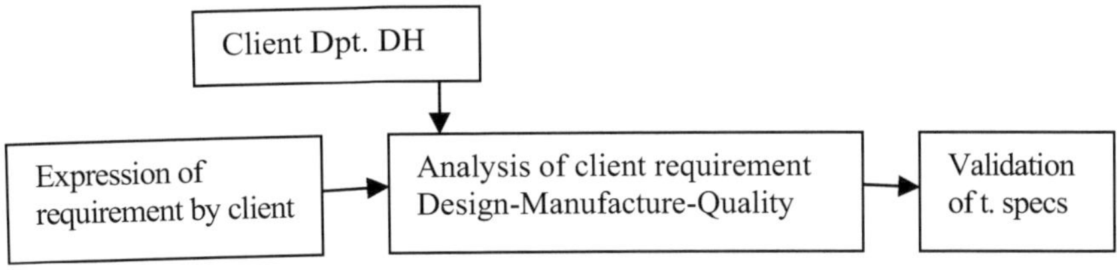

To satisfy the requirement expressed above, we use PERT (Program Evaluation and Review Technique) – i.e. the spider chart shown below. The service functions describe the client's requirements, for which the designer needs to cater.

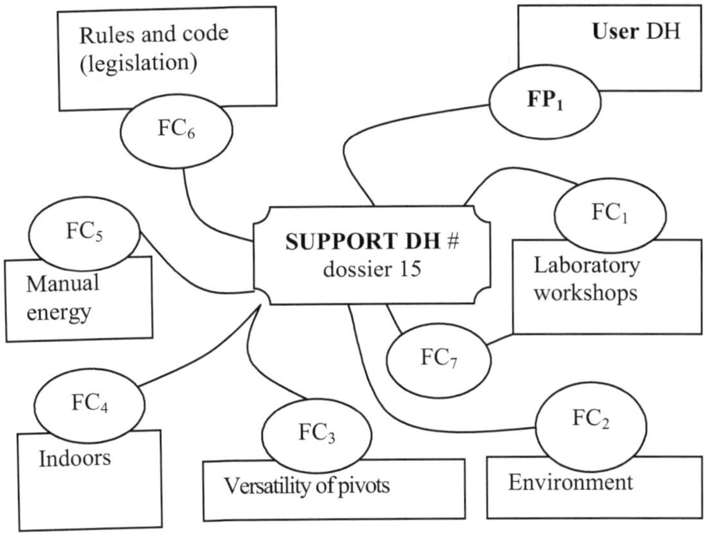

Figure 11.19. *Spider diagram based on the APTE principle of jaw support*

Where:

– FP_1 is the primary function of the support: to enable the hygienist to perform demonstrations by mounting the jaws on stem 3 (between two media);

– FC_1: the first corollary function is to withstand the manipulation (guidance) around nylon 6/6 spheres, and thus easily pivot the skull and jaws as required for the demonstrations (see Figure 11.19);

– FC_2: to respect the environment and the total mass of the mechanical system;

– FC_3: to be able to easily maneuver the jaws (versatility) of different sizes;

– FC_4: to withstand the forces of friction and wedging (jamming);

– FC_5: to be lightweight, malleable, safe and reliable;

– FC_6: to conform to the applicable norms and other codes;

– FC_7: to be comfortable and ergonomic.

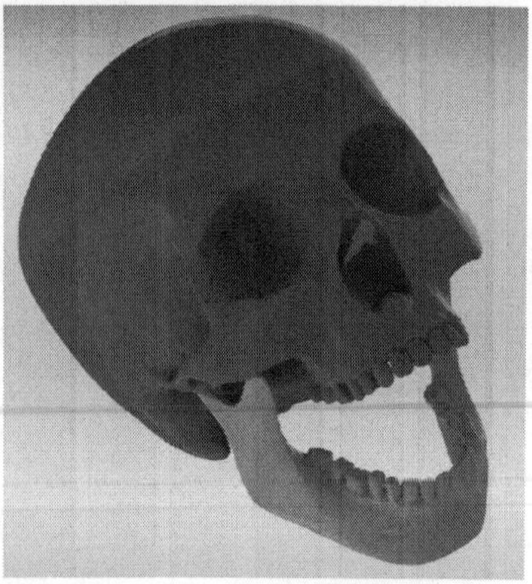

Figure 11.20. *Skull and jaws scanned and drawn in 3D. For a color version of this figure, see www.iste.co.uk/grous/design.zip*

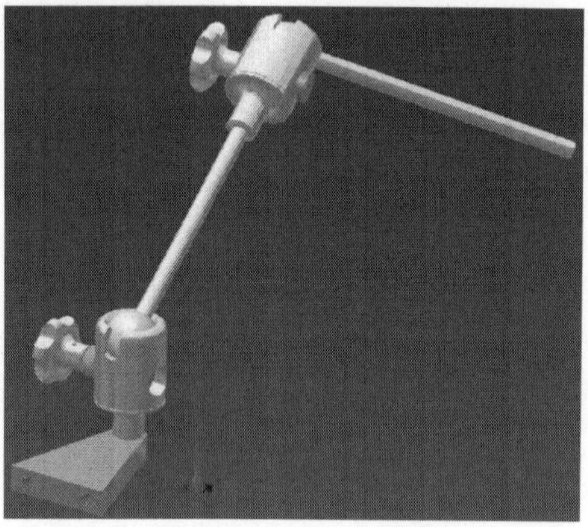

Figure 11.21. *Pivoting support: mechanical jaw system. For a color version of this figure, see www.iste.co.uk/grous/design.zip*

The jaws shown in Figure 11.19, which fit to the skull, are affixed to the stem (3). The rod (1) holds a sheet of metal which fits perfectly to the shape of its housing. The intermediary stem (2), which fits to the sphere (nylon 6/6), is free to pivot however the user deems fit. The same is true for the sphere on stem (3). The principle is similar to the use of a support for the dial-type comparators used in machining metrology. The jaws can be opened and closed by the dental hygienist.

The logical steps in the process are:

– *states*: the states correspond to the operation, assembly and maintenance;

– *environment*: the elements orbiting the system (the support mechanism) are functional because their tolerance values are designed in keeping with the materials and standards of GPS (geometrical product specifications). The energy used is manual, and an appropriate choice of materials for the mechanism optimizes the friction, for greater reliability/quality;

– *primary functions*: the dummy is easy and ergonomically safe to operate. The system is versatile, because different size jaws can be fitted to it, depending on the needs of the DH labs and workshops.

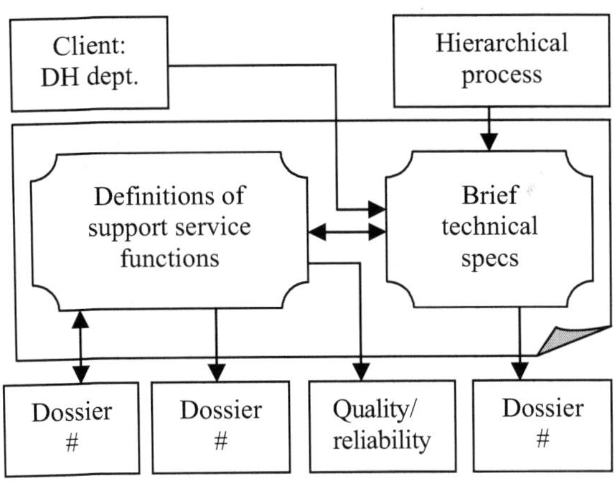

Figure 11.22. *Detail of dossier # 15. Technical specifications*

The needs analysis, here, feeds into the technical specifications, approved by the dental hygiene department (the client). Thus, we move on to the feasibility study.

Feasibility study		
Feasibility	Dossier XX	
Client	Department	
Date of expression	....	**Dossier 25**

Table 11.6. *Dossier for feasibility study*

There is no "recipe" or book to guide designers in realizing a project. In the specific case of interest to us here, rather than setting out to "reinvent" an already-existing mechanism, we need to follow the client's requests in order to cater adequately to their needs, with emphasis placed on the specific details inherent to the use of the system by the students and their teachers. The aim, at this stage, is to put forward usable solutions. Tables 11.7 and 11.8 also show a part of the proposed solution which was not accepted, for the support.

Before seeking to create any solutions, we define the assessment criteria.

Proposed solution(s) dossier		
Technical specifications	Dossier #	
Client	DH department	
Date of expression	....	Computer-aided drawing → Dossier 25-1
Selection criteria		Computer-aided design → Dossier 25-2
FAST method		Graphs → Dossier 25-3
FAST		Calculations → Dossier 25-4
Dossier of solutions not adopted		
Application 1	Reason: too heavy for the proposed materials…	
Application 2	Reason: lightweight (plastic) but not reliable …	
Date of expression	....	**Dossier 30**

Table 11.7. *Proposed solution dossiers*

11.7.1. *Support clamped to the lab bench in the dental hygiene department*

A user could quite easily apply a force of 20 lbf on the end of the handle:

1) What would the torsion stress on the clamp be, and where would it be applied?

2) What is the pressure P that can be handled by the internal thread?

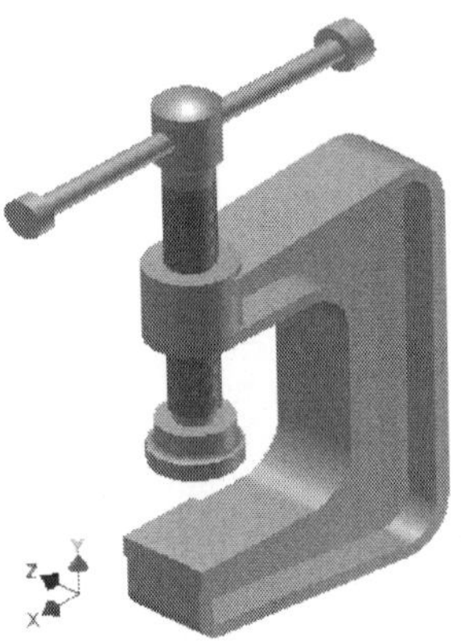

Initial data:

Outside diameter $D_e = 0.5$ in
Root diameter R_r=0.4001 in
Root surface Sr = 0.1257 in^2
Friction coefficient, f= 0.120
Friction coefficient of $f_c = 0.250$
Mean radius of press spires – collar:
$R_c = 0.25$ in
Outside-manual load W = 1000 *lb*
The thread is in accordance with US standard:
13 threads per inch (single-threaded)

Figure 11.23. *Clamp to hold the dental hygiene dummy.*
For a color version of this figure, see www.iste.co.uk/grous/design.zip

SOLUTION.– Calculation of the mean radius, $R_m = [0.50 + 0.4001]/4 = 0.225$ in.

Calculation of the helix angle if the step of the helix is $p = 1/13$; then →

$$\tan(\alpha) = step/2\pi R_m = 0.0544 \rightarrow then\ \alpha = \arctan(0.0544) = 3.114°$$

We consider $\theta_n = 30$ deg, because the helix angle α is reasonably small, so Cos $(\theta_n) = 0.866$. For: W = 1000 *lb*; tan (α) = 0.544; f = 0.120; $f_c = 0.25$; $R_c = 0.25$ and $R_m = 0.225$:

$$\begin{cases} Torque(W) = \dfrac{W}{2}\left(R_m \dfrac{\tan(\alpha) + f/\cos(\theta_n)}{1 - [f\tan(\alpha)/\cos(\theta_n)]} + f_c R_c\right) = 114.308\ lbfin = 12.915\ Nm \\ T_W = \dfrac{1000}{2}\left(0.225\left(0.544 + \dfrac{0.12}{0.866}\right)\Big/\left(1 - \dfrac{0.12 \times 0.0544}{0.866}\right) + 0.25 \times 0.25\right) = 108.622\ lbfin \end{cases}$$

To develop that moment with the load of 20 lb, we need **L**? Let the radius be r = 0.200 in, and calculate the quadratic moment of inertia J:

$$J = \pi r^4/2 = 2.513\times10^{-3}\left[in^4\right]$$

Calculate the torsion stress for: l = 6 in $\tau = Torque\times r/J = 8.644\times10^3\left[psi\right]$.

Calculate the bending stress for: l = 6 in; r = 0.2 in.

The bending moment is: $M = F\times l = 120\ in.lb$. Knowing the quadratic moment of inertia J_1 = 1.257×10^{-3} in^4, calculate the bending stress as follows:

For: $J_1 = \pi r^4/4 = 1.257\times10^{-3}\ in^4\ then\ M = Mr/J_1 = 19100\ psi = 131.69\ MPa$

The maximum torsion stress: $\tau_{\max} = \sqrt{(\sigma/2)^2 + \tau^2} = 12880\ MPa = 12.88\ psi$.

Consider the cross-section subject to a twisting moment, calculated as follows for r = 0.2 in:

$$M = Wf_c r_c/J = 19890\ MPa = 4970\ psi$$

The compressive stress for the area A = 0.1257 in^2 would be M = W/A = 7955 *psi*. Thus, we conclude that the maximum stress on the section A-A is:

$$n = 13\ threads;\ R_m = 0.225\ in;\ h = 0.050\ in \rightarrow P = W/2\pi n R_m h = 1088\ psi = 7.5\ MPa$$

The length of the threaded stem = 1 in; the pitch = 1/13. This proves that the number of threads per inch (N) is indeed 13.

N = N$_b$ of threads per inch/pitch = 1[1/3] = 13 threads per inch (*QED*) → For: r_0 = 0.25 in; r_i = 0.2 in, the height of the thread would be: $h = r_0 - r_i = 0.05\ in$ (*QED*).

CONCLUSION.– The clamp is well suited for the required forces of pressure, and the characteristics of the threading (material) are tried and tested, and subject to rigorous standards. For reasons of efficiency, the solutions not used are not discussed here.

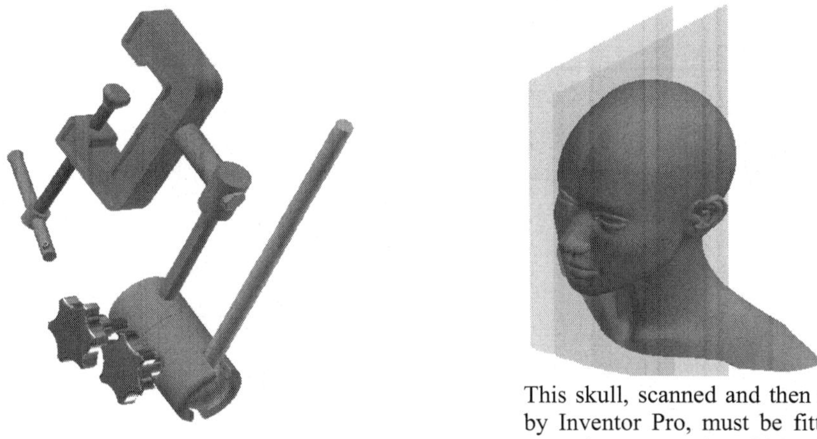

This skull, scanned and then simulated by Inventor Pro, must be fitted to the stem shown opposite without breaking.

Figure 11.24. *System for clamping the dummy, designed by the author and his team. For a color version of this figure, see www.iste.co.uk/grous/design.zip*

Dossier of rejected solutions		
Solution 1 rejected	All parts made of 6061 steel and brass (?) → Heavy Reasons: system heavy and costly. Expensive machining process. Activity chart shows lengthy hours, etc.	
Solution 2	If actuators were pneumatic: system would perform well but would be cumbersome. Costly to maintain. Added weight, actuator less easy to use. High price. Less machining.	
Solution 3	Present version adopted (assembly illustrated in Figure 11.25) Reasons: client approved the technical specifications of the design	
Date of expression	….. ….. …..	**Dossier 30**
Application 1	Mechanical system – simple, flexible, versatile and lightweight	
Principles of calculations	Calculation dossier (CAD): GPS and SOM	
Plan and definition	CAD dossier (parts and assembly)	
Manufacture	Machining dossier	
Quality/Reliability	Dimensional metrology dossier	
Date of expression	….. ….. …..	**Dossier 35**

Mechanical production dossier		
Machining 1	Dossier #	
Metrology/Quality		
Date of expression	....	**Dossier 40**
Test array dossier – Assembly and Quality/Reliability		
Application	Dossier # assembly of manufactured elements	
Date of expression	....	**Dossier 45**
Development and final delivery of the project		
Project development	Dossier #	
Demonstration		
Delivery of project		
Date of expression	....	**Dossier 50**

Table 11.8. *Dossiers for the design of a dummy for use in a dental hygiene lab*

11.7.2. *Case studies of a complete block and crank link*

A crank, as a cylinder (sphere) excentered from its rotational axis, can act either as a power transmission system or a tightening system. It is the latter scenario on which this case study focuses.

STATEMENT OF THE PROBLEM.– Consider a sphere of nylon 66 and another sphere of brass. It acts as a multiplier of the pressing force, the equation for which is given below:

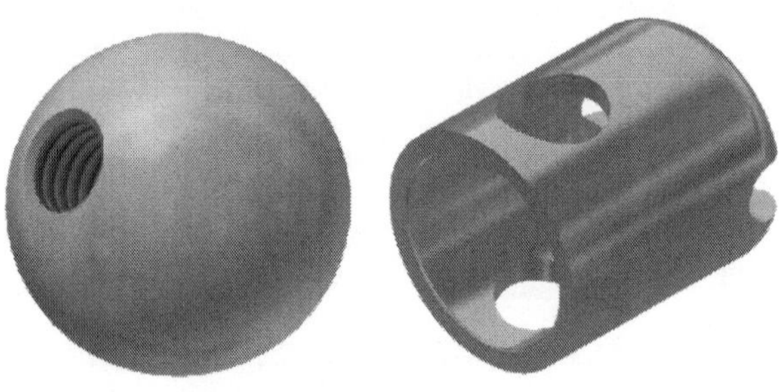

Figure 11.25. *Nylon 66 sphere and its housing (also see Figure 11.25). For a color version of this figure, see www.iste.co.uk/grous/design.zip*

For: l = 4 mm = 0.157 in; L = 12 mm = 0.472 in; d = 8 mm = 0.315 in; F_{m1} = 5 N = 1.124 *lbf*; E_1 = 16 MPa = 2.321×10^3 and E_3 = 9 MPa = 1.305×10^3.

SOLUTION.– From the equilibrium condition, we isolate the reaction R_{31} as follows:

$$R_{31}l = F_{m1}L \rightarrow R_{31} = F_{m1}\left(L/l\right) = 15\ N = 3.372\ lbf$$

The result is conclusive. The resistance condition is ensured by: $R_{31} > F_{m1}$, meaning that: 15 N > 5 N. From Hertz's law, we derive the maximum contact pressure:

$$P_{\max} = 0.81\times\sqrt{R_{31}E_1E_3/Ld\left(E_1+E_3\right)} = 0.768\ MPa = 111.452\ psi$$

The P_{max} found does not exceed the elastic limit, the blocking is reliable and will take place:

$$From\ 2\times\varepsilon \le \mu\times d\ ;\ for\,\mu = 0.25 \rightarrow \varepsilon \le \mu d/2 = 1\ mm = 0.039\ in$$

11.7.3. *Explanatory photographic definition of the final product*

It would take an entire book to present all the graphical and numerical results found in this project. For that reason, we choose only to present the graphical results, with a few accompanying examples of the calculation, in order to *spur on* the users (students and teachers) in reproducing similar project models. The dummy was designed and built in the TGM department, and was shown on the television channel of radio Canada Ottawa-Gatineau in 2015 – the link to watch the video is below (French language):

http://ici.radio-canada.ca/regions/ottawa/2015/12/03/009-innovation-capitale-cegep-outaouais-genie-mecanique-hygiene-dentaire-mannequins.shtml

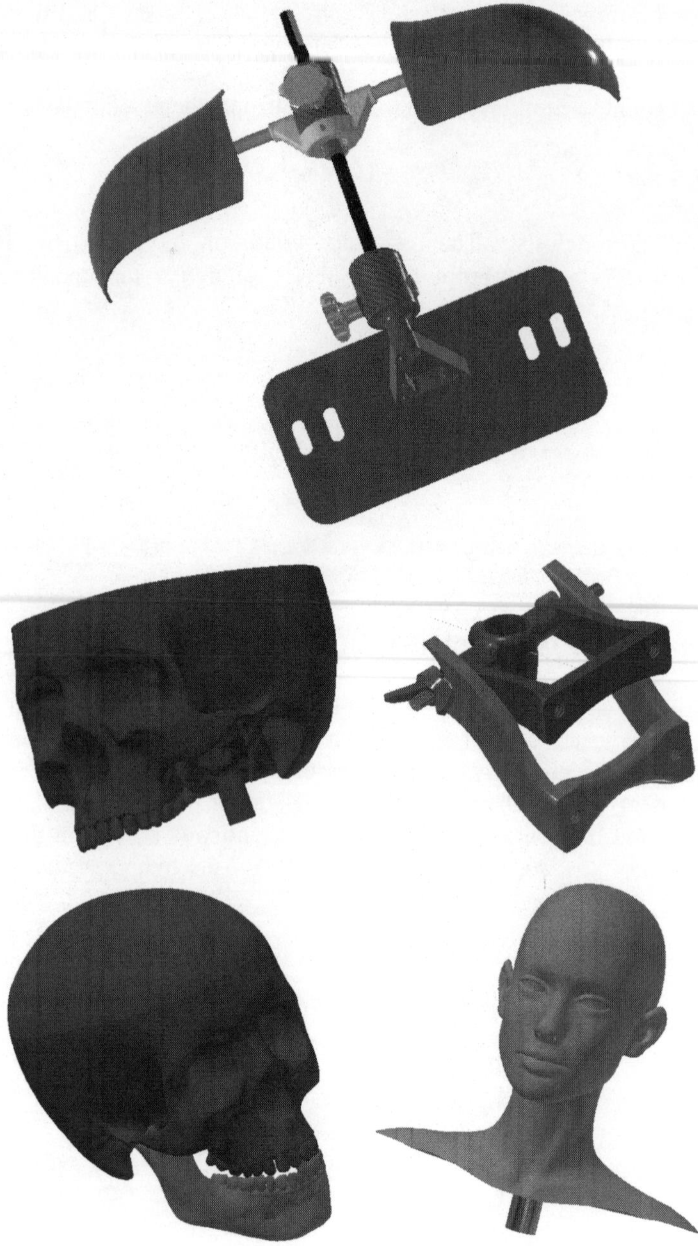

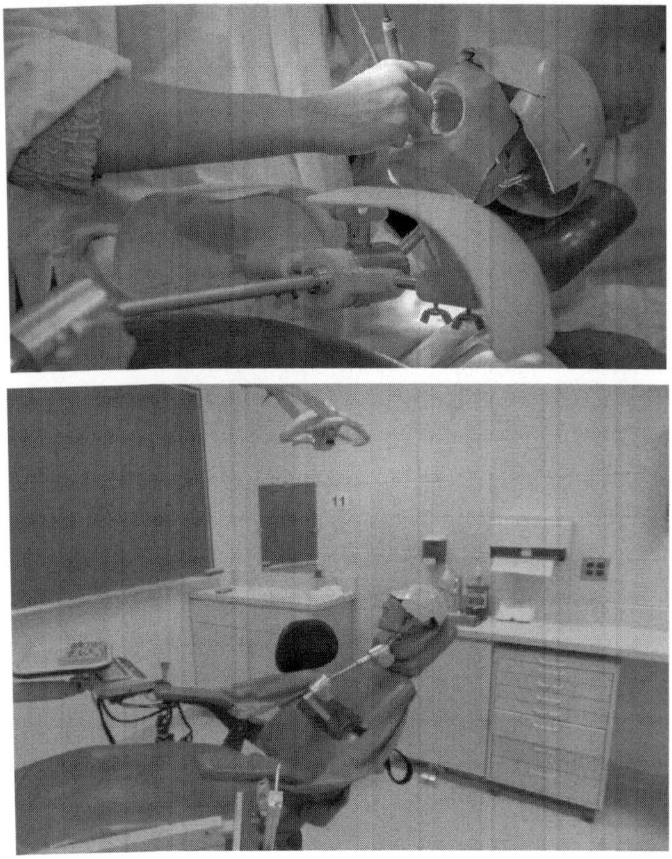

Figure 11.26. *Real photographs of the dummy in a dental hygiene lab class. Dummy designed with Inventor Pro, used in the dental hygiene department. For a color version of this figure, see www.iste.co.uk/grous/design.zip*

Conclusion

Recap and Reflections on Applied Mechanical Design

Parallels could be drawn between summarizing a book on mechanical design and summarizing a wide array of disciplines relating to engineering sciences. In this case, we would see a series of brief and digestible overviews of subjects such as applied physics; materials science; strength of materials; dimensional and geometric specifications of machine elements; theory of mechanisms and machines; applied metrology; drafting and assembly techniques; etc.

In writing this book, I made practical choices based on my own introductory classes to mechanical design, which also provided the case studies presented, along with their solutions. Thus, the book opened with the methodology, consisting of presenting the steps of design, from an analysis of the client's requirements to the constitutive dossiers of the project preliminaries. Then, we chose industrial case studies where solved problems are linked to the techniques of mechanical engineering, and sometimes civil engineering, with a significant role also given to the materials. Whether it is a question of a power transmission mechanism, a piece of equipment (shaft, beam, gear system, pulley, etc.), a good grasp of materials sciences is at the very heart of designers' needs.

The problems discussed here involve calculations to find quantitative solutions. In many cases, design students have to make qualitative judgments based on the technical specifications. The primary goal was to better equip learners and teachers by offering examples with solutions. This strategy is aimed at instilling a formative virtue in the teaching of the elements of applied mechanical design. The secondary goal was to delve deep into the choice of materials and the physical and mechanical characteristics to optimize the design.

Without materials science, mechanical design would be nothing more than a mathematical exercise, devoid of structural analysis. It often happens that the question relates to the best choice of materials, rather than to the mathematical approach taken to solve an algebraic search. The laws of strength of materials and of mechanism and machine theory have been solidly established in the specialized literature. The computational methods are well known.

It is sometimes hard to predict how materials will respond to the stresses of assembly and natural stresses, or environmental pressures. For this reason, it is absolutely crucial to gather all possible data about the behavior of the materials used in design. The usage conditions of the materials need to be duly and clearly set out. It is uncommon for one material alone to guarantee all the optimal properties we want. Take the simple example of a material which is strong but not ductile: we need to find an acceptable compromise in order to choose the best properties to serve the objective of the design – i.e. the client's objective.

Nowadays, design software packages are very well-equipped with calculation tools. Therefore, students focus more on grasping the concepts of mechanical assembly and simulation of stress analysis. For example, when we choose a material (however simple the choice may be), the software tells us all about that material's physical and mechanical properties. It is the designer's *duty* to have a good grasp of these characteristics in order to bring the design project to fruition for a reasonably low cost.

In many engineering schools, designs are presented as assembled drafts and drawings. In my humble opinion, this approach is over-simplistic, because design is, first and foremost, a noble question of art and genuine craft. It should not – indeed *must* not – be squeezed into the domain of simple graphic representation. In the final analysis, it must be remembered that the field of design is so vast that it is difficult – and sometimes over-simplistic – to present the best solution to a given project. Besides, there is no single right solution; there are many solutions. Therefore, this design book must be viewed as a guide, rather than an encyclopedic coverage of the subject.

Appendix

Lexicon, Glossary and Standards

A.1. Introduction

The calculation principles include combinations of properties to achieve the best possible performances from a project. Modeling gives us exact values for the stresses, strains and component failures, which are taken into account in the operational limits. In this book, we have given concrete examples with geometric forms subjected to various types of loading. The formulae presented herein are drawn from the specialized literature, cited throughout the chapters. The tables, standards, formulae and other examples are presented here to serve as guidelines. Under no circumstances should what follows be viewed as an exhaustive guide. Readers are encouraged to refer to other, more specialized publications.

A.2. Abbreviations used in this book

CSA	Canadian Standards Association
SAE	Society of Automotive Engineers
AISI	American Iron and Steel Institute
ASME	American Society of Mechanical Engineers
ASTM	American Society of Testing Materials
CAD	Computer-Aided Design and Drafting
CAM	Computer-Aided Manufacturing
CAD/CAM	Computer-Aided Design and Manufacturing

A.3. A few international standards on surface state and metrology

ISO standards on this subject are frequently updated. For a fuller understanding, it is advisable to refer to the latest version of the following.

ISO standards	Méthode du profil ISO 25178 Techniques de filtrage. See http://www.digitalsurf.fr/en/guide-metrology-standards.html, replaced by EN https://www.iso.org/standard/50176.html
ISO 3274:1998	Geometrical Product Specifications (GPS) — Surface texture: Profile method — Nominal characteristics of contact (stylus) devices https://www.iso.org/obp/ui/#iso:std:iso:3274:ed-2:v1:cor:1:v1:en
ISO 12085:1998	Geometrical Product Specifications (GPS) — Surface texture: Profile method — Motif parameters https://www.iso.org/obp/ui/#iso:std:iso:12085:ed-1:v1:cor:1:v1:en
ISO 1:2002	Geometrical Product Specifications (GPS) — Standard reference temperature for geometrical product specification and verification https://www.iso.org/standard/28086.html
ISO 1101:2013	Geometrical product specifications (GPS) — Geometrical tolerancing — Tolerances of form, orientation, location and run-out https://www.iso.org/obp/ui/#iso:std:iso:1101:ed-3:v1:cor:1:v1:en
ISO 8785:1999	Geometrical Product Specification (GPS) — Surface imperfections — Terms, definitions and parameters https://www.iso.org/standard/16210.html
ISO 14253-1:2013	Geometrical product specifications (GPS) — Inspection by measurement of workpieces and measuring equipment — Part 1: Decision rules for verifying conformity or nonconformity with specifications https://www.iso.org/standard/70137.html
ISO/IEC Guide 98-3:2014	Uncertainty of measurement — Part 3: Guide to the expression of uncertainty in measurement (GUM:1995) https://www.iso.org/standard/50461.html
ISO/IEC Guide 99:2011	International vocabulary of metrology — Basic and general concepts and associated terms (VIM) https://www.iso.org/standard/45324.html

Table A.1. *ISO standards on surface texture employed in this book*

A.3.1. *Standard SAE grades of steel (ASTM- SAE, USA, Canada)*

Grade number	Standard SAE labels
1xxx	Carbon steels
2xxx	Nickel steels
3xxx	Nickel-chromium steels
4xxx	Molybdenum steels
5xxx	Chromium steels
6xxx	Chromium-vanadium steels
7xxx	Tungsten steels

8xxx	Nickel-chromium-molybdenum steels
9xxx	Silicon-manganese steels
Representative example: SAE 1040 → 1 = Carbon steel. 0 = No significant content of other elements besides carbon. 40 = 0.45 % carbon content.	

Table A.2. *Material grades according to the SAE*

A.3.2. *Reference table giving some friction coefficients (μ)*

Material 1	Material 2	Dry friction coefficient		Greased friction coefficient	
		Static	Slip	Static	Slip
Aluminum	Aluminum	1.05-1.35	1.4	0.3	
Brake (leather)	Cast iron	0.4			
Brass	Cast iron		0.3		
Bronze	Steel			0.16	
Cast iron	Cast iron	1.1	0.15		0.07
Copper	Copper	1.0		0.08	
Copper	Mild steel	0.53	0.36		0.18
Leather	Wood	0.3-0.4			
Polystyrene	Steel	0.3-0.35		0.3-0.35	
Steel	Aluminum	0.45			
Steel	Cast iron	0.4		0.21	
Teflon	Steel	0.04		0.04	0.04

Table A.3. *Friction coefficients (μ)*

A.3.3. *Mechanical properties of wood*

Source (Material)	E, [MPa] × 10^3 Elasticity modulus for bending	σ_t [MPa] Compressive strength	σ_C [MPa] Traction strength
Birch	13300 →16200	42 ... 60	130 ... 140
Cherry tree	9500 →11000	44 ... 55	98
Chestnut	8200 → 8800	40 ... 52	115 ... 142
Oak	10500 → 14500	52 ... 64	88 ... 110
Spruce	10000 → 12000	40 ... 50	80 ... 90
Maple	9100 → 12000	46 ... 62	80 ... 140
Elm	10800	45 ... 55	78

Source (Material)	σ_F [MPa] Bending strength	τ_C [MPa] Shear strength	σ_{Shock} Nm/cm^2 Shock resistance	τ_C [MPa] Brinell hardness r_{ball} = 10-12%	
				// to fibers	⊥ to fibers

Birch	120-144	11.8-14.2	7.5-10	48	21-34
Cherry tree	83-110	-	-	51-58	28-31
Chestnut	63-79	7.8-9.3	5.5-5.9	32-37	15-23
Oak	86-108	9.3-11.5	5-7.4	50-65	23-42
Spruce	65-77	5-7.5	4-5	31	12-14
Maple	85-135	8.5-11	6.2-6.6	48-61	27-34
Elm	72-105	6.8	5.9	58-63	37

Table A.4. *Examples of mechanical properties of wood. Source: Sell and Kropf*

We measure the *shock resistance* by the work needed to break a bar with the *cross-section area (A)*. It is characteristic of the *resilience*. The *fragility* of a wood is an important property. This parameter is influenced by irregularities in the growing of the wood. This is the main reason for the broad range of averages observed. The *hardness* of a wood surface, subjected to local compressive stress, is often determined by the Brinell method.

A.4. Recap of some calculation formulae used in design

In design, mathematical modeling and properly scaled sketches are two complementary parts of the process, and are absolutely ineluctable.

Ref.	**Elastic deformation**	**Plastic deformation**
Uni-axial	$\varepsilon_1 = \sigma_1 / E$	$\sigma_1 \geq R_e$
General	$\varepsilon_1 = \dfrac{\sigma_1}{E} - \dfrac{\nu(\sigma_2 + \sigma_3)}{E}$	$\sigma_1 \succ \sigma_2 \succ \sigma_3$ *Tresca criterion* $\rightarrow \sigma_1 - \sigma_3 = R_e$ *von Mises criterion* $\rightarrow$ $\sigma_e^1 = \dfrac{(\sigma_1 - \sigma_3)^2 + (\sigma_2 - \sigma_3)^2 + (\sigma_3 - \sigma_1)^2}{2}$
	Creep	**Fracture mechanics**
Uni-axial	$\dfrac{\varepsilon_1}{dt} = \dfrac{\varepsilon_0}{dt}\left(\dfrac{\sigma}{\sigma_0}\right)^n$	$\sigma_1 = \dfrac{C_{perf} K_{IC}}{\sqrt{\pi a}}, C_{perf}^{trac} \cong 1; C_{perf}^{Comp} \cong 15$
General	$\dfrac{\varepsilon_1}{dt} = \dfrac{\varepsilon_0}{dt}\left(\dfrac{\sigma^{n-1}}{\sigma_0^n}\right)\left(\sigma_1 - \dfrac{\sigma_2 - \sigma_3}{2}\right)$	$\sigma_1 = \dfrac{C_{perf} K_{IC}}{\sqrt{\pi a}}, (\sigma_1 \succ \sigma_2 \succ \sigma_3)$

Table A.5. *Stresses and strains (deformations). Source: [ASH 13]*

A.4.1. *Contact stresses according to Hertzian theory*

Contact theory (Hertz) comes into play when surfaces come into contact by touching at one or more discrete points. Depending on the loads applied to the components, the points of contact are flattened elastically. The areas of contact may increase, to the extent that failures sometimes occur. Plastic failure occurs under shear stress. R is the radius of the contact zone and (δ) is the deformation.

Descriptions	**Contact radius**	**Displacement, δ (m)**
F is the force (load) in N ε is the contact radius in (m) (δ) is the displacement in (m) σ is the applied stress in MPa σ_e is the elastic limit in MPa ν are the Poisson ratios R_1 and R_2 are the radii of the spheres in m E_1, E_2 are the elastic moduli in MPa Equations [A.1]	$\varepsilon = 7/10\sqrt[3]{FR/E}$ $\nu_{Poisson} = 1/3$	$\delta = 1.0 \times \sqrt[3]{F^2/E^2R}$ $\nu_{Poisson} = 1/3$
	$\varepsilon = \sqrt[3]{\frac{3}{4}\frac{F \times R_1 \times R_2}{E^*\left(R_1 + R_2\right)}}$	$\delta = \sqrt[3]{\frac{9}{16}\frac{F^2\left(R_1 + R_2\right)}{\left(E^*\right)^2 R_1 R_2}}$
	Stresses σ in MPa	
	$\sigma_{cmax} = \frac{3F}{2\pi a^2}$; $\sigma_{smax} = \frac{F}{2\pi a^2}$; $\sigma_{\tau max} = \frac{F}{6\pi a^2}$	

Table A.6. *Contact stresses according to Hertzian theory*

$$E^* = 1\bigg/\left\{\frac{1-\nu_1^2}{E_1} + \frac{1-\nu_2^2}{E_2}\right\} \text{ and } 3\sigma_e = F/\pi a^2 \qquad \text{[A.2]}$$

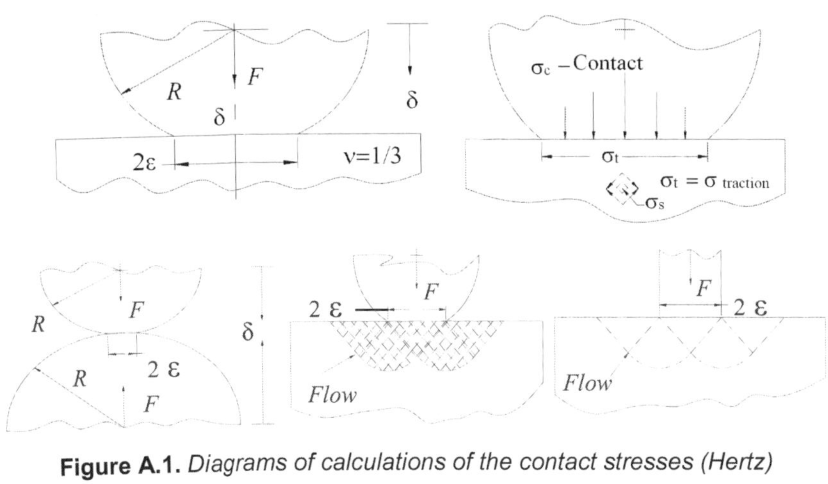

Figure A.1. *Diagrams of calculations of the contact stresses (Hertz)*

A.4.2. *Buckling of columns (Euler) and other homogeneous plates*

For slender forms, an elastic column under compressive stress is represented by $F_{critical}$. The expressions are as follows:

$$\begin{cases} P_{crit} = F_{critical} = n^2 \dfrac{EJ\pi^2}{l^2} \text{ or indeed} \rightarrow F_{critical} / A = n^2 \dfrac{E\pi^2}{(l/r)^2} \\ P_{crit} = \dfrac{EJ\pi^2}{l^2} - \dfrac{M^2}{4EJ} \text{ or } F_{crit} = n^2 \dfrac{EJ\pi^2}{l^2} + \dfrac{\kappa l^2}{n^2} \text{ and } P_{crit} = \dfrac{3EJ}{R_{mean}{}^3} \end{cases} \quad [A.3]$$

where:

– (n) is the constant assimilable to a half wavelength in conditions of buckling;

– κ is the stiffness per unit depth in $N/m^2 = Pa$;

– γ is the local lateral deformation in m or in mm;

– J is the moment of inertia of the section in m^4 or in mm^4;

– P is the pressure exerted (see figures below) $\rightarrow P = \kappa \times \gamma$ in $N/m^2 = Pa$;

– E is the elasticity modulus (Young's modulus) in Pa;

– M is the moment in N.m;

– r is the gyradius $r = \sqrt{J/A}$ in m or mm;

– F is the force in (N) and A is the area in m^2.

The effect of (γ) is to increase the critical force $F_{critical}$. The stresses considered for each couple (2 by 2) are presented below.

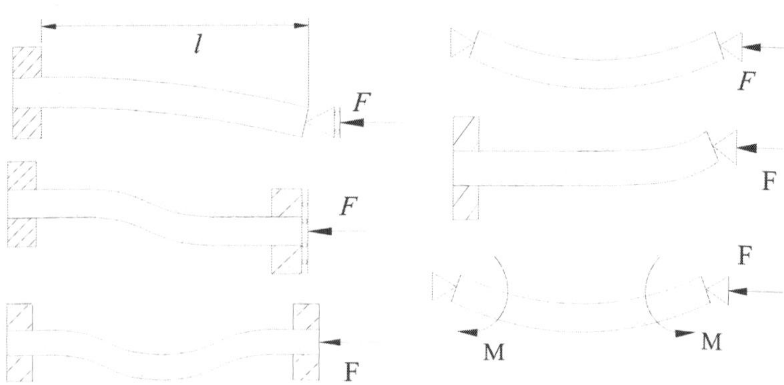

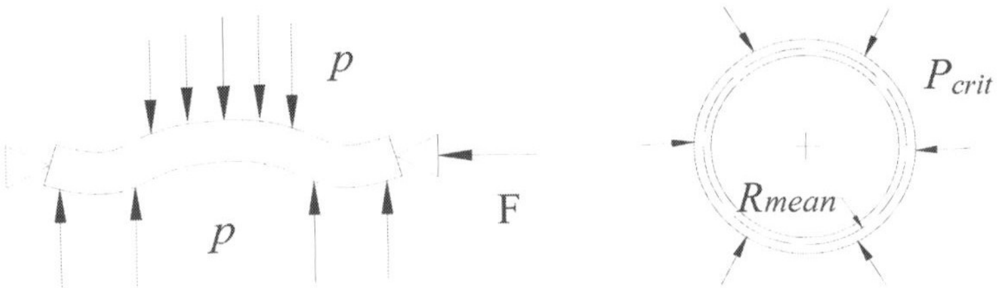

Figure A.2. *Diagrams of calculations for buckling columns and plates*

A.4.3. *Shafts subject to torsion (twisting)*

The twisting torque, applied to the extremities of an *isotropic* material having a uniform cross-section, acting perpendicular to the axis of the bar (shaft), produces a torsion angle (θ). In conditions of torsion, it is the stiffness (G) which is considered. It is represented by its shear modulus (G). In the case of circular pipes, *K* replaces the polar moment of inertia (*J*). In other cases, *K* is significantly less than *J*. Here are the main expressions of the calculations:

$$\begin{cases} \theta = l \times T_\tau / KG \leftarrow \text{torsion angle in degrees} = 0^\text{o} \swarrow \\ \textit{start of plasticity } \tau_f = K\sigma_\gamma / d_0 \text{; } \textit{in fragile failure } \tau_f = 2K\sigma_\gamma / d_0 \end{cases} \quad [A.4]$$

In cases where *there is absolutely no elasticity*, we say that there is *failure* of the shaft, because the maximum stress is higher than the elastic limit (R_e) of the material:

$$\tau = (C_\tau r)/K = (G\theta r)/l \textit{ and } \tau_{\max} = \sigma_{\max} = (C_\tau d_0)/2K = (G\theta d_0)/2l \quad [A.5]$$

The Tresca criterion applies if τ_{max} is higher than $R_e/2$. The same is true if σ_{max} is higher than $R_e/2$:

– (T_τ) is the twisting torque in N.m;

– θ is the torsion angle in degrees. *l* is the length in m;

– σ_f is the failure modulus in Pa = N/m^2;

– σ_γ is the elastic limit in Pa;

– G is the stiffness modulus (shear) in Pa.

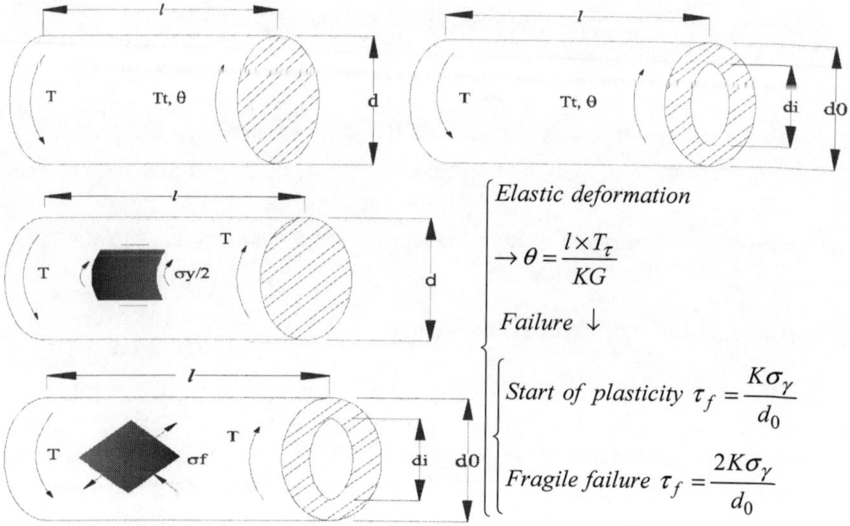

$$\begin{cases} \text{Elastic deformation} \\ \to \theta = \dfrac{l \times T_\tau}{KG} \\ \text{Failure} \downarrow \\ \begin{cases} \text{Start of plasticity } \tau_f = \dfrac{K\sigma_\gamma}{d_0} \\ \text{Fragile failure } \tau_f = \dfrac{2K\sigma_\gamma}{d_0} \end{cases} \end{cases} \quad \text{[A.6]}$$

Figure A.3. *Diagrams of calculations for shafts under torsion*

A.4.4. *Static and rotating discs: consider a homogeneous disc*

$$\begin{cases} \text{Simple discs} \to \delta = \dfrac{3}{4}\left(1-\nu^2\right)\dfrac{\Delta pR^4}{Ee^3} \text{ and } \sigma_{\max} = \dfrac{3}{8}(3+\nu)\dfrac{\Delta pR^2}{e^2} \\ \text{Pressed discs} \to \delta = \dfrac{3}{16}\left(1-\nu^2\right)\dfrac{\Delta pR^4}{Ee^3} \text{ and } \sigma_{\max} = \dfrac{3}{8}(3+\nu)\dfrac{\Delta pR^2}{e^2} \end{cases} \quad \text{[A.7]}$$

$$\begin{cases} \text{Discs} \to \delta = \dfrac{\pi\rho e\omega R^4}{4} \text{ and } \sigma_{\max} = \dfrac{1}{8}(3+\nu)\dfrac{\rho\omega^2 R^2}{e^2} \\ \text{Annuli} \to \delta = \pi\rho e\omega^2 R^3 x \text{ and } \sigma_{\max} = \rho\omega^2 R^2 \end{cases} \quad \text{[A.8]}$$

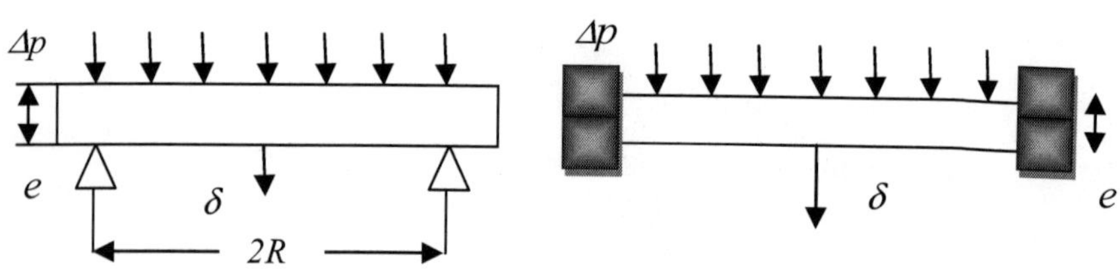

Figure A.4. *Diagrams of calculations for static and rotating discs*

A.4.5. *Stress intensification factor (SIF)*

In mechanical design, we often find ourselves dealing with stresses and deformations which are intensified around holes, grooves, cracks, etc. This occurs on variations of area in the elastic body. In the vicinity of these sections, we see sites of cracking under fatigue, which grow over time. The stress ratio is expressed as:

$$\text{Stress ratio: } \frac{\sigma_{\max}}{\sigma_{nom}} = 1 + \alpha \times \sqrt{\frac{e_{characteristic}}{\rho}} \quad \text{[A.9]}$$

The stress intensification resulting from the variation of the section "*tapers off*" – indeed sometimes disappears altogether – over distances according to Saint-Venant's principle. The *maximum intensification is limited to plastic failure. The increase in deformation* goes hand in hand with the *start of plasticity*. The *stress intensification* remains essentially *constant*. At a medium distance, the deformation intensification becomes a very considerable value. Then, Saint-Venant's principle no longer applies:

– F is the force in N;

– A_{min} is a minimum area in m^2;

– $e_{characteristic}$ is a characteristic distance in m;

– ρ is the radius of curvature in m and $\alpha = 2.00$ under torsion and $\alpha = 0.5$ under traction;

– σ_{nom} is the nominal stress = F/A_{min};

– σ_{max} is the maximum stress = F/A_{max}.

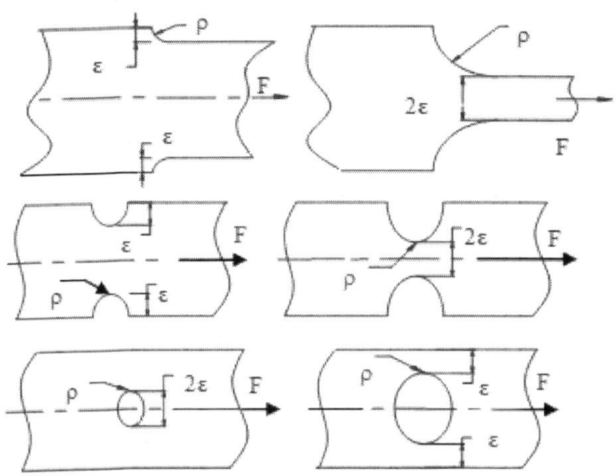

Figure A.5. *Diagrams of calculations of stress intensification factors*

A.4.6. *Cracking. Penetrating cracks*

$$\text{Local stress: } \sigma_l = C \times \left(\sigma\sqrt{\pi a}\right) / \sqrt{2\pi r}\text{, ſlc } K - C\sigma\sqrt{\pi a} \quad \text{[A.10]}$$

$$\text{Failure occurs when: } K_1 \geq K_{IC}\left[\sqrt[3]{m}\right], K_1 = C\sigma\sqrt{\pi a}:$$

$$\begin{cases} \text{Local loading} \rightarrow \ \sigma \geq F/2ab \\ \text{Stress} \rightarrow \ \sigma = 6M/be \\ \text{Bending at three different points} \rightarrow \ \sigma \geq 3Fl/2be \end{cases} \quad \text{[A.11]}$$

A.4.7. *Cracking: pressurized tanks (or vats)*

Membranes and thin walls under pressure are modeled in the same way as tanks. It should, however, be noted that this approximation is made for a thickness $\varepsilon < (a/4)$. In the case of thick walls, the stresses vary depending on the radial distance (***r***). These stresses are all the more intense *between* the inside and outside surfaces. They are stronger on the inside (see the equations below). When the von Mises stresses surpass the elastic limit R_e, the plasticity is predominant in pressurized tanks. Failure occurs when the greatest traction stress surpasses the failure stress:

$$\sigma_f \rightarrow \sigma_f = C\left(K_{IC}/\sqrt{\pi a}\right) \quad \text{[A.12]}$$

where:

– a is the outside radius in m and b is the inside radius in m;

– r is the radial coordinate in m and e is the thickness of the wall in m;

– p is the pressure in N/m^2 = Pa;

– K_{IC} is the toughness of the material. C is a constant.

$$\begin{cases} Cylinder: -p = \sigma_\theta = \sigma_r \\ Sphere: -p = \sigma_\theta = \sigma_r = \sigma_\phi \end{cases}; \text{ failure stress: } \sigma_f = C\frac{K_{IC}}{\sqrt{\pi a}} \quad \text{[A.13]}$$

$$-\ Cylinder: \left\{\sigma_\theta = \frac{pb}{e};\ \sigma_r = \frac{-p}{2};\ \sigma_z = \frac{pb}{2e}\right\} \quad \text{[A.14]}$$

– *Sphere:* $\left\{\sigma_\theta = \sigma_\phi = \dfrac{pb}{2e} \text{ et } \sigma_r = \dfrac{-p}{2}\right\}$ [A.15]

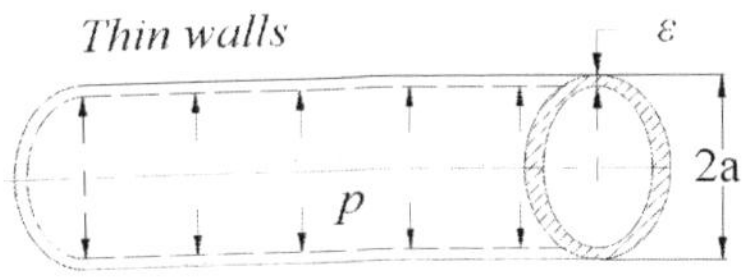

$$\begin{Bmatrix} \sigma_\theta = pa/\varepsilon \\ \sigma_r = -p/2 \\ \sigma_z = pa/2\varepsilon \end{Bmatrix} \quad [A.16]$$

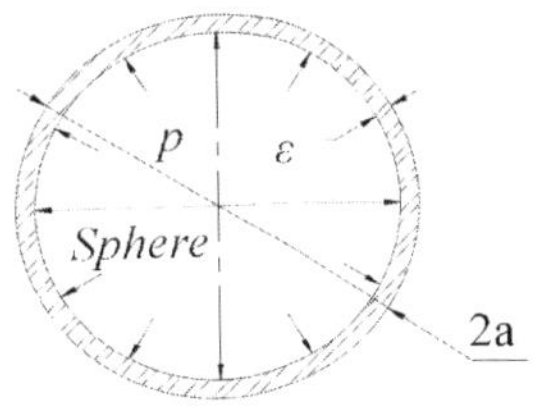

$$\begin{Bmatrix} \sigma_\theta = \sigma_\phi = pa/2\varepsilon \\ \text{and } \sigma_r = -p/2 \end{Bmatrix}$$

$$\begin{Bmatrix} \sigma_\theta = \dfrac{pb^2}{r^2}\left(\dfrac{a^2 - r^2}{a^2 - b^2}\right) \\ \sigma_r = \dfrac{pb^2}{r^2}\left(\dfrac{a^2 + r^2}{a^2 - b^2}\right) \end{Bmatrix} [A.17]$$

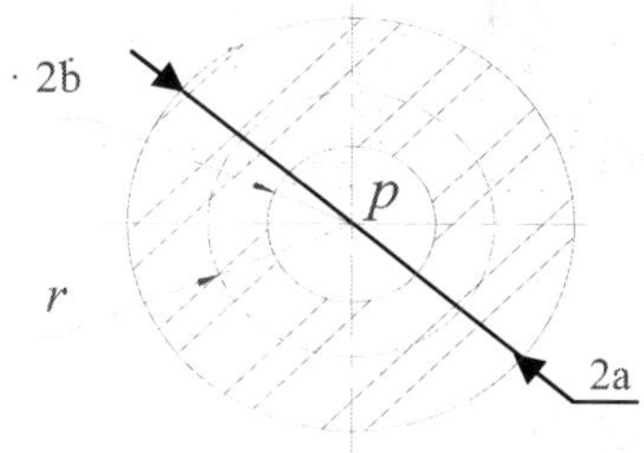

$$\begin{Bmatrix} \sigma_\theta = \dfrac{pb^3}{2r^3}\left(\dfrac{a^3 + 2r^3}{a^3 - b^3}\right) \\ \sigma_r = \dfrac{pb^3}{r^3}\left(\dfrac{a^3 - r^3}{a^3 - b^3}\right) \end{Bmatrix}$$

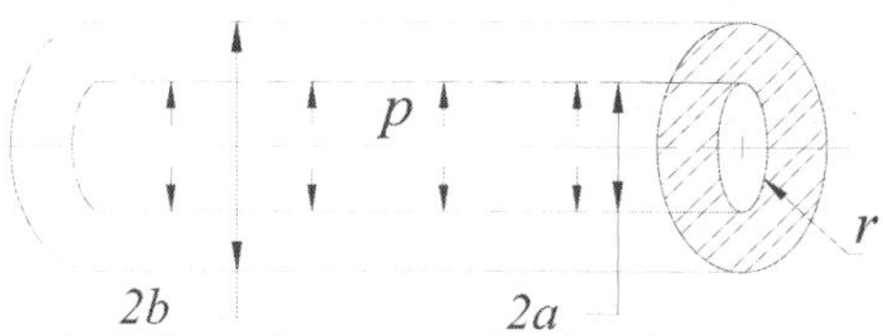

Figures A.6. *Diagrams of calculations for pressurized tanks*

A.4.8. *Creep, and creep fracture mechanics*

The phenomenon of creep and creep fracture is dependent on the temperatures, occurring at a third of the melting temperature of the materials in question. Creep is encountered when materials are under stress, at a given temperature T. The evolution of the strain rate, in the permanent regime, obeys an exponential power law as a function of the temperature.

$$\left\{\begin{array}{l}\text{Strain rate} \rightarrow \dot{\varepsilon} = A\left(\dfrac{\sigma}{\sigma_0}\right)^n Exp - \left(\dfrac{Q}{RT}\right) \\ \text{Const. temperature } \dot{\varepsilon} = \dot{\varepsilon}_0\left(\sigma/\sigma_0\right)^n;\ with\ prolonged\ creep\ \dot{\varepsilon} = C_{\text{constant}}^{\text{material}}/\tau_f\end{array}\right\} \quad [A.18]$$

Q is the activation energy and A is the kinetic energy. R is taken from ideal gases. It is a constant, sometimes referred to as the Mendeleev–Clapeyron constant. At a constant temperature, we read the previous equation of $\dot{\varepsilon}$ where $(\dot{\varepsilon}, \sigma_0\ and\ n)$ are constants pertaining to creep. In cases of prolonged creep, we see the accumulation of damage, leading to failure after a period of creep expressed thus:

$$\tau_f \times \dot{\varepsilon} = C \quad [A.19]$$

C is a constant of the material. For ductile materials experiencing creep, C ranges between 0.1 and 0.5. For fragile materials, the constants are smaller than 0.01.

Variables used in creep:

– σ is the applied stress, in N/m^2 or Pa. F is the force applied in N;

– α and β are the respective radii (inside and outside) in m;

– l, b and e are the dimensions represented in the figures;

– C_1 and C_2 are constants;

– ρ is the relative density;

– $\dot{\varepsilon}$, $\dot{\delta}$ and $\dot{u}$ are the strain rates in m/S;

– n, $\dot{\varepsilon}_0$ and $\dot{\sigma}_0$ are constants pertaining to creep.

The equations drawn from the existing body of literature, expressing the strain gradients, are as follows:

$$\left\{\begin{array}{l}\dot{\varepsilon} = \dot{\varepsilon}_0\left(\dfrac{\sigma}{\sigma_0}\right)^n \text{ and } \dot{u} = C_1\dot{\varepsilon}_0\sqrt{A}\left(\dfrac{C_2F}{A\sigma_0}\right)^n; \dot{\rho} = 2\dot{\varepsilon}_0\dfrac{\rho(1-\rho)}{\left(1-\sqrt{(1-\rho)}\right)^n}\left(\dfrac{2\left(p_{ext}-p_{\text{int}}\right)}{n\sigma_0}\right)^n \\ \dot{\delta} = \dfrac{2\dot{\varepsilon}_0}{n+2}\left(\dfrac{(2n+1)}{n\sigma_0}\dfrac{E}{be}\right)^n\left(\dfrac{\sigma}{\sigma_0}\right)^n l\left(\dfrac{l}{e}\right)^{n+1}; \dot{\rho} = \dfrac{3}{2}\dot{\varepsilon}_0\dfrac{\rho(1-\rho)}{\left(1-\sqrt{(1-\rho)}\right)^n}\left(\dfrac{3\left(p_{ext}-p_{\text{int}}\right)}{2n\sigma_0}\right)^n\end{array}\right. \quad [A.20]$$

– The relative density (ρ) is expressed by: $\left(b^3 - a^3\right)/b^3 = \rho$

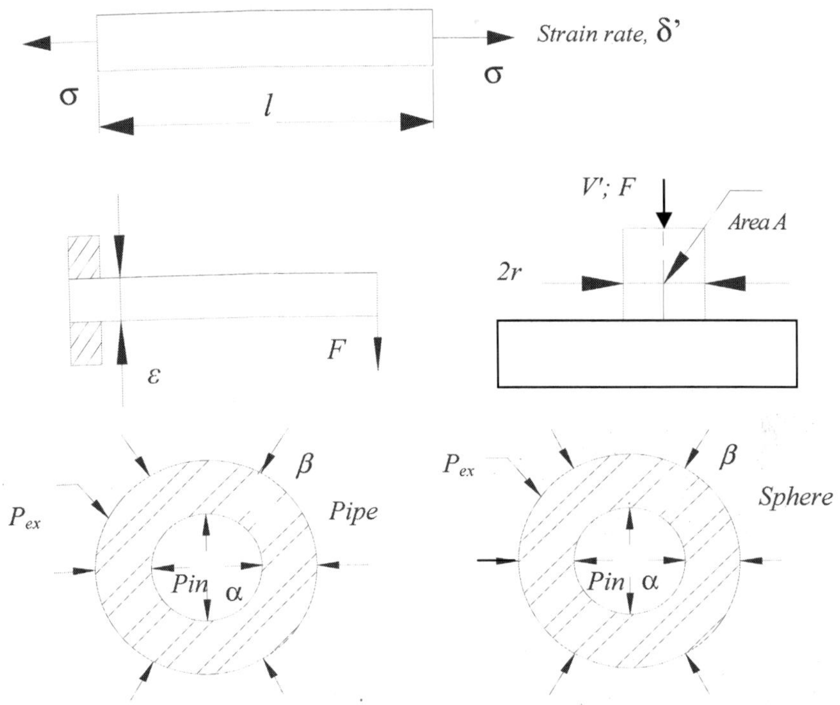

Figure A.7. *Diagrams of calculations for pipes and spheres experiencing creep failure*

A.5. Properties of regular sections

WARNING.– The formulae for the properties of sections presented here are reproduced from published documentation. We shall recap the main sections which have been studied thus far.

Diagram	Description – unit – formula [t = ε]
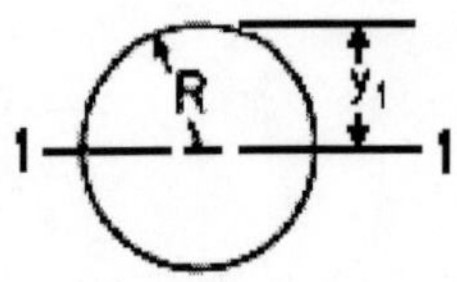	Area of the section in² or mm² $A = \pi R^2$
	Distance from the centroid to the extremities $y_1 = R$
	Moment of inertia in⁴ or mm⁴ $J_1 = \pi R^4/4$
	Gyradius with respect to the central axes in or mm $r_1 = R/2$
Solid circle section	Plastic section modulus $W_2 = W_1$, $W_1 = 1.333R^3$
	Form factor $\varphi_1 = 1.698$
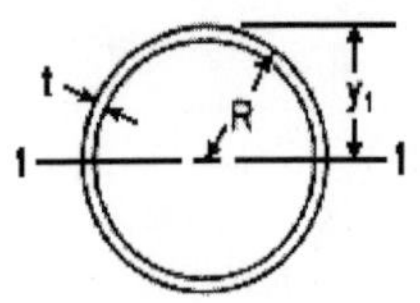	Area of the section in² or mm² $A = 2\pi R\varepsilon$
	Distance from the centroid to the extremities $y_1 = R$
	Moment of inertia in⁴ or mm⁴ $J_1 = \pi R^3\varepsilon$
Thin annulus	Gyradius with respect to the central axes in or mm $r_1 = 0.707R$
Diagram	**Description – unit – formula**
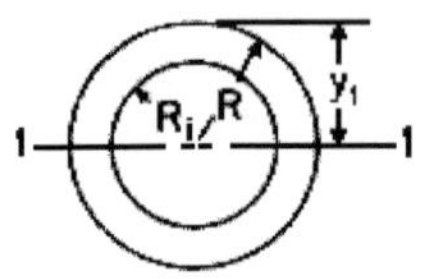	Area of the section in² or mm² $A = \pi\left(R_{ex}^2 - R_{in}^2\right)$
	Distance from the centroid to the extremities $y_1 = R_{ex}$
	Moment of inertia in⁴ or mm⁴ $J_1 = \left(R_{ex}^2 - R_{in}^2\right)\pi/4$
	Gyradius with respect to the central axes in or mm $r_1 = \sqrt{R_{ex}^2 + R_2^2}/2$
	Plastic section modulus $W_1 = 1.333\left(R_{ex}^2 - R_{in}^2\right)$
Hollow circle	Form factor: $\varphi_1 = 1.698\left(R_{ex}^4 - R_{in}^3 R_{ex}\right)/\left(R_{ex}^4 - R_{in}^4\right)$

Diagram	Description – unit – formula
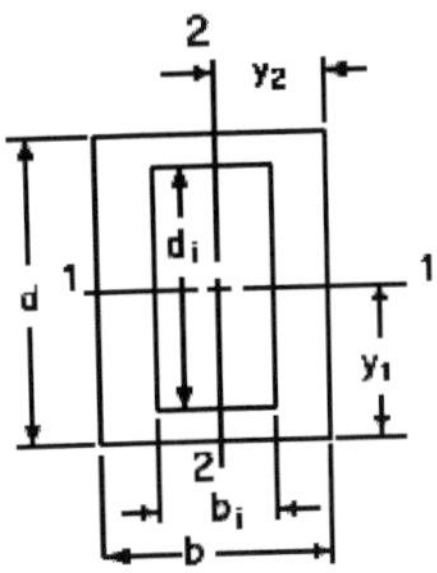	Area of the section in^2 or mm^2 $A = b_{ex} d_{ex} - b_{in} d_{in}$
	Distance from the centroid to the extremities $y_1 = d/2$ $y_2 = b/2$
	Moment of inertia in^4 or mm^4 $J_1 = \left(b_{ex} d_{ex}^3 - b_{in} d_{in}^3\right)/12 \quad J_2 = \left(d_{ex} b_{ex}^3 - d_{in} b_{in}^3\right)/12$
	Gyradius with respect to the central axes in or mm $r_1 = \sqrt{J_1/A}\ ;\ r_2 = \sqrt{J_2/A}$
	Plastic section modulus $W_1 = \left(b_{ex} d_{ex}^2 - b_{in} d_{in}^2\right)/4 \quad W_2 = \left(d_{ex} b_{ex}^2 - d_{in} b_{in}^2\right)/4$
Hollow rectangle	Form factor: $\varphi_1 = W_1 d_{ex}/2J_1\ ;\ \varphi_2 = W_2 b_{ex}/2J_2$
Diagram	Description – unit – formula
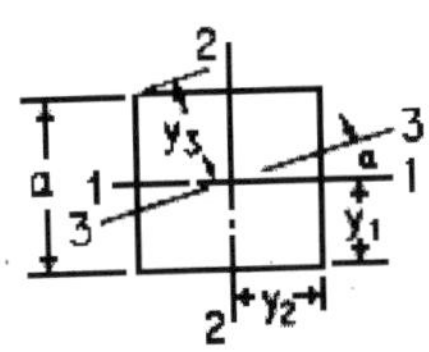	Area of the section in^2 or mm^2 $A = 2\pi R\varepsilon$
	Distance from the centroid to the extremities $y_3 = 0.707a\cos(45° - \alpha)\ \ y_2 = y_1\ \ y_1 = a/2$
	Moment of inertia in^4 or mm^4 $J_3 = J_2 = J_1;\ J_1 = a^4/12$
	Gyradius with respect to the central axes in or mm $r_3 = r_2 = r_1\ ;\ r_1 = 0.2887a$
	Plastic section modulus $W_2 = W_1$ $W_1 = 0.25a^3$
Square	Form factor: $\varphi_2 = \varphi_1\ ;\ \varphi_1 = 1.5$

Diagram	Description – unit – formula [t = ε] and [$t_w = \varepsilon_w$]
	Area of the section in² or mm² $A = b\varepsilon + \varepsilon_w d$
	Distance from the centroid to the extremities $$y_1 = \frac{b\varepsilon^2 + \varepsilon_w d(2\varepsilon + d)}{2(\varepsilon b + \varepsilon_w d)}\ ;\ y_2 = b/2$$
	Moment of inertia in⁴ or mm⁴ $J_2 = \frac{\varepsilon b^3}{12} + \frac{d\varepsilon_w^3}{12}$ $$J_1 = \frac{b}{3}(d+\varepsilon) - \frac{d^3}{3}(b - \varepsilon_w) - A(d + \varepsilon - y_1)^2$$
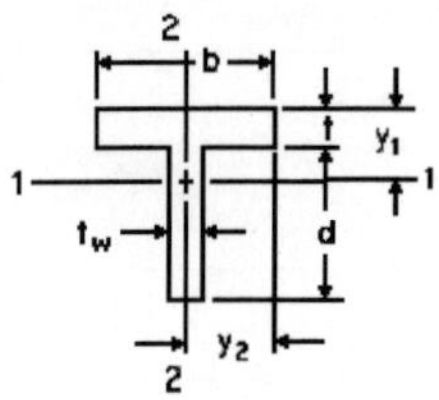	Gyradius with respect to the central axes in or mm $r_1 = \sqrt{J_1/A}\ ;\ r_2 = \sqrt{J_2/A}$
	Plastic section modulus $W_2 = W_2 b/2J_2$ $$W_1 = (\varepsilon_w d \prec b\varepsilon)\left(\frac{d^2\varepsilon_w}{4} - \frac{b^2\varepsilon^2}{4\varepsilon_w} + \frac{b\varepsilon(d+\varepsilon)}{2}\right) + \ldots$$ $$\ldots + (\varepsilon_w d \succ b\varepsilon)\left(\frac{d^2\varepsilon_w}{4} + \frac{b\varepsilon(d + \varepsilon - d\varepsilon_w/2)}{2}\right)$$
	Form factor: $\varphi_1 = W_1(d + \varepsilon - y_1)/J_1 \quad \varphi_2 = W_2 b/2J_2$
Tee section	Position of axis of the neutral fiber: $$y_p = (\varepsilon_w d \prec b\varepsilon)\left(\frac{b\varepsilon}{2\varepsilon_w} + \frac{d}{2}\right) + (d\varepsilon_w \succ b\varepsilon)\left(\frac{d\varepsilon}{2b} + \frac{\varepsilon}{2}\right)$$
Diagram	Description – unit – formula
	Area of the section in² or mm² $A = b \times d$
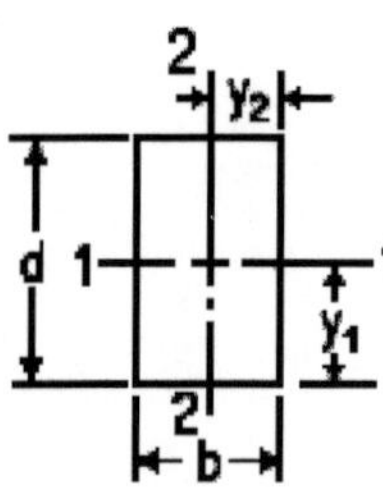	Distance from the centroid to the extremities $y_1 = d/2$ $y_2 = b/2$
	Moment of inertia in⁴ or mm⁴ $If\ d \succ b \rightarrow J_1 = bd^3/12; J_2 = db^3/12$
	Gyradius with respect to the central axes in or mm $r_1 = 0.288d\ ;\ r_2 = 0.288b$
Rectangle	Plastic section modulus $W_1 = 0.25bd^2\ ;\ W_2 = 0.25db^2$
	Form factor: $\varphi_1 = \varphi_2 = 1.5$

Diagram	Description – unit – formula [t = ε]
	Semi-axes (a_{in}, a_{ex}) *and* (b_{in}, b_{ex}) $a_{ex} = a_{in} + \varepsilon/2 \;\; b_{ex} = b_{in} + \varepsilon/2$
	Thickness in or mm $t = \varepsilon = a_{ex} - a_i$ $\varepsilon_{\max} = \left(\frac{a_{ex}}{b_{ex}} \geq 1\right)\frac{2b_{ex}^2}{a_{ex}} + \left(\frac{a_{ex}}{b_{ex}} \prec 1\right)\frac{2a_{ex}^2}{b_{ex}}$
Stresses $\frac{2}{10} \prec \frac{a_{ex}}{b_{ex}} \prec 5$ 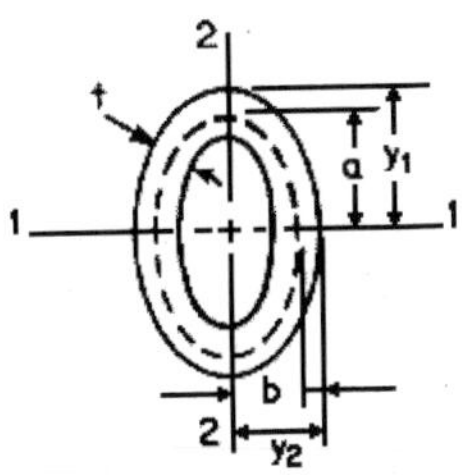 	Constants K_i: $K_1 = 0.2464 + 0.002222\left(\frac{a_{ex}}{b_{ex}} + \frac{b_{ex}}{a_{ex}}\right)$ $K_2 = 0.1349 + 0.1279\frac{a_{ex}}{b_{ex}} - 0.01284\left(\frac{a_{ex}}{b_{ex}}\right)^2$ $K_3 = 0.1349 + 0.1279\frac{b_{ex}}{a_{ex}} - 0.01284\left(\frac{b_{ex}}{a_{ex}}\right)^2$ $K_4 = 0.1835 + 0.895\frac{a_{ex}}{b_{ex}} - 0.00978\left(\frac{a_{ex}}{b_{ex}}\right)^2$ $K_5 = 0.1835 + 0.895\frac{b_{ex}}{a_{ex}} - 0.00978\left(\frac{b_{ex}}{a_{ex}}\right)^2$
Hollow ellipse; constant wall thickness	Area of the section in^2 or mm^2 $A = \pi\varepsilon(a_{ex} + b_{ex})\left(1 + K_1\left(\frac{a_{ex} - b_{ex}}{a_{ex} + b_{ex}}\right)^2\right)$ Distance from the centroid to the extremities $y_1 = a_{ex} + \frac{\varepsilon}{2}, y_2 = b_{ex} + \frac{\varepsilon}{2}$
[t = ε]	Moment of inertia in^4 or mm^4 $J_1 = \frac{\pi}{4}\varepsilon a_{ex}^2(a_{ex} + 3b_{ex})\left(1 + K_2\left(\frac{a_{ex} - b_{ex}}{a_{ex} + b_{ex}}\right)^2\right) +$ $\ldots + \frac{\pi}{16}\varepsilon^3(3a_{ex} + b_{ex})\left(1 + K_3\left(\frac{a_{ex} - b_{ex}}{a_{ex} + b_{ex}}\right)^2\right)$

	$J_2 = \frac{\pi}{4}\varepsilon b_{ex}^2 (b + 3a_{ex})\left(1 + K_3\left(\frac{b_{ex} - a_{ex}}{b_{ex} + a_{ex}}\right)^2\right) + \ldots + \frac{\pi}{16}\varepsilon^3 (3b_{ex} + a_{ex})\left(1 + K_2\left(\frac{b_{ex} - a_{ex}}{b_{ex} + a_{ex}}\right)^2\right)$
	Plastic section modulus $W_1 = 1.333\varepsilon a_{ex}(a_{ex} + 2b_{ex})\left(1 + K_4\left(\frac{a_{ex} - b_{ex}}{a_{ex} + b_{ex}}\right)^2\right) + \frac{\varepsilon^3}{3}$ $W_2 = 1.333\varepsilon b_{ex}(b_{ex} + 2a_{ex})\left(1 + K_5\left(\frac{b_{ex} - a_{ex}}{b_{ex} + a_{ex}}\right)^2\right) + \frac{\varepsilon^3}{3}$
Diagram	Description – unit – formula [t = ε]
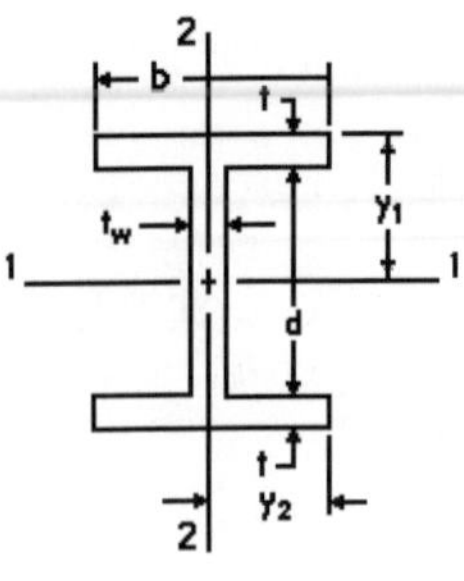 I-beam *Wide-flange beam with equal flanges*	Area of the section in² or mm² $A = 2b\varepsilon + \varepsilon_w d$
	Distance from the centroid to the extremities $y_1 = b/2 + \varepsilon$; $y_2 = b/2$
	Moment of inertia in⁴ or mm⁴ $J_1 = \frac{b(d + 2\varepsilon_w)^3}{12} - \frac{(b - \varepsilon_w)d^3}{12}$; $J_2 = \frac{b^3\varepsilon}{6} + \frac{\varepsilon_w^3 d}{12}$
	Gyradius with respect to the central axes in or mm $r_1 = \sqrt{J_1/A}$; $r_2 = \sqrt{J_2/A}$
	Plastic section modulus $W_1 = \frac{\varepsilon_w d^2}{4} + b\varepsilon(d + \varepsilon)$; $W_2 = \frac{\varepsilon b^2}{2} + \frac{d\varepsilon_w^2}{4}$
	Form factor: $\varphi_1 = W_1 y_1/J_1$ $\varphi_2 = W_2 y_2/J_2$

Table A.7. *Parameters in geometric sections: [BEE 14, OBE 16, ROA 11]*

A.6. Glossary and definitions of mechanical characteristics of plastic materials

Notches: notches or acute angles lead to stress intensification, which can cause the fracture and premature failure of the part.

Notch sensitivity: notch sensitivity refers to the propensity for propagation of a crack through a plastic material from a pre-existing area of stress intensification (acute angles, grooves, perforations, sudden changes in transverse sections).

Thermal ageing: plastic materials deteriorate after prolonged exposure to high temperatures.

Isotropy: an *isotropic* material retains all of its characteristics irrespective of the direction in which they are measured.

Ductility: the characteristic of plastic materials whereby they can be stretched, tensed or rolled without compromising their integrity.

Slip: typical failure of plastic materials results from slip, where their molecules slip past one another. Materials with a high slip coefficient tend to exhibit a low friction coefficient, either with themselves or with a different material, and have no propensity for scuffing. The slip capacity applies to the characteristics of resistance to loads in relative motion.

Plasticity: the plasticity of a material is its ability to keep its shape and original size after its first transformation. Plasticity is seen when the stress exceeds the plastic resistance threshold on the stress/strain curve for the plastic material.

Warping: warping is the deformation of plastic parts, resulting from internal stresses. The complexities of the configuration of the finished parts mean they have different contraction rates depending on the dimensions of the molded part.

A.7. Glossary for metrology and quality control

Here we present the fundamentals of interest pertaining to design, control and processes. The VIM (*Vocabulaire international de métrologie* – Internal Metrology Lexicon) is an absolutely invaluable, and ineluctable reference [GRO 11, GRO 13].

Terms or acronyms	**Meaning in the context of project management for a design/manufacture project**
ACTIVITY	An activity corresponds to the tasks needing to be done for a project.
ADVANCEMENT	The advancement/status represents the progress of the activity. For instance, 100 % indicates that the project has been completed.
CAPABILITY	The capability is a system's ability to make products which conform to the pre-established requirements. The aim is to stay abreast of the quality of the products, keeping them free of faults outside the tolerance range.

CRITICAL PATH	The path, constituting a series of activities which, ultimately, determines the length of the project. Any delay encountered will have an impact on the final duration of the project.
FAST method	*Function Analysis System Technique* is a method aimed at ordering functions and breaking down tasks into chunks according to a logic which produces the appropriate technical solutions.
FMECA	Failure mode, effects, and criticality analysis
GANTT chart	Representation over time, with an equivalent segment
GPS, in metrology and tolerancing	Geometrical Product Specifications
KANBAN	In Japanese, KANBAN = label. Circulation of labels representing the automated triggering of orders (flow)
KAIZEN	In Japanese, KAI = to study and ZEN = to improve
LOAD	The load conditions the duration of the project.
MILESTONE	The milestone corresponds to an important deadline (key date) for the project. It has no duration as such. It is represented in graphs by a lozenge symbol.
NON-CONFORMITY	Failure to satisfy the specified requirements
PERT	*Program Evaluation and Review Technique* defines the critical paths of tasks for a project
POKA YOKE	In Japanese, POKA = error and YOKERY = to avoid or ERROR–PROOFING system
PROCEDURE	Reference documents determining the planning, oversight or improvement of different processes.
PROCESS	Mode of technology-based transformations
PROCESSES	Activity system: handling input functions to obtain the output functions, by giving them an added value.
PROJECT	SPECIFIC methodological approach, whereby users can PROGRESSIVELY monitor a REALITY which has no equivalent prior model (control)
PSM	Problem-Solving Methodology
SIMOGRAM	Graph of times taken for the production of the final products
SPC	Statistical Process Control
TIME	Production time is the amount of time having passed between the expression of the requirement and the actual date of delivery of the finished product. Manufacturing time expresses the technical component of the production time including the amounts of time given over to each phase.

	Productive time is the time when we are actually producing (machining) and also performing checks (metrology). NON-productive time is the time given over to the blips in the production process, such as unforeseen stoppages, repairs or changeover. Preparation time is the execution time, including the following: – T_t = technology time → time when the machine is operating, depending on the slice conditions chosen (Taylor); – T_m = manual time → general human work; – T_z = hidden time → additional time invested in another activity during the normal operation of the machine; – T_{tm} = technical-manual time → joint time between the machine and the human operator; – T_S = serial time → establishment of the phases of manufacture, assembly/disassembly and cleaning.
TRACEABILITY	Method (not an enquiry) for historically locating an entity using recorded means to identify the plot.
WILSON method	Mathematical simplifying method used to calculate the economical quantity for product orders for manufacture.

Table A.8. *Lexicon of terms employed in this book [GRO 11, GRO 13]*

Bibliography

[ASH 91] ASHBY M.F., "Materials and Shape", *Acta Metall. Mater.*, vol. 39, pp. 1025–1039, 1991.

[ASH 13] ASHBY M.F., *Choix des matériaux en conception mécanique*, Dunod, Paris, 2013.

[ASS 64] ASSOCIATED SPRING CORPORATION, *Handbook of Mechanical Spring Design*, Associated Spring, Bristol, 1964.

[AVA 06] AVALLONE E.A., BAUMEISTER III T., SADEGH A.M., *Marks' Standard Handbook for Mechanical Engineers*, 11th ed., McGraw-Hill Professional, New York, 2006.

[BAU 67] BAUMEISTER III T., *Marks' Standard Handbook for Mechanical Engineers,* 7th ed., McGraw-Hill Book Company, New York, 1967.

[BEE 14] BEER F., JOHNSTON JR E.R., DEWOLF J. *et al.*, *Mechanics of Materials*, McGraw-Hill Education, 2014.

[BUD 79] BUDINSKI K., *Engineering Materials: Properties and Selection*, Prentice Hall, Englewood Cliffs, 1979.

[CEB 94] CEBON D., ASHBY M.F., "Materials selection for precision instruments", *Measurement Sciences and Technologies*, vol. 5, pp. 296–306, 1994.

[CHE 87] CHETWYND D.G., "Selection of structural materials for precision devices", *Precision Engineering,* vol. 9, no. 1, pp. 3–6, 1987.

[COT 64] COTTRELL A.H., *Mechanical Properties of Matter*, Wiley, New York, 1964.

[DEJ 97] DEJEANS M., LEHU H., SACQUEPEY D. *et al.*, *Précis de construction mécanique, tome 3 : projets-calculs, dimensionnement, normalisation*, Nathan/Afnor, Paris, 1997.

[ENS 89] ENSINGER INC., Catalogue des propriétés des matières plastiques, 1989.

[FAN 14] FANCHON J.L., *Guide de Mécanique*, Nathan, Paris, 2014.

[FAR 89] FARAG M.M., *Selection of Materials and Manufacturing Processes for Engineering Design,* Prentice Hall, Englewood Cliffs, 1989.

[FRE 85] FRENCH M.J., *Conceptual Design for Engineering,* 2nd ed., Design Council, London and Springer, Berlin, 1985.

[GER 85] GERE G.M., TIMOSHENKO S.P., *Mechanics of Materials*, 2nd ed., Wadsworth International, London, 1985.

[GOE 95] GOEDKOOP M.J., DEMMERS M., COLLIGNON M.X., Eco-Indicator 95, Manual, Netherlands Agency for Energy and Environment, Amersfort, 1995.

[GRO 94] GROUS A., Étude probabiliste du comportement des structures en croix soudées, PhD Thesis, Université des Hautes-Alpes, 1994.

[GRO 11] GROUS A., *Applied Metrology for Manufacturing Engineering*, ISTE Ltd, London and John Wiley & Sons, New York, 2011.

[GRO 13a] GROUS A., *Contrôle de qualité appliquée – Études de cas et nouvelle organisation du travail,* Hermès-Lavoisier, Paris, 2013.

[GRO 13b] GROUS A., *Fiabilité mécanique appliquée,* Hermès-Lavoisier, Paris, 2013.

[HAY 90] HAYES M., "Materials Update 2: Springs", *Engineering*, p. 42, May 1990.

[HER 89] HERTZBERG R.W., *Deformation and Fracture Mechanics of Engineering Materials*, Wiley, New York, 1989.

[ISO 06] ISO: INTERNATIONAL ORGANIZATION FOR STANDARDIZATION, Rolling bearings: Static load ratings, Standard ISO 76:2006, 2006.

[ISO 07] ISO: INTERNATIONAL ORGANIZATION FOR STANDARDIZATION, Rolling bearings: Dynamic load ratings and rating life, Standard ISO 281:2007, 2007.

[JEN 36] JENNINGS C.H., "Welding Design", *Trans. ASME*, vol. 58, p. 497, 1936.

[LEW 90] LEWIS G., *Selection of Engineering Materials,* Prentice Hall, Englewood Cliffs, 1990.

[LUN 07] LUNDBERG A., PALMGREN A., "Dynamic Capacity of Rolling Bearings", *Acta Polytechnica*, vol. 1, no. 3, p. 7, 1947.

[MIS 65] MISCHKE C.R., "Bearing Reliability and Capacity", *Machine Design*, vol. 37, pp. 139–40, 1965.

[OBE 16] OBERG E., JONES F., HORTON H. *et al.*, *Machinery's Handbook*, 30th ed., Industrial Press Inc., South Norwalk, 2016.

[PET 74] PETERSON R.E., *Stress Contribution Factor*, Wiley, New York, 1974.

[PHE 70] PHELAN R.M., *Fundamentals of Mechanical Design*, McGraw-Hill Education, London, 1970.

[PIO 06] PIOTROWSKI J., *Shaft Alignment Handbook*, 3rd ed., CRC Press, Boca Raton, 2006.

[POP 10] POPOV V.L., *Contact Mechanics and Friction : Physical Principles and Applications,* Springer-Verlag, Berlin, 2010.

[RAB 95] RABINOWICZ E., *Friction and Wear of Material*, 2nd ed., Wiley, New York, 1995.

[RAD 92] RADZEVICH S.P., *Dudley's Handbook of Practical Gear Design and Manufacturing*, CRC Press, Boca Raton, 1992.

[ROA 12] ROARK R., YOUNG W., SADEGH A.M., *Formulas for Stress and Strain*, 8th ed., McGraw-Hill Education, 2012.

[ROT 73] ROTHFUSS N.B., "Design and Application of Flexible Diaphragm Couplings to Industrial-Marine Gas Turbines", *ASME 1973 International Gas Turbine Conference and Products Show*, Washington DC, USA, April 8-12, 1973.

[SHA 93] SHARP A., *Bicycles and Tricycles. An Elementary Treatise on Their Design and Construction*, MIT Press, Cambridge, 1993.

[SIN 14] SINNIAH I., MONTERRUBIO L., YUSUKE M., *The Rayleigh-Ritz Method for Structural Analysis*, ISTE Ltd, London and John Wiley & Sons, New York, 2014.

[SOD 30] SODERBERG C.R., "Factor of Safety and Working Stress", *Trans. ASME*, vol. 52, pp. 13–28, 1930.

[SOD 35] SODERBERG C.R., "Working Stresses", *Trans. ASME*, vol. 57, pp. 106–108, 1935.

[STI 07] STRIBECK R., "Ball Bearings for Various Loads", *Trans. ASME*, vol. 29, pp. 420–463, 1907.

[SVE 76] SVETLICKIJ V.A., *Vibrations aléatoires des systèmes mécaniques,* Technique et Documentation Lavoisier, Paris, 1980.

[TIM 51] TIMOSHENKO S., GOODIER J.N., *Theory of Elasticity*, 2nd ed., McGraw-Hill Book Company, New York, 1951.

[TIM 58] TIMOSHENKO S.P., *Résistance des Matériaux*, Van Nostrand Reinhold Company, New York, 1958.

[TIM 67] TIMOSHENKO S.P., *Vibrations dans le domaine technique*, MIR, Moscow, 1967.

[TRO 99] TROTIGNON J.P., VERDU J., DOBRACZYNSKI A. *et al.*, *Matières plastiques : Structures-propriétés, mise en œuvre, normalisation,* Nathan/Afnor, Paris, 1999.

[TSU 15] TSUBAKI, Cam Clutch and Mechanical Components, Catalogue 4, Japan, 2015.

[ULL 92] ULLMAN D.G., *The Mechanical Design Process*, McGraw-Hill, New York, 1992.

[WAH 29] WAHL A.M., "Stresses in Heavy Closely Coiled Helical Springs", *Trans. ASME*, vol. 51, 1929.

[WAH 63] WAHL A.M., *Mechanical Springs*, 2nd ed., McGraw-Hill Book Company, New York, 1963.

[WHI 85] WHITT R.R., WILSON D.G., *Bicycling Science*, 2nd ed., MIT Press, Cambridge, 1985.

Index

D

E

F

O, P

R

S

T

V, W

Other titles from

ISTE

in

Systems and Industrial Engineering – Robotics

2018

MARÉ Jean-Charles
Aerospace Actuators 3: European Commercial Aircraft and Tiltrotor Aircraft

SIMON Christophe, WEBER Philippe, SALLAK Mohamed
Data Uncertainty and Important Measures
(Systems Dependability Assessment Set – Volume 3)

2017

ANDRÉ Jean-Claude
From Additive Manufacturing to 3D/4D Printing 1: From Concepts to Achievements
From Additive Manufacturing to 3D/4D Printing 2: Current Techniques, Improvements and their Limitations
From Additive Manufacturing to 3D/4D Printing 3: Breakthrough Innovations: Programmable Material, 4D Printing and Bio-printing

ARCHIMÈDE Bernard, VALLESPIR Bruno
Enterprise Interoperability: INTEROP-PGSO Vision

CAMMAN Christelle, FIORE Claude, LIVOLSI Laurent, QUERRO Pascal
Supply Chain Management and Business Performance: The VASC Model

FEYEL Philippe
Robust Control, Optimization with Metaheuristics

MARÉ Jean-Charles
Aerospace Actuators 2: Signal-by-Wire and Power-by-Wire

POPESCU Dumitru, AMIRA Gharbi, STEFANOIU Dan, BORNE Pierre
Process Control Design for Industrial Applications

RÉVEILLAC Jean-Michel
Modeling and Simulation of Logistics Flows 1: Theory and Fundamentals
Modeling and Simulation of Logistics Flows 2: Dashboards, Traffic Planning and Management
Modeling and Simulation of Logistics Flows 3: Discrete and Continuous Flows in 2D/3D

2016

ANDRÉ Michel, SAMARAS Zissis
Energy and Environment
(Research for Innovative Transports Set - Volume 1)

AUBRY Jean-François, BRINZEI Nicolae, MAZOUNI Mohammed-Habib
Systems Dependability Assessment: Benefits of Petri Net Models (Systems Dependability Assessment Set - Volume 1)

BLANQUART Corinne, CLAUSEN Uwe, JACOB Bernard
Towards Innovative Freight and Logistics (Research for Innovative Transports Set - Volume 2)

COHEN Simon, YANNIS George
Traffic Management (Research for Innovative Transports Set - Volume 3)

MARÉ Jean-Charles
Aerospace Actuators 1: Needs, Reliability and Hydraulic Power Solutions

REZG Nidhal, HAJEJ Zied, BOSCHIAN-CAMPANER Valerio
Production and Maintenance Optimization Problems: Logistic Constraints and Leasing Warranty Services

TORRENTI Jean-Michel, LA TORRE Francesca
Materials and Infrastructures 1 (Research for Innovative Transports Set - Volume 5A)
Materials and Infrastructures 2 (Research for Innovative Transports Set - Volume 5B)

WEBER Philippe, SIMON Christophe
Benefits of Bayesian Network Models
(Systems Dependability Assessment Set – Volume 2)

YANNIS George, COHEN Simon
Traffic Safety (Research for Innovative Transports Set - Volume 4)

2015

AUBRY Jean-François, BRINZEI Nicolae
Systems Dependability Assessment: Modeling with Graphs and Finite State Automata

BOULANGER Jean-Louis
CENELEC 50128 and IEC 62279 Standards

BRIFFAUT Jean-Pierre
E-Enabled Operations Management

MISSIKOFF Michele, CANDUCCI Massimo, MAIDEN Neil
Enterprise Innovation

2014

CHETTO Maryline
Real-time Systems Scheduling
Volume 1 – Fundamentals
Volume 2 – Focuses

DAVIM J. Paulo
Machinability of Advanced Materials

ESTAMPE Dominique
Supply Chain Performance and Evaluation Models

FAVRE Bernard
Introduction to Sustainable Transports

GAUTHIER Michaël, ANDREFF Nicolas, DOMBRE Etienne
Intracorporeal Robotics: From Milliscale to Nanoscale

MICOUIN Patrice
Model Based Systems Engineering: Fundamentals and Methods

MILLOT Patrick
Designing Human–Machine Cooperation Systems

NI Zhenjiang, PACORET Céline, BENOSMAN Ryad, RÉGNIER Stéphane
Haptic Feedback Teleoperation of Optical Tweezers

OUSTALOUP Alain
Diversity and Non-integer Differentiation for System Dynamics

REZG Nidhal, DELLAGI Sofien, KHATAD Abdelhakim
Joint Optimization of Maintenance and Production Policies

STEFANOIU Dan, BORNE Pierre, POPESCU Dumitru, FILIP Florin Gh., EL KAMEL Abdelkader
Optimization in Engineering Sciences: Metaheuristics, Stochastic Methods and Decision Support

2013

ALAZARD Daniel
Reverse Engineering in Control Design

ARIOUI Hichem, NEHAOUA Lamri
Driving Simulation

CHADLI Mohammed, COPPIER Hervé
Command-control for Real-time Systems

DAAFOUZ Jamal, TARBOURIECH Sophie, SIGALOTTI Mario
Hybrid Systems with Constraints

FEYEL Philippe
Loop-shaping Robust Control

FLAUS Jean-Marie
Risk Analysis: Socio-technical and Industrial Systems

FRIBOURG Laurent, SOULAT Romain
Control of Switching Systems by Invariance Analysis: Application to Power Electronics

GROSSARD Mathieu, REGNIER Stéphane, CHAILLET Nicolas
Flexible Robotics: Applications to Multiscale Manipulations

GRUNN Emmanuel, PHAM Anh Tuan
Modeling of Complex Systems: Application to Aeronautical Dynamics

HABIB Maki K., DAVIM J. Paulo
Interdisciplinary Mechatronics: Engineering Science and Research Development

HAMMADI Slim, KSOURI Mekki
Multimodal Transport Systems

JARBOUI Bassem, SIARRY Patrick, TEGHEM Jacques
Metaheuristics for Production Scheduling

KIRILLOV Oleg N., PELINOVSKY Dmitry E.
Nonlinear Physical Systems

LE Vu Tuan Hieu, STOICA Cristina, ALAMO Teodoro,
CAMACHO Eduardo F., DUMUR Didier
Zonotopes: From Guaranteed State-estimation to Control

MACHADO Carolina, DAVIM J. Paulo
Management and Engineering Innovation

MORANA Joëlle
Sustainable Supply Chain Management

SANDOU Guillaume
Metaheuristic Optimization for the Design of Automatic Control Laws

STOICAN Florin, OLARU Sorin
Set-theoretic Fault Detection in Multisensor Systems

2012

AÏT-KADI Daoud, CHOUINARD Marc, MARCOTTE Suzanne, RIOPEL Diane
Sustainable Reverse Logistics Network: Engineering and Management

BORNE Pierre, POPESCU Dumitru, FILIP Florin G., STEFANOIU Dan
Optimization in Engineering Sciences: Exact Methods

CHADLI Mohammed, BORNE Pierre
Multiple Models Approach in Automation: Takagi-Sugeno Fuzzy Systems

DAVIM J. Paulo
Lasers in Manufacturing

DECLERCK Philippe
Discrete Event Systems in Dioid Algebra and Conventional Algebra

DOUMIATI Moustapha, CHARARA Ali, VICTORINO Alessandro, LECHNER Daniel
Vehicle Dynamics Estimation using Kalman Filtering: Experimental Validation

GUERRERO José A, LOZANO Rogelio
Flight Formation Control

HAMMADI Slim, KSOURI Mekki
Advanced Mobility and Transport Engineering

MAILLARD Pierre
Competitive Quality Strategies

MATTA Nada, VANDENBOOMGAERDE Yves, ARLAT Jean
Supervision and Safety of Complex Systems

POLER Raul *et al.*
Intelligent Non-hierarchical Manufacturing Networks

TROCCAZ Jocelyne
Medical Robotics

YALAOUI Alice, CHEHADE Hicham, YALAOUI Farouk, AMODEO Lionel
Optimization of Logistics

ZELM Martin *et al.*
Enterprise Interoperability –I-EASA12 Proceedings

2011

CANTOT Pascal, LUZEAUX Dominique
Simulation and Modeling of Systems of Systems

DAVIM J. Paulo
Mechatronics

DAVIM J. Paulo
Wood Machining

GROUS Ammar
Applied Metrology for Manufacturing Engineering

KOLSKI Christophe
Human–Computer Interactions in Transport

LUZEAUX Dominique, RUAULT Jean-René, WIPPLER Jean-Luc
Complex Systems and Systems of Systems Engineering

ZELM Martin, *et al.*
Enterprise Interoperability: IWEI2011 Proceedings

2010

BOTTA-GENOULAZ Valérie, CAMPAGNE Jean-Pierre, LLERENA Daniel, PELLEGRIN Claude
Supply Chain Performance / Collaboration, Alignement and Coordination

BOURLÈS Henri, GODFREY K.C. Kwan
Linear Systems

BOURRIERES Jean-Paul
Proceedings of CEISIE'09

CHAILLET Nicolas, REGNIER Stéphane
Microrobotics for Micromanipulation

DAVIM J. Paulo
Sustainable Manufacturing

GIORDANO Max, MATHIEU Luc, VILLENEUVE François
Product Life-Cycle Management / Geometric Variations

LOZANO Rogelio
Unmanned Aerial Vehicles / Embedded Control

LUZEAUX Dominique, RUAULT Jean-René
Systems of Systems

VILLENEUVE François, MATHIEU Luc
Geometric Tolerancing of Products

2009

DIAZ Michel
Petri Nets / Fundamental Models, Verification and Applications

OZEL Tugrul, DAVIM J. Paulo
Intelligent Machining

PITRAT Jacques
Artificial Beings

2008

ARTIGUES Christian, DEMASSEY Sophie, NERON Emmanuel
Resources–Constrained Project Scheduling

BILLAUT Jean-Charles, MOUKRIM Aziz, SANLAVILLE Eric
Flexibility and Robustness in Scheduling

DOCHAIN Denis
Bioprocess Control

LOPEZ Pierre, ROUBELLAT François
Production Scheduling

THIERRY Caroline, THOMAS André, BEL Gérard
Supply Chain Simulation and Management

2007

DE LARMINAT Philippe
Analysis and Control of Linear Systems

DOMBRE Etienne, KHALIL Wisama
Robot Manipulators

LAMNABHI Françoise *et al.*
Taming Heterogeneity and Complexity of Embedded Control

LIMNIOS Nikolaos
Fault Trees

2006

FRENCH COLLEGE OF METROLOGY
Metrology in Industry

NAJIM Kaddour
Control of Continuous Linear Systems

Printed and bound by CPI Group (UK) Ltd, Croydon, CR0 4YY